NUTRIENTS IN NATURAL WATERS
Herbert E. Allen and James R. Kramer, Editors

pH AND pION CONTROL IN PROCESS AND WASTE STREAMS
F. G. Shinskey

INTRODUCTION TO INSECT PEST MANAGEMENT
Robert L. Metcalf and William H. Luckman, Editors

OUR ACOUSTIC ENVIRONMENT
Frederick A. White

ENVIRONMENTAL DATA HANDLING
George B. Heaslip

THE MEASUREMENT OF AIRBORNE PARTICLES
Richard D. Cadle

ANALYSIS OF AIR POLLUTANTS
Peter O. Warner

ENVIRONMENTAL INDICES
Herbert Inhaber

URBAN COSTS OF CLIMATE MODIFICATION
Terry A. Ferrar, Editor

CHEMICAL CONTROL OF INSECT BEHAVIOR: THEORY AND APPLICATION
H. H. Shorey and John J. McKelvey, Jr.

MERCURY CONTAMINATION: A HUMAN TRAGEDY
Patricia A. D'Itri and Frank M. D'Itri

POLLUTANTS AND HIGH RISK GROUPS
Edward J. Calabrese

SULFUR IN THE ENVIRONMENT, Parts I and II
Jerome O. Nriagu, Editor

ENERGY UTILIZATION AND ENVIRONMENTAL HEALTH
Richard A. Wadden, Editor

METHODOLOGICAL APPROACHES TO DERIVING ENVIRONMENTAL AND OCCUPA-
TIONAL HEALTH STANDARDS
Edward J. Calabrese

ZINC IN THE ENVIRONMENT

Part II: Health Effects

ZINC IN THE ENVIRONMENT

Part II: Health Effects

Edited by

JEROME O. NRIAGU

Canada Centre for Inland Waters
Burlington, Ontario, Canada

A WILEY-INTERSCIENCE PUBLICATION
JOHN WILEY & SONS
New York • Chichester • Brisbane • Toronto

Library of Congress Cataloging in Publication Data
Main entry under title:

Zinc in the environment.

   (Environmental science and technology)
   "A Wiley-Interscience publication."
   Includes index.
   CONTENTS: pt. 1. Ecological cycling.–pt. 2. Health
effects.
   1. Zinc–Environmental aspects. I. Nriagu, Jerome O.
TD196.Z56   574.5'222   79-19257
ISBN 0-471-05888-2 (v. 1)
ISBN 0-471-05889-0 (v. 2)

Printed in the United States of America

10 9 8 7 6 5 4 3 2 1

SERIES PREFACE

Environmental Science and Technology

The Environmental Science and Technology Series of Monographs, Textbooks, and Advances is devoted to the study of the quality of the environment and to the technology of its conservation. Environmental science therefore relates to the chemical, physical, and biological changes in the environment through contamination or modification, to the physical nature and biological behavior of air, water, soil, food, and waste as they are affected by man's agricultural, industrial, and social activities, and to the application of science and technology to the control and improvement of environmental quality.

The deterioration of environmental quality, which began when man first collected into villages and utilized fire, has existed as a serious problem under the ever-increasing impacts of exponentially increasing population and of industrializing society. Environmental contamination of air, water, soil, and food has become a threat to the continued existence of many plant and animal communities of the ecosystem and may ultimately threaten the very survival of the human race.

It seems clear that if we are to preserve for future generations some semblance of the biological order of the world of the past and hope to improve on the deteriorating standards of urban public health, environmental science and technology must quickly come to play a dominant role in designing our social and industrial structure for tomorrow. Scientifically rigorous criteria of environmental quality must be developed. Based in part on these criteria, realistic standards must be established and our technological progress must be tailored to meet them. It is obvious that civilization will continue to require increasing amounts of fuel, transportation, industrial chemicals, fertilizers, pesticides, and countless other products; and that it will continue to produce waste products of all descriptions. What is urgently needed is a total systems approach to modern civilization through which the pooled talents of scientists and engineers, in cooperation with social scientists and the medical profession, can be focused on the development of order and equilibrium in the presently disparate segments of the human environment. Most of the skills and tools that are needed are already in existence. We surely have a right to hope a technology that has created such manifold environmental problems is also capable of solving them. It is our hope that this Series in Environmental Sciences and Technology will not only serve to make this challenge more explicit to the established professionals, but that it also will help to stimulate the student toward the career opportunities in this vital area.

Robert L. Metcalf
Werner Stumm

PREFACE

This book presents a comprehensive review of the current knowledge about the role of zinc in the environment. It contains articles written by experts from many scientific disciplines, all focusing on the chemical behavior and biological effects of zinc in the biosphere, hydrosphere, and geosphere. While the focus is on zinc, many authors have also related their topics to the broader framework of metal contaminants in the environment.

Zinc is a key element in our health and life-styles. It has been used by humans since time immemorial and today finds myriad applications in our industrialized society. Zinc is essential for the normal activity of DNA polymerase and for protein synthesis, and thus plays a vital role in the healthy development of many life forms. Excessive amounts of zinc, however, may be toxic, especially to the aquatic biota. This volume endeavors to interface the various biological, chemical, geological, and clinical studies on zinc and represents perhaps the first attempt at looking holistically at the biogeochemistry of this element. Some overlapping of material between chapters is inevitable in dealing with different environments in which diverse processes are closely interlinked.

The chapters have been grouped into two parts. Part I (this volume) covers the sources, distribution, behavior, and flow of zinc in the environment, and emphasizes the pathways of environmental zinc to man. Part II deals with the biological, ecological, and health effects of zinc, with particular attention to zinc deficiency in humans, animals, and plants. *Zinc in the Environment* thus should be of interest to scientists, advanced students, public health officials, veterinarians, and anyone involved in pollution research and regulation or environmental management. The broad aspects of zinc biochemistry and toxicology covered should make the volume a fundamental reading for occupational hygienists and bioinorganic scientists.

It is a pleasure to acknowledge the tremendous effort and cooperation of our distinguished group of contributors. My sincere appreciation and gratitude also go to Wiley-Interscience for invaluable editorial assistance.

JEROME O. NRIAGU

Burlington, Ontario, Canada
October 1979

CONTENTS

1. Epidemiological Aspects of Human Zinc Deficiency   1
   Clare E. Casey and K. Michael Hambidge

2. Manifestations of Zinc Abnormalities in Human Beings   29
   Ananda S. Prasad

3. Manifestations of Zinc Abnormalities in Animals   61
   W. Jack Miller and Milton W. Neathery

4. Biochemical Changes of Hormones and Metalloenzymes in Zinc
   Deficiency   71
   M. Kirchgessner and H. -P. Roth

5. Zinc Metabolism in Human Beings   105
   Herta Spencer, Carol Ann Gatza, Lois Kramer, and Dace Osis

6. Cellular and Molecular Aspects of Mammalian Zinc Metabolism and
   Homeostasis   121
   Robert J. Cousins and Mark L. Failla

7. Blood Zinc in Health and Disease   137
   John D. Bogden

8. Blood Zinc and Lead Poisoning   171
   Morris M. Joselow

ix

9.   Zinc and Pregnancy   183
     Sten Jameson

10. Tumorgenesis and Zinc   199
     Jerry L. Phillips and Mary K. Kindred

11. Zinc in Wound Healing   215
     Andre M. van Rij and Walter J. Pories

12. Zinc Dental Cements under Physiological Conditions   237
     Ralph G. Silvey and George E. Myers

13. Accumulation of Zinc by Marine Biota   259
     Ronald Eisler

14. Zinc Speciation and Toxicity to Fish   353
     Gordon K. Pagenkopf

15. Zinc and Plants in Rivers and Streams   363
     Brian A. Whitton

16. Zinc Uptake and Accumulation by Agricultural Crops   401
     P. M. Giordano and J. J. Mortvedt

17. Zinc Tolerance by Plants   415
     Werner Mathys

18. Zinc Transport and Metabolism in Microorganisms   439
     Mark L. Failla and Eugene D. Weinberg

Index   467

CONTENTS
PART I

1. *Production and Uses of Zinc*
    V. Anthony Cammarota, Jr.

2. *Zinc in Soils*
    Larry M. Shuman

3. *Zinc Interaction with Soil and Sediment Components*
    William F. Pickering

4. *Zinc in the Atmosphere*
    Jerome O. Nriagu and Cliff I. Davidson

5. *Biological Indicators of Atmospheric Zinc Dispersal and Deposition*
    Suzanne S. Groet

6. *Distribution of Zinc in Natural Waters*
    John H. Martin, George A. Knauer, and A. Russell Flegal

7. *Speciation of Zinc in Natural Waters*
    T. Mark Florence

8. *Zinc in Urban Storm- and Wastewaters*
    William G. Wilber, Joseph V. Hunter, and Joel Balmat

9. *Zinc in the Limestone Cycle*
    Nicholas E. Pingitore, Jr.

10. *Cycling of Zinc in the Nearshore Marine Environment*
    David R. Young, Tsu-Kai Jan, and G. Patrick Hershelman

11. *Zinc Pollution and the Ecology of the Freshwater Environment*
    Alan H. Weatherley, Philip S. Lake, and Stephen C. Rogers

12. *Zinc Distribution and Cycling in Forest Ecosystems*
    Robert I. Van Hook, Dale W. Johnson, and Brian P. Spalding

Index

ZINC IN THE ENVIRONMENT

Part II: Health Effects

1

EPIDEMIOLOGICAL ASPECTS OF HUMAN ZINC DEFICIENCY

Clare E. Casey
K. Michael Hambidge

Department of Pediatrics, University of Colorado Medical Center, Denver, Colorado

1. **Introduction**   **2**
2. **Zinc as an Essential Nutrient**   **2**
   2.1. Body content and metabolism   2
   2.2. Biochemical and physiological role   3
   2.3. Availability   4
   2.4. Requirements   4
   2.5. Intakes   5
   2.6. Measurement of nutritional status   5
3. **Zinc Deficiency in Man**   **6**
   3.1. Etiology   6
   3.2. Clinical manifestations   8
   3.3. Diagnosis and treatment   9
4. **Epidemiological Aspects of Zinc Deficiency**   **10**
   4.1. Fetus   10
   4.2. Infants   10
   4.3. Preschool children   12
   4.4. Older children   13
   4.5. Adolescents   14
   4.6. Adults   15
   4.7. Pregnancy and lactation   15

4.8. Total parental nutrition   16
4.9. Acrodermatitis enteropathica   17
4.10. Secondary zinc deficiency   17
5. Scope of the Problem   **18**
Acknowledgments   **19**
References   **20**

1. INTRODUCTION

Zinc was first shown to be an essential nutrient for mammals more than 45 years ago (Todd et al., 1934). As with many other trace elements later shown to be essential, interest in zinc metabolism in man focused largely on its toxic properties: it was considered that the ubiquity of zinc in the environment made human deficiencies unlikely. However, in 1958 Prasad suggested that the syndrome of dwarfism and hypogonadism, seen in adolescent males in Iran, might be due to a nutritional deficiency of zinc (Prasad et al., 1961). Since then, there have been many reports of zinc deficiency occurring in groups of people in widely differing circumstances and in many different countries. Thus, although zinc is ubiquitous in the environment, so too appears to be human zinc deficiency.

2. ZINC AS AN ESSENTIAL NUTRIENT

2.1. Body Content and Metabolism

The adult human body contains about 2.5 g of zinc. The highest concentrations (100 to 200 ppm) occur in the eye, hair, bone, and male reproductive organs, with intermediate concentrations (40 to 50 ppm) in the liver, kidney, and muscle (Underwood, 1977). In the blood about 80% of the total zinc is in the red cells. The mean plasma zinc level is about 90 μg/100 ml, serum zinc being about 10% higher. About one half of the plasma zinc is in a freely exchangeable form, loosely bound to albumin. Most of the remainder is tightly bound to α_2-macroglobulin, with 7% of the total bound to amino acids, the most important of which are histidine and cysteine (Beisel et al., 1976).

Absorption of zinc takes place mainly in the duodenum. The mechanism appears to be an active transport involving a small binding ligand of pancreatic origin. This ligand facilitates the uptake of zinc from the lumen of the small intestine into the epithelial cells. The element is then transferred to a binding site on the basolateral membrane, which functions as a zinc donor to plasma albumin (Evans, 1976). The percentage of zinc absorbed from the diet has been reported to vary from 20 to 80%, most values being between 20 and 30%. Zinc absorption is dependent partially on the level of intake and on zinc nutritional status, a higher percentage being absorbed in a deficient state.

The principal route of excretion is via the intestine. Fecal zinc consists of unabsorbed dietary zinc with a small amount of endogenous origin, mainly from pancreatic exocrine secretions. Urinary zinc losses are small, amounting to 0.3 to 0.5 mg/24 hr in the normal, healthy adult (Hambidge and Walravens, 1975). Sweat losses are normally low but may become important in climates or situations that produce copious sweating. Under such conditions, losses of 4 mg Zn/day may occur (Prasad et al., 1963a).

2.2. Biochemical and Physiological Role

More than 20 different zinc metalloenzymes have been identified, including carbonic anhydrase, alkaline phosphatase, and alcohol dehydrogenase. Many other enzymes are known to be activated by zinc (Parisi and Vallee, 1969). Zinc has an important role in the metabolism of proteins and nucleic acids. It is apparently essential for the synthesis of DNA and ribosomal RNA; RNA polymerase is a zinc metalloenzyme (Fernandez-Madrid et al., 1973). Zinc also has a role in mitotic cell division and in the utilization of amino acids in protein synthesis, particularly of collagen in skin and scar tissue (McClain et al., 1973). Although little is known of the relation between the biochemical role of zinc and the clinical features of zinc deficiency, some of these features may arise from disturbances of nucleic acid and protein metabolism.

Numerous experimental studies with rats have shown an absolute requirement for zinc for growth, and its deficiency has severe effects on all stages of reproduction and of tissue proliferation in the young (Hurley and Schrader, 1972). Reproductive performance in both males and females is impaired by zinc deficiency. The effect on the young depends on the duration and timing of the zinc deprivation. In early gestation, zinc deficiency may cause severe congenital abnormalities. Later in gestation, it can cause growth retardation and impairment of brain growth, leading to impaired behavioral development after birth (McKenzie et al., 1975). Feeding a low-zinc diet to lactating dams produces symptoms of zinc deficiency, including impaired growth, in suckling pups (Mutch and Hurley, 1974).

Zinc deficiency is not yet recognized as a cause of human congenital malformations, but this possibility has been suggested on the basis of epidemiology. Imparied sexual maturation in adolescent males and poor growth in young children have been observed in human zinc deficiency.

Zinc has also been shown to have a role in the growth of hair, skin, and bone

(Calhoun et al., 1974) and in vitamin A metabolism (Smith et al., 1974), probably through its role in protein metabolism. A number of other physiological processes, including taste acuity, hormone metabolism, and immune function, also appear to involve zinc.

2.3. Availability

Zinc is less available for absorption from plant foods than from animal foods. One of the factors responsible may be the high phytate (inositol hexaphosphate) content of many plant foods, particularly grains. Phytate binds strongly to zinc, markedly decreasing its availability (Halsted et al., 1974). Recent studies indicate that fiber, particularly in unleavened breads, binds strongly with zinc, making it unavailable for absorption (Reinhold et al., 1976). Other plant constituents such as hemicelluloses and amino acid-carbohydrate complexes may also be involved.

2.4. Requirements

The precise minimal requirements for zinc for optimal health and growth are not known. These depend, in part, on the composition of the diet. The source of protein may affect the zinc requirement as much as fourfold, the variation arising from the lower availability of zinc in plant foods. Various physiological

Table 1. Recommended Daily Dietary Allowances for Zinc

Age	Amount (mg)	
	World Health Organization[a]	National Research Council
0–4 months	6	3
5–12 months	6	5
1–10 years	8	10
Males		
11–17 years	14	15
18+ years	11	15
Females		
10–13 years	13	15
14+ years	11	15
Pregnancy	15	20
Lactation	27	25

[a]Availability of 20%

conditions—for example, pregnancy, lactation, and periods of rapid growth and development in infants, children, and adolescents—are associated with a relatively high zinc requirement.

The World Health Organization (1973) has published provisional values for requirements of dietary zinc. These values (Table 1) were estimated on a factorial basis which included retention, urinary and sweat excretion, and change in lean body mass with growth of infants and children. Requirements are given for levels of available zinc in the diet of 10, 20 and 40%.

The Food and Nutrition Board (United States) included zinc in its recommended dietary allowances in 1974 (National Research Council, 1974). These values (Table 1) are based on the results of balance studies. No allowances are made for the variations in availability, but a recommendation that "the zinc intake should come from a balanced diet contaning sufficient animal protein" is made.

2.5. Intakes

In recent years a number of papers have been published giving values for zinc in individual foodstuffs and in diets. The distribution of zinc in the various food groups is similar in most reports, but values vary with the source, reports from some European countries (Schlettwein-Gsell and Mommsen-Straub, 1970) and New Zealand (Guthrie, 1975) giving generally lower levels than U.S. reports (Freeland and Cousins, 1976; Haeflein and Rasmussen, 1977). The best sources of zinc are meats, fish, shellfish, whole-grain cereals, and legumes. The refining of cereals decreases the zinc content but increases its availability (Reinhold et al., 1976).

Most of the total diets analyzed for zinc content have been institutional. Values reported for daily intakes of zinc range from 7.3 mg in a hospital diet for renal patients (Brown et al., 1976) up to 21 mg for a high-protein, high-calorie diet (Osis et al., 1972). However, most reported daily intakes, for both institutionalized and free-living adults, range from 8 to 14 mg Zn (Waslien, 1976). Values for the daily intake obtained by calculation may be somewhat higher than those obtained by analysis. One study found that there could be an overestimation of 35% when zinc intake was calculated from food tables (Brown et al., 1976).

2.6. Measurement of Nutritional Status

A number of parameters may be used to describe nutritional status with respect to zinc. The most common are zinc concentrations in plasma or serum and in hair; zinc levels in whole blood, red cells, saliva (parotid or mixed), and urine, and 24-hr urinary output, are also frequently used. Additional information may

Table 2. Normal Adult Values for Some Parameters Used in Assessing Zinc Status

Parameter	Value
Plasma zinc	68–110 μg/100 ml
Urine zinc	100–700 μg/24 hr
Hair zinc	$>$105 μg/g
Erythrocyte zinc	10.1–13.4 μg/ml
Serum alkaline phosphatase	150–510 IU/l
Serum ribonuclease	6440–9940 units/ml

be provided by measuring the activities of some zinc-dependent enzymes such as ribonuclease, alkaline phosphatase, and carbonic anhydrase. Results, however, are not always easily interpreted. Table 2 gives some normal values for adults obtained in this laboratory (Walravens and Hambidge, 1978).

Measurements of plasma zinc (or serum zinc, which is usually 5 to 10 μg/100 ml higher) are readily made and are the most useful for indicating short-term changes in zinc nutrition. Hair concentrations may provide useful information on long-term zinc status. Some workers have found a relationship between hair levels and plasma levels (Klevay, 1970), but others have not (McBean et al., 1971). An attempt to relate zinc levels in hair to those in tissues at autopsy in a small group of adults was unsuccessful (McKenzie, 1974). However, hair zinc concentrations do reflect dietary intake, and consistently low values over a period of time can provide a sensitive index of zinc deficiency (Hambidge and Walravens, 1975).

3. ZINC DEFICIENCY IN MAN

In 1938 Eggleton estimated that the daily zinc intake of the poorest class in south China was less than 6 mg. It was suggested at the time that zinc may be limiting at this level (Eggleton, 1939). Nonetheless, the occurrence of zinc deficiency in man was not proven until 1963 (Prasad et al., 1963a, 1963b, 1963c). In the past 15 years an increasing number of zinc-responsive disorders have been recognized, and some degree of zinc depletion appears to be quite common in children and adolescents.

3.1. Etiology

Most nutritional deficiencies may arise through a variety of different causes, operating either alone or in combination. Zinc is no exception: Table 3 outlines

Table 3. Etiological Factors Contributing to Zinc Deficiency

Factor	Examples
Inadequate dietary intake	Protein-calorie malnutrition Low-income diets Old age
Decreased availability	High fiber/phytate diets Infant formulas
Decreased absorption	Malabsorption syndromes Steatorrhea
Excessive losses	Hyperzincuria Surgery Burns Increased sweating
Increased requirement	Rapid growth Pregnancy Lactation Tissue anabolism
Iatrogenic	Chelating drugs TPN
Genetic defect	Acrodermatitis enteropathica

the various factors that may contribute to zinc deficiency. Simple nutritional deficiency due to marginal zinc intake may be quite common even in countries such as the United States (Sandstead, 1973). Inadequate intakes of zinc have been reported for groups of infants and children receiving otherwise sufficient diets (Hambidge, 1977). Institutional and hospital diets and poor economic status have also been associated with low zinc intakes. Synthetic diets and solutions used for total parenteral nutrition (TPN), when not supplemented with zinc, generally supply grossly inadequate amounts of this element (Hauer and Kaminski, 1978), and severe zinc deficiency has been observed in patients on such regimens (van Rij and McKenzie, 1977).

In rural areas of Egypt and Iran, where human zinc deficiency was first identified, dietary intakes of zinc appeared initially to be quite adequate (Sarram et al., 1969). However, the diets were subsequently found to contain a number of substances, such as phytate and fiber, that bind with zinc and make it largely unavailable for absorption (Reinhold et al., 1976). Thus decreased availability may contribute to the development of a nutritional zinc deficiency. This may also be the cause of the poorer zinc status of some formula-fed infants as compared with breast-fed infants; the zinc in various cow's milk and soy-based formulas is less bioavailable than that in human milk (Johnson and Evans, 1978).

High urinary losses of zinc over an extended period may lead to zinc depletion. Urinary excretion of zinc appears to vary directly with the amount of the element bound to the plasma amino acid pool. Conditions that increase the circulating amino acids, particularly histidine and cysteine, also cause an increase in the renal clearance of zinc (Morrison et al., 1978). Increased urinary excretion has been reported in a number of conditions, including alcoholism and post-alcoholic cirrhosis (Sullivan and Lankford, 1962) and other liver diseases. Hyperzincuria occurs in any condition associated with increased catabolism, such as surgery, burns, multiple injuries, major fractures, diabetes mellitus, protein deprivation, and starvation (Spencer et al., 1976). Excessive urinary excretion also results from abnormal excretion of molecules to which zinc binds and which prevent it from being reabsorbed, for example, albumin in nephrotic syndrome and chelating agents such as ethylenediamine tetraacetate (EDTA) and penicillamine (Sandstead et al., 1976).

Intestinal malabsorption associated with steatorrhea is the most common mechanism of zinc depletion in patients with gastrointestinal disease. Protein-losing enteropathies, zinc complexing with the protein, and massive loss of intestinal secretions may also contribute to the lowered zinc status of such patients. Increased losses of zinc due to chronic blood loss and excessive sweating were thought to be factors contributing to the zinc deficiency of nutritional dwarfism (Prasad et al., 1963a).

A number of physiological conditions are known to increase the requirement for zinc, including pregnancy and lactation. Rapid growth, such as the catch-up growth of premature infants, and rehabilitation after starvation or trauma also increase nutritional requirements. Poor zinc status may be associated with several genetic diseases, including thalassemia and sickle-cell anemia (Prasad et al., 1975a), but only one inherited disorder of zinc metabolism causing a deficiency syndrome has been described. This is acrodermatitis enteropathica (AE), an autosomal recessively inherited disease, the symptoms of which are consistent with severe zinc deficiency (Neldner and Hambidge, 1975).

3.2. Clinical Manifestations

The signs and symptoms of zinc deficiency depend, to some extent, on age, acuteness of onset, duration and severity of the zinc depletion, and the circumstances in which the zinc deficiency occurs. Many of the features observed in man are similar to those seen in zinc-deficient animals. Chronic deficiency in pediatric and adolescent age groups causes retarded growth. In adolescence, sexual maturation is delayed. Anorexia, pica, impaired taste acuity, and mental lethargy have all been reported to occur. Other disturbances may include rough, dry skin, impaired wound healing, and increased susceptibility to infection.

Abnormal glucose tolerance has been observed in the Middle East (Sandstead et al., 1967), and impaired secretion of luteinizing hormone has been reported. Acute zinc deficiency can occur in patients on long-term TPN, on penicillamine therapy, and in AE. The predominant clinical features are skin lesions and alopecia; diarrhea, mental lethargy and depression, and eye lesions have also been reported.

3.3. Diagnosis and Treatment

Zinc deficiency may be suspected in the presence of the clinical features described, particularly when they are associated with a suggestive nutritional history or a condition known to predispose to zinc deficiency. Laboratory data provide support for the diagnosis but must be interpreted with care. Plasma zinc levels are usually, but not always, low in zinc deficiency. However, hypozincemia also occurs in a number of conditions as a result of altered zinc metabolism, rather than zinc deficiency; such conditions include acute infections and hypoproteinemia, as well as the use of oral contraceptives (Beisel and Pekarek, 1972). Similarly, there is no evidence to indicate that the hypozincemia observed in Down's syndrome, pernicious anemia, and various cancers is associated with a deficiency state.

Hair zinc concentrations reflect past dietary intake but may be complicated by a decline in the rate of hair growth. Low hair zinc levels may occur in the absence of other signs of zinc deficiency, particularly in infants and preschool children in the United States. These low concentrations probably result from a marginal intake which may provide sufficient zinc to the tissues and organs that take up zinc rapidly but which is insufficient to maintain normal levels in tissues, like hair, with slow uptake. Measurements of zinc concentrations in other tissues and fluids may also be helpful in detecting or confirming zinc deficiency. These include erythrocyte and parotid saliva zinc levels and urinary zinc excretion. Demonstration of reduced activity of a zinc-dependent enzyme would also provide a biological index of zinc depletion. Unfortunately, the zinc metalloenzymes most senssitive to deficiency include pancreatic carboxypeptidase and liver alcohol dehydrogenase, which are not readily accessible. Serum alkaline phosphatase may be useful, however, as activity is frequently depressed in zinc-deficient states and increases after supplemention.

In most circumstances oral administration of 1 mg Zn^{2+}/kg body weight per day is adequate to treat zinc deficiency (Hambidge and Walravens, 1975). This amount is within the range of normal daily intakes of zinc, and no side effects have been reported. Zinc is equally well absorbed as the sulfate or the acetate, although the former may cause gastrointestinal irritation, which is reduced if the sulfate is taken with meals. Both forms are soluble and thus can be administered in a liquid, which is more suitable for children. Intravenous requirements for patients maintained on TPN appear to be about 20 to 40 μg Zn^{2+}/kg per day. In acute zinc deficiency syndromes, improvement in the patient's condition may occur very rapidly with the institution of zinc therapy. Mood changes have been observed within a few hours, diarrhea may be completely controlled within 3 to 4 days, and skin lesions may clear up in 2 weeks (Kay and Tasman-Jones, 1975).

4. EPIDEMIOLOGICAL ASPECTS OF ZINC DEFICIENCY

4.1. Fetus

Animal studies have shown the importance of adequate maternal zinc nutrition for normal fetal growth and development (Underwood, 1977). Hurley and Swenerton (1971) found that bone and liver zinc in pregnant rats was not mobilized even under conditions of deficiency that would cause teratogenesis. This may also be the case in the human mother, so that poor dietary intake during pregnancy not only will cause the mother to become zinc depleted but may also adversely affect her fetus. Several dietary surveys have indicated that zinc intakes during pregnancy are often poor (Sandstead, 1973). Maternal zinc deficiency has also been suggested as a factor in the high incidence of anencephaly found in Alexandria, Egypt, and Shiraz, Iran (Sever, 1973). A high incidence of congenital malformations has been observed in adult women suffering from AE (Hambidge et al., 1975). Of the seven reported pregnancies in such patients, one resulted in spontaneous abortion, and one in delivery of an anencephalic stillbirth with multiple severe skeletal abnormalities. The malformations of the organ systems involved were among the most common that have been reported in the offspring of zinc-deficient rats. The congenital malformations could not be attributed to maternal drug therapy but did appear to be related to the severity of the disease and hence possibly to the severity of the maternal zinc deficiency.

4.2. Infants

The newborn infant does not appear to have a store of zinc like the reserves of iron and copper, and tissue concentrations are similar to adult levels (Casey and Robinson, 1978). Large negative zinc balances, arising partly from high urinary losses, have been reported in the first weeks after birth (Cavell and Widdowson, 1964). There is some evidence that zinc can be mobilized from bone in the young animal during zinc deficiency (Calhoun et al., 1978); nonetheless, the neonate very quickly becomes dependent on a dietary supply of zinc. Recommended daily intakes have been estimated as 3 mg (National Research Council, 1974) and 6 mg (World Health Organization, 1973) for infants 0 to 6 months of age and as 5 mg for 7 to 12 months of age (World Health Organization, 1973; National Research Council, 1974). Such intakes are generally not attained by the breast-fed infant. Mature human milk contains 0.1 to 4 μg Zn/ml (Picciano and Guthrie, 1976), which supplies a maximum of about 2.5 mg Zn/day. Full-term infants fed solely on human milk do not, however, show signs of zinc depletion, and plasma levels at 3 months of age are the same as those found in healthy adults. Several studies have suggested that the availability of zinc from breast milk is rather higher than that from other milks (Johnson and Evans, 1978), and absorption and retention of zinc by low-birth-weight infants is greater from human milk than from a cow's milk formula (Widdowson et al., 1974). Further evidence for the greater bioavailability of zinc in human milk is provided by its efficacy in AE; breast milk may improve the clinical status of some patients, and the onset of symptoms is usually delayed until weaning in breast-fed infants. The consumption of colostrum, which has a zinc content ranging up to 20 μg/ml, may help to maintain the zinc status of breast-fed infants. Mice deprived of their mother's colostrum were found to develop signs of severe zinc deficiency (Nishimura, 1953).

The zinc content of cow's milk is about 3 to 4 μg/ml, similar to that of human milk during the first months of lactation (Casey, 1977), but the bioavailability is somewhat lower. Most of the zinc in milk is associated with the protein fraction. The protein content of cow's milk is usually reduced in the manufacture of infant milk formulas to correspond to the level in human milk. There is a corresponding reduction in the zinc content, and the final concentration of zinc in unsupplemented formulas is generally less than 2 μg/ml. Although soy-based formulas may have a higher zinc content, bioavailability is lower than that in cow's milk. Mean zinc concentrations in plasma (Walravens and Hambidge, 1976) and in hair (Strain et al., 1967) of formula-fed infants in the United States have been reported to be considerably lower than those of older children and adults and of breast-fed infants.

Higher zinc levels, similar to those in adults, have been reported for infants in Germany (Hellwege, 1971), Sweden (Berfenstam, 1952), and New Zealand (Douglas et al., 1976). A report from Japan indicated that breast-fed infants tended to have a higher serum zinc level at 3 months than formula-fed infants of the same age (Ohtake, 1977). Plasma zinc levels in young infants have been found to be dependent on dietary intakes, so that the low concentrations observed in American infants could be explained by suboptimal zinc nutrition. Hair levels of zinc in these infants also showed a sharp decline in the first year, which was not seen in Thai or bottle-fed British infants (Hambidge, 1974).

Further evidence that nutritional zinc deficiency severe enough to retard growth occurs in the United States was provided by the results of a recent investigation. This was a double-blind study in which the infants in the control group were fed a commercial infant milk formula providing 1.8 mg Zn/l and the test group received the same formula with additional zinc to provide 5.8 mg/l. At 6 months of age the height and weight increments of the males in the test group were significantly greater than those of the control group. The test infants also had higher plasma zinc levels (Walravens and Hambidge, 1976). Recent experience at the University of Colorado Medical Center indicates that zinc deficiency may be an etiological factor in some cases of failure to thrive.

Premature infants may have a greater zinc requirement because of catch-up growth, but a lower intake than full-term infants. Longer periods of negative balance have also been reported in the premature neonate (Tkachenko, 1970), and retention of zinc from cow's milk preparations is poor (Widdowson et al., 1974). The premature infant is therefore at greater risk of becoming zinc depleted. However, little information is available on the zinc status of breast-fed or formula-fed premature infants.

4.3. Preschool Children

Preschool children (1 to 4 years) have received considerable attention in dietary surveys and nutrition intervention programs, but only recently has zinc status been included in such studies. A study in Denver found that mean hair zinc levels of preschool children from middle-income families were lower than those of older children and adults (Hambidge et al., 1972). Laboratory data suggest that zinc nutrition may also be poor in preschoolers from low-income families. mean hair and plasma values in these children are lower than the levels in middle-income children, and abnormally low plasma zinc values (<68 μg/100 ml) were present in one third of this group. These children were enrolled in the Denver Head Start program and were selected on the basis of their low height percentiles (Hambidge et al., 1976). It is possible that poor zinc nutrition may be one of the environmental factors responsible for the high incidence of low growth centiles seen in this population group and similar ones. Low levels of hair zinc have also been reported for normal preschool children from middle- and upper-income groups in Chandigarh, India (Hambidge et al., 1974), but not in some other countries, including Thailand, Peru (Bradfield et al., 1969), and Panama (Klevay, 1970).

Plasma zinc concentrations have been measured in children suffering from protein-calorie malnutrition (PCM) in Cairo (Sandstead et al., 1965), Pretoria (Smit and Pretorius, 1964), Cape Town (Hansen and Lehmann, 1969), and Hyderabad (Kumar and Rao, 1973). In all locations, levels were low on admission and remained low in the Cairo and Pretoria groups after serum proteins had returned to normal. Hair zinc levels in children with PCM in Peru (Bradfield et al., 1969) and Chandigarh (Hambidge et al., 1974) were similar to those of normal children of the same age. These results are difficult to interpret because the rate of hair growth is adversely affected by protein deprivation; the Chandigarh group had significantly lower hair zinc levels on rehabilitation than during the acute illness. However, zinc deficiency appears to be associated with PCM. A recent study in Jamaica (Golden et al., 1978) suggested that zinc deficiency is a cause of the immunoincompetence seen in malnutrition, specifically impairing the cell-mediated system.

4.4. Older Children

In a survey of the trace element levels in the hair of a group of normal children (4 to 17 years) from middle- and upper-income American families, 10 were found to have zinc concentrations of less than 70 μg/g (Hambidge et al., 1972). Of these 10 children, 8 had low height centiles not attributable to other causes, and 5 of 6 tested showed consistent evidence of objective hypoguesia. These children also had a history of poor appetitie, particularly a low consumption of meats, which are the best source of available zinc. Several months of zinc supplementation (0.2 to 0.4 mg/kg · day) led to an increase in hair zinc levels and an improvement, to normal, of taste acuity. It was concluded that these otherwise normal children were suffering from a dietary deficiency of zinc, sufficient to impair taste acuity and possibly also to disturb the appetite and limit the growth rate.

A syndrome similar to that seen in adolescents with nutritional dwarfism has been described in school-age children in Iran (Eminians et al., 1967). Clinical features include anemia, hepatosplenomegaly, and growth retardation. Children (both boys and girls) with this syndrome were found to have mean serum zinc levels significantly lower than those of normal village and suburban children of the same age. Hair zinc levels of the affected girls were also lower than those of the normal groups. It seems likely that zinc deficiency contributed to the poor growth of these children, particularly as their diets, which contain large amounts of phytate and fiber, are known to be low in available zinc. The syndrome is most frequently seen in the small rural communities in which there is also a high incidence of adolescent nutritional dwarfism.

Serum zinc levels of normal school-age children have been reported from West Germany (Kasperek et al., 1977) and New Zealand (Lines et al., 1977). Values were generally not different from adult levels. The New Zealand 12-year-olds had significantly lower zinc levels than other age groups, but there was no other evidence of zinc depletion in these children. Low serum zinc concentrations have also been reported to occur in children on synthetic diets, particularly for phenylketonuria (PKU) (Kasperek et al., 1977), although the New Zealand study did not find this.

4.5. Adolescents

Zinc deficiency was first recognized in 1961 as being a major factor in the syndrome of nutritional dwarfism seen in adolescents in rural areas of Iran and Egypt (Prasad et al., 1961). About 3% of the adolescent population in these areas may be affected, and a similar syndrome is found in other countries, including Turkey, Tunisia, Morocco, Portugal, and Panama (Halsted et al, 1974). Several cases have also been reported in the United States, including one early in this century (Caggiano et al., 1969; Lemann, 1910). The syndrome was first described in males but has subsequently been found to occur in females also (Ronaghy and Halsted, 1975). The age at diagnosis is generally between 12 and 21 years.

The clinical picture varies somewhat with geographical location, severity of the zinc deficiency, etiological factors involved, and occurrence of other nutritional deficiencies. The latter frequently include iron deficiency and sometimes inadequate protein nutrition. The principal features resulting from the zinc deficiency are extreme growth retardation (dwarfism) and lack of sexual development. Other features attributed to the zinc deficiency include roughened, hyperpigmented skin, lethargy, poor appetite, and hepatosplenomegaly (Prasad et al., 1963a, 1963b). Geophagia, a common feature of this syndrome in Iran (Ronaghy, 1974), was thought previously to be an etiological factor contributing to zinc deficiency, but it now appears that pica, including geophagia, may result from the zinc deficiency (Hambidge and Silverman, 1973). Less severe cases of zinc deficiency in these adolescents are characterized by moderate growth failure and delayed sexual maturation in otherwise normal boys. Laboratory evidence for zinc deficiency in this syndrome include low levels of zinc in plasma, hair, red cells, urine, and sweat. The metabolic abnormalities caused by the zinc deficiency and responsible for the retarded growth and delayed sexual maturation have not been elucidated. One factor contributing to the severe zinc deficiency in the Middle East is the widespread use of unleavened breads as a major staple food. These breads are made from flours of very high extraction rate and contain considerable amounts of phytate and fiber. Thus, although intakes of zinc may exceed the recommended allowances by a wide margin, the element is largely unavailable for absorption (Reinhold et al., 1976). Other etiological factors involved may include excessive losses of zinc in sweat and, in Egypt, gastrointestinal blood loss caused by infestation with schistosomiasis and hookworm (Sandstead et al., 1967).

Zinc supplementation produces an increase in growth rate and sexual maturation; controlled studies have demonstrated that this effect is specific for zinc (Halsted et al., 1972). In the uncontrolled village situation the response to zinc therapy is slower than that produced in the controlled studies, and a higher level of supplementation is required, apparently because of the factors in the local diet that inhibit zinc absorption (Ronaghy et al., 1974).

4.6. Adults

Once growth and development have stopped, relative zinc requirements are lower and the effects of zinc deficiency are less severe. However, a number of factors suggest that suboptimal zinc nutrition may be present in certain groups of adults such as the elderly. One study of elderly men and women living in an institution for the aged in the United States found a daily intake of zinc of 7 to 9 mg. Of the 65 subjects, 5% had hair zinc levels below 75 μg/g (Greger, 1977). Free-living elderly people participating in an urban feeding program in Indiana obtained about 10 mg Zn/day in their diet. Taste acuity was generally less in these subjects than in younger adults, but none had hair zinc levels below 70 μg/g (Greger and Sciscoe, 1977). Taste acuity in elderly subjects does not appear to be improved by zinc supplementation (Greger and Geissler, 1978). Thus it is difficult to separate the effects of aging on zinc status from those of marginal zinc nutrition.

Reinhold and co-workers (1966) reported that village women in Iran had lower hair zinc levels than women living in cities. About 16% of the adults in these rural areas have low plasma zinc concentrations (Sarram et al., 1969). Like the zinc-deficient adolescents, they subsist mainly on unleavened wholemeal breads, which considerably reduce the availability of dietary zinc.

Wound healing is impaired in patients with low plasma zinc levels and may be accelerated by zinc therapy (Pories and Strain, 1974). Improvement has also been noted in the rate of healing of venous leg ulcers, but only in patients with hypozincemia before the therapy. There is some indication that idiopathic hypoguesia may also benefit from zinc supplementation (Henkin et al., 1971).

4.7. Pregnancy and Lactation

During the second and third trimesters of pregnancy about 750 μg Zn/day is required for the growth of the conceptus (Prasad, 1978). The increased recommended daily allowance of 20 mg is rarely met by the dietary intake, even in well-to-do Western communities. Animal studies suggest that bone zinc cannot be utilized (Hurley and Swenerton, 1971), so the mother may become zinc depleted, particularly if she has had repeated, close pregnancies.

Plasma zinc concentrations fall throughout pregnancy; however, this decrease may be attributed to increased body levels of estrogen, as women on oral contraceptives also have low plasma zinc levels (Prasad et al., 1975b), and may also reflect maternal hemodilution. Hair zinc levels were found to be lower at 38 weeks than at 16 weeks in a group of healthy, pregnant women with a mean parity of 1.0 (Hambidge and Droegemueller, 1974). This decline probably represents some degree of zinc depletion. Similarly, another study in the United States found a significant decline in hair zinc levels with increasing parity (Baumslag et al., 1974). Sarram and co-workers (1969) found pregnant village women in Iran had lower plasma and hair levels of zinc than women living in similar socioeconomic conditions in urban areas.

Lactation greatly increases the requirement for zinc by the mother. Recommended daily allowances have been set at 27 mg (World Health Organization, 1973) and 25 mg (National Research Council, 1974). Very few unsupplemented diets provide such high levels. Although a considerable number of lactating mothers may be at risk of marginal zinc deficiency, there have been few studies of zinc nutrition during this period. One study, conducted in Teheran, Iran, found a zinc intake of 11 mg by low-income lactating women in the third month postpartum, and of 12 mg by middle-income women. There was no evidence of overt zinc deficiency in any of the mothers, but biochemical parameters of zinc status were not measured (Geissler et al., 1978).

4.8. Total Parental Nutrition

Severe, acute zinc deficiency in patients receiving long-term TPN without zinc supplements has been reported by a number of workers. Symptoms are similar to those seen in AE and include skin rashes, alopecia, diarrhea and depression. The onset of symptoms, usually 5 to 10 weeks after the start of TPN, is related to the severity of zinc depletion in the patient. Large amounts of zinc may be lost in wound exudates; high urinary concentrations, possibly related to tissue catabolism, are also seen in postsurgical patients. The amount of zinc lost in the urine can also be increased by the type of amino acid infusion used: D-amino acids and heat sterilization of sugar-amino acid mixtures have been implicated in excessive zincuria (van Rij and McKenzie, 1978).

An extensive study of zinc deficiency in adult patients on TPN was reported from New Zealand (Kay et al., 1976). In one of this group of patients a plasma zinc level as low as 8 μg/100 ml and a urinary excretion of 22.6 mg/day were recorded before zinc therapy was instituted. The lowest recorded plasma level in all patients was below 20 μg/100 ml, and their response to intravenously administered zinc was striking. Other cases have been reported from New Zealand (van Rij and McKenzie, 1977) and Japan (Okada et al., 1975).

There have also been a number of reports of infants on prolonged TPN developing zinc deficiency. Most such cases involved premature infants. Low plasma zinc levels have been reported in these patients in the United States (Michie and Wirth, 1978) and France (Ricour et al., 1975). Overt signs of zinc deficiency were reported in two Japanese infants 20 days after commencing TPN (Arakawa et al., 1976) and in two premature infants in the United States who developed symptoms after the cessation of prolonged TPN (Sivasubramanian and Henkin, 1978).

4.9. Acrodermatitis Enteropathica

Acrodermatitis enteropathica is a familial disease with an autosomal recessive inheritance. The disorder is characterized by skin lesions, diarrhea, and alopecia, symptoms consistent with severe zinc deficiency. In the clincially active state, plasma zinc levels are considerably lower than normal. Hair levels, however, may be only slightly lower, severe zinc deficiency causing imparied hair growth rather than depressed zinc concentration. The disorder is effectively treated with oral zinc therapy (Neldner and Hambidge, 1975). The basic molecular defect has not yet been identified but appears to be in the intestine, where it causes a partial block in zinc absorption (Lombeck et al., 1975a).

The disorder is rare but appears to occur with more frequency in individuals of Mediterranian descent (Italian, Armenian, Iranian). Recent cases have been reported in the United States (Hambidge et al., 1978), West Germany (Lombeck et al., 1975b), England (Barnes and Moynahan, 1973), South Africa (Rubin et al., 1978), and Canada (Bohane et al., 1977).

4.10. Secondary Zinc Deficiency

Zinc deficiency can complicate a number of conditions affecting the gastrointestinal tract, including celiac disease, cystic fibrosis, regional enteritis, and disaccharide intolerance. The zinc deficiency is presumably due to impaired absorption and possibly excessive losses in duodenal secretions (Sandstead et al., 1976). In a case of nutritional dwarfism reported in the United States the zinc deficiency was secondary to regional enteritis of long standing (Ronaghy, 1974). Hyperzincuria with consequent poor zinc nutritional status has been found in chornic alcoholism and in cirrhosis and other liver disorders (Sullivan, 1974). Low plasma zinc levels have been reported to occur in a wide variety of conditions, including viral and bacterial infections, cancers, blood disorders, rheumatoid arthritis, schizophrenia, and cardiovascular disease (National Academy of Sciences, 1978). It is not known, however, whether these lowered levels reflect an increased requirement for zinc or an alteration in the distribution and metabolism of the element in response to the disorder.

5. SCOPE OF THE PROBLEM

In the 15 years since human zinc deficiency was demonstrated to occur, it has been observed in a wide variety of geographical areas and economic circumstances. Reported cases arise from a variety of causes and range in severity from mild depletion, with few visible symptoms, to the acute syndrome seen in TPN and AE patients and the chronic deficiency of nutritional dwarfism. The areas of the world from which zinc deficiency has been reported are shown in Figure 1; no attempt has been made to illustrate the number or severity of the cases involved. Unfortunately, at present this distribution represents awareness in diagnosing overt cases, and interest in and facilities for searching for more subtle manifestations, rather than the actual worldwide incidence of zinc deficiency.

Marginal zinc deficiency appears to occur more readily in certain groups, the most "at risk" being infants and young children. Many reports of deficiency in this age group have come from the United States. This apparently higher incidence may represent a greater amount of research in this country than elsewhere, although reports from Sweden, West Germany, and New Zealand suggest

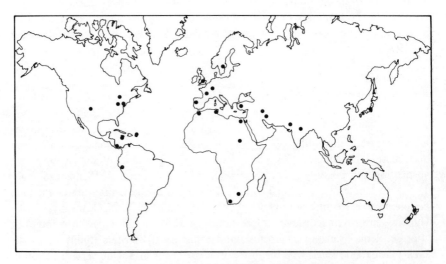

Figure 1.   Areas in which zinc deficiency has been reported.

that children in these countries enjoy better zinc status. Other occurrences of low zinc status in young children have been observed in conjunction with PCM. It seems likely that most children with PCM have some degree of zinc depletion, but its effects are masked by the general picture of malnutrition.

Old age, pregnancy, lactation, and alcoholism are also associated with higher incidence of poor zinc nutrition. Again, such observations have been made mainly in the United States; but as workers elsewhere become aware of the problem, it will undoubtedly be recognized in other countries as well.

Dietary surveys of groups of healthy adults in a number of countries have shown that mean zinc intakes are frequently less than the recommended daily allowances (Waslien, 1976). Intakes of less than half the recommended values are not uncommon, and it has been suggested from such results that substantial portions of the U.S. population may be at risk of zinc deficiency (Sandstead, 1973; Mertz, 1974). However, care must be taken in using dietary intake as an indicator of risk of depletion, as other dietary and environmental factors may have significant effects on zinc status. For example, zinc deficiency occurs in rural Iran despite relatively high dietary intakes because other dietary factors make ingested zinc largely unavailable. Conversely, Tokelau Islanders, whose diet consists mainly of coconut and fish, have a very low zinc intake (4.5 mg/day from food). However, their zinc status, as measured by hair and serum levels and urinary excretion, appears to be good (McKenzie et al., 1978).

More severe zinc deficiency appears to be prevalent in some areas of the Middle East and North Africa. It has been estimated that up to 3% of adolescents in rural areas show varying degrees of deficiency, and the condition has also been reported in school children. Probably some degree of zinc depletion is present in most people in such areas, particularly in infants and pregnant and lactating women. The zinc deficiency seen in these countries is generally associated with the consumption of unrefined cereals as a major part of the diet. It may be suspected to occur, therefore, in other underdeveloped areas where the staple foodstuff is a cereal.

It is obvious from Figure 1 that poor zinc nutrition is a widespread problem. However, there are insufficient data to allow an estimation of the incidence of zinc deficiency.

ACKNOWLEDGMENTS

This work was supported by U.S. Public Health Service Grant R01-AM-12432 from the National Institute of Arthritis and Metabolic Disease, by Grant RR-69 from the General Clinical Research Centers Program of the Division of Research Resources, National Institutes of Health, and by a grant from the Thrasher Research Fund.

REFERENCES

Arakawa, Ts., Tamura, T., Igarashi, Y., Suzuki, H., and Standstead, H. H. (1976). "Zinc Deficiency in Two Infants during Total Parenteral Alimentation for Intractable Diarrhoea," *Am. J. Clin. Nutr.*, **29**, 197–204.

Barnes, P. M. and Moynahan, E. J. (1973). "Zinc Deficiency in Acrodermatitis Enteropathica," *Proc. R. Soc. Med.*, **66**, 327–329.

Baumslag, N., Yeager, D., Levin, L., and Petering, H. G. (1974). "Trace Metal Content of Maternal and Neonate Hair," *Arch. Environ. Health*, **29**, 186–191.

Beisel, W. R. and Pekarek, R. S. (1972). "Acute Stress and Trace Element Metabolism," In C. C. Pfeiffer, Ed., *Neurobiology of the Trace Metals Zinc and Copper*. Academic Press, New York, pp. 53–82.

Beisel, W. R., Pekarek, R. S., and Wannemacher, R. W. (1976). "Homeostatic Mechanisms Affecting Plasma Zinc Levels in Acute Stress." In A. S. Prasad, Ed., *Trace Elements in Human Health and Disease*, Vol. 1. Academic Press, New York, pp. 87–106.

Berfenstam, R. (1952). "Studies on Blood Zinc," *Acta Paediatr. (Scand.)*, **41**, Supple. 87, 5–105.

Bohane, T. D., Cutz, E., Hamilton, J. R., and Gall, D. G. (1977). "Acrodermatitis Enteropathica, Zinc, and the Paneth Cell," *Gastroenterology*, **73**, 587–592.

Bradfield, R. B., Yee, T. and Baertl, J. M. (1969). "Hair Zinc Levels of Andean Indian Children during Protein-Calorie Malnutrition," *Am. J. Clin. Nutr.*, **22**, 1349–1353.

Brown, E. D. McGuckin, M. A., Wilson, M., and Smith, J. C. (1976). "Zinc in Selected Hospital Diets," *J. Am. Diet. Assoc.*, **69**, 632–635.

Caggiano, V., Schnitzler, R., Strauss, W., Baker, R. K., Carter, A. C., Josephson, A. S., and Wallach, S. (1969). "Zinc Deficiency in a Patient with Retarded Growth, Hypogonadism, Hypogammaglobulinemia, and Chronic Infection," *Am. J. Med. Sci.*, **257**, 305–319.

Calhoun, N. R., Smith, J. C., and Becker, K. L. (1974). "The Role of Zinc in Bone Metabolism," *Clin. Orthopaed.*, **103**, 212–234.

Calhoun, N. R., McDaniel, E. G., Howard, M. P., and Smith, J. C. Jr. (1978). "Loss of Zinc from Bone during Deficiency State," *Nutr. Rep. Int.*, **17**, 299–306.

Casey, C. E. (1977). "The Content of Some Trace Elements in Infant Milk Foods and Supplements Available in New Zealand," *N.Z. Med. J.*, **85**, 275–278.

Casey, C. E. and Robinson, M. F. (1978). "Copper, Manganese, Zinc, Nickel, Cadmium and Lead in Human Foetal Tissues," *Br. J. Nutr.*, **39**, 639–646.

Cavell, P. A. and Widdowson, E. M. (1964). "Intakes and Excretions of Iron, Copper, and Zinc in the Neonatal Period," *Arch. Dis. Child.*, **39**, 496–501.

Douglas, B. S., Lines, D. R., and Tse, C. A. (1976). "Serum Zinc Levels in New Zealand Children," *N.Z. Med. J.*, **83**, 192–194.

Eggleton, W. G. E. (1939). "The Zinc Content of Epidermal Structures in Beriberi," *Biochemistry*, **33**, 403–406.

Eminians, J., Reinhold, J. G., Kfoury, G. A., Amirhakimi, G. H., Sharif, H., and Ziai, M. (1967). "Zinc Nutrition of Children in Fars Province of Iran," *Am. J. Clin. Nutr.*, **20**, 734–742.

Evans, G. W. (1976). "Zinc Absorption and Transport." In A. S. Prasad, Ed., *Trace Elements in Human Health and Disease*, Vol. 1. Academic Press, New York, pp. 181–187.

Fernandez-Madrid, F., Prasad, A. S., and Oberleas, D. (1973). "Effect of Zinc Deficiency on Nucleic Acids, Collagen, and Non-collagenous Protein of the Connective Tissue," *J. Lab. Clin. Med.*, **82**, 951–961.

Freeland, J. H. and Cousins, R. J. (1976). "Zinc Content of Selected Foods," *J. Am. Diet. Assoc.*, **68**, 526–529.

Geissler, C., Calloway, D. H., and Margen, S. (1978). "Lactation and Pregnancy in Iran. II: Diet and Nutritional Status," *Am. J. Clin. Nutr.*, **31**, 341–354.

Golden, M. H. N., Golden, B., Harland, P.S.E.G., and Jackson, A. A. (1978). "Zinc and Immunocompetence in Protein-Energy Malnutrition," *Lancet*, i, 1226–1227.

Greger, J. L. (1977). "Dietary Intake and Nutritional Status in Regard to Zinc of Institutionalized Aged," *J. Gerontol.*, **32**, 549–553.

Greger, J. L. and Geissler, A. H. (1978). "Effect of Zinc Supplementation on Taste Acuity of the Aged," *Am. J. Clin. Nutr.*, **31**, 633–637.

Greger, J. L. and Sciscoe, B. S. (1977). "Zinc Nutriture of Elderly Participants in an Urban Feeding Program," *J. Am. Diet. Assoc.*, **70**, 37–41.

Guthrie, B. E. (1975). "Chromium, Manganese, Copper, Zinc and Cadmium Content of New Zealand Foods," *N.Z. Med. J.*, **82**, 418–424.

Haeflein, K. A. and Rasmussen, A. I. (1977). "Zinc Content of Selected Foods," *J. Am. Diet. Assoc.*, **70**, 610–616.

Halsted, J. A., Ronaghy, H. A., Abadi, P., Haghshenass, M., Amirhakemi, G. H., Barakat, R. M., and Reinhold, J. G. (1972). "Zinc Deficiency in Man," *Am. J. Med.*, **53**, 277–284.

Halsted, J. A., Smith, J. C., and Irwin, M. I. (1974). "A Conspectus of Research on Zinc Requirements of Man," *J. Nutr.*, **104**, 345–378.

Hambidge, K. M. (1974). "The Clinical Significance of Trace Element Deficiencies in Man," *Proc. Nutr. Soc.*, **33**, 249–255.

Hambidge, K. M. (1977). "Trace Elements in Pediatric Nutrition," *Adv. Pediatr.* **24**, 191–231.

Hambidge, K. M. and Droegemueller, W. (1974). "Changes in Plasma and Hair Concentrations of Zinc, Copper, Chromium and Manganese during Pregnancy," *Obstet. Gynecol.*, **44**, 666–672.

Hambidge, K. M. and Silverman, A. (1973). "Pica with Rapid Improvement after Dietary Zinc Supplementation," *Arch. Dis. Child.*, **48**, 567–568.

Hambidge, K. M. and Walravens, P. (1975). "Trace Elements in Nutrition," *Prac. Pediatr.*, **1**, 1–40.

Hambidge, K. M., Hambidge, C., Jacobs, M., and Baum, J. D. (1972). "Low Levels of Zinc in Hair, Anorexia, Poor Growth and Hypogeusia in Children," *Pediatr. Res.*, **6**, 868–874.

Hambidge, K. M., Walravens, P., Kumar, V., and Tuchinda, C. (1974). "Chromium, Zinc, Manganese, Copper, Nickel, Iron and Cadmium Concentrations in the Hair of Residents of Chandigarh, India, and Bangkok, Thailand." In D. D. Hemphill, Ed., *Trace Substances in Environmental Health*—VIII. University of Missouri, Columbia, pp. 39–44.

Hambidge, K. M., Neldner, K. H., and Walravens, P. A. (1975). "Zinc Acrodermatitis Enteropathica, and Congenital Malformations," *Lancet*, i, 577–578.

Hambidge, K. M., Walravens, P. A., Brown, R. M., Webster, J., White, S., Anthony, M., and Roth, M. L. (1976). "Zinc Nutrition of Preschool Children in the Denver Head Start Program," *Am. J. Clin. Nutr.*, **29**, 734–738.

Hambidge, K. M., Neldner, K. H., Walravens, P. A., Weston, W. L., Silverman, A., Sabol, J. L., and Brown, R. M. (1978). "Zinc in Acrodermatitis Enteropathica." In K. M. Hambidge and B. L. Nichols, Eds., *Zinc and Copper in Clinical Medicine*. Spectrum, New York, pp. 81–98.

Hansen, J. D. L. and Lehmann, B. H. (1969). "Serum Zinc and Copper Concentrations in Children with Protein-Calorie Malnutrition," *S. Afr. Med. J.*, **43**, 1248–1251.

Hauer, E. C. and Kaminski, M. V. (1978). "Trace Metal Profile of Parenteral Nutrition Solutions," *Am. J. Clin. Nutr.*, **31**, 264–268.

Hellwege, H. H. (1971). "Der Serumzinkspiegel und seine Veranderunger bei einigen Frankheiten im Kindersalter," *Monatsschr. Kinderheilkd.*, **119**, 37–41.

Henkin, R. I., Schechter, P. J. Hoye, R. and Mattern, C. F. T. (1971). "Idiopathic Hypogeusia with Dysgeusia, Hyposmia and Dysosmia," *J. Am. Med. Assoc.*, **217**, 434–440.

Hurley, L. S. and Schrader, R. E. (1972). "Congenital Malformations of the Nervous System in Zinc-Deficient Rats." In C. C. Pfeiffer, Ed., *Neurobiology of the Trace Metals Zinc and Copper*. Academic Press, New York, pp. 7–51.

Hurley, L. D. and Swenerton, H. (1971). "Lack of Mobilization of Bone and Liver Zinc under Teratogenic Conditions of Zinc Deficiency in Rats," *J. Nutr.*, **101**, 597–604.

Johnson, P. E. and Evans, G. W. (1978). "Relative Zinc Availability in Human Breast Milk, Infant Formulas, and Cow's Milk," *Am. J. Clin. Nutr.*, **31**, 416–421.

Kasperek, K., Feinendegen, L. E., Lombeck, I., and Bremer, H. J. (1977). "Serum Zinc Concentration during Childhood," *Eur. J. Pediatr.*, **126**, 199–202.

Kay, R. G. and Tasman-Jones, C. (1975). "Acute Zinc Deficiency in Man during Intravenous Alimentation," *Aust. N.Z. J. Surg.*, **45**, 325–330.

Kay, R. G., Tasman-Jones, C., Pybus, J., Whiting, R., and Black, H. (1976). "A Syndrome of Acute Zinc Deficiency during Total Parenteral Alimentation in Man," *Ann. Surg.*, **183**, 331–340.

Klevay, L. M. (1970). "Hair as a Biopsy Material. I: Assessment of Zinc Nutriture," *Am. J. Clin. Nutr.*, **23**, 284–289.

Kumar, S. and Rao, K. S. J. (1973). "Plasma and Erythrocyte Zinc Levels in Protein-Calorie Malnutrition," *Nutr. Metab.*, **15**, 364–371.

Lemann, I. I. (1910). "A Study of the Type of Infantilism in Hookworm Disease," *Arch. Int. Med.*, **6**, 139–146.

Lines, D. R., Bell, E. B., and Pybus, J. (1977). "Zinc Levels in Childhood Health and Disease," *Proc. Nutr. Soc. N.Z.*, **2**, Part 3, 31–37.

Lombeck, I., Schnippering, H. G., Ritzl, F., Feinendegen, L. E., and Bremer, H. J. (1975a). "Absorption of Zinc in Acrodermatitis Enteropathica," *Lancet*, i, 885.

Lombeck, I., Schnippering, H. G., Kasperek, K., Ritzl, F., Kästner, H., Feinendegen, L. E., and Bremer, H. J. (1975b). "Akrodermatitis enteropathica— eine Zinkstoffwechselstörung mit Zink Malabsorption," *Z. Kinderheilkd.*, **120**, 181–189.

McBean, L. D., Mahloudji, M., Reinhold, J. G., and Halsted, J. A. (1971). "Correlation of Zinc Concentrations in Human Plasma and Hair," *Am. J. Clin. Nutr.*, **24**, 506–509.

McClain, P. E., Wiley, E. R., Beecher, G. R., Anthony, W. L., and Hsu, J. M. (1973). "Influence of Zinc Deficiency on Synthesis and Crosslinking of Rat Skin Collagen," *Biochem. Biophys. Acta*, **304**, 457–465.

McKenzie, J. M. (1974). "Tissue Concentrations of Cadmium, Zinc and Copper from Autopsy Samples," *N.Z. Med. J.*, **79**, 1016–1019.

McKenzie, J. M., Fosmire, G. J., and Sanstead, H. H. (1975). "Zinc Deficiency during the Latter Third of Pregnancy: Effects on Fetal Rat Brain, Liver, and Placenta," *J. Nutr.*, **105**, 1466–1475.

McKenzie, J. M., Guthrie, B. E., and Prior, I. A. M. (1978). "Zinc and Copper Status of Polynesian Residents in the Tokelau Islands," *Am. J. Clin. Nutr.*, **31**, 422–428.

Mertz, W. (1974). "The Effects of Zinc in Man: Nutritional Considerations." In W. J. Pories, W. H. Strain, J. M. Hsu, and R. L. Woosley, Eds., *Clinical Applications of Zinc Metabolism*. Charles C Thomas, Springfield, Ill., pp. 93–100.

Michie, D. D. and Wirth, F. H. (1978). "Plasma Zinc Levels in Premature Infants Receiving Parenteral Nutrition," *J. Pediatr.*, **92**, 798–800.

Morrison, S. A., Russell, R. M., Carney, E. A., and Oaks, E. V. (1978). "Zinc Deficiency: a Cause of Abnormal Dark Adaptation in Cirrhotics," *Am. J. Clin. Nutr.,* **31**, 276–281.

Mutch, P. and L. S. Hurley (1974). "Effect of Zinc Deficiency during Lactation on Postnatal Growth and Development of Rats," *J. Nutr.,* **104**, 828–842.

National Academy of Sciences (1978). *Zinc.* Report of a Subcommittee of the National Research Council, Washington, D. C.

National Research Council (1974). *Recommended Dietary Allowances,* 8th ed. Food and Nutrition Board, National Academy of Sciences, Washington, D.C.

Neldner, K. H. and Hambidge, K. M. (1975). "Zinc Therapy of Acrodermatitis Enteropathica," *N. Engl. J. Med.,* **292**, 879–882.

Nishimura, H. (1953). "Zinc Deficiency in Suckling Mice Deprived of Colostrum," *J. Nutr.,* **49**, 79–97.

Ohtake, M. (1977). "Serum Zinc and Copper Levels in Healthy Japanese Infants," *Tohoku J. Exp. Med.,* **123**, 265–270.

Okada, A., Takagi, Y., Itakura, F., Satani, M., Manabe, H., and Iida, Y. (1975). "Zinc Deficiency during Intravenous Alimentation." In *Proceedings of Xth International Congress of Nutrition, Kyoto, Japan,* p. 236.

Osis, D., Kramer, L., Wiatrowski, E., and Spencer, H. (1972). "Dietary Zinc Intake in Man," *Am. J. Clin. Nutr.,* **25**, 582–588.

Parisi, A. F. and Vallee, B. L. (1969). "Zinc Metalloenzymes: Characteristics and Significance in Biology and Medicine," *Am. J. Clin. Nutr.,* **22**, 1222–1239.

Picciano, M. F. and Guthrie, H. A. (1976). "Copper, Iron, and Zinc Contents of Mature Human Milk," *Am. J. Clin. Nutr.,* **29**, 242–254.

Pories, W. J. and Strain, W. H. (1974). "Zinc Sulphate Therapy in Surgical Patients." In W. J. Pories, W. H. Strain, J. M. Hsu, and R. L. Woosley, Eds., *Clinical Applications of Zinc Metabolism.* Charles C Thomas, Springfield, Ill., pp. 139–157.

Prasad, A. S. (1978). "Zinc Deficiency in Man." In K. M. Hambidge and B. L. Nichols, Eds., *Zinc and Copper in Clinical Medicine.* Spectrum, New York, pp. 1–14.

Prasad, A. S., Halsted, J. A., and Nadimi, M. (1961). "Syndrome of Iron Deficiency Anemia, Hepatosplenomegaly, Hypogonadism, Dwarfism and Geophagia," *Am. J. Med.,* **31**, 532–546.

Prasad, A. S., Miale, A., Farid, Z., Sanstead, H. H., and Schulert, A. R. (1963a). "Zinc Metabolism in Patients with the Syndrome of Iron Deficiency Anemia, Hepatosphenomegaly, Dwarfism, and Hypogonadism," *J. Lab. Clin. Med.,* **61**, 534–548.

Prasad, A. S., Schulert, A. R., Miale, A., Farid, Z., and Sandstead, H. H. (1963b). "Zinc and Iron Deficiencies in Male Subjects with Dwarfism and Hypogonadism but without Ancylostomiasis, Schistosomiasis or Severe Anemia," *Am. J. Clin. Nutr.*, **12**, 437–444.

Prasad, A. S., Miale, A., Farid, Z., Sandstead, H., Schulert, A. R., and Darby, W. J. (1963c). "Further Biochemical Studies in Patients with Syndrome of Iron Deficiency Anemia, Hepatosplenomegaly, Dwarfism and Hypogonadism," *Arch. Int. Med.*, **111**, 407–428.

Prasad, A. S., Schoomaker, E. B., Ortega, J., Brewer, G. J. , Oberleas, D., and Oelshlegel, F. J. (1975a). "Zinc Deficiency in Sickle Cell Disease," *Clin. Chem.*, **21**, 582–587.

Prasad, A. S., Oberleas, D., Lei, K. Y., Moghissi, K. S., and Stryker, J. C. (1975b). "Effect of Oral Contraceptive Agents on Nutrients. 1: Minerals," *Am. J. Clin. Nutr.*, **28**, 377–384.

Reinhold, J. G., Kfoury, G. A., Ghalambor, M. A., and Bennett, J. C. (1966). "Zinc and Copper Concentrations in Hair of Iranian Villagers," *Am. J. Clin. Nutr.*, **18**, 294–300.

Reinhold, J. G., Faradji, B., Abadi, P., and Ismail-Beigi, F. (1976). "Binding of Zinc to Fiber and Other Solids of Wholemeal Bread." In A. S. Prasad, Ed., *Trace Elements in Human Health and Disease*, Vol. 1. Academic Press, New York, pp. 163–180.

Ricour, C., Gros, J., Maziere, B., and Comar, D. (1975). "Trace Elements in Infants on T.P.N." In *Proceedings of the Xth International Congress of Nutrition, Kyoto, Japan*, p. 236.

Ronaghy, H. A. (1974). "Dwarfism and Delayed Sexual Maturation Caused by Zinc Deficiency." In W. J. Pories, W. H. Strain, J. M. Hsu and R. L. Woosley, Eds., *Clinical Applications of Zinc Metabolism*. Charles C Thomas, Springfield, Ill., pp. 119–129.

Ronaghy, H. A. and Halsted, J. A. (1975). "Zinc Deficiency Occurring in Females: Report of Two Cases," *Am. J. Clin. Nutr.*, **28**, 831–836.

Ronaghy, H. A., Reinhold, J. G., Mahloudji, M., Ghavami, P., Spivey Fox, M. R., and Halsted, J. A. (1974). "Zinc Supplementation of Malnourished Schoolboys in Iran: Increased Growth and Other Effects," *Am. J. Clin. Nutr.*, **27**, 112–121.

Rubin, I. L., Hansen, J. D. L., Goldberg, B., Presbury, D. G. C., and Kilroe-Smith, T. A. (1978). "Acrodermatitis Enteropathica–a Zinc Deficiency State," *S. Afr. Med. J.*, **53**, 497–498.

Sandstead, H. H. (1973). "Zinc Nutrition in the United States," *Am. J. Clin. Nutr.*, **26**, 1251–1260.

Sandstead, H. H., Shukry, A. S., Prasad, A. S., Gabr, M. K., Hifney, A. E., Mokhtar, N., and Darby, W. J. (1965). "Kwashiorkor in Egypt," *Am. J. Clin. Nutr.*, **17**, 15–26.

Sandstead, H. H., Prasad, A. S., Schulert, A. R., Farid, Z., Miale, A., Bassilly, S., and Darby, W. J. (1967). "Human Zinc Deficiency, Endocrine Manifestations and Response to Treatment," *Am. J. Clin. Nutr.*, **20**, 422–442.

Sandstead, H. H., Vo-Khactu, K. P., and Solomons, N. (1976). "Conditioned Zinc Deficiencies." In A. S. Prasad and D. Oberleas, Eds., *Trace Elements in Human Health and Disease,* Vol. 1. Academic Press, New York, pp. 33–49.

Sarram, M., Younessi, M., Khorvash, P., Kfoury, G. A., and Reinhold, J. G. (1969). "Zinc Nutrition in Human Pregnancy in Fars Province, Iran," *Am. J. Clin. Nutr.,* 22, 726–732.

Schlettwein-Gsell, D. and Mommsen-Straub, S. (1970). Übersicht Spurenelemente in Lebensmittel. I: Zink," *Int. Z. Vit. Forsch.,* 40, 659–692.

Sever, L. S. (1973). "Zinc Deficiency in Man," *Lancet,* i, 887.

Sivasubramanian, K. N. and Henkin, R. I. (1978). "Clinical Changes and Low Serum Zinc and Copper in Premature Infants Corrected by Exogenous Zinc," *Fed. Proc.,* 37, 324.

Smit, Z. M. and Pretorius, P. J. (1964). "Studies in Metabolism of Zinc," *J. Trop. Pediatr.,* 9, 105–112.

Smith, J. E., Brown, E. D., and Smith, J. C. (1974). "The Effect of Zinc Deficiency on the Metabolism of Retinol-Binding Protein in the Rat," *J. Lab. Clin. Med.,* 84, 692–697.

Spencer, H., Osis, D., Krammer, L., and Norris, C. (1976). "Intake, Excretion, and Retention of Zinc in Man." In A. S. Prasad and D. Oberleas, Eds., *Trace Elements in Human Health and Disease,* Vol. 1. Academic Press, New York, pp. 345–361.

Strain, W. H., Lascari, A., and Pories, W. J. (1967). "Zinc Deficiency in Babies." In *Proceedings of the VIIth International Congress of Nutrition,* Vol. 5. Pergamon Press, Oxford, pp. 759–765.

Sullivan, J. F. (1974). "Zinc Deficiency and Chronic Alcoholism." In W. J. Pories, W. H. Strain, J. M. Hsu, and R. L. Woosley, Eds., *Clinical Applications of Zinc Metabolism.* Charles C Thomas, Springfield, Ill., pp. 113–118.

Sullivan, J. F. and Lankford, H. G. (1962). "Urinary Excretion of Zinc in Alcoholism and Postalcoholic Cirrhosis," *Am. J. Clin. Nutr.,* 10, 153–157.

Tkachenko, S. K. (1970). "Intake and Excretion of Some Trace Elements in Premature Infants," *Pediatriya,* 49, 10–15.

Todd, W. R., Elvehjem, C. A., and Hart, E. B. (1934). "Zinc in the Nutrition of the Rat," *Am. J. Physiol.,* 107, 146–156.

Underwood, E. J. (1977). *Trace Elements in Human and Animal Nutrition,* 4th ed., Academic Press, New York, pp. 196–242.

van Rij, A. M. and McKenzie, J. M. (1977). "Zinc in Parenteral Nutrition," *Proc. Nutr. Soc. N.Z.,* 2, Part 3, 87–94.

van Rij, A. M. and McKenzie, J. M. (1978). "Hyperzincuria and Zinc Deficiency in Total Parenteral Nutrition." In M. Kirchgessner, Ed., *Trace Element Metabolism in Man and Animals—3.*ATW, Freising-Weihenstephan, pp. 288–291.

Walravens, P. A. and Hambidge, K. M. (1976). "Growth of Infants Fed a Zinc Supplemented Formula," *Am. J. Clin. Nutr.,* 29, 1114–1121.

Walravens, P. A. and Hambidge, K. M. (1978). "Zinc Nutrition and Deficiency in Pediatrics." In K. M. Hambidge and B. L. Nichols, Eds., *Zinc and Copper in Clinical Medicine.* Spectrum, New York, pp. 49–58.

Waslien, C. I. (1976). "Human Intake of Trace Elements." In A. S. Prasad, Ed., *Trace Elements in Human Health and Disease,* Vol. II. Academic Press, New York, pp. 347–370.

Widdowson, E., Dauncey, J. and Shaw, J. C. L. (1974). "Trace Elements in Foetal and Early Postnatal Development," *Proc. Nutr. Soc.* **33**, 275–284.

World Health Organization (1973). *Trace Elements in Human Nutrition.* Technical Report Series 532, Geneva, pp. 9–13.

2

MANIFESTATIONS OF ZINC ABNORMALITIES IN HUMAN BEINGS

Ananda S. Prasad

Division of Hematology, Department of Medicine, Wayne State University School of Medicine and Harper Hospital, Detroit, Michigan, and Veterans Administration Hospital, Allen Park, Michigan

1.	**Introduction**	**30**
2.	**Causes of Zinc Deficiency**	**30**
	2.1. Nutritional causes	30
	2.2. Nutritional zinc deficiency in children in the United States	32
	2.3. Zinc nutrition and deficiency in infants	33
	2.4. Alcohol	33
	2.5. Liver disease	34
	2.6. Gastrointestinal disorders	34
	2.7. Renal diseases	34
	2.8. Neoplastic diseases	35
	2.9. Burns and skin disorders	35
	2.10. Parasitic infestations	36
	2.11. Iatrogenic causes	36
	2.12. Diabetes	36
	2.13. Collagen diseases	37
	2.14. Pregnancy	37
	2.15. Genetic disorders	38
	Sickle-cell disease	38
	Acrodermatitis enteropathica	38
	Miscellaneous genetic disorders	39

3. Clinical Manifestations of Zinc Deficiency   39
   3.1.   Experimental production of zinc deficiency in human beings   42
   3.2.   Metabolism and biochemistry   45
4. Hyperzincemia   51
5. Toxicity   51
6. Summary   52
Acknowledgments   53
References   53

1. INTRODUCTION

Recently many biochemical, physiological, and clinical roles of zinc in human beings have been reported. It has been known for many years that zinc is essential for the growth of microorganisms, plants, and animals (Bertrand and Bhattacherjee, 1934; Todd et al, 1934), but its essential nature for human beings was recognized only recently (Prasad et al. 1961, 1963a, 1963b). Human zinc deficiency was suspected for the first time in 1961 (Prasad et al. 1961) and was confirmed in 1963 (Prasad et al. 1963). In later studies from Iran (Halsted et al. 1972; Ronaghy et al., 1974) it was clearly demonstrated that zinc is a principal limiting nutrient in children and adolescents, a fact that probably accounts for the growth retardation so commonly seen there. Recent reports indicate that marginal deficiency of zinc in human beings is probably widespread (Sandstead et al., 1976). It is now evident that not only nutritional but also conditioned zinc deficiency may complicate many disease states.

2. CAUSES OF ZINC DEFICIENCY

2.1: Nutritional Causes

In 1958 a 21-year-old patient at Saadi Hospital, Shiraz, Iran, was brought to my attention. He looked like a 10-year-old boy. Besides dwarfism and hypogonadism, he had hepatosplenomegaly, rough and dry skin, mental lethargy, and geophagia (Prasad et al., 1961). He ate only bread made of wheat flour, and his intake of animal protein was negligible. He also consumed nearly 1 lb of clay daily; geophagia (clay eating) is common in the villages in Iran. During a short period of time we found 10 similar cases.

Although there was no evidence of blood loss, the patient was anemic, and it was determined after extensive investigation that the anemia was related to a nutritional deficiency of iron. It was concluded that the iron deficiency was due to a lack of iron availability from the bread, greater loss of iron caused by excessive sweating in the hot climate, and the adverse effect of geophagia on iron absorption. In every case the anemia was completely corrected by administration of oral iron.

Inasmuch as growth retardation and testicular atrophy are not seen in iron-deficient experimental animals, the possibility that zinc deficiency may have been a complicating factor was considered. Zinc deficiency is known to produce growth retardation, testicular atrophy, and skin changes in animals (Todd et al., 1934). Since heavy metals may form insoluble complexes with phosphates, it was considered possible that the factors responsible for decreased iron availability in these patients may also adversely affect zinc availability.

Subsequently, zinc deficiency was documented in similar patients in Egypt (Prasad et al., 1963a, 1963b; Prasad, 1966). This conclusion was based on (1) decreased zinc concentrations in plasma, red cells, and hair; and (2) radioactive ^{65}Zn studies, which revealed that the plasma zinc turnover rate was greater, the 24-hr exchangable pool smaller, and the excretion of ^{65}Zn in stool and urine less in the patients than in control subjects.

Further studies in Egypt showed that the growth rate was greater in patients who received zinc than in those who received iron or only an animal protein diet (Prasad, 1966; Sandstead et al., 1967). Genitalia size became normal, pubic hair appeared, and other secondary sexual characteristics developed within 12 to 24 weeks in all subjects receiving zinc. No such changes were observed in a comparable length of time in the iron-supplemented group or in the group on an animal protein diet alone. Thus the growth retardation and gonadal hypofunction in these subjects were related to zinc deficiency.

In the Middle East excessive intake of phytate, which is known to decrease zinc availability for absorption, was considered a principal factor responsible for zinc deficiency. Excessive loss of zinc by sweating and blood loss due to parasitic infestation in Egypt were regarded as additional factors accounting for zinc deficiency in human subjects.

Further studies in Iran demonstrated that zinc is a principal limiting factor in the nutrition of children (Ronaghy et al., 1974). It was also evident from several studies conducted in the Middle East that the requirement for zinc under different dietary conditions varied widely. For instance, in the studies reported by Prasad et al. (1963a, 1963b) and Sandstead et al. (1967), 18 mg of supplemental zinc, with adequate animal protein and caloric diet, was sufficient to produce a definite response with respect to growth and gonads, but in other studies where the subjects continued to eat the village diet up to 40 mg of zinc supplement was required to show some growth effect (Ronaghy et al., 1974).

Halsted et al. (1972) published results of a study on a group of 15 men, who were rejected at the Iranian army induction center because of "malnutrition," and 2 women. All subjects were 19 or 20 years of age, and their clinical features were similar to those of zinc-deficient dwarfs reported earlier by Prasad et al. (1961, 1963a, 1963b). They were studied for 6 to 12 months. One group was given a well-balanced, nutritious diet containing ample animal protein plus a placebo capsule. A second group received the same diet without additional medication for 6 months, followed by the diet plus zinc for another 6-month period. The development in subjects receiving the diet alone was slow, whereas the effect on height increment and onset of sexual function was strikingly enhanced in those receiving zinc. The zinc-supplemented subjects gained considerably more height than those receiving the ample protein diet alone and showed evidence of early onset of sexual function, as defined by nocturnal emission in males and menarche in females. The two women described in this report were from a hospital clinic and represented the first cases of dwarfism in females due to zinc deficiency (Halsted et al., 1972).

Brief mention should be made of the prevalence of nutritional zinc deficiency in human populations throughout the world. Clinical features similar to those reported by Prasad and co-workers in zinc-deficient dwarfs have been observed in many countries, such as Turkey, Portugal, and Morocco. Also, zinc deficiency should be prevalent in other countries where primarily cereal proteins are consumed by the population.

A clinical syndrome similar to that of "adolescent nutritional dwarfism" has been identified in younger children in Iran, although failure of sexual maturation was not evident prior to adolescence. This syndrome is most common in small rural communities where there is also a high incidence of adolescent nutritional dwarfism.

2.2. Nutritional Zinc Deficiency in Children in the United States

In 1972 a number of Denver children were reported to show evidence of symptomatic zinc deficiency (Hambidge and Walravens, 1976). These children were identified as the result of a survey of trace element concentrations in the hair of apparently normal children from middle- and upper-income families. There was no apparent cause of the relatively poor growth in the majority of children, most of whom also had a history of poor appetite.

It has been calculated recently that substantial sections of the U.S. population are at risk from suboptimal zinc nutrition. Those at particular risk include persons whose zinc requirements are relatively high (e.g., at times of rapid growth) and also people subsisting on low-income diets. Although severe or moderately severe zinc deficiency on a nutritional basis is unlikely to be seen in the developed countries, marginal deficiency may occur frequently.

2.3. Zinc Nutrition and Deficiency in Infants

Hair and plasma zinc levels are exceptionally low in infants in the United States compared with other age groups, including the neonate, older children, and adults. It appears unlikely that these low levels can be regarded as entirely normal for this age, as levels are higher in other countries where comparable data have been obtained (Walravens and Hambidge, 1977). Similar considerations apply equally to the low plasma zinc levels reported for infants in this country; plasma concentrations in infants in Sweden and Germany have been reported to be no lower than those for adults in these countries.

Several factors, including difficulty in achieving a positive zinc balance in early postnatal life and a "dilutional" effect of rapid growth, may contribute to zinc depletion in infants. A unique factor in the United States that may contribute to zinc deficiency is the low concentration of this metal in certain popular infant milk formulas.

The majority of infants with low levels of zinc in plasma and hair do not show any detectable signs of zinc deficiency. However, those at the lower end of a spectrum of zinc depletion, as manifested by low levels of hair and plasma zinc, and perhaps those who remain moderately depleted for a prolonged period of time seem to develop symptomatic zinc deficiency (Walravens and Hambidge, 1977). In the original Denver study, 8 of 93 infants and children aged less than 4 years had hair zinc levels below 30 ppm, and 6 of these manifested declining growth percentiles and poor appetitie. Experience with a number of infants who had low levels of zinc in plasma and hair and who responded favorably to zinc supplementation indicates that some cases of failure to thrive in infancy may be caused by zinc deficiency.

2.4. Alcohol

Alcohol induces hyperzincuria (Gudbjarnason and Prasad, 1969; Prasad, 1976); the mechanism is unknown. A direct effect of alcohol on the renal tubular epithelium may be responsible. Acute ingestion of alcohol did not induce zincuria in some experiments (Sullivan, 1962), however, Gudbjarnason and Prasad (1969) reported increased urinary zinc excretion after alcohol intake. This effect was evident when the complete urine collection was analyzed for zinc during the first and second 3-hr periods following ingestion of 6 oz of chilled vodka.

The serum zinc level of the alcoholic subjects tends to be lower than that of controls. An absolute increase in the renal clearance of zinc in the alcoholic, demonstrable at both normal and low serum zinc concentrations, has been observed (Allan et al., 1975). Thus the measurement of renal clearance of zinc may be utilized clinically for etiological classification of chronic liver disease due to alcohol in different cases. Excessive ingestion of alcohol may lead to severe deficiency of zinc (Weismann et al., 1976).

2.5. Liver Disease

Vallee et al (1956) initially described the abnormal zinc metabolism that occurs in patients with cirrhosis of the liver. These investigators demonstrated that patients with cirrhosis had low serum zinc, diminished hepatic zinc, and, para-doxically, hyperzincuria. These observations led them to suggest that zinc deficiency in alcoholic cirrhotic patients may be a conditioned state related in some way to alcohol ingestion. These observations have since been confirmed by other investigators. Saldeen and Brunk (1967) have shown that parenteral zinc salts protect rat liver from damage by carbon tetrachloride, suggesting that zinc exerts a protective effect on the liver.

2.6. Gastrointestinal Disorders

Zinc deficiency has been reported in patients with steatorrhea (Sandstead et al., 1976; Love et al. 1978). In an alkaline environment, zinc would be expected to form insoluble complexes with fat and phosphate analogous to those formed by calcium and magnesium. Thus fat malabsorption due to any cause should result in increased loss of zinc in the stool.

Exudation of large amounts of zinc-protein complexes into the intestinal lumen may also contribute to the decrease in plasma zinc concentrations that occur in patients with inflammatory disease of the bowel. It seems likely that protein-losing enteropathy due to other causes may also impair zinc homeostasis. Another potential cause of a negative zinc balance is a massive loss of intestinal secretions.

2.7. Renal Diseases

The potential causes of conditioned deficiency of zinc in patients with renal diseases include proteinuria and failure of tubular reabsorption (Mansouri et al., 1970; Halsted and Smith, 1970; Lindeman et al., 1977). In patients with renal failure the occurrence of conditioned zinc deficiency may result from a combi-nation of factors that at present are ill defined. If 1,25-dihydroxychlolecalciferol plays a role in the intestinal absorption of zinc, an impairment in its formation by the diseased kidney would be expected to result in malabsorption of this element. It seems likely that plasma and soft tissue concentrations of zinc may be "protected" in some individuals with renal failure by the dissolution of bone, which occurs as a result of increased parathyroid activity in response to low serum calcium. In experimental animals, calcium deficiency has been shown to cause zinc release from bone. In some patients successfully treated for hyperphosphatemia and hypocalcemia, the plasma zinc concentration may be expected to decline because of deposition of zinc, along with calcium, in bone. Thus, in the hypocalcemic group in particular, a diet low in protein and high in refined cereal products and fat (i.e., low in zinc) is likely to contribute to a conditioned zinc deficiency. The patients reported by Mansouri et al. (1970), who were treated with a diet containing 20 to 30 g of protein daily and who had low plasma concentrations of zinc, appear to represent such a clinical instance. Presumably the patients of Halsted and Smith (1970) were similarly restricted in dietary protein. Further studies are required to establish the role of zinc in chronic renal diseases.

2.8. Neoplastic Diseases

The occurrence of conditioned zinc deficiency in patients with neoplastic diseases will depend upon the nature of the neoplasm. Anorexia and starvation, plus avoidance of foods rich in available zinc, are probably important conditioning factors. Increased zinc excretion subsequent to its mobilization by leukocyte endogenous mediator (LEM) in response to tissue necrosis may be another factor.

2.9. Burns and Skin Disorders

The causes of zinc deficiency in patients with burns include losses in exudates. Starvation of burned patients is a well-recognized cause of morbidity and mortality. The contribution of conditioned zinc deficiency to the morbidity of burned patients has not been defined. Limited studies indicate that epithelialization of burns may be improved by zinc treatment. This finding is consistent with the beneficial effect of zinc in the treatment of leg ulcers and the well-defined requirement of zinc for collagen synthesis (Proies and Strain, 1966; Fernandez-Madrid et al., 1973).

In psoriasis the loss of large numbers of skin cells may possibly result in zinc depletion. The skin contains approximately 20% of body zinc. Thus, if the loss of epithelial cells is great enough, it is conceivable that the massive formation of new cells by the skin could lead to conditioned deficiency. Although low levels of plasma zinc have been reported in some patients with extensive psoriasis (Greaves and Boyde, 1967; Greaves, 1972), others have not been able to confirm these findings (Portnoy and Molokhia, 1971, 1972).

2.10. Parasitic Infestations

Blood loss due to parastitic diseases may contribute to conditioned zinc deficiency; this appears to have been the case in the zinc-responsive "dwarfs" reported from Egypt (Prasad et al. 1963a, 1963b). As red blood cells contain 12 to 14 μg Zn/ml, hookworm and/or schistosomiasis infections that are severe enough to cause iron deficiency will probably contribute to zinc deficiency as well.

2.11. Iatrogenic Causes

Possible iatrogenic causes of conditioned zinc deficiency include the use of antimetabolites and antianabolic drugs. Treatment with some of these drugs makes patients feel ill. They become anorectic and may starve. With catabolism of body mass, urinary excretion of zinc is increased. Moreover, commonly used intravenous fluids are relatively zinc-free. Thus, under the usual circumstances, a negative zinc balance could occur in patients who are given antimetabolites, antianabolic agents, or prolonged intravenous therapy.

Failure to include zinc in fluids for total parenteral nutrition (TPN) can also lead to iatrogenically induced conditioned zinc deficiency. A decline in plasma zinc has been observed in several patients given TPN fluids containing less than 1.25 mg Zn daily (Greene, 1977). In some cases urinary zinc loss is excessive (Greene, 1977). Some patients on TPN have developed a clinical picture resembling that of acrodermatitis enteropathica. Typical severe zinc deficiency occurred in one patient following penicillamine therapy for Wilson's disease (Klingberg et al., 1976).

2.12. Diabetes

Some patients with diabetes mellitus have been found to have increased urinary losses of zinc (Pidduck et al., 1970). The mechanism is unknown. Presumably, some of them may become zinc deficient, although the plasma zinc generally is not affected.

2.13. Collagen Diseases

In patients with inflammation, such as occurs in rheumatoid arthritis, lupus erythematosus, infection, or injury, two factors—loss of zinc from catabolized tissue (Cuthbertson et al., 1972) and mobilization of zinc by LEM (Pekarek et al., 1972) to the liver and its subsequent excretion in the urine—may account for conditioned zinc deficiency. Recently, a beneficial effect of zinc therapy in patients with rheumatoid arthritis has been reported (Simkin, 1977).

2.14. Pregnancy

The plasma concentration of zinc decreases during human pregnancy. Presumably, the decrease reflects, in part, the uptake of zinc by the fetus and other products of conception. It has been estimated that a pregnant woman must retain approximately 750 μg Zn/day for growth of the products of conception during the last two trimesters. Thus, when zinc deficiency occurs in pregnancy, a conditioning factor is the demand of the fetus for zinc. Studies in the rat suggest that the placenta actively provides zinc to the fetus (Henkin et al., 1971). If the diet of the pregnant woman does not include liberal amounts of animal protein, the likelihood of conditioned deficiency of zinc is increased, as the element is probably less available from food derived from grains and other plants.

The possible importance of zinc deficiency in human pregnancy is implied by the observations of Hurley (1976), Caldwell et al. (1970), and Halas et al (1976). Zinc deficiency in pregnant rats was shown to cause fetal abnormalities, behavioral impairment in the offspring, and maternal difficulty in parturition. In view of the important role of zinc in nucleic acid metabolism, Hurley and Shrader (1972) proposed that impaired deoxyribonucleic acid (DNA) synthesis in zinc-deprived embryos prolongs the mitotic cycle and reduces the number of normal neural cells, leading to malformations of the central nervous system. It is tempting to speculate that the exceptionally high rates of congenital malformations of the central nervous system reported from the Middle East (Damyanov and Dutz, 1971) might be caused by maternal zinc deficiency. Caldwell et al. (1970) were the first to show that in both prenatal and postnatal nutrition even mild zinc deficiency in rats had a profound influence on behavior potential, despite an apparently adequate protein level in the diet.

Plasma zinc is known to decrease after use of oral contraceptive agents (Halsted et al., 1968; Prasad et al., 1975a). Our recent data indicate that, whereas the plasma zinc may decline, the zinc content of the red blood cells increases as a result of oral contraceptive administration. This phenomenon may merely mean a redistribution of zinc from the plasma pool to the red cells. Alternatively, oral contraceptives may enhance carbonic anhydrase (a zinc metallo-enzyme) synthesis, thus increasing the red cell zinc content.

2.15. Genetic Disorders

Sickle-Cell Disease

Recently, zinc deficiency in sickle-cell disease has been recognized (Prasad et al., 1975b, 1977). Since zinc is an important constituent of erythrocytes, it is possible that long-continued hemolysis in patients with sickle-cell disease may lead to a zinc-deficient state.

In a study reported by Prasad et al. (1975b, 1977), zinc in plasma, erythrocytes, and hair was decreased and urinary zinc excretion was increased in sickle-cell anemia patients, as compared to controls. Erythrocyte zinc and daily urinary zinc excretion were inversely correlated in the anemia patients ($r = .71, p < .05$), suggesting that hyperzincuria may have caused their zinc deficiency. Carbonic anhydrase, a zinc metalloenzyme, correlated significantly with erythrocyte zinc ($r = + .94, p < .001$). Plasma ribonuclease (RNase) activity was significantly greater in anemia subjects than in controls, consistently with the hypothesis that sickle-cell anemia patients are zinc deficient.

In spite of the tissue zinc depletion in the sickle-cell anemia patients, the mean urinary excretion of zinc was higher than in the controls. This may have been a direct result of increased filtration of zinc by the glomeruli owing to continued hemolysis, or there may have been a defect in tubular reabsorption of zinc somehow related to sickle-cell anemia—a possibility that cannot yet be exluded. Continued hyperzincuria may have been responsible for tissue depletion of zinc, as suggested by a significant negative correlation between the values for 24-hr urinary zinc excretion and erythrocyte zinc. At this stage, however, one cannot rule out additional factors such as predominant dietary use of cereal proteins and other nutritional factors that affect zinc availability adversely, thus accounting for zinc deficiency. Further work is warranted to properly elucidate the pathogenesis of zinc deficiency in sickle-cell anemia.

Acrodermatitis Enteropathica

In 1973 Barnes and Moynahan (1973) studied a 2-year-old girl with severe acrodermatitis enteropathica, who was being treated with diiodohydroxyquinolone and a lactose-deficient synthetic diet. The clinical response to this therapy was not satisfactory, and the physicians sought to identify contributory factors. Finding that the concentration of zinc in the patient's serum was profoundly reduced, they administered oral zinc. The skin lesions and gastrointestinal symptoms cleared completely, and the patient was discharged from the hospital.

When zinc was inadvertently omitted from the child's regimen, she suffered a relapse, but promptly responded when oral zinc was reinstituted.

In their initial reports Barnes and Moynahan attributed zinc deficiency in this patient to the synthetic diet. Later it was recognized that zinc might be fundamental to the pathogenesis of this rare inherited disorder, and that the clinical improvement reflected improvement in zinc status. Support for the zinc-deficiency hypothesis came from the observation that a close resemblance existed between the symptoms of zinc deficiency in animals and human beings, as reported earlier (Prasad, 1966), and those of acrodermatitis enteropathica, particularly with respect to skin lesions, growth pattern, and gastrointestinal symptoms.

Zinc supplementation to these patients led to complete clearance of skin lesions and restoration of normal bowel function, which had previously resisted various dietary and drug regimens. This original observation was quickly confirmed in other cases with equally good results. The underlying mechanism of the zinc deficiency in these patients is probably malabsorption. The cause of the poor absorption is obscure, but abnormality of Paneth's cells may be involved.

Miscellaneous Genetic Disorders

Low levels of plasma zinc have been noted in patients with Down's syndrome (Pradsad, 1966, 1976). The mechanism is unknown.

Congenital hypoplasia of the thymus gland in cattle may be an example of zinc deficiency on a genetic basis (Brummerstedt et al., 1971). It is not known whether thymus hypoplasia in human beings is somehow related to zinc deficiency.

3. CLINICAL MANIFESTATIONS OF ZINC DEFICIENCY

Growth retardation, hypogonadism in males, poor appetite, mental lethargy, and skin changes were the classic clinical features of chronic zinc deficiency reported from the Middle East by Prasad et al. (1961, 1963a, 1963b). Hepato-splenomegaly was consistently present in the zinc-deficient dwarfs, who improved after zinc supplementation. The mechanism of spleen and liver enlargement in this syndrome, however, is not well understood. All the features mentioned above were corrected by zinc supplementation.

Recently, Morrison et al. (1978) reported abnormal dark adaptation in six stable alcoholic cirrhotics, who also had low serum zinc levels. Zinc administration to these patients resulted in improvement of dark adaptation. This interesting clinical observation needs further confirmation. The effect of zinc on the retina may be mediated by retinene reductase, which is a zinc-dependent enzyme.

Some of the clinical features of cirrhosis of the liver, such as loss of body hair, testicular hypofunction, poor appetitie, mental lethargy, impairment of wound healing, and night blindness, are probably related to the secondary zinc-deficient state in this disease. In the future, careful clincial trials with zinc supplementation must be carried out to determine whether zinc is beneficial to patients with chronic liver disease.

Several publications have focused on a distinctive pattern of abnormalities known as "fetal alcohol syndrome," occurring in the offspring of alcoholic mothers. It is characterized by prenatal and postnatal growth deficiency, microcephaly, short palpebral fissures, epicanthal folds, cleft palate, micrognathia, joint anomalies, cardiac and renal anomalies, and anomalies of the external genitalia (Jones and Smith, 1973; DeBeukelaer et al., 1977). Many of these features are similar to those reported in rat fetuses when zinc intake was restricted in the mothers during the crucial stage of gestation (Hurley, 1976). As excessive alcohol intake may deplete the body zinc store and as zinc plays a vital role in DNA synthesis and cell division, one could speculate that maternal zinc deficiency may be responsible for the "fetal alcohol syndrome." A careful clinical study is warranted to test this hypothesis.

In sickle-cell disease, delayed onset of puberty and hypogonadism in males, characterized by decreased facial, pubic, and axillary hair, short stature, low body weight, rough skin, and poor appetitie, have been noted and related to a secondary zinc-deficient state (Prasad et al., 1975b, 1977). Many patients with sickle-cell anemia develop chronic leg ulcers that do not heal, and a beneficial effect of zinc supplementation in such cases has been reported (Serjeant et al., 1970; Prasad et al., 1975b). In one study, zinc sulfate (660 mg/day was administered orally to seven men and two women with sickle-cell anemia. Two 17-year-old males grew 10 cm in height during 18 months of therapy, and all but one patient gained weight. Five of the males showed increased growth of pubic, axillary, facial, and body hair, and in one a leg ulcer healed in 6 weeks on zinc. In two others some beneficial effect of zinc therapy on the healing of ulcers was noted. Further controlled clinical trials are needed to establish the effect of zinc therapy on this condition.

In limited noncontrolled studies, zinc appears to have been effective in decreasing the symptoms and crisis of sickle-cell anemia patients (Brewer et al., 1975). The therapeutic rationale is based on the effects of zinc on the red cell membrane. It decreases hemoglobin and calcium binding and improves deformability; this may result in decreased trapping of skickle cells in the capillaries, where the pain cycle is normally initiated. Undoubtedly, more thorough evaluation of zinc therapy in sickle-cell disease is needed.

Zinc deficiency may play an important role in the functional activity of the enterocyte in celiac disease, and some "nonresponsive" celiac disease patients may be profoundly zinc deficient (Love et al., 1978). A recent report described six patients who failed to respond to diet, steroids, and nutritional supplements but made a remarkable recovery when zinc was administered (Love et al., 1978).

They gained weight, and their *d*-xylose absorption tests and steatorrhea improved after zinc therapy.

Zinc therapy in a few subjects with malabsorption syndrome (other than celiac disease) seemed to produce beneficial results with respect to growth retardation, hypogonadism in the males, mental lethargy, skin changes, and loss of hair (Sandstead et al., 1976). One should be aware, therefore, of zinc deficiency occurring as a possible complication of malabsorption syndrome, as this is easily correctable.

In 1966 Pories and his collaborators (Pories and Strain, 1966; Pories et al., 1967) reported that oral administration of zinc to military personnel with marsupialized pilonidal sinuses was attended by a twofold increase in the rate of reepithelialization. The authors' conclusion that zinc can promote the healing of cutaneous sores and wounds has been a subject of controversy in recent years. Clinical investigations by Cohen (1968), Husain (1969), Greaves and Skillen (1970), and Serjeant et al. (1970) have substantiated the beneficial effects of zinc on wound healing, whereas studies by Barcia (1970), Myers and Cherry (1970), Clayton (1972), and Greaves and Ive (1972) have failed to demonstrate any therapeutic benefit. Hallbook and Lanner (1972) found that the reepithelialization rate of venous leg ulcers was enhanced by zinc in patients who initially had diminished concentrations of serum zinc, but observed no benefit in patients whose initial measurements of serum zinc were within the normal range.

Studies in experimental animals have demonstrated that (1) healing of incised wounds is impaired in rats with dietary zinc deficiency; (2) collagen and non-collagen proteins are reduced in skin and connective tissues from rats with dietary zinc deficiency; (3) zinc supplementation does *not* augment wound healing in normal rats; and (4) zinc supplementation *does* augment wound healing in chronically ill rats (Oberleas et al., 1971). These data provide evidence that zinc supplementation may promote wound healing in zinc-deficient patients.

Abnormalities of taste have been related to a deficiency of zinc in human beings and animals (Henkin and Bradley, 1969; Henkin et al. 1976; Catalanotto, 1978). Animal studies with various drugs seem to indicate that thiols, as well as agents that deplete trace metals such as zinc, cause increased intake of certain solutions distinguished primarily by their taste. This suggests that a change in taste function may have occurred because of zinc deficiency (Catalanotto, 1978). Decreased taste acuity (hypogeusia) has been observed in zinc-deficient human subjects, such as patients with liver disease and malabsorption syndrome, after thermal burns and after the administration of penicillamine or histidine (Henkin and Bradley, 1969). A double-blind study, however, failed to show the effectiveness of zinc in the treatment of hypogeusia in various patients (Henkin et al., 1976). This may suggest that zinc depletion leads to decreased taste acuity, but not all cases of hypogeusia are related to zinc deficiency. The role of zinc in hypogeusia needs to be delineated further, and careful clinical studies are warranted.

Severe and relatively acute deficiency of zinc has been observed in patients with acrodermatitis enteropathica (Barnes and Moynahan, 1973), after penicillamine therapy in a patient with Wilson's disease (Klingberg et al., 1976), after TNP (Greene, 1977), and after excessive ingestion of alcohol (Weismann et al., 1976).

In acrodermatitis enteropathica the dermatological manifestations include progressive bullous-pustular dermatitis of the extremities and the oral, anal, and genital areas, combined with paronychia and generalized alopecia. Infection with *Candida albicans* is a frequent complication. Ophthalmic signs may include blepharitis, conjunctivitis, photophobia, and corneal opacities. Gastrointestinal disturbances are usually severe and include chronic diarrhea, malabsorption, steatorrhea, and lactose intolerance. Neuropsychiatric signs include irritability, emotional disorders, tremor, and occasional cerebellar ataxia. The patients generally have retarded growth, and males exhibit hypogonadism. Zinc therapy has been shown to produce remarkable improvements and is considered to be a life-saving measure for these subjects.

A very similar clinical picture has been reported in a patient who received penicillamine therapy for Wilson's disease. After TPN or excessive ingestion of alcohol, the clinical manifestations of zinc deficiency resemble those of acrodermatitis enteropathica, and such cases may be considered examples of acquired acrodermatitis enteropthica. Once the condition is recognized, zinc therapy is imperative.

3.1. Experimental Production of Zinc Deficiency in Human Beings

Recently, marginal deficiency of zinc has been successfully produced in human volunteers by dietary means, and changes in several zinc-dependent parameters have been documented (Prasad et al., 1978). Four male volunteers (1, 2, 3, and 4) ranging in age from 55 to 65 years participated in this study. The physical examination revealed no abnormal features, and their zinc status was within normal limits. A semipurified diet based on texturized soy protein, purchased from General Mills Co., Minneapolis, Minnesota (Bontrae Products) and Worthington Foods Co., Division of Miles Laboratory, Elkhart, Indiana, was developed for this study. Soy protein isolate, which was used as soy flour in the baked goods, was purchased from General-Biochemicals (Teklad Mills, Chagrin Falls, Ohio). The texturized soy meals utilized were hamburger granules, chicken and turkey slices, and chicken chunks. The texturized soy protein and soy protein isolate were washed twice with ethylenediamine tetraacetate (EDTA) and then rinsed three times with deionized water, boiled for 30 min, and kept frozen until use. The diet was supplemented with vitamins, mineral mix (except zinc), and protein supplement to meet the recommended daily requirements. The daily intake of zinc was 2.7 mg for patients 1 and 2, and 3.5 mg for patients 3 and 4.

The patients were kept under strict metabolic conditions. The first group (subjects 1 and 2) received the hospital diet for 2 weeks, then the experimental diet with 10 mg supplemental Zn (as zinc acetate) orally for 6 weeks. Thereafter they were given only the experimental diet (daily zinc intake, 2.7 mg) for 24 weeks. At the end of this phase, while continuing the experimental diet, these two subjects received 30 mg supplemental Zn (as zinc acetate) daily, orally, for 12 weeks. Finally, they were maintained on the hospital diet with a total daily intake of 10 mg Zn plus 30 mg supplemental Zn (as zinc acetate) orally for 8 weeks. The hospital diet provided the same number of calories and amount of protein as the experimental diet. Thus these two subjects were observed for a total period of 52 weeks.

Subjects 3 and 4 (the second group) received the hospital diet (10 mg Zn intake daily) for 3 weeks, followed by the experimental diet, with 30 mg oral Zn supplement (as zinc acetate) for 5 weeks. They were then given the experimental diet only (3.5 mg Zn intake daily) for 40 weeks. The repletion phase was begun by giving 30 mg Zn (as zinc acetate) orally while maintaining the same experimental diet; this continued for a total of 8 weeks, after which the hospital diet replaced the experimental diet. Oral zinc supplement (30 mg as zinc acetate) was continued along with the hospital diet for 8 weeks. Altogether, these two subjects were observed for 64 weeks.

Body weight decreased in all four subjects as a result of dietary zinc restriction, but the weight loss was more pronounced in the first group. Body weight correlated highly with subscapular thickness in the two subjects in whom these data were obtained. An approximate calculation revealed that the weight loss could be accounted for as follows: 50% fat, 30% water, and 20% other.

In the first group of subjects, during the zinc restriction phase (2.7 mg Zn daily intake) the apparent negative balance for zinc ranged from 1 to 4 mg/day, whereas in the second group, the apparent negative balance for zinc was 1 to 2 mg/day. After supplementation with 30 mg Zn the positive zinc balance ranged from 11 to 22 mg/day in the first group of two subjects. In the second group, during the baseline period when the daily zinc intake was 33.5 mg daily the positive balance for zinc was 3 to 4 mg/day. On the other hand, these subjects, on the same level of zinc intake (33.5 mg daily) showed after the zinc depletion phase a positive zinc balance of 14 to 16 mg/day.

Plasma zinc decreased significantly in all four subjects as a result of zinc restriction and increased after zinc supplementation. The changes were more marked for the patients on 2.7 mg than for those on 3.5 mg Zn daily intake. The red cell zinc decreased significantly in the first group (subjects 1 and 2), although the decrease was not evident until 12 weeks on restricted zinc intake. In the second group, although the red cell zinc did not decrease significantly during the zinc-restricted period, it showed a marked increase after zinc supplementation. The leukocyte zinc decreased significantly as a result of zinc restriction in the second group of subjects, in whom this parameter was measured.

Plasma alkaline phosphatase was monitored carefully in the second group. In both subjects the activity slowly declined as a result of zinc restriction; after supplementation with zinc, the activity nearly doubled in 8 weeks. In all four subjects plasma ribonuclease activity was almost twice as great during the zinc-restricted period as in the zinc-supplemented phase. Surprisingly, it was observed that plasma ammonia levels increased during zinc restriction and decreased after zinc supplementation in the second group of two subjects in whom this was monitored.

Urinary excretion of zinc decreased in three of the four subjects as a result of zinc restriction. In one case a decrease in urinary zinc excretion was not seen because of the diuretic therapy he received for mild hypertension during the study.

Total protein, total collagen, ribonucleic acid (RNA), DNA, and deoxythymidine kinase activity were measured in the connective tissue obtained after implantation subcutaneously of a small sponge during the zinc restriction and zinc supplementation phases in the first group. A marked increase in the total protein, total collagen, and RNA/DNA was observed in the sponge connective tissue as a result of zinc supplementation. Whereas the activity of deoxythymidine kinase was not measurable in the connective tissue during the zinc restriction phase, it became nearly normal after zinc supplementation.

Thus changes in the zinc concentrations of plasma, erythrocytes, leukocytes, and urine, and changes in the activities of zinc-dependent enzymes such as alkaline phosphatase, and RNase in the plasma and deoxythymidine kinase in the connective tissue during the zinc restriction phase, appear to have been induced specifically by a mild deficiency of zinc in the volunteers. One unexpected finding involved the plasma ammonia level, which appeared to increase during the zinc-restricted period. We have recently reported a similar finding in zinc-deficient rats (Rabbani and Prasad, 1978). This may have important health implications concerning zinc deficiency in human beings, inasmuch as in liver disease hyperammonemia is believed to affect the central nervous system adversely.

The changes observed in body weight were related to dietary zinc intake. Increased loss of fat, as determined by subscapular thickness, and normal absorption of fat during the zinc restriction phase suggest that zinc deficiency may have led to hypercatabolism of fat in these subjects. In experimental animals an increase in free fatty acids has been observed as a result of zinc deficiency (Underwood, 1977). Indeed, more studies are required in human subjects to document increased fat catabolism due to zinc restriction.

Changes in the plasma zinc concentration were observed early (within 4 to 6 weeks) and were correlated with the severity of the dietary zinc restriction. Thus plasma zinc may be very useful in assessing human zinc status, provided that infections, myocardial infarction, intravascular hemolysis, and acute stress can be ruled out. As a result of infections, myocardial infarction, and acute stress, zinc from the plasma compartment may redistribute to other tissues, thus making an assessment of zinc status in the body a difficult task. Intravascular hemolysis also spuriously increases the plasma zinc level, inasmuch as the zinc content of red cells is much higher than that of the plasma.

Changes in the red cell zinc were slow to appear, as expected; on the other hand, the leukocyte zinc appeared more sensitive to changes in zinc intake. Urinary excretion of zinc decreased as a result of dietary zinc restriction, suggesting that renal conservation of this element may be important for the hemostatic control mechanism in human beings. Thus determination of zinc in 24-hr urine may be of additional help in diagnosing zinc deficiency, provided that cirrhosis of the liver, sickle-cell disease, and chronic renal disease, are ruled out. These conditions are known to produce hyperzincuria and associated zinc deficiency.

Our data indicate that, after the zinc-deficient state, the four subjects showed a greater positive balance for zinc. This would suggest that a test based on oral challenge of zinc and subsequent plasma zinc determination may be able to distinguish between zinc-deficient and zinc-sufficient states in human subjects.

3.2. Metabolism and Biochemistry

The zinc content of a normal 70-kg male is approximately 1.5 to 2.0 g. Liver, kidney, bone, retina, prostate, and muscle appear to be rich in zinc. In man the zinc contents of testes and skin have not been determined accurately, although clinically it appears that these tissues are sensitive to zinc depletion.

Zinc in plasma is mostly present as bound to albumin, but other proteins, such as α_2-macroglobulin, transferrin, ceruloplasmin, haptoglobin, and gamma globulins, also bind a significant amount of zinc (Prasad and Oberleas, 1970). Besides the protein-bound fraction, a small proportion of zinc (2 to 3% of oral Zn) in the plasma exists as an ultrafilterable fraction, mostly bound to amino acids but with a smaller amount in ionic form.

Approximately 20 to 30% of ingested dietary zinc is absorbed. Data on both the site(s) of absorption in human beings and the mechanism(s) of absorption, whether active, passive, or facultative transport, are meager. Zinc absorption is variable and highly dependent on a variety of factors. Zinc is more available for absorption from animal proteins. Among other factors that may affect zinc absorption are body size, level of zinc in the diet, and presence in the diet of other potentially interfering substances, such as calcium, phytate, fiber, and the chelating agents. Recently it has been shown that prostaglandin E_2 not only binds zinc but also facilitates its transport across the intestinal mucosa in the rat (Song and Adham, 1978).

Normal zinc intake in a well-balanced American diet with animal protein is approximately 12 to 15 mg/day. Urinary zinc loss is approximately 0.5 mg/day. Loss of zinc by sweat may be considerable under certain climatic conditions; under normal conditions approximately 0.5 mg Zn may be lost daily by sweating. Endogenous zinc loss in the gastrointestinal tract may amount to 1 to 2 mg/day.

The daily requirement of zinc for human subjects has not been established. In view of the fact that several dietary constituents may affect the availability of zinc, however, dietary requirements must vary greatly from one region to another, depending on the food habits of the population.

The major functions of zinc in human and animal metabolism appear to be enzymatic; over 70 metalloenzymes are now known to require zinc for their functions (Riordan, 1976). Zinc enzymes are known to participate in a wide variety of metabolic processes. The metal is present in several dehydrogenases, aldolases, peptidases, and phosphatases.

A deficiency of zinc in *Euglena gracilis* has been shown to adversely affect all phases of the cell cycle (G_1, S, G_2, and mitosis), indicating that zinc is required for the biochemical processes essential for cells to pass from G_2 to mitosis, from S to G_2, and from G_1 to S (Riordan, 1976). The effect of zinc on the cell cycle is undoubtedly due to its vital role in DNA synthesis (Prasad and Oberleas, 1974; Kirchgessner et al., 1976).

Many studies have shown that zinc deficiency in animals impairs the incorporation of labeled thymidine into DNA; this effect has been detected within a few days after instituting the zinc-deficient diet. Prasad and Oberleas (1974) showed that decreased activity of deoxythymidine kinase may be responsible for this early reduction in DNA synthesis. As early as 6 days after rats were placed on the dietary regimen, the activity of deoxythymidine kinase was reduced in rapidly regenerating connective tissue of the zinc-deficient animals, as compared to pair-fed controls. These results have recently been confirmed by Dreosti and Hurley (1975). The activity of deoxythymidine kinase in 12-day-old fetuses taken from females exposed to dietary zinc deficiency during pregnancy was significantly lower than in *ad libitum* and restricted-fed controls. Activity of the enzyme was not restored by *in vitro* addition of zinc, and addition of copper severely affected the enzyme activity adversely.

Zinc has been shown to be an essential constituent for the DNA polymerase of *Escherichia coli* (Slater, et al., 1971). Whether this enzyme is affected adversely in an animal model by zinc deficiency is not known. Livers from zinc-deficient rats incorporated less ^{32}P into the nucleotides of RNA than livers from pair-fed controls, and DNA-dependent RNA polymerase has been shown to be a zinc-dependent enzyme (Terhune and Sandstead, 1972; Scrutton et al., 1971). The fact that the activity of RNase is increased in zinc-deficient tissues (Prasad and Oberleas, 1973) suggests that the catabolism of RNA may be regulated by zinc.

From the above discussion it appears that zinc may have its primary effect on zinc-dependent enzymes that regulate the biosynthesis and catabolic rate of RNA and DNA. In addition, zinc may also play a role in the maintenance of polynucleotide conformation. Sandstead et al. (1971) observed abnormal polysome profiles in the liver of zinc-deficient rats and mice. Acute administration of zinc appeared to stimulate polysome formation both *in vivo* and *in vitro*. This finding is supported by the data of Fernandez-Madrid et al. (1973), who noted a decrease in the polyribosome content of zinc-deficient connective tissue from rats and a concomitant increase in inactive monosomes.

During the past decade it was shown that the activity of various zinc-dependent enzymes was reduced in the testes, bones, esophagus, and kidneys of zinc-deficient rats in comparison to their pair-fed controls (Prasad et al., 1967). These results correlated with the decreased zinc content in the same tissues of the zinc-deficient rats.

In several studies the activity of alkaline phosphatase was found to be reduced in bones from zinc-deficient rats, pigs, chicks, turkey poults, and quails (Kirchgessner et al., 1976). The activity of alkaline phosphatase may be reduced also in the intestine, kidney and stomach of experimental animals because of zinc deficiency. Not only may there be a loss of activity due to a lack of sufficient zinc for maintaining the enzyme activity, but also the amount of apoenzyme present appears to be diminished because of either decreased synthesis or increased degradation. Inasmuch as a reduction in the activity of alkaline phosphatase in intestinal tissue and the plasma is observed before any sign of a lowered food intake is evident, it is concluded that the loss of enzyme activity is directly attributable to zinc deficiency.

Two important zinc metalloenzymes in protein digestion are pancreatic carboxypeptidase A and B. A loss of activity of pancreatic carboxypeptidase A in zinc deficiency is a consistent finding (Prasad and Oberleas, 1971; Prasad et al., 1971). According to some investigators, within 2 days of instituting a zinc-deficient diet in rats the enzyme lost 24% of its activity, and within 3 days of dietary zinc repletion the activity of pancreatic carboxypeptidase was restored to normal levels (Kirchgessner et al., 1976). These results indicate that the level of food intake has no influence and that the decreased activity of carboxypeptidase A in the pancreas was related specifically to a dietary lack of zinc. As with alkaline phosphatase, the amount of carboxypeptidase A apoenzyme appears to be diminished in zinc-deficient pancreas.

Reduced activity of carbonic anhydrase, another zinc metalloenzyme, has been reported in gastric and intestinal tissues and in erythrocytes when the activity of the enzyme was expressed per unit of erythrocytes (Kirchgessner et al., 1976).

Recently, in patients with sickle-cell disease, an example of a conditioned zinc-deficient state, the content of carbonic anhydrase in the red cells was found to be decreased, correlating with the zinc content of the red cells (Prasad et al., 1975). Inasmuch as the technique measured the apoenzyme content, it appears that zinc may have a specific effect on the synthesis of this protein by some mechanism yet to be elucidated.

Several investigators have now shown that zinc deficiency lowers the activity of alcohol dehydrogenase in the liver, bones, testes, kidneys and esophagus of rats, and pigs (Prasad et al., 1967, 1971; Prasad and Oberleas, 1971; Kirchgessner et al., 1976). In another study, alcohol dehydrogenase was assayed in subcellular fractions of liver and retina from zinc-deficient and control rats, using retinol and ethanol as substrates (Huber and Gershoff, 1975). The activity of alcohol dehydrogenase was significantly decreased as a result of zinc deficiency in growing animals. In older rats, although no changes in liver zinc and activity of alcohol dehydrogenase were found, the retina was sensitive to the lack of zinc. These data show that zinc is required for the metabolism of vitamin A, as well as the catabolism of ethanol. An attempt to demonstrate accumulation of apo-enzyme of alcohol dehydrogenase in the zinc-deficient tissues was unsuccessful.

Recently, the role of zinc in gonadal function was investigated in rats (Lei et al., 1976). The increases in luteinizing hormone (LH), follicle-stimulating hormone (FSH), and testosterone were assayed after intravenous administration of synthetic LH-releasing hormone (LH-RH) to zinc-deficient and restricted-fed control rats. Body weight gain, zinc content of testes and weights of testes were significantly lower in the zinc-deficient than in the control rats. The serum LH and FSH responses to LH-RH administration were higher in the zinc-deficient rats, but the serum testosterone response was lower than than in the restricted-fed controls. These data indicate a specific effect of zinc on testes and suggest that gonadal function in the zinc-deficient state is affected through some alteration of testicular steroidogenesis.

Zinc prevents induced histamine release from mast cells (Chvapil, 1976). It is believed that this effect of zinc is due to its action on the cell membrane. Platelets are also affected by zinc ions. Collagen-induced aggregation of dog platelets and collagen- or epinephrine-induced release of [14C] serotonin were significantly inhibited by zinc. Supplementation of zinc in dogs effectively decreased the aggregatability of platelets, as well as the magnitude of [14C] serotonin release.

Zinc may form mercaptides with thiol groups of proteins, possibly linking to the phosphate moiety of phospholipids or interacting with carboxyl groups of sialic acid or proteins on plasma membranes, resulting in changes in the fluidity and stabilization of the membranes (Chvapil, 1976).

Several receptors at the plasmatic membrane presumably function as a gate for transmitting information in intracellular space. In the case of mast cells, histamine-releasing agents seem to work through specific receptors at the membrane. The masking of such a receptor site by membrane-impermeable zinc-8-hydroxyquinolone would thus explain the inhibition of the release reaction.

The role of Ca^{2+} in the function of the cell microskeleton, represented by microtubules and microfilament, has been well documented. The contractile elements of this system are responsible in some way for the mobility of microorganelles, and the transport of granules to the membrane, as well as the excitability of the plasma membrane itself. Zinc may compete with calcium and thereby inhibit the calcium effect.

Zinc has been shown to improve the filterability through a $3.0-\mu$ Nucleopore filter of sickle cells *in vitro* (Dash et al., 1974). Improvement in filterability at low zinc concentration suggests that the process of formation of irreversibly sickled cells involves the cell membrane. Calcium and/or hemoglobin binding may promote the formation of irreversibly sickled cells, thus hindering the filterability of such cells. Zinc may act favorably on the filterability of sickled cells by blocking the proposed calcium and/or hemoglobin binding to the membrane. The beneficial effect of zinc on the sickling process has been demonstrated both *in vitro* and *in vivo* (Brewer et al., 1977). It has been suggested that irreversibly sickled cells are stabilized by abnormal interactions between the spectrin and/or actin components of the membrane skeleton, and that disruption of these bonds by zinc may allow the skeleton to resume a normal shape (Lux and John, 1978).

Zinc is known to compete with Cd, Pb, Cu, Fe, and Ca for similar binding sites (Hill, 1976). In future, a potential use of zinc may be to alleviate the toxic effects of cadmium and lead in human subjects. The use of zinc as an antisickling agent is an example of its antagonistic effect on calcium, which is known to produce irreversible sickle cells by its action on red cell membrane.

The therapeutic use of zinc produces hypocupremia in human subjects (Prasad et al., 1978). Whether zinc could be utilized to decrease the copper load in Wilson's disease remains to be demonstrated.

Zinc may also intervene in nonenzymic, free radical reactions (Editorial, 1978); in particular, zinc is known to protect against iron-catalyzed free radical damage. It has been known that free radical oxidation (auto-oxidation) of polyunsaturated lipids is most effectively induced by the interaction of inorganic iron, oxygen, and various redox couples, and recent work suggests that this interaction underlies the pathological changes and clinical manifestations of iron toxicity. Iron-catalyzed free radical oxidation is inhibited by zinc, ceruloplasmin, metalloezymes (catalase, peroxidases, and superoide dismutase), and free radical scavenging antioxidants such as vitamin E.

Carbon tetrachloride-induced liver injury is another animal model for studying free radical injury to tissues. Animals maintained on a high-zinc regimen resist this type of biochemical injury, suggesting that zinc may protect against free radical injury. More recent studies have shown that zinc also inhibits the analogous metronidazole-dependent free radical sequence.

Zinc-deficient animals are known to be susceptible to infections (Beisel, 1976). Reduced resistance to infections is seen in severely deficient calves with lethal trait A 46, commonly known as Adema disease, as well as in human beings with acrodermatitis enteropathica. The primary cause of death in both conditions is sepsis or pneumonia. After zinc replacement, calves with Adema disease and patients with acrodermatitis enteropathica recover, and the reduced resistance to infections normalizes.

Absolute lymphopenia has been noted in patients with cirrhosis of the liver (an example of a conditioned zinc-deficient state) and in some species of animals made zinc deficient. Frost et al. (1976) reported that patients with sickle-cell anemia (also a conditioned zinc-deficient state) have an increased number of null cells. In addition, these patients, who show normal mitogen responsiveness *in vitro*, demonstrated an impaired cellular immune response *in vivo,* as measured by skin tests.

Administration of zinc in incubation media stimulates DNA synthesis of lymphocytes within 6 to 7 days (Rhul et al., 1971). Zinc must be present in the media for the entire culture period to produce maximal stimulation of [^{3}H] thymidine incorporation in DNA lymphocytes.

Decreased ability to develop cell-mediated immune response *in vitro* has been demonstrated in zinc-deficient calves and rats (Weismann, 1976). Fraker et al. (1977) reported that animals immunized with keyhole limpet hemocyanin-P-azophenylarsonate on the first day of a zinc-deficient diet and subsequently at 10–day intervals up to 30 days showed a severe depression in their antibody response to this antigen. These findings implied that both the primary immune and the memory immune responses are impaired in zinc-deficient mice. Recent studies in rats indicate that zinc deficiency causes a selective suppression of lymphoid organ weight and abnormalities in the immune response to sheep erythrocytes (Frost et al., 1977); the overall response to this antigen is markedly depressed and delayed. Thus, although it appears that zinc is involved in immune responses, further studies are required to define its role more precisely.

Serum zinc is known to decline during infection in human beings or after endotoxin administration to experimental animals (Beisel, 1976). A decrease in serum zinc may be accounted for to a large extent by an accelerated flux of zinc from plasma to liver mediated by LEM, which is liberated by phagocytizing cells. Leukocyte endogenous mediator is a heat-labile trace protein of low molecular weight and acts on the liver to stimulate and accelerate the flux of iron and zinc into hepatic cells and cause an accelerated synthesis of ceruloplasmin (Beisel, 1976). The possible purposes or values to the host of these changes in the plasma concentrations of trace elements remain to be elucidated.

4. HYPERZINCEMIA

Recently, an example of familial hyperzincemia has been reported (Smith et al., 1976). In five of seven members of a family and two of three second-generation individuals, the plasma zinc ranged from 250 to 435 μg %. Zinc levels in the red cells, hair, and bone were unremarkable, however, and no clinical effects of hyperzincemia were observed.

5. TOXICITY

Three types of toxic reactions to zinc have been reported in human beings (Prasad, 1976). First "metal fume fever," characterized by pulmonary manifestations, fever, chills, and gastroenteritis, has been reported to occur in industrial workers exposed to fumes. The second type of toxicity was observed in a 16-year-old male who ingested 12 g of zinc sulfate over a period of 2 days. The third type of acute zinc toxicity involved a patient with renal failure after hemodialysis. (The water for hemodialysis was stored in a galvanized tank.) The patient suffered from nausea, vomiting, fever, and severe anemia.

Many of the toxic effects attributed to zinc in the past may actually be due to other contaminating elements, such as Pb, Cd, or As. Zinc is noncumulative and the proportion absorbed is thought to be inversely related to the amount ingested. Vomiting, a protective phenomenon, occurs after ingestion of large quantities of zinc; in fact, 2 g of zinc sulfate has been recommended as an emetic.

The symptoms of human zinc toxicity include dehydration, electrolyte imbalance, abdominal pain, nausea, vomiting, lethargy, dizziness, and muscular incoordination. Acute renal failure caused by zinc chloride poisoning has been reported. The symptoms occurred within hours of ingesting large quantities of zinc. Death is reported to have occurred after ingestion of 45 g of zinc sulfate. This dose would be regarded as massive, since the daily human requirement of zinc is considered to be in the range of 15 to 20 mg/day.

Recently, gastrointestinal bleeding was observed in a patient after ingestion of zinc sulfate, 220 mg twice daily, for treatment of acne (Moore, 1978). The zinc level in the plasma was not reported. In our experience, this has not occurred in our patients. With prolonged use of zinc, however, we have observed hypocupremia in sickle-cell anemia patients (Prasad et al., 1978). Zinc acetate or zinc gluconate may be preferable for oral use, as gastric discomfort has not been a problem in our patients receiving this treatment.

In rats, ingestion of 0.5 to 1.0% Zn results in reduced growth, anemia, poor reproduction, and decreased activity of liver catalase and cytochrome oxidase.

The enzyme effects are reversed by copper administration, indicating that excessive intake of zinc may induce copper deficiency. In a study in which five to six times the normal level of dietary zinc intake was provided to pregnant rats, a higher rate of resorption of fetuses was observed (Kumar, 1976). The same author also reported that 100-mg supplements of zinc sulfate given during the third trimester of pregnancy in human subjects resulted in three premature births and one stillbirth in four consecutive subjects. Unfortunately, critical dietary and biochemical information with respect to zinc studies were not provided in either the rat or the human study by the investigator, so a proper evaluation of these results cannot be made. By contrast, other investigators have not observed this effect with excess zinc supplementation to pregnant rats (Hurley, 1968), and during the past few years in the United States many vitamin supplements containing comparable amounts of zinc have been used by pregnant women without any untoward effects.

In view of the long-term clinical usage of zinc in therapeutic dosages for several clinical conditions in human beings, one must remain alert for possible toxic effects. There may be other toxic effects of high-dosage zinc administration for long periods of time which have not yet been recognized. In general, however, zinc appears to be much less toxic than other trace elements.

6. SUMMARY

A deficiency of zinc in the growing age period results in growth retardation. Hypogonadism in males, skin changes, poor appetite, mental lethargy, and delayed wound healing are some of the recognized clinical manifestations of chronic zinc deficiency in human beings. A severe deficiency of zinc, such as occurs in acrodermatitis enteropathica and after TPN, is characterized by progressive bullous-pustular dermatitis, alopecia, severe diarrhea, and emotional disorders, and may be fatal if untreated.

The major functions of zinc in human and animal metabolism appear to be enzymatic; over 70 metalloenzymes are now known to require zinc for their functions. Zinc is required in all stages of the cell cycle, and zinc deficiency affects DNA and protein synthesis adversely. The activities of many enzymes are known to be decreased in zinc-deficient tissues of experimental animals. Thymidine kinase, alkaline phosphatase, and carboxypeptidase are affected adversely within 6 days of instituting a zinc-deficient diet in experimental animals.

There is evidence to suggest that zinc may play a significant role in the stabilization of biomembrane structure and polynucleotide conformation. Inasmuch as zinc appears to protect against hepatic cellular damage induced by carbon tetrachloride poisoning, it is reasonable to suggest that this element may have a direct effect on free radicals.

Zinc deficiency affects testicular functions adversely in human beings and animals. This effect of zinc operates at the end-organ level, and it appears that this element is essential for spermatogenesis and testosterone steroidogenesis.

Zinc is know to compete with Cd, Pb, Cu, Fe, and Ca for similar binding sites. The use of zinc as an antisickling agent is an example of its antagonistic effect on calcium, which produces irreversible sickle cells by its action on red cell membranes. The therapeutic use of zinc is known to produce hypocupremia in human subjects. Whether zinc could be utilized to decrease the copper load in Wilson's disease remains to be demonstrated.

ACKNOWLEDGMENTS

This research was supported in part by a grant from the National Institute of Health, NIAMD (AM-19338), a contract from the Food and Drug Administration, a comprehensive sickle-cell center grant from the U.S. Public Health Service, (HL-16008) to Wayne State University, and a grant from Meyer Laboratories, Inc., Fort Lauderdale, Florida.

REFERENCES

Alan, J. G., Fell, G. S., and Russell, R. I. (1975). "Urinary Zinc in Hepatic Cirrhosis," *Scot. Med. J.,* **109**, 109–111.

Barcia, P. J. (1970). "Lack of Acceleration of Healing with Zinc Sulfate," *Ann. Surg.,* **172**, 1048–1050.

Barnes, P. M. and Moynahan, E. J. (1973). "Zinc Deficiency in Acrodermatitis Enteropathica: Multiple Dietary Intolerance Treated with Synthetic Diet," *Proc. R. Soc. Med.,* **66**, 327–329.

Beisel, W. R. (1976). "Trace Elements in Infectious Processes," *Med. Clin. N. Am.,* **60**, 831–849.

Bertrand, G. and Bhattacherjee, R. C. (1934). "L'Action Combinée du Zinc et des Vitamines dans l'Alimentation des Animaux," *C. R. Acad. Sci.,* **198**, 1823–1827.

Brewer, G. J., Oelshlegel, F. J., Jr., and Prasad, A. S. (1975). "Zinc in Sickle Cell Anemia." In G. Brewer, Ed., *Erythrocyte Structure and Function.* Alan R. Liss, New York, pp. 417–435.

Brewer, G. J., Brewer, L. F., and Prasad, A. S. (1977). "Suppression of Irreversibly Sickled Erythrocytes by Zinc Therapy in Sickle Cell Anemia," *J. Lab. Clin. Med.,* **90**, 549–554.

Brummerstedt, E., Flagstad, T., and Andresen, E. (1971). "The Effect of Zinc on Calves with Hereditary Thymus Hypoplasia (Lethal Tract A 46)," *Acta Pathol. Microbiol. Scand.*, Section A, **79**, 686–687.

Caldwell, D. F., Oberleas, D., Clancy, J. J., and Prasad, A. S. (1970). "Behavioral Impairment in Adult Rats Following Acute Zinc Deficiency," *Proc. Soc. Exp. Biol. Med.*, **133**, 1417–1421.

Catalanotto, F. A. (1978). "The Trace Metal Zinc and Taste," *Am. J. Clin. Nutr.*, **31**, 1098–1103.

Chvapil, M. (1976). "Effect of Zinc on Cells and Biomembranes," *Med. Clin. N. Am.*, **60**, 799–812.

Clayton, R. J. (1972). "Double-Blind Trial of Oral Zinc Sulfate in Patients with Leg Ulcers," *Br. J. Clin. Pract.*, **26**, 368–370.

Cohen, C. (1968). "Zinc Sulfate and Bedsores," *Br. Med. J.*, **2**, 561.

Cuthbertson, D. P., Fell, G. S., Smith, C. M., and Tolstone, W. J. (1972). "Metabolism after Injury. I: Effects of Severity, Nutrition, and Environmental Temperature on Protein, Potassium, Zinc, and Creatine," *Br. J. Surg.*, **59**, 926–931.

Damyanov, I. and Dutz, W. (1971). "Anencephaly in Shiraz, Iran," *Lancet*, **1**, 82.

Dash, S., Brewer, G. J., and Oelshlegel, F. J., Jr. (1974). "Effect of Zinc on Hemoglobin Binding by Red Blood Cell Membranes," *Nature*, **250**, 251–252.

DeBeukelaer, M. M., Randall, C. L., and Stroud, D. R. (1977). "Renal Anomalies in the Fetal Alcohol Syndrome," *J. Pediatr.*, **91**, 759–760.

Dreosti, I. E. and Hurley, L. S. (1975). "Depressed Thymidine Kinase Activity in Zinc-Deficient Rat Embryos," *Proc. Soc. Exp. Biol. Med.*, **150**, 161–165.

Editorial (1978). "A Radical Approach to Zinc," *Lancet*, **1**, 191.

Fernandez-Madrid, F., Prasad, A. S., and Oberleas, D. (1973). "Effect of Zinc Deficiency on Nucleic Acids, Collagen, and Noncollagenous Protein of the Connective Tissue," *J. Lab. Clin. Med.*, **82**, 951–961.

Fraker, P. J., Haas, S. M., and Luecke, R. W. (1977). "The Effect of Zinc Deficiency on the Young Adult A/J Mouse," *J. Nutr.*, **107**, 1889–1895.

Frost, P., Chen, J. C., Amjad, H., and Prasad, A. S. (1976). *Clin Res.*, **24**, 570A (abstr.).

Frost, P., Chen, J. C., Rabbani, P., Smith, J., and Prasad, A. S. (1977). "The Effect of Zinc Deficiency on the Immune Response." In G. J. Brewer and A. S. Prasad, Eds., *Zinc Metabolism: Current Aspects in Health and Disease.* Alan R. Liss, New York, pp. 143–153.

Greaves, M. W. (1972). "Zinc and Copper in Psoriasis," *Br. J. Dermatol.*, **86**, 439–440.

Greaves, M. W. and Boyde, T. R. C. (1967). "Plasma Zinc Concentrations in Patients with Psoriasis, Other Dermatosis, and Venous Ulcerations," *Lancet*, **2**, 1019–1020.

Greaves, M. W. and Ive, F. A. (1972). "Double-Blind Trial of Zinc Sulphate in the Treatment of Chronic Venous Leg Ulceration," *Br. J. Dermatol.*, 87, 632–634.

Greaves, M. W. and Skillen, A. W. (1970). "Effects of Long-Continued Ingestion of Zinc Sulphate in Patients with Venous Leg Ulceration," *Lancet*, 2, 889–891.

Greene, H. L. (1977). "Trace Metals in Parenteral Nutrition." In G. Brewer and A. Prasad, Eds., *Zinc Metabolism: Current Aspects in Health and Disease.* Alan R. Liss, New York, pp. 87–97.

Gudbjarnason, S. and Prasad, A. S. (1969). "Cardiac Metabolism in Experimental Alcoholism." In V. Sardesai, Ed., *Biochemical and Clinical Aspects of Alcohol Metabolism.* Charles C Thomas, Springfield, Ill., pp. 61–70.

Halas, E. S., Rowe, M. C., Orris, R. J., McKenzie, J. M., and Sandstead, H. H. (1976). "Effects of Intrauterine Zinc Deficiency on Subsequent Behavior." In A. Prasad, Ed., *Trace Elements in Human Health and Disease*, Vol. I. Academic Press, New York, pp. 327–343.

Hallbook, T. and Lanner, E. (1972). "Serum Zinc and Healing of Venous Leg Ulcers," *Lancet*, 2, 780–782.

Halsted, J. A. and Smith, J. C., Jr. (1970). "Plasma-Zinc in Health and Disease," *Lancet*, 1, 322–324.

Halsted, J. A., Hackley, B. M., and Smith, J. C., Jr. (1968). "Plasma-Zinc and Copper in Pregnancy and after Oral Contraceptives," *Lancet*, 2, 278–279.

Halsted, J. A., Ronaghy, H. A., Abadi, P., Haghshenass, M., Amirhakimi, G. H., Barakat, R. M., and Reinhold, J. G. (1972). "Zinc Deficiency in Man: The Shiraz Experiment," *Am. J. Med.*, 53, 277–284.

Hambidge, K. M. and Walravens, P. A. (1976). "Zinc Deficiency in Infants and Preadolescent Children." In A. S. Prasad, Ed., *Trace Elements in Human Health and Disease*, Vol. I. Academic Press, New York, pp. 21–23.

Henkin, R. I. and Bradley, D. F. (1969). "Regulation of Taste Acuity by Thiols and Metal Ions," *Proc. Natl. Acad. Sci.*, 62, 30–37.

Henkin, R. I., Marshell, J. R., and Meret, S. (1971). "Maternal-Fetal Metabolism of Copper and Zinc at Term," *Am. J. Obstet. Gynecol.*, 110, 131–134.

Henkin, R. I., Scheichler, P. J., Friedewald, W. T., Demets, D. L., and Raff, M. (1976). "A Double-Blind Study of the Effects of Zinc Sulfate on Taste and Smell Function," *Am. J. Med. Sci.*, 272, 285–299.

Hill, C. H. (1976). Mineral Interrelationships. In A. Prasad, Ed., *Trace Elements in Human Health and Disease*, Vol. I. Academic Press, New York, pp. 281–299.

Huber, A. M. and Gershoff, S. N. (1975). "Effects of Zinc Deficiency on the Oxidation of Retinol and Ethanol in Rats," *J. Nutr.*, 105, 1486–1490.

Hurley, L. S. (1968). "Approaches to the Study of Nutrition in Mammalian Development," *Fed. Proc.*, 27, 193–198.

Hurley, L. S. (1976). "Perinatal Effects of Trace Element Deficiencies." In A.

Prasad, Ed., *Trace Elements in Human Health and Disease,* Vol. I. Academic Press, New York, pp. 301–314.

Hurley, L. S. and Shrader, R. E. (1972). "Congenital Malformations of the Nervous System of Zinc Deficient Rats," *Int. Rev. Neurobiol.,* Suppl. 1, 7–51.

Husain, S. L. (1969). "Oral Zinc Sulfate in Leg Ulcers," *Lancet,* 1, 1069–1071.

Jones, K. L. and Smith, D. W. (1973). "Recognition of the Fetal Alcohol Syndrome in Early Infancy," *Lancet,* 2, 999–1001.

Kirchgessner, M., Roth, H. P., and Weigand, E. (1976). "Biochemical Changes in Zinc Deficiency." In A. S. Prasad, Ed., *Trace Elements in Human Health and Disease,* Vol I. Academic Press, New York, pp 189–219.

Klingberg, W. G., Prasad, A. S., and Oberleas, D. (1976). "Zinc Deficiency Following Penicillamine Therapy." In A. Prasad, Ed., *Trace Elements in Human Health and Disease,* Vol. I. Academic Press, New York, pp. 51–65.

Kumar, S. (1976). "Effect of Zinc Supplementation on Rats during Pregnancy," *Nutr. Rep. Int.,* 13, 33–36.

Lei, K. Y., Abbasi, A., and Prasad, A. S. (1976). "Function of Pituitary Axis in Zinc-Deficient Rats," *Am. J. Physiol.,* 230, 1730–1732.

Lindeman, R. D., Baxter, D. J., Yunice, A. A., King, R. W., and Kraikit, S. (1977). "Zinc Metabolism in Renal Disease and Renal Control of Zinc Excretion." In G. Brewer and A. Prasad, Eds., *Zinc Metabolism: Current Aspects in Health and Disease.* Alan R. Liss, New York, pp. 193–209.

Love, A. H. G., Elmes, M. Golden, M. K., and McMaster, D. (1978). "Zinc Deficiency and Coeliac disease." In M. Kirchgessner, Ed., *Proceedings of the 3rd International Symposium on Trace Element Metabolism in Man and Animals, Freising, West Germany,* pp. 357–358.

Lux, S. E. and John, K. M. (1978). *Pediatr. Res.,* 12, 630 (abstr.).

Mansouri, K., Halsted, J., and Gombos, E. A. (1970). "Zinc, Copper, Magnesium, and Calcium in Dialysed and Non-Dialysed Uremic Patients," *Arch. Int. Med.,* 125, 88–93.

Moore, R. (1978). "Bleeding Gastric Erosion after Oral Zinc Sulphate," *Brit. Med. J.,* 1, 754–755.

Morrison, S. A., Russell, R. M., Carney, E. A., and Oaks, E. V. (1978). "Zinc Deficiency: A Cause of Abnormal Dark Adaptation in Cirrhotics," *Am. J. Clin. Nutr.,* 31, 276–281.

Myers, M. B. and Cherry, G. (1970). "Zinc and the Healing of Chronic Leg Ulcers," *Am. J. Surg.,* 120, 77–81.

Oberleas, D., Seymour, J. K., Lenaghan, R., Hovanesian, J., Wilson, R. F., and Prasad, A. S. (1971). "Effect of Zinc Deficiency on Wound Healing in Rats," *Am. J. Surg.,* 121, 566–568.

Pekarek, R. S., Wannemacher, R. W., and Beisel, W. R. (1972). "The Effect of Leukocyte Endogenous Mediator (LEM) on the Tissue Distribution of Zinc and Iron," *Proc. Soc. Exp. Biol. Med.,* 140, 685–688.

Pidduck, H. G., Wren, P. J. J., and Price Evans, D. A. (1970). "Plasma Zinc and Copper in Diabetes Mellitus," *Diabetes,* **19**, 234–239.

Pories, W. J. and Strain, W. H. (1966). "Zinc and Wound Healing." In A. S. Prasad, Ed., *Zinc Metabolism.* Charles C Thomas, Springfield, Ill., pp. 378–394.

Pories, W. J., Henzel, J. H., Rob, C. G., and Strain, W. H. (1967). "Acceleration of Wound Healing in Man with Zinc Sulfate Given by Mouth," *Lancet,* **1**, 121–124.

Portnoy, B. and Molokhia, M. M. (1971). "Zinc and Copper in Psoriasis," *Br. J. Dermatol.,* **85**, 597.

Portnoy, B. and Molokhia, M. M. (1972). "Zinc and Copper in Psoriasis," *Br. J. Dermatol.,* **86**, 205.

Prasad, A. S. (1966). "A Century of Research on the Metabolic Role of Zinc." In A. Prasad, Ed., *Zinc Metabolism.* Charles C Thomas, Springfield, Ill., pp. 250–302.

Prasad, A. S., Ed. (1976). "Deficiency of Zinc in Man and Its Toxicity." In *Trace Elements in Human Health and Disease,* Vol. 1. Academic Press, New York, pp. 1–20.

Prasad, A. S. and Oberleas, D. (1970). "Binding of Zinc to Amino Acids and Serum Proteins *in vitro,*" *J. Lab. Clin. Med.,* **76**, 416–425.

Prasad, A. S. and Oberleas, D. (1971). "Changes in Activity of Zinc-Dependent Enzymes in Zinc-Deficient Tissues of Rats," *J. Appl. Physiol.,* **31**, 842–846.

Prasad, A. S. and Oberleas, D. (1973). "Ribonuclease and Deoxyribonuclease Activities in Zinc-Deficient Tissues," *J. Lab. Clin. Med.,* **82**, 461–466.

Prasad, A. S. and Oberleas, D. (1974). "Thymidine Kinase Activity and Incorporation of Thymidine into DNA in Zinc-Deficient Tissue," *J. Lab. Clin. Med.,* **83**, 634–639.

Prasad, A. S., Halsted, J. A., and Nadimi, M. (1961). "Syndrome of Iron Deficiency Anemia, Hepatosplenomegaly, Hypogonadism, Dwarfism and Geophagia," *Am. J. Med.,* **31**, 532–546.

Prasad, A. S., Miale, A., Farid, Z., Sandstead, H. H., and Darby, W. J. (1963a). "Biochemical Studies on Dwarfism, Hypogonadism and Anemia," *Arch. Int. Med.,* **111**, 407–428.

Prasad, A. S., Miale, A., Farid, Z., Schulert, A., and Sandstead, H. H. (1963b). "Zinc Metabolism in Patients with the Syndrome of Iron Deficiency Anemia, Hypogonadism and Dwarfism," *J. Lab. Clin. Med.,* **61**, 537–549.

Prasad, A. S., Oberleas, D., Wolf, P., and Horwitz, J. P. (1967). "Studies on Zinc Deficiency: Changes in Trace Elements and Enzyme Activities in Tissues of Zinc-Deficient Rats," *J. Clin. Invest.,* **46**, 549–557.

Prasad, A. S., Oberleas, D., Miller, E. R.., and Luecke, R.W. (1971). "Biochemical Effects of Zinc Deficiency: Changes in Activities of Zinc-Dependent Enzymes and Ribonucleic Acid and Deoxyribonucleic Acid Content of Tissues," *J. Lab. Clin. Med.,* **77**, 144–152.

Prasad, A. S., Oberleas, D., Lei, K. Y., Moghissi, K. S., and Stryker, J. C. (1975a). "Effect of Oral Contraceptive Agents on Nutrients. I: Minerals," *Am. J. Clin. Nutr.*, 28, 377–384.

Prasad, A. S., Schoomaker, E. B., Ortega, J., Brewer, G. J., Oberleas, D., and Oelshlegel, F. J. (1975b). "Zinc Deficiency in Sickle Cell Disease," *Clin. Chem.*, 21, 582–587.

Prasad, A. S., Abbasi, A., and Ortega, J. (1977). "Zinc Deficiency in Man: Studies in Sickle Cell Disease." In G. Brewer and A. Prasad, Eds., *Zinc Metabolism: Current Aspects in Health and Disease.* Alan R. Liss, New York, pp. 211–236.

Prasad, A. S., Brewer, G. J., Schoomaker, E. B., and Rabbani, P. (1978a). *J. Am. Med. Assoc.* 240, 2166–2168.

Prasad, A. S., Rabbani, P., Abbasi, A., Bowersox, E., and Fox, M. R. S. (1978b). *Ann. Int. Med.*, 89, 482–490.

Rabbani, P., and Prasad, A. S. (1978). *Am. J. Physiol.*, 235(2), E 203–E206.

Riordan, J. F. (1976). "Biochemistry of Zinc," *Med. Clin. N. Am.*, 60, 661–674.

Ronaghy, H. A., Reinhold, J. G., Mahloudji, M., Ghavami, P., Fox, M. R. S., and Halsted, J. A. (1974). "Zinc Supplementation of Malnourished Schoolboys in Iran: Increased Growth and Other Effects," *Am. J. Clin. Nutr.*, 27, 112–121.

Ruhl, H., Kirchner, H., and Bochert G. (1971). "Kinetics of the Zn^{2+} Stimulation of Human Peripheral Lymphocytes *in vitro*," *Proc. Soc. Exp. Biol. Med.*, 137, 1089–1092.

Saldeen, T., and Bounk, U. 1967. "Enzyme histochemical investigations of the inhibitory effect of zinc on the injurious action of carbon tetrachloride on the liver," Frankf. Z. Pathol., 76, 419–426.

Sandstead, H. H., Prasad, A. S., Schulert, A. R., Farid, Z., Miale, A., Jr., Bassily, S., and Darby, W. J. (1967). "Human Zinc Deficiency, Endocrine Manifestations and Response to Treatment," *Am. J. Clin. Nutr.*, 20, 422–442.

Sandstead, H. H., Hollaway, W. L., and Baum, V. (1971). "Zinc Deficiency: Effect on Polysomes," *Fed. Proc. Fed. Am. Soc. Exp. Biol.*, 30, 517.

Sandstead, H. H., Vo-Khactu, K. P., and Solomons, N. (1976). "Conditioned Zinc Deficiences." In A. S. Prasad, Ed., *Trace Elements in Human Health and Disease,* Vol. I. Academic Press, New York, pp. 33–46.

Scrutton, M. C., Wu, C. W., and Goldthwait, D. A. (1971). "The Presence and Possible Role of Zinc in RNA Polymerase Obtained from *Escherichia coli,*" *Proc. Natl. Acad. Sci.*, 68, 2497–2501.

Serjeant, G. R., Galloway, R. E., and Gueri, M. C. (1970). "Oral Zinc Sulphate in Sickle-Cell Ulcers," *Lancet*, 2, 891–892.

Simkin, P. A. (1977). "Zinc Sulphate in Rheumatoid Arthritis." In G. Brewer and A. Prasad, Eds., *Zinc Metabolism: Current Aspects in Health and Disease.* Alan R. Liss, New York, pp. 343–351.

Slater, J. P., Mildvan, A. S., and Loeb, L. A. (1971). "Zinc in DNA Polymerase," *Biochem. Biophys. Res. Commun.*, **44**, 37–43.

Smith, J. C., Jr., Zeller, J. A., Brown, E. D., and Ong, S. C. (1976). "Elevated Plasma Zinc: A Heritable Anomaly," *Science*, **193**, 496–498.

Song, M. K. and Adham, N. F. (1978). "Role of Prostaglandin E_2 in Zinc Absorption in the Rat," *Am. J. Physiol.*, **234**, E99–E105.

Sullivan, J. F. (1962). "Effect of Alcohol on Urinary Zinc Excretion," *Quart. J. Stud. Alcohol*, **23**, 216–220.

Terhune, M. W. and Sandstead, H. H. (1972). "Decreased RNA Polymerase Activity in Mammalian Zinc Deficiency," *Science*, **177**, 68–69.

Todd, W. R., Elvehjem, C. A., and Hart, E. B. (1934). "Zinc in the Nutrition of the Rat," *Am. J. Physiol.*, **107**, 146–156.

Underwood, E. J. (1977). "Zinc." In E. Underwood, Ed., *Trace Elements in Human and Animal Nutrition*. Academic Press, New York, pp. 196–242.

Vallee, B. L., Wacker, W. E. C., Bartholomay, A. F., and Robin, E. D. (1956). "Zinc Metabolism in Hepatic Dysfunction. I: Serum Zinc Concentrations in Laennec's Cirrhosis and Their Validation by Sequential Analysis," *N. Engl. J. Med.*, **255**, 403–408.

Walravens, P. A. and Hambidge, K. M. (1977). "Nutritional Zinc Deficiency in Infants and Children." In G. Brewer and A. Prasad, Eds., *Zinc Metabolism: Current Aspects in Health and Disease*. Alan R. Liss, New York, pp. 61–70.

Weismann, K., Roed-Petersen, J., Hjorth, N., and Kopp, H. (1976). "Chronic Zinc Deficiency Syndrome in a Beer Drinker with a Billroth II Resection," *Int. J. Dermatol.*, **15**, 757–761.

3

MANIFESTATIONS OF ZINC ABNORMALITIES IN ANIMALS

W. Jack Miller
Milton W. Neathery

Animal and Dairy Science Department, University of Georgia, Athens, Georgia

1. Introduction   61
2. Performance, Clinical, and Behavioral Effects of Zinc Deficiency   62
3. Effects of Excess Dietary Zinc   65
References   65

1. INTRODUCTION

The essential nature of zinc in numerous life processes is widely recognized (Underwood, 1977). Likewise, the ability of animals to perform normally over a wide range of zinc intake levels has been firmly established (Miller, 1970, 1971, 1973, 1975). This is due to homeostatic control mechanisms, which permit animals to adapt effectively to a wide range of zinc intake.

In the practical feeding of animals, the zinc intake often falls outside the range in which health and performance are optimum. When too little zinc is consumed, deficiency occurs; if the intake is sufficiently excessive, there is toxicity.

Zinc deficiency was experimentally produced and described in rats and mice in the 1930s (Miller, 1971; Todd et al., 1934). At that time and for many years thereafter it was widely accepted by nutritionists that zinc deficiency would never be a practical problem in any species of animals (Miller, 1970). Even so,

61

parakeratosis (zinc deficiency) in swine was a major economic problem, costing as much as $100 million annually (Miller, 1970, 1971). Since Tucker and Salmon (1955) established that swine parakeratosis is a zinc-deficiency problem, the manifestations of zinc deficiency have constituted an active research area. Zinc deficiency has been produced and described in numerous species, including cattle (Kirchgessner and Schwarz, 1975; Legg and Sears, 1960; Miller and Miller, 1960, 1962; Mills et al., 1967; Ott et al., 1965), goats (Groppel and Hennig, 1971; Miller et al., 1964; Neathery et al., 1973), sheep (Mills et al., 1967; Ott et al., 1964) dogs (Robertson and Burns, 1963), guinea pigs (Alberts et al., 1977; McBean et al., 1972), squirrel monkeys (Macapinlac et al., 1967), chickens— chicks and laying hens (Kienholz et al., 1961; O'Dell and Savage, 1957; O'Dell et al., 1958), and Japanese quail (Fox and Harrison, 1964). Although there are some differences in the manifestations of zinc deficiency among species, many similarities also exist.

2. PERFORMANCE, CLINICAL, AND BEHAVIORAL EFFECTS OF ZINC DEFICIENCY

One of the first effects of zinc deficiency in animals is reduced feed consumption (Miller 1970, 1971; Mills et al., 1969). This is due not just to decreased appetite, but also to fundamental changes that reduce the rate of feed utilization (Miller, 1971; Quarterman et al., 1970). Forced feeding of normal amounts of feed to zinc-deficient animals can be fatal (Miller, 1971; Mills et al., 1969).

When zinc-deficient and control animals eat the same amount of feed, the controls grow more rapidly (Miller et al., 1965b; Neathery et al., 1973). The feed is digested just as efficiently by deficient animals, but digested nutrients are utilized less efficiently (Hiers et al., 1968; Miller et al., 1966; Somers and Underwood, 1969). Nitrogen and sulfur balance studies show that less protein is deposited and urinary loss of nitrogen and sulfur is greater in zinc-deficient animals than in pair-fed controls (Somers and Underwood, 1969).

The passage rate of feed in the gastrointestinal tract is slower in zinc-deficient animals (Lantzsch et al., 1977; Pate et al., 1970; Quarterman et al., 1970). When fed *ad libitum,* the amount of feed consumed by deficient rats follows a cyclic pattern, with a frequency of 3.5 to 4 days (Mills and Chesters, 1970; Mills et al., 1969; Williams and Mills, 1970).

Reduced growth is another major manifestation of inadequate zinc nutrition (Miller, 1970, 1971). The extent of the reduction can vary from a very small percentage that is not readily detectable to an actual loss in weight, depending on the degree of the deficiency.

An increase in the amount of feed consumed is one of the first changes observed when adequate zinc is fed. The response can be quite dramatic within only a few hours (Miller, 1971).

"Parakeratosis" is a term often used virtually synonomously with "zinc deficiency." The skin lesions responsible for this name generally are not evident with a borderline inadequacy of zinc (Blackmon et al., 1967; Miller and Miller, 1962; Ott et al., 1965; Underwood, 1977). Thus the manifestations of zinc deficiency are quite dependent on the severity of the deficiency.

With very severe zinc deficiency, skin lesions are usually among the most prominent clinical signs (Miller and Miller, 1962). These may include loss of hair; hard, dry, scaly skin; thickening, cracking, and inflammation of the skin; and raw, bleeding areas and cracks that may become deep fissures around hooves (Miller, 1970; Miller and Miller, 1962). The occurrence and the location of parakeratotic skin appear to depend on secondary factors such as mild trauma and rubbing (Kirchgessner and Schwarz, 1975; Miller et al., 1965a). Histologically, parakeratotic skin is thickened or keratinized, with retention of the nuclei of the epithelial cells (Miller and Miller, 1962; Underwood, 1977).

In some species, such as the chick, rat, swine, and squirrel monkey, parakeratosis of the esophagus is often observed in severe zinc deficiency (Barney et al., 1967, 1968; Diamond et al., 1971; Follis et al., 1941; O'Dell et al., 1958). In other species, including cattle, parakeratotic lesions in the esophagus have not been found (Blackmon et al., 1967; Miller and Miller, 1962).

In birds zinc deficiency causes retarded and abnormal feather development (Berg et al., 1963; O'Dell et al., 1958). Often the feathers are described as frizzled (O'Dell et al., 1958).

Reproduction is adversely affected in both males and females by inadequate dietary zinc (Berg et al., 1963; Egan, 1972; Hoekstra et al., 1967; Millar et al., 1958; Pitts et al., 1966; Underwood and Somers, 1969). In young male sheep the zinc requirement was found to be substantially higher for normal testicular growth and sperm production than for maximum body growth (Underwood and Somers, 1969). In mature male goats libido and spermatogenesis were severely reduced by zinc deficiency (Neathery et al., 1973).

Several studies have shown that very severe zinc deficiency during pregnancy seriously interferes with reproduction in female rats, mice, and rabbits (Apgar, 1968, 1970, 1971, 1972, 1973; Hurley and Mutch, 1973; Hurley and Swenerton, 1966; Hurley et al., 1971; Warkany and Petering, 1972). Congenital malformations can be an important manifestation of such a deficiency. The exact nature of the defects and reproductive problems depends on both the time during pregnancy when the deficiency exists and its extent (Apgar, 1972, 1973). Often with a borderline deficiency the only obvious effect may be lower reproductive efficiency, such as the birth of fewer lambs (Egan, 1972). Parturition can be extremely stressful in severely deficient rats (Apgar 1973).

In birds zinc deficiency reduces egg production and hatchability (Berg et al., 1963; Kienholz et al., 1961). Likewise, often the chicks are hatched with symptoms of zinc deficiency, including weakness, labored breathing, abnormal feathering, and grossly impaired skeletal development (Kienholz et al., 1961).

In zinc-deficient calves there is a rapid increase in the number of bacteria in the mouth (Mills et al., 1967). This is alleviated by adequate dietary zinc. Excessive salivation, which may be transitory, is observed in zinc-deficient calves (Miller, 1970; Miller and Miller, 1962; Mills et al., 1967) and goats (Neathery et al., 1973).

Lethargy is one of the most obvious manifestations of zinc deficiency in animals such as cattle (Miller, 1970; Miller and Miller, 1962). Affected cattle often show extreme lethargy when severely deficient. Major improvement may be observed within as little as 1 day when dietary zinc is fed to a deficient calf (Miller, 1970). Zinc-deficient rats appear to have decreased taste acuity (Catalanotto and Lacy, 1977), and learning ability may be impaired (Caldwell et al., 1970).

Often the mortality rate is substantially higher in zinc-deficient animals (Blackmon et al., 1967; Miller, 1971). Much of the increase is associated with greater susceptibility to nonspecific infections. The generally reduced resistance to infections that appears to be responsible may be caused, at least in part, by a decreased immune response associated with thymus atrophy and impairment of T-cell helper function (Fraker et al., 1977; Luecke et al., 1978; Miller et al., 1968).

A reduced rate of wound healing has been observed in zinc-deficient animals of several species, including calves (Miller et al., 1965a) and rats (Oberleas et al., 1971; Sandstead et al., 1970). In one study, however, wounds appeared to heal normally in zinc-deficient cows (Schwarz and Kirchgessner, 1975).

Hair and wool growth rates are reduced in zinc-deficient animals (Miller et al., 1965c; Mills et al., 1967). Likewise, wool from zinc-deficient lambs is brittle and lacking in crimp (Mills et al., 1967).

The manifestations of zinc deficiency vary among species of animals. For example, bone abnormalities are much more evident in young chickens and swine than in young cattle. In chicks and swine zinc deficiency causes a marked shortening and thickening of the bones (Miller et al., 1968, O'Dell et al., 1958). In contrast, relatively few, if any, differences were observed in the bones of zinc-deficient calves, even after several weeks of severe deficiency (Blackmon et al., 1967; Miller and Miller, 1962). Stiff and/or swollen joints, variable degrees of leg weakness, arthritic-like defects, and an unsteady gait were evident in several species (Nielsen et al., 1968; O'Dell et al., 1958). Pathological defects occur in the epiphyseal cartilage of zinc-deficient chicks (Westmoreland and Hoekstra, 1969). Hoof elongation and horn abnormalities have been noted in zinc-deficient sheep and goats (Mills et al., 1967; Neathery et al., 1973).

Most of the effects of zinc deficiency are reversible (Miller, 1970, 1978). However, the new growth of hair may be gray in places where hair was lost in cattle with zinc-deficiency parakeratosis (Miller and Miller, 1962). Likewise, defects in young animals born of mothers that were deficient during pregnancy generally are not corrected by subsequent administration of sufficient zinc.

In Dutch-Friesian cattle a hereditary zinc deficiency occurs in calves fed a diet containing a normal amount of zinc (Andresen et al., 1970; Kroneman et al., 1975). Apparently the defect is due to a single recessive gene.

3. EFFECTS OF EXCESS DIETARY ZINC

Although animals are able to tolerate far more dietary zinc than the minimum amount needed to permit optimum performance, a sufficiently high excess results in adverse effects on performance and health.

The first observed abnormalities due to excess zinc include reduced feed intake, growth rate, and feed efficiency (Berg and Martinson, 1972; Brink et al., 1959: Ott et al., 1966b). There may be a depraved appetite, denoted by excessive salt consumption and wood chewing (Ott et al., 1966b). Prolonged feeding of excess zinc can cause death (Ott et al., 1966a). Often, but not invariably, anemia or low hemoglobin accompanies zinc toxicity (Brink et al., 1959; Gasoway and Buss, 1972; Magee and Spahr, 1964). With zinc poisoning a dramatic fall in milk production has been observed (Allen, 1968).

With extremely high levels of dietary zinc, diarrhea, drowsiness, and paralysis have been observed in cattle (Van Ulsen, 1973). Likewise, acute enteritis, rapid emaciation, black scouring, and death have been noted when cattle were zinc poisoned (Allen, 1968).

In chicks, reduced bone ash is observed with zinc toxicity (Berg and Martinson, 1972). When turkey poults were fed 4000 to 10,000 ppm Zn, weight gains were depressed, but there was no mortality (Vohra and Kratzer, 1968).

In the mallard duck, zinc toxicity caused a reduction in the size of the pancreas and gonads relative to body weight (Gasoway and Buss, 1972). Diarrhea, leg paralysis, weight loss, anemia, and high mortality were also observed in the ducks.

Zinc toxicity symptoms in the pig, in addition to reduced rate of gain, feed intake, and feed efficiency, included arthritis, congestion of the mesentery, and gastritis (Brink et al., 1959). Likewise, there were catarrhal enteritis, and hemorrhages in the axillary spaces, brain, lymph nodes, and spleen. Hemoglobin values were unaffected, but death often occurred (Brink et al., 1959).

REFERENCES

Alberts, J. C., Lang, J. A., Reyes, P. S., and Briggs, G. M. (1977). "Zinc Requirement of the Young Guinea Pig," *J. Nutr.,* **107**, 1517–1527.

Allen, G. S. (1968). "An Outbreak of Zinc Poisoning in Cattle," *Vet. Rec.,* **83**, 8–9.

Andresen, E., Flagstad, T., Basse, A., and Brummerstedt, E. (1970). "Evidence of a Lethal Trait, A46, in Black Pied Danish Cattle of Friesian Descent," *Nord. Vet. Med.,* **22**, 473–485.

Apgar, J. (1968). "Effect of Zinc Deficiency on Parturition in the Rat," *Am. J. Physiol.,* **215**, 160–163.

Apgar, J. (1970). Effect of Zinc Deficiency on Maintenance of Pregnancy in the Rat," *J. Nutr.,* **100**, 470–476.

Apgar, J. (1971). "Effect of a Low Zinc Diet during Gestation on Reproduction in the Rabbit," *J. Anim. Sci.,* **33**, 1255–1258.

Apgar, J. (1972). "Effect of Zinc Deprivation from Day 12, 15, or 18 of Gestation on Parturition in the Rat," *J. Nutr.,* **102**, 343–347.

Apgar, J. (1973)." Effect of Zinc Repletion Late in Gestation on Parturition in the Zinc-Deficient Rat," *J. Nutr.,* **103**, 973–981.

Barney, G. H., Macapinlac, M. P., Pearson, W. N., and Darby, W. J. (1967). "Parakeratosis of the Tongue—A Unique Mistopathologic Lesion in the Zinc-Deficient Squirrel Monkey," *J. Nutr.,* **93**, 511–517.

Barney, G. H., Orgebin-Crist, M. C., and Macapinlac, M. P. (1968). "Genesis of Esophageal Parakeratosis and Histologic Changes in the Testes of the Zinc-Deficient Rat and Their Reversal by Zinc Repletion," *J. Nutr.,* **95**, 526–534.

Berg, L. R. and Martinson, R. D. (1972). "Effect of Diet Composition on the Toxicity of Zinc for the Chick," *Poult. Sci.,* **51**, 1690–1694.

Berg, L. R., Bearse, G. E., and Merrill, L. H. (1963). "Evidence for a High Zinc Requirement at the Onset of Egg Production," *Poult. Sci.,* **42**, 703–707.

Blackmon, D. M., Miller, W. J., and Morton, J. D. (1967). "Zinc Deficiency in Ruminants: Occurrence, Effects, Diagnosis, Treatments," *Vet. Med.,* **62**, 265–270.

Brink, M. F., Becker, D. E., Terrill, S. W., and Jensen, A. H. (1959). "Zinc Toxicity in the Weanling Pig," *J. Anim. Sci.,* **18**, 836–842.

Caldwell, D. F., Oberleas, D., Clancy, J. J., and Prasad, A. S. (1970). "Behavioral Impairment in Adult Rats Following Acute Zinc Deficiency," *Proc. Soc. Exp. Biol. Med.,* **133**, 1417–1421.

Catalanotto, F. A. and Lacy, P. (1977). "Effects of a Zinc Deficient Diet upon Fluid Intake in the Rat," *J. Nutr.,* **107**, 436–442.

Diamond, I., Swenerton, H., and Hurley, L. S. (1971). "Testicular and Esophageal Lesions in Zinc-Deficient Rats and Their Reversibility," *J. Nutr.,* **101**, 77–84.

Egan, A. R. (1972). "Reproductive Responses to Supplemental Zinc and Manganese in Grazing Dorset Horn Ewes," *Aust. J. Exp. Agric. Anim. Husb.,* **12**, 131–135.

Follis, R. H., Jr., Day, H. G., and McCollum, E. V. (1941). "HIstological Studies of the Tissues of Rats Fed a Diet Extremely Low in Zinc," *J. Nutr.,* **22**, 223–237.

Fox, M. R. S. and Harrison, B. N. (1964). "Use of Japanese Quail for the Study of Zinc Deficiency," *Proc. Soc. Exp. Biol.Med.,* **116**, 256–259.

Fraker, P. J., Haas, S. M., and Luecke, R. W. (1977). "Effect of Zinc Deficiency on the Immune Response of the Young Adult A/J Mouse," *J. Nutr.*, **107**, 1889–1895.

Gasaway, W. C. and Buss, I. O. (1972). "Zinc Toxicity in the Mallard Duck," *J. Wildl. Manage.*, **36**, 1107–1117.

Groppel, B. and Hennig, A. (1971). "Zinc Deficiency in Ruminants," *Arch. Exp. Veterinärmed.*, **25**, 817–821.

Hiers, J. M., Jr., Miller, W. J., and Blackmon, D. M. (1968). "Effect of Dietary Cadmium and Ethylenediaminetetraacetate on Dry Matter Digestibility and Organ Weights in Zinc Deficient and Normal Ruminants," *J. Dairy Sci.*, **51**, 205–209.

Hoekstra, W. G., Faltin, E. C., Lin, C. W., Roberts, H. F., and Grummer, R. H. (1967). "Zinc Deficiency in Reproducing Gilts Fed a Diet High in Calcium and Its Effect on Tissue Zinc and Blood Serum Alkaline Phosphatase," *J. Anim. Sci.*, **26**, 1348–1357.

Hurley, L. S. and Mutch, P. B. (1973). "Prenatal and Postnatal Development after Transitory Gestational Zinc Deficiency in Rats," *J. Nutr.*, **103**, 649–656.

Hurley, L. S. and Swenerton, H. (1966). "Congenital Malformations Resulting from Zinc Deficiency in Rats," *Proc. Soc. Exp. Biol. Med.*, **123**, 692–696.

Hurley, L. S., Gowan, J., and Swenerton, H. (1971). "Teratogenic Effects of Short-Term and Transitory Zinc Deficiency in Rats," *Teratology*, **4**, 199–204.

Kienholz, E. W., Turk, D. E., Sunde, M. L., and Hoekstra, W. G. (1961). "Effects of Zinc Defieicny in the Diets of Hens," *J. Nutr.*, **75**, 211–221.

Kirchgessner, M. and Schwarz, W. A. (1975). "Relationships between Clincial Zinc Deficiency Symptoms and Zinc Status in Lactating Cows," *Zbl. Vet. Med.*, **22**, 572–582.

Kroneman, J., Mey, G. J. W., and Helder, A. (1975). "Hereditary Zinc Deficiency in Dutch Friesian Cattle," *Zbl. Vet. Med.*, **22**, 201–208.

Lantzsch, H. J., Schenkel, H., and Menke, K. H. (1977). "The Use of Chelating Agents for Characterization of the Zinc Nutrition State. 2: Relationship between Zn Intake, Zn Retention, and Zn Excretion in Urine after a Single i.p. EDTA Injection in the Rat," *Z. Tierphysiol., Tierernähr. Futtermittelkd.*, **38**, 106–118.

Legg, S. P. and Sears, L. (1960). "Zinc Sulphate Treatment of Parakeratosis in Cattle," *Nature*, **186**, 1061–1062.

Luecke, R. W., Simonel, C. E., and Fraker, P. J. (1978). "The Effect of Restricted Dietary Intake on the Antibody Mediated Response of the Zinc Deficient A/J Mouse," *J. Nutr.*, **108**, 881–887.

Macapinlac, M. P., Barney, G. H., Pearson, W. N., and Darby, W. J. (1967). "Production of Zinc Deficiency in the Squirrel Monkey (*Saimiri sciureus*)," *J. Nutr.*, **93**, 499–510.

Magee, A. C. and Spahr, S. (1964). "Effects of Dietary Supplements on Young Rats Fed High Levels of Zinc," *J. Nutr.*, **82**, 209–216.

McBean, L. D., Smith, J. C., Jr., and Halsted, J. A. (1972). "Zinc Deficiency in Guinea Pigs," *Proc. Soc. Exp. Biol. Med.*, **140**, 1207–1209.

Millar, M. J., Fischer, M. I., Elcoate, P. V., and Mawson, C. A. (1958). "The Effects of Dietary Zinc Deficiency on the Reproductive System of Male Rats," *Can. J. Biochem. Physiol.*, **36**, 557–569.

Miller, E. R., Luecke, R. W., Ullrey, D. E., Baltzer, B. V., Bradley, B. L., and Hoefer, J. A. (1968). "Biochemical, Skeletal and Allometric Changes Due to Zinc Deficiency in the Baby Pig," *J. Nutr.*, **95**, 278–286.

Miller, J. K. and Miller, W. J. (1960). "Development of Zinc Deficiency in Holstein Calves Fed a Purified Diet," *J. Dairy Sci.*, **43**, 1854–1856.

Miller, J. K. and Miller, W. J. (1962). "Experimental Zinc Deficiency and Recovery of Calves," *J. Nutr.*, **76**, 467–474.

Miller, W. J. (1970. "Zinc Nutrition of Cattle: A Review," *J. Dairy Sci.*, **53**, 1123–1135.

Miller, W. J. (1971). "Zinc Metabolism in Farm Animals." In *Trace Mineral Studies with Isotopes in Domestic Animals.* Proceedings of panel organized by the Joint FAO/IAEA Division of Atomic Energy in Food and Agriculture, Vienna, Sept. 28 to Oct. 3, 1970, pp. 23–41.

Miller, W. J. (1973). "Dynamics of Absorption Rates, Endogenous Excretion, Tissue Turnover, and Homeostatic Control Mechanisms of Zinc, Cadmium, Manganese, and Nickel in Ruminants," *Fed. Proc.*, **32**, 1915–1920.

Miller, W. J. (1975). "New Concepts and Developments in Metabolism and Homeostasis of Inorganic Elements in Diary Cattle: A Review," *J. Dairy Sci.*, **58**, 1549–1560.

Miller, W. J. (1978). *Dairy Cattle Feeding and Nutrition.* Academic Press, New York (in press).

Miller, W. J., Pitts, W. J., Clifton, C. M., and Schmittle, S. C. (1964). "Experimentally Produced Zinc Deficiency in the Goat," *J. Dairy Sci.*, **47**, 556–559.

Miller, W. J., Morton, J. D., Pitts, W. J., and Clifton, C. M. (1965a). "Effect of Zinc Deficiency and Restricted Feeding on Wound Healing in the Bovine," *Proc. Soc. Exp. Biol. Med.*, **118**, 427–430.

Miller, W. J., Pitts, W. J., Clifton, C. M., and Morton, J. D. (1965b). "Effects of Zinc Deficiency per se on Feed Efficiency, Serum Alkaline Phosphatase, Zinc in Skin, Behavior, Greying, and Other Measurements in the Holstein Calf," *J. Dairy Sci.*, **48**, 1329–1334.

Miller, W. J., Powell, G. W., Pitts, W. J., and Perkins, H. F. (1965c). "Factors Affecting Zinc Content of Bovine Hair," *J. Dairy Sci.*, **48**, 1091–1095.

Miller, W. J., Powell, G. W., and Hiers, J. M., Jr. (1966). "Influence of Zinc Deficiency on Dry Matter Digestibility in Ruminants," *J. Dairy Sci.*, **49**, 1012–1013.

Mills, C. F. and Chesters, J. K. (1970). "Problems in the Execution of Nutri-

tional and Metabolic Experiments with Trace Element Deficient Animals."
In C. F. Mills, Ed., *Trace Element Metabolism in Animals*, E. & S. Livingstone, Edinburgh and London, pp. 39–50.

Mills, C. F., Dalgarno, A. C., Williams, R. B., and Quarterman, J. (1967). "Zinc
Deficiency and the Zinc Requirements of Calves and Lambs," *J. Nutr.*, 21,
751–768.

Mills, C. F., Quarterman, J., Chesters, J. K., Williams, R. B., and Dalgarno, A. C.
(1969). "Metabolic Role of Zinc," *Am. J. Clin. Nutr.*, 22, 1240–1249.

Neathery, M. W., Miller, W. J., Blackmon, D. M., Pate, F. M., and Gentry, R. P.
(1973). "Effects of Long Term Zinc Deficiency on Feed Utilization, Reproductive Characteristics, and Hair Growth in the Sexually Mature Male Goat,"
J. Dairy Sci., 56, 98–105.

Nielsen, F. H., Sunde, M. L., and Hoekstra, W. G. (1968). "Alleviation of the
Leg Abnormality in Zinc-Deficient Chicks by Histamine and by Various
Anti-arthritic Agents," *J. Nutr.*, 94, 527–533.

Oberleas, D., Seymour, J. K., Lenaghan, R., Hovanesian, J., Wilson, R. F., and
Prasad, A. S. (1971). "Effect of Zinc Deficiency on Wound-Healing in Rats,"
Am. J. Surg., 121, 566–568.

O'Dell, B. L. and Savage, J. E. (1957). "Potassium, Zinc and Distiller's Dried
Solubles as Supplements to a Purified Diet," *Poult. Sci.*, 36, 459–460.

O'Dell, B. L., Newberne, P. M., and Savage, J. E. (1958). "Significance of
Dietary Zinc for the Growing Chicken," *J. Nutr.*, 65, 503–523.

Ott, E. A., Smith, W. H., Stob, M., and Beeson, W. M. (1964). "Zinc Deficiency
Syndrome in the Young Lamb," *J. Nutr.*, 82, 41–50.

Ott, E. A., Smith, W. H., Stob, M., Parker, H. E., and Beeson, W. M. (1965).
"Zinc Deficiency Syndrome in the Young Calf," *J. Anim. Sci.*, 24, 735–741.

Ott, E. A., Smith, W. H., Harrington, R. B., and Beeson, W. M. (1966a). "Zinc
Toxicity in Ruminants. I: Effect of High Levels of Dietary Zinc on Gains,
Feed Consumption and Feed Efficiency of Lambs," *J. Anim. Sci.*, 25, 414–
418.

Ott, E. A., Smith, W. H., Harrington, R. B., and Beeson, W. M. (1966b). "Zinc
Toxicity in Ruminants. II: Effect of High Levels of Dietary Zinc on Gains,
Feed Consumption and Feed Efficiency of Beef Cattle," *J. Anim. Sci.*, 25,
419–423.

Pate, F. M., Miller, W. J., Blackmon, D. M., and Gentry, R. P. (1970). "[65]Zn
Absorption Rate Following Single Duodenal Dosing in Calves Fed Zinc-
Deficient or Control Diets," *J. Nutr.*, 100, 1259–1266.

Pitts, W. J., Miller, W. J., Fosgate, O. T., Morton, J. D., and Clifton, C. M.
(1966). "Effect of Zinc Deficiency and Restricted Feeding from Two to
Five Months of Age on Reproduction in Holstein Bulls," *J. Dairy Sci.*, 49,
995–1000.

Quarterman, J., Humphries, W. R., and Florence, E. (1970). "Changes in Appetite and Alimentary Muco-Substances in Zinc Defieicency." In C. F.

Mills, Ed., *Trace Element Metabolism in Animals*. E. & S. Livingstone, Edinburgh and London, pp. 167–169.

Robertson, B. T. and Burns, M. J. (1963). "Zinc Metabolism and the Zinc-Deficiency Syndrome in the Dog," *Am. J. Vet. Res.*, 24, 997–1002.

Sandstead, H. H., Lanier, V. C., Jr., Shephard, G. H., and Gillespie, D. D. (1970). "Zinc and Wound Healing: Effects of Zinc Deficiency and Zinc Supplementation," *Am. J. Clin. Nutr.*, 23, 514–519.

Shwarz, W. A. and Kirchgessner, M. (1975). "Experimental Zinc Deficiency in Lactating Dairy Cows," *Vet. Med. Rev.*, No. 1/2, pp. 19–41.

Somers, M. and Underwood, E. J. (1969). "Studies of Zinc Nutrition in Sheep. II: The Influence of Zinc Deficiency in Ram Lambs upon the Digestibility of the Dry Matter and the Utilization of the Nitrogen and Sulphur of the Diet," *Aust. J. Agric. Res.*, 20, 899–903.

Todd, W. R., Elvehjem, C. A., and Hart, E. B. (1934). "Zinc in the Nutrition of the Rat," *Am. J. Physiol.*, 107, 146–156.

Tucker, H. F. and Salmon, W. D. (1955). "Parakeratosis or Zinc Deficiency Disease in the Pig," *Proc. Soc. Exp. Biol. Med.*, 88, 613–616.

Underwood, E. J. (1977). *Trace Elements in Human and Animal Nutrition*, 4th ed. Academic Press, New York.

Underwood, E. J. and Somers, M. (1969). "Studies of Zinc Nutrition in Sheep. I: The Relation of Zinc to Growth, Testicular Development, and Spermatogenesis in Young Rams," *Aust. J. Agric. Res.*, 20, 889–897.

Van Ulsen, F. W. (1973). "Cattle and Zinc," *Tijdschr. Diergeneeskd.*, 98, 543–546.

Vohra, P. and Kratzer, F. H. (1968). "Zinc, Copper and Manganese Toxicities in Turkey Poults and Their Alleviation by EDTA," *Poult. Sci.*, 47, 699–704.

Warkany, J. and Petering, H. G. (1972). "Congenital Malformations of the Central Nervous System in Rats Produced by Maternal Zinc Deficiency," *Teratology*, 5, 319–334.

Westmoreland, N. and Hoekstra, W. G. (1969). "Pathological Defects in the Epiphyseal Cartilage of Zinc-Deficient Chicks," *J. Nutr.*, 98, 76–82.

Williams, R. B. and Mills, C. F. (1970). "The Experimental Production of Zinc Deficiency in the Rat," *Br. J. Nutr.*, 24, 989–1003.

4

BIOCHEMICAL CHANGES OF HORMONES AND METALLOENZYMES IN ZINC DEFICIENCY

M. Kirchgessner
H. -P. Roth

Institut für Ernährungsphysiologie der Technischen Universität München, Freising-Weihenstephan, West Germany

1.	**Introduction**	**72**
2.	**Effects on Hormones•**	**72**
	2.1. Insulin	72
	Glucose tolerance	73
	Insulin	75
	2.2. Adrenocorticotropin and growth hormones	77
	2.3. Sex hormones	79
3.	**Effect on Zinc Metalloenzymes**	**81**
	3.1. Changes in enzyme activities	81
	Alkaline phosphatase	81
	Carbonic anhydrase	85
	Carboxypeptidase A and B	86
	Dehydrogenases	86
	RNA and DNA polymerase	87
	Ribonuclease	88
	Thymidine kinase	89

3.2. Biochemical aspects   89
       Differences in the responses of individual zinc
       metalloenzymes to zinc depletion   89
       Relationships between reduced enzyme activities and zinc-
       deficiency symptoms   91
References   94

1. INTRODUCTION

Zinc deficiency has been demonstrated in a number of birds and mammals, including human beings. Among the most frequently reported deficiency symptoms are loss of appetite; retardation or even cessation of growth; skin defects, including parakeratotic lesions; hair loss; impaired wound healing; and defects leading to reproductive failure (Prasad, this volume).

Poor appetite and failure to grow are commonly noted as the earliest conspicuous responses to zinc deficiency, although not in adult animals. In dairy cows, for example, dietary zinc depletion did not affect food intake or milk production at all, but did induce dermal lesions on udder, legs, and tail, which were reversible by supplementing the diet with adequate levels of zinc (Schwarz and Kirchgessner, 1975).

Little is known about the metabolic defects responsible for the zinc-deficiency symptoms, but much scientific effort has been invested, especially in recent years, to elucidate the biochemical changes in the human or animal body suffering from an inadequate supply of zinc (Kirchgessner et al., 1976a; Roth and Kirchgessner, 1978a; Underwood, 1977).

2. EFFECTS ON HORMONES

Relationships between zinc and hormones not only are known for insulin but are also considered for other hormones, such as glucagon, growth hormone, and sex hormones. The importance of the zinc status on the growth and sexual development of males is well documented for various species, including human beings (Underwood, 1977; Sandstead et al., 1967). Trace elements can affect hormones at various action sites, such as their secretion, activity, and tissue binding sites. Conversely, hormones also can affect the metabolism of the trace elements at various sites, for example, their excretion and transport sites.

2.1. Insulin

Zinc is present in the endocrine pancreas at a relatively high concentration. The pancreas is one of the most sensitive tissues with regard to zinc metabolism and shows rapid accumulation and turnover of the zinc retained. Accordingly, this tissue responds to dietary zinc deficiency with rapid and severe zinc loss (Roth and Kirchgessner, 1975) and with a reduction in the activity of the digestive enzymes synthesized there, specifically carboxypeptidases A and B (Roth and Kirchgessner, 1974c). Since the discovery of Scott (1934) that crystalline insulin contains considerable amounts of zinc, this close functional and morphological relationship between zinc and insulin has been confirmed in numerous studies. The importance of insulin for the regulation of carbohydrate, fat, and protein metabolism has long been known. Although th etiology of disturbances of this endocrine system has been studied extensively, there are only a few studies on the involvement of dietary factors in insulin metabolism, such as the influence of the dietary zinc content. Changes in the zinc status of an animal, however, are likely to influence the synthesis, storage, secretion, and hormonal activity of insulin.

Glucose Tolerance

One of the best-known functions of insulin is to lower the blood glucose level. As early as 1937, Hove et al. published their first studies on the glucose tolerance of zinc-deficient rats. They noted, after oral glucose doses had been administered, only minor differences in the glucose tolerance curves between zinc-deficient and *ad libitum*-fed control rats. Hendricks and Mahoney (1972) found no difference between zinc-deficient and zinc-supplemented rats in the ability to metabolize orally administered glucose.

However, when glucose was injected intraperitoneally into rats fasted overnight after a long period of dietary treatment, as was done by Quarterman et al. (1966), the glucose tolerance of zinc-deficient animals was depressed compared to that of pair-fed controls. This finding was confirmed in studies by Boquist and Lernmark (1969) using Chinese hamsters and by Hendricks and Mahoney (1972) and Huber and Gershoff (1973) using rats. Figure 1 shows glucose tolerance curves obtained by Roth et al. (1975) for zinc-deficient rats in comparison to pair-fed and *ad libitum*-fed control animals. In these studies, rats that had been depleted by being fed a semisynthetic zinc-deficient casein diet (2 ppm Zn) for 34 days received an intramuscular injection of 80 mg glucose/100 g body weight after they had been fasted for 12 hr. The zinc-depleted rats, which had the same initial plasma glucose concentration as the pair-fed and *ad libitum*-fed control animals, had significantly lower glucose tolerance. Since the pair-fed animals exhibited an even better glucose tolerance than the *ad libitum*-fed controls, the lowered glucose tolerance of the zinc-deficient animals cannot be attributed to inanition.

In contrast to these findings are studies by Macapinlac et al. (1966), who were unable to demonstrate that zinc deficiency affects the tolerance for intraperitoneally injected glucose, and by Quarterman and Florence (1972), who used meal-eating and continuously eating pair-fed control rats. Quarterman and Florence suggested that the reduced glucose tolerance of zinc-deficient rats

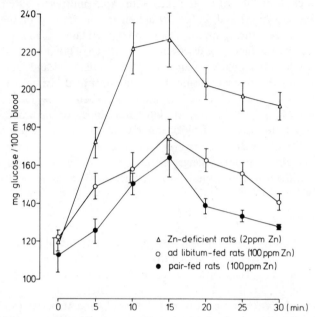

Figure 1. Glucose tolerance curves of zinc-deficient rats in comparison to ad libitum-fed and pair-fed control animals. Vertical bars represent the standard errors of the mean of six animals.

was merely the result of different patterns in food intake, because zinc-deficient animals eat slowly and continuously throughout the day, whereas their meal-fed pair-mates consume their dietary allowance for the day in a rather short time. Thus these authors consider the amount of food consumed on the day before the glucose tolerance test to be the decisive factor. The conflicting results obtained for glucose tolerance after oral dosing, on the one hand, and after intraperitoneal or intravenous injection, on the other hand, may be explained by a greater stimulation of insulin secretion by glucose given orally (McIntyre et al., 1965; Fasel et al., 1970).

After an intraperitoneal glucose injection into zinc-deficient and pair-fed control rats, within 10 to 15 min the serum zinc levels of the pair-fed animals rose significantly above the initial levels (Roth and Kirchgessner, 1979a). It was precisely within this period of time of the high zinc concentration that the highest blood glucose values also were measured. After 20 min the serum zinc content returned to normal. The serum zinc concentration of the depleted animals was reduced by about half in comparison to that of the pair-fed controls and remained largely unchanged because of the glucose injection.

Conversely, the administration of insulin to rats brought about a significant rise in the zinc contents of the cerebellum, pons, pancreas, and other organs (Ribas et al., 1978). The chromium content of the serum also showed, as with zinc, a curvelike increase during the glucose tolerance test in the case of the pair-fed control animals, reaching its maximum after 10 min (Roth and Kirchgessner, 1979a). In contrast to zinc, however, there was also a rise in the serum chromium content in the case of the deficient animals, especially toward the end of the glucose tolerance test, whereby the serum chromium contents of the zinc-deficient rats were generally somewhat lower than those of the control animals.

Insulin

The reasons for the poorer glucose tolerance of zinc-deficient animals observed in several studies can be discussed in connection with insulin. Quarterman et al. (1966) demonstrated that zinc-deficient rats exhibit a reduced concentration of plasma insulin compared to pair-fed controls. They postulated that the rate of insulin secretion in response to glucose stimulation is reduced in zinc deficiency. Furthermore, the zinc-depleted animals were less sensitive to coma and convulsions when soluble zinc-free insulin was injected intraperitoneally, although there were no differences in the blood glucose levels. Similarly, Huber and Gershoff (1973) noted that the serum of zinc-deficient rats contained less immunoreactive insulin than that of *ad libitum* control animals; there was no difference, however, in comparison with pair-fed controls. Total serum insulin-like activity measured by *in vitro* adipose tissue assay was significantly lower in the zinc-deficient groups than in the pair-fed and *ad libitum* control rats. Huber and Gershoff demonstrated *in vitro* that pancreas from zinc-deficient rats, incubated with glucose as stimulant, released less immunoreactive insulin, as well as insulin-like activity. Quarterman and Florence (1972) showed that zinc addition enhances the potency of insulin in increasing glucose uptake by adipose tissue.

In a further experiment Quarterman and Florence (1972) found no difference in plasma insulin levels between zinc-deficient rats and their zinc-supplemented, continuously fed or meal-fed pair-mates. Similarly, in four studies conducted by Roth and Kirchgessner (1975), the serum or plasma insulin levels differed significantly only once between the zinc-deficient and the pair-fed rats, although they were consistently lower than those of the *ad libitum*-fed controls. In more recent studies (Roth and Kirchgessner, 1979b) the basal levels of immunoreactive serum insulin were no different between zinc-deficient animals and pair-fed controls. After an intraperitoneal glucose injection, the serum insulin levels rose significantly during the glucose tolerance test in the pair-fed control animals, whereas they remained unchanged in the zinc-deficient animals. However, 10 to 15 min after the glucose injection, when the blood glucose values were highest, the total insulin-like activity was higher in the serum of the zinc-deficient rats than in the control animals (Roth and Kirchgessner, 1979b). The proinsulin contents of the serum did not differ between the two experimental groups either before or after the glucose injection.

Since, on the one hand, zinc deficiency in growing animals leads to loss of appetite, which means a lower food intake, and since, on the other hand, a low or no food intake involves a greatly reduced insulin level, it is difficult to determine whether the decreased serum insulin contents result primarily from the zinc deficiency or from the depressed food intake. The latter possibility could, however, be ruled out by studies on mature dairy cows, in which voluntary food consumption was not depressed despite distinct zinc-deficiency symptoms, whereas the insulin level was significantly reduced but could be raised again by high zinc supplementation (Kirchgessner et al., 1976c). Here the elevated liver glutathione contents found by Hsu (1976) in zinc-deficient rats could play a causal role, since glutathione is a coenzyme of glutathione-insulin transhydrogenase, an enzyme that promotes the degradation of insulin into its A and B chains by reductive hydrolysis of the insulin molecule.

Boquist and Lernmark (1969) did not find reduced serum insulin concentration before or after the intravenous administration of glucose to zinc-deficient hamsters, although they observed lowered glucose tolerance. Since a similar reduction in glucose tolerance was found after pancreatectomy (Boquist, 1967) and after the administration of alloxan (Boquist, 1968), Boquist and Lernmark believe that zinc deficiency causes a "prediabetic" condition. Furthermore, light and electron microscopic studies showed that the beta cells of the pancreas exhibit reduced granulation and, possibly, reduced insulin content. In rabbits with alloxan-induced or dithizone-induced diabetes, the zinc concentration of the islet tissue was also markedly lower than in normal control animals (Lazaris et al., 1971). Engelbart and Kief (1970) found that acute stimulation of insulin secretion in rats also reduces the zinc content in the beta cells of the pancreas. Since it can be assumed that zinc participates in the synthesis and storage of insulin in the beta cells, the amount of insulin stored during zinc deficiency may well be lower. Coombs et al. (1971), using equilibrium dialysis, demonstrated that porcine proinsulin aggregates to soluble polymers by binding 5 mol Zn^{2+}/ mol, whereas porcine insulin binds only 1 mol Zn^{2+}/mol and precipitates from solution. It may therefore be assumed that zinc is also of importance *in vivo* for the poor solubility of insulin and can thereby play a role in the mechanism of its release. Zinc must be dissolved before the insulin can be released from the insulin-zinc complex. In this respect, metabolites of the stimulated beta cells, such as citrate, oxalate, or organic phosphor compounds, which are strong zinc-complexing agents, could play a role (Maske, 1960). On the other hand, Hendricks and Mahoney (1972) postulated that the reduced glucose tolerance of zinc-deficient animals is caused by an increased rate of insulin degradation. This could also explain the increased insulin resistance of zinc-depleted rats observed by Quarterman et al. (1966). To clarify the relationship between zinc deficiency and insulin, further research must consider the synthesis, function, and degradation of active insulin.

2.2. Adrenocorticotropin and Growth Hormones

Homan et al. (1954) demonstrated that the addition of zinc salts increases and prolongs the physiological potency of corticotropin preparations. In *in vitro* studies with human cell cultures, adrenal steroid hormones with glucocorticoid activity increased the uptake of Zn^{2+} (Cox and Ruckenstein, 1971). In studies by Flynn et al. (1972a) the *in vitro* effect of zinc on the stimulation of corticosterone synthesis by adrenocorticotropin hormone (ACTH) was tested with isolated adrenal glands. It was found that ACTH does not stimulate corticosterone synthesis in the presence of a zinc-chelating agent in the incubation medium. If, however, zinc is added in excess to the chelating agent, the activity of ACTH is restored. This observation shows that ACTH is functionally dependent on zinc. If the ACTH activity in the intact animal depends on the extracellular zinc, reduced corticosterone synthesis should be evident in zinc-deficient rats.

In zinc-deficient patients in Iran and Egypt, accurate analysis of their endocrine functions of the anterior hypophysis revealed that these were in part suboptimal (Sandstead et al., 1967). Growth failure, including dwarfism, and hypogonadism were the most striking features. More than half of a group of human zinc-deficient dwarfs under investigation showed a reduced hypophyseal ACTH reserve and responded to ACTH injection with an abnormal delay of the renal output of 17-hydroxysteroids. After prolonged treatment with zinc, these patients responded to ACTH injection with a normal excretion pattern of renal steroids. However, this could also be an indirect effect of the zinc application, inasmuch as the zinc therapy resulted in a general improvement of the patients' health.

Henkin et al. (1969) and Lifschitz and Henkin (1971) also reported on changes in the zinc metabolism of several patients with abnormalities in the adrenocorticotropin metabolism. Patients with an isolated deficiency of growth hormone (GH), having plasma GH levels below the analytical limit of detection, showed significantly higher serum zinc concentrations and lower renal zinc excretions than control persons (Henkin, 1974b). After treatment with exogenous GH the serum zinc concentration fell again, while the renal zinc excretion rose. Conversely, an increased concentration of circulating GH, as in untreated acromegaly, leads to a reduction of the serum zinc and an elevation of the renal zinc excretion (Henkin, 1974b). Again, treatment of these patients by surgical hypophysectomy or with X-rays brings about a decrease in the circulating GH level, an increase in the serum zinc concentration, and a reduction in the renal zinc excretion. In adrenalectomized or hypophysectomized cats (Henkin, 1974a) and rats (Prasad et al., 1969b), the serum zinc concentration also increased while the renal zinc excretion decreased. These results demonstrate that the GH level in plasma is inversely proportional to the serum zinc content and directly proportional to the renal zinc excretion.

These changes are presumably the consequence of a direct and/or indirect influence of GH on the binding of zinc to macromolecular and micromolecular ligands in the blood, so that renal excretion of the zinc bound by micromolecular ligands is affected (Giroux and Henkin, 1972; Henkin 1974c, 1976). By contrast, the circulating zinc ions are primarily bound to the histidine residues of albumin, the major macromolecular ligand. When there is an increase in the concentration of circulating histidine, the major micromolecular ligand, the zinc shifts from albumin to the amino acid histidine, which, because of its size, permeates the renal glomerular membrane, whereas albumin normally cannot do so. This view is supported by clinical observations at elevated contents of circulating GH that brought about a higher cellular turnover rate and an increased content of serum and urine amino acids, including histidine (Henkin, 1974b). Consequently, an intact pituitary-adrenal cortex system is required to maintain normal circulating zinc levels and to mobilize body zinc depots (Flynn et al., 1972b). In zinc-deficient rats, however, Reeves et al. (1977) found that the serum corticosterone concentration depended neither on the dietary zinc content nor on the zinc status of the animals. Consequently, there was no response if the previously zinc-depleted animals were repleted. Furthermore, zinc depletion also had no effect on the ACTH-induced serum corticosterone levels. It must be assumed therefore, that short-term zinc supplementation of depleted animals does not influence the serum corticosterone level. Nor did the administration of ACTH have any influence on the serum zinc concentration, whether or not the animals had been given an adequate zinc supply.

Treating zinc-deficient rats with bovine GH did not improve weight gains (Macapinlac et al., 1966). Similarly, Ku (1971) reported that GH given to zinc-deficient pigs did not improve growth and food intake, nor did it influence their serum zinc levels, serum alkaline phosphatase activity, or parakeratotic lesions. The administration of bovine GH to zinc-deficient, nonhypophysectomized rats in the studies by Prasad et al. (1969b) also failed to enhance growth, whereas growth rates greatly increased after zinc supplementation. The growth rates of hypophysectomized rats, however, responded to both hormone and zinc supplementation, regardless of zinc status. Here the effects of the hormone and the zinc were additive but independent of each other.

2.3. Sex Hormones

It had been shown by Bischoff (1936, 1938) and Maxwell (1934) that the activities of the hyophyseal follicle-stimulating hormone (FSH) and luteinizing hormone (LH) could be increased when zinc salts were added to extracts of the anterior pituitary before they were injected into sexually immature or hypophysectomized rats. Injections of gonadotropin and testosterone stimulated the growth of all the accessory sex organs under zinc deficiency but did not prevent tubular atrophy of the testes, which is considered to be a typical zinc-deficiency symptom (Millar et al., 1960). On the other hand, few if any, changes were evident in the serum or urinary zinc after estrogen administration to rats, while the serum zinc concentration decreased after progesterone administration (Sato and Henkin, 1973). Briggs et al. (1971) reported a significant decrease in the serum zinc in women after estrogen therapy. Many other authors (Halsted et al., 1968; Halsted and Smith, 1970; Schenker et al., 1971; Briggs et al., 1971; Prasad et al., 1975b; McBean et al., 1971) reported a similar response of the plasma or serum zinc levels to the use of oral contraceptives.

Apgar (1970) was able to maintain pregnancy to term in about 50% of zinc-deficient rats by the administration of progesterone and estrone. According to studies by Pories et al. (1976), progesterone is regarded as the factor that may be responsible for the mobilization of zinc from tissues. They found an increase in serum zinc content during pregnancy in women and a decrease after parturition; this may be attributed to the progesterone activity (Flynn et al., 1973). Also, the effects of estrogen during pregnancy must be considered as an additional factor influencing serum zinc contents. The changes in the zinc contents in the serum and organs during the course of pregnancy and lactation are different, however, depending on the zinc nutrition (Kirchgessner and Schneider, 1978; Schneider and Kirchgessner, 1978).

Gombe et al. (1973) observed that the LH content in pooled pituitaries of zinc-deficient female rats was no different from that of pair-fed and *ad libitum*-fed control animals. The levels of LH and progesterone, however, were reduced in the plasma of the zinc-deficient and also the restricted-fed animals, in comparison to that of the *ad libitum*-fed controls. The lower plasma LH levels of zinc-deficient rats and their restricted-fed mates do not seem to be due to a lack of the LH-releasing factor, since its level was comparable in all three groups. Lactating dairy cows, in which experimental zinc depletion resulted in the appearance of distinct deficiency symptoms without their feed consumption being depressed, showed unaltered basal levels of LH and FSH in their serum, compared to repleted animals (Kirchgessner et al., 1976b). Also, in human blood donors selected according to low and high serum zinc contents, the serum testosterone correlated with the serum zinc content only between the ages of 36 and

60, while there was no relationship to the FSH and LH (Hartoma, 1977). This is evidence that, contrary to previous beliefs, a slight zinc deficiency does not reduce the hypophyseal gonadotropins, but rather affects the testicular contents. Lei et al. (1976) found in zinc-deficient rats an increased response of the serum LH and FSH to an intravenous injection of LH-releasing factor, whereas the serum testosterone response was reduced in comparison with that of restrictively fed control rats. This also indicates that the role of zinc in the male reproductive system concerns mainly the testicular site.

In patients with sickle-cell anemia who suffer, in part, from zinc deficiency (Prasad et al., 1974, 1975a; Brewer et al., 1976), the average basal levels of serum testosterone, dehydrotestosterone, and androtestosterone were significantly lower than in normal persons (Abbasi et al., 1976). After stimulation with gonadotropin-releasing hormone (Gn-RH), the rise in the serum LH and FSH was higher in the patients with sickle-cell anemia, whereas the testosterone response to Gn-RH was sluggish (Prasad et al., 1977). The basal serum LH and FSH values were also higher than in the control patients. The androgen deficiency, always a characteristic symptom in patients with sickle-cell anemia, appears to be primarily the consequence of testicular rather than hypophyseal atrophy. Alternatively, it could be assumed that the androgen catabolism within the testes is increased during zinc deficiency. The basic abnormalities responsible for the defective testicular function have not yet been fully clarified.

Acrodermatitis enteropathica, an autosomal recessive disease, is characterized by abnormal metabolism of the essential fatty acids (Neldner et al., 1974; Cash and Berger, 1969; White and Montalvo, 1973) and by disorders in zinc absorption (Moynahan, 1974; Neldner and Hambidge, 1975; Lombeck et al., 1975). Evans et al. (1975) found a low-molecular zinc-binding ligand in human milk to be essential for the maintenance of normal zinc absorption. This ligand was identified by Song and Adham (1976) as prostaglandin. In cow's milk this zinc-prostaglandin complex could not be detected, a finding that may explain the positive effect of human breast milk in the treatment of patients with acrodermatitis. The reduced arachidonic acid levels found in the serum of patients with acrodermatitis enteropathica (Neldner et al, 1974; Cash and Berger, 1969; White and Montalvo, 1973) indicate a disturbed prostaglandin synthesis, because arachidonic acid serves as the precursor of prostaglandins (Evans and Johnson, 1977; Weismann and Flagstad, 1976). As shown in studies on rats, the essential role of prostaglandin E_2 in zinc absorption lies not only in its chelation of zinc but, more importantly, in an increase in the zinc transport across the intestinal mucosa (Song and Adham, 1977, 1978). These results show that the symptoms of acrodermatitis enteropathica result from inability to synthesize prostaglandin, which stimulates intestinal zinc absorption. Pathological symptoms similar to those seen in zinc deficiency were evident in female rats given aspirin at the end of pregnancy (O'Dell et al., 1977). Aspirin inhibits prostaglandin biosynthesis in various tissues and hence results in impaired zinc absorption in rats (Evans and Johnson, 1977). Because of the similarity of the defects that can be brought about by a prostaglandin inhibitor such as aspirin and by zinc deficiency in pregnant rats, one could speculate that zinc participates in the biosynthesis or function of prostaglandin (O'Dell et al., 1977). However, additional studies are needed to show whether or not zinc and the prostaglandins are related by a common metabolic pathway that, when disturbed, leads to the similarity in pathology.

In summary, it may be said, in reference to zinc and hormones, that there is a great need for further research on the role of zinc in hormone metabolism, especially in respect to its functions in the synthesis and secretion of various hormones. Specifically, additional studies must clairfy the involvement of zinc in the biopotency of these hormones.

3. EFFECT ON ZINC METALLOENZYMES

The essential nature of zinc for the living system is fundamentally based on its role as an integral part of a number of metalloenzymes and as a cofactor for regulating the activity of specific zinc-dependent enzymes. The level of zinc in cells can therefore govern many metabolic processes, specifically carbohydrate, fat, and protein metabolism and nucleic acid synthesis or degradation, through the initiation and/or regulation of the activity of zinc-dependent enzymes. The decrease in the activity of a particular enzyme in response to deficient zinc nutrition depends on how tightly the zinc cation is bound to the protein (thermodynamic stability) or how fast the rate of exchange of the ligands is (kinetic stability). These are also the reasons why only a few of the known zinc metalloenzymes (see Table 1) respond sensitively and rapidly to a deficient zinc supply.

3.1. Changes in Enzyme Activities

Alkaline Phosphatase

Some metalloenzymes contain concrete metal-binding sites that are important for catalytic function and essential for structural stability. Thus alkaline phosphatase from *Escherichia coli,* for example, contains 4 g-atoms Zn/mol; 2 are essential for catalytic activity (Plocke et al., 1962), while the additional 2 zinc atoms stabilize the protein structure (Simpson and Vallee, 1968). Accordingly, alkaline phosphatase exhibits quick loss of activity in experimental zinc deficiency. In the serum of rats, Roth and Kirchgessner (1974a) found the activity of alkaline phosphatase to decrease by as much as 25% after just 2 days

Table 1. Zinc Metalloenzymes in the Animal and Human Body

Enzyme	EC Number	Activity Change[a]	Examples of Occurrence
Alcohol dehydrogenase	1.1.1.1	–	Liver, adipose tissue, lung
Glutamate dehydrogenase	1.4.1.3	0	Liver, cerebrum, kidneys, cardiac muscle, etc.
Malate dehydrogenase	1.1.1.37	0	Brain, adipose tissue, liver, pancreas, etc.
Lactate dehydrogenase	1.1.1.27	0	Brain, adipose tissue, cardiac muscle, kidneys, liver pancreas, etc.
Glyceraldehydephosphate dehydrogenase	1.2.1.13		Muscle
RNA polymerase	2.7.7.6	–	Liver, kidneys
DNA polymerase	2.7.7.7	–	Liver, kidneys
Alkaline phosphatase	3.1.3.1	–	Bone, mucosa, serum, kidneys
Leucine aminopeptidase	3.4.1.1		Kidneys
Carboxypeptidase A	3.4.2.1	–	Pancreas
Carboxypeptidase B	3.4.2.2	–	Pancreas
Dipeptidase	3.4.3	–	Kidneys
AMP aminohydrolase	3.5.4.6		Muscle
Carbonic anhydrase	4.2.1.1	–	Erythrocytes
δ-Aminolevulinic acid dehydrogenase	4.2.1.24		Liver

[a]In zinc deficiency: 0 = unchanged in comparison with pair-fed controls, – = reduced in comparison with pair-fed controls.

of dietary zinc depletion and by 50% after 4 days. This loss of activity was not due to reduced food intake; in fact, the activity of this serum enzyme had already diminished before growth and food intake noticeably decreased. Furthermore, there was no difference in the activity between restrictively fed and *ad libitum*-fed control groups throughout the experimental period. Also, the serum alkaline phosphatase was greatly decreased in zinc-deficient lactating cows that did not restrict their feed intake (Kirchgessner et al., 1975a).

The activity of alkaline phosphatase returned nearly to the level of that in control animals or in animals having adequate zinc within 3 days after zinc had been added to the diet or injected (Roth and Kirchgessner, 1974a, 1978b) (see also Figure 2). Although preincubation of serum from zinc-deficient rats with zinc *in vitro* enhanced the activity of the alkaline phosphatase, the level of control serum was not reached because there the activity was also stimulated by zinc addition. This demonstrates that the decrease in alkaline phosphatase activity in

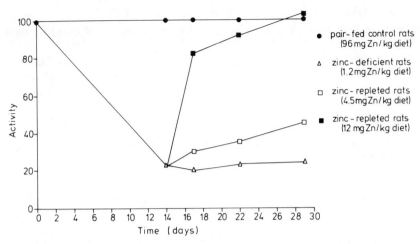

Figure 2. Activities of alkaline phosphatase in serum of depleted, repleted and control rats (pair-fed control rats = 100).

zinc-deficient status is the result of lowered enzyme concentration, caused by reduced enzyme synthesis or enhanced degradation (Roth and Kirchgessner, 1978b).

Under controlled experimental conditions the activity of alkaline phosphatase thus is a good indicator of the zinc supply status. Since, however, under field conditions the activity of alkaline phosphatase is subject to large individual variation and to many other influences, the problem of nominal or normal values was circumvented in a study on the diagnosis of zinc deficiency by determination of the activity before and 3 days after zinc injection. As is evident from Table 2, an increase in the activity of alkaline phosphatase could be brought about only in animals that were supplied more or less inadequately with zinc (Roth and Kirchgessner, 1978b). Hence the activity of the alkaline phosphatase in serum and its response to zinc injection constitute a simple and sensitive system to diagnose zinc status, because at optimum zinc supply additional doses of the element fail to bring about an increment in activity.

Henkin et al. (1976) found the activity of the alkaline phosphatase in the leukocytes of 106 patients with gustatory and olfactory disturbances, a zinc-deficiency symptom, to lie significantly below normal values. Similarly, in volunteers in whom a slight zinc deficiency was produced experimentally for the first time with human beings (Prasad et al., 1978), the activity of the alkaline phosphatase in serum slowly decreased after zinc restriction. After zinc supplementation of the diet, all the other dietary constituents remaining the same except for zinc intake, the activity of the alkaline phosphatase doubled within 8 weeks. This also proves that this biochemical change in enzyme activity is to be attributed

Table 2.   Increase in Alkaline Phosphatase Activity and Zinc Content in Serum of Rats with Different Dietary Zinc Supplies after Injection of Zinc (0.8 mg Zn per animal)

Supply	Dietary Zinc Content (ppm)	Increase after Zinc Injection (Initial Level = 100)	
		Zinc Content in Serum	Alkaline Phosphatase Activity
Deficient	1.3	292	660
	4	265	415
Suboptimal	8	200	159
	10	183	139
Optimal	20	133	142
	100	105	103

solely to zinc deficiency. In patients with acrodermatitis enteropathica, a lethal autosomal, recessive disease that is probably due to a disorder in zinc absorption (Lombeck et al., 1975), the rate of renal zinc excretion and the activity of the serum alkaline phosphatase were also greatly reduced, in addition to the zinc contents in plasma and skin (Hambidge et al., 1977; Neldner and Hambidge, 1975). This indicates a severe state of zinc deficiency. An oral supply of 22 to 44 mg Zn twice daily (see Table 3) brought about rapid and complete clinical remission in all four patients within 2 weeks and the return of biochemical indices such as plasma zinc content, renal zinc excretion, and alkaline phosphatase activity of serum to normal values (Neldner and Hambidge, 1975).

The clinical effect of zinc therapy was equally fast on skin lesions, diarrhea, lack of appetite, and emotional depression. This immediate clinical response to zinc therapy before the biochemical parameters of the zinc supply status became

Table 3.   Biochemical Data on Four Patients with Acrodermatitis Enteropathica before and after Zinc Therapy[a]

Biochemical Index	Before Zinc Therapy	After Zinc Therapy
Plasma zinc (μg/100 ml)	25 ± 10	112 ± 28
Renal zinc excretion (μg Zn/24 hr)	47 ± 11	714 ± 384
Alkaline phosphatase in serum (IU/ℓ)	57 ± 14	181 ± 18

[a] Hambidge et al. (1977).

fully normal shows that the nutritive effect of zinc supplementation is to be attributed to an amelioration of the zinc deficiency and not to a pharmacological effect of excessive zinc (Hambidge et al., 1977). Ten patients with parenteral nutrition developed skin lesions similar to those seen in acrodermatitis enteropathica, and showed low serum zinc levels and reduced activity of the alkaline phosphatase in serum (Van Vloten and Bos, 1978). In Friesian calves with Adema disease, a malady with symptoms paralleling those of acrodermatitis enteropathica of man, it was consistently found that the serum zinc contents were subnormal and the alkaline phosphatase activity reduced, and that both increased immediately after zinc therapy (Weismann and Flagstad, 1976; Kroneman et al., 1975; Stöber, 1971).

In arthritis rheumatica, an increasingly disabling disease of unknown etiology, there also are indications of deficient zinc status (Simkin, 1976). After 12 weeks of oral zinc sulfate therapy, the activity of the alkaline phosphatase in serum had increased significantly, whereby the serum zinc content was closely correlated with the activity of the alkaline phosphatase. Recent studies on patients with sickle-cell anemia showed that zinc plays an important role in this disease and that some of these patients suffer from zinc deficiency (Prasad et al., 1974; Prasad, 1976; Brewer et al., 1976); zinc therapy therefore has proved to be beneficial. In the Middle East dietary supplementation with zinc augmented the activity of the alkaline phosphatase in the serum of zinc-deficient persons showing dwarfism (Prasad et al., 1961, 1963). Determination of the activity of the alkaline phosphatase in serum or plasma before and after zinc supplementation is therefore a useful approach to the diagnosis of zinc deficiency in human beings (Roth and Kirchgessner, 1978b).

Carbonic Anhydrase

The carbonic anhydrase of erythrocytes also requires zinc for its physiological functions. The first zinc metalloenzyme detected by Keilin and Mann (1940a, 1940b), it contains 1 g-atom Zn/mol. In rats the activity of carbonic anhydrase was reduced by about 20 and 40% within 2 and 4 days, respectively, after feeding them a zinc-deficient diet (Roth and Kirchgessner, 1974b). However, at the end of the 30-day experiment, there was no longer a difference in the activities of the deficient and the control animals. The number of erythrocytes per cubic millimeter of blood had increased in this time by 40%. The body thus strives to maintain the activity of this essential enzyme by raising the erythrocyte concentration in the blood. When the activity is expressed per unit of erythrocytes, a reduction in the activity of the carbonic anhydrase in the blood can be demonstrated soon after the start of zinc depletion and later in the stage of extreme zinc deficiency (Kirchgessner et al., 1975b). Also, in the intestine and stomach of zinc-deficient rats carbonic anhydrase was reduced by 33 and 47%, respectively (Iqbal, 1971). In patients with sickle-cell anemia the activity of the carbonic anhydrase in the red blood cells closely correlated with the zinc content (Prasad et al., 1975a).

Carboxypeptidase A and B

Two additional zinc metalloenzymes important in protein digestion are the pancreatic carboxypeptidases A and B. Each possesses 1 g-atom Zn/mol and has a molecular weight of about 34,000. In zinc deficiency studies carboxypeptidase A showed a reduced activity in the pancreas of rats (Hsu et al., 1966; Mills et al., 1967; Prasad and Oberleas, 1971; Roth and Kirchgessner, 1974c) and pigs (Prasad et al., 1971). The activity of this enzyme was about 25% lower in the pancreas of rats after two days of zinc depletion (Roth and Kirchgessner, 1974c). In repletion studies the activity of carboxypeptidase A returned to nearly normal values within 3 days after zinc supplementation (Figure 3). Also, in the case of pancreatic carboxypeptidase B, a reduction of its activity by roughly 50% was found in zinc-deficient rats compared with pair-fed and *ad libitum*-fed control animals (Roth and Kirchgessner, 1974c).

Dehydrogenases

The lactate, malate, alcohol, and glutamate dehydrogenases are other zinc metalloenzymes that differ in their molar contents of zinc. They show, depending on

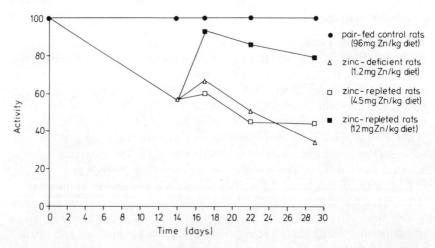

Figure 3. Activities of pancreatic carboxypeptidase A of depleted, repleted and control rats (pair-fed control rats = 100).

species and tissue, unchanged or only slightly reduced activities in response to zinc deficiency (Kirchgessner et al., 1976a). In experimental zinc deficiency in human beings (Prasad et al., 1978) the activity of the lactate dehydrogenase in the plasma of all subjects correlated with the phases of zinc restriction and zinc repletion. During zinc restriction lactate dehydrogenase activity was decreased; during zinc repletion a significant elevation of the activity was observed.

RNA and DNA Polymerase

In 1959 Wacker and Valle showed that zinc occurred in several highly purified preparations of RNA and DNA from very different sources. On the basis of these observations many experimental studies were undertaken to investigate whether this element plays a role in nucleic acid metabolism and in protein synthesis (Wacker, 1962). Several years ago zinc became known as an essential constituent of the DNA-dependent DNA and RNA polymerase. Using highly purified preparations of DNA-dependent RNA polymerase from *Escherichia coli*, Scrutton et al., (1971) determined its zinc content as 2 g-atoms/mol of enzyme (mol wt, 370,000). A homogeneous DNA polymerase was isolated by Slater et al. (1971) from *E. coli* and sea urchins and determined to contain 2 and 4 g-atoms Zn/mol of enzyme. Definite proof that the DNA polymerase of *E. coli* is a metalloenzyme was established by Springgate et al. (1973a, 1973b). The RNA-dependent DNA polymerase, that is, the reverse transcriptase, also is a zinc metalloenzyme (Auld et al., 1974a, 1974b) and occurs in a number of viruses. These transcription enzymes have a key position in nucleic acid metabolism and hence also in protein biosynthesis. Lieberman and Ove (1962) found in *in vitro* experiments with kidney cells from rats that zinc ions are necessary for DNA synthesis. Removal of zinc from the growth medium resulted in inhibition of both DNA synthesis and DNA polymerase and thymidine kinase activity. Terhune and Sandstead (1972) studied the influence of zinc deficiency on the activity of the DNA-dependent RNA polymerase in the nuclei of liver cells of suckling rats born of zinc-deficient mothers. The activity of this enzyme steadily decreased in the young from the 10th day of life, demonstrating that zinc is necessary for the activity of the nuclear DNA-dependent RNA polymerase in mammalian liver. In a similar study on the brains of prenatal zinc-deficient rats, reduced RNA polymerase activity was also found, in addition to smaller brain size and diminished DNA synthesis (Sandstead et al., 1972). Likewise, the activity of DNA polymerase was significantly lower in embryos of zinc-deficient rats than in those of control animals from the 9th to 12th days of pregnancy.

On the basis of these observations it may be concluded that zinc-deficient organisms show impaired nucleic acid synthesis. Although the exact mechanism responsible for the defective protein synthesis during zinc deficiency cannot yet be stated with certainty, it appears likely that it also results from impaired nucleic acid synthesis.

Ribonuclease

Fernandez-Madrid et al. (1973) and Somers and Underwood (1969) expressed the opinion that Zn^{2+} can inhibit ribonuclease, whereby RNA degradation would be lowered. Also, enhanced activity of this enzyme was observed in zinc-deficient tissues of experimental animals (Prasad and Oberleas, 1973). Hence the zinc content in cells regulates the activity of ribonuclease, so that the rate of RNA catabolism also depends on the tissue zinc concentration. Thus the testes of zinc-deficient rats contained less zinc, RNA, DNA, and protein, but at the same time they exhibited an elevated ribonuclease activity (Somers and Underwood, 1969). The primary defect is also the increased ribonuclease activity, which leads to increased protein catabolism. In similar experiments (Prasad and Oberleas, 1973), the activities of ribonuclease and deoxyribonuclease were studied in the testes, kidneys, bones, and thymus of zinc-deficient rats. Whereas the deoxyribonuclease activity did not differ between the zinc-deficient and pair-fed control rats, the ribonuclease activity was elevated in the zinc-deficient tissues. The authors believe that the increased ribonuclease activity is responsible for the lowered RNA/DNA ratio observed in previous investigations with zinc-deficient pigs (Prasad et al., 1971), as well as for the lowered protein synthesis and the growth depressions observed regularly in many species, including human beings.

In the plasma of patients with sickle-cell anemia a reduced zinc concentration and an elevated ribonuclease activity were demonstrated providing further evidence that these subjects have a zinc-deficient status (Prasad et al., 1975a). After zinc therapy, the plasma zinc content increased and the activity of the plasma ribonuclease decreased. Also, in a child with acrodermatitis entero-pathica, Hambidge et al. (1977) found an elevated ribonuclease activity in serum, besides a reduced vitamin A content; both the vitamin A content in the serum and the ribonuclease activity returned to normal after zinc therapy was begun. In volunteers in whom a mild zinc deficiency was produced experimentally, the activity of the plasma ribonuclease was nearly twice as high during the phase of zinc restriction as during the zinc repletion phase (Prasad, 1978). These results are similar to those obtained with zinc-deficient rats (Prasad and Oberleas, 1974).

Determination of the activity of a zinc-dependent enzyme, such as the ribonuclease in plasma, could therefore be an additional parameter useful for the diagnosis and assessment of the zinc status in human beings. Opposed to this conclusion stands a recent report by Chesters and Will (1978), who state that ribonuclease activity, total ribonuclease concentration, and free ribonuclease inhibitor concentration in liver, kidneys, and testes of zinc-deficient rats were not altered in comparison to those of pair-fed control animals. They assume that the effect of a dietary zinc deficiency on the ribonuclease activities of tissues is a general response to the altered growth rate rather than a direct effect of zinc on the enzyme system.

Thymidine Kinase

More recent studies have showed that zinc is also required for the activity of thymidine kinase, an enzyme essential for DNA synthesis and hence also for cell division. Thymidine kinase could therefore be the enzyme that is responsible for the symptoms of early zinc depletion. Lieberman et al. (1963) had indicated that zinc-deficient mammalian cells had a lower thymidine kinase activity after *in vitro* incubation than did control cells. In zinc-deficient rats the activity of thymidine kinase in fast-regenerating connective tissue was, after 6 days of zinc-deficient nutrition, significantly reduced below the values for pair-fed or *ad libitum* control animals (Prasad and Oberleas, 1974). Similar results were obtained after 13 days, and after 17 days of zinc depletion the thymidine kinase was no longer detectable. The incorporation of $[^{14}C]$-thymidine into the DNA of connective tissue was also reduced after 6 days on the zinc-deficient diet. In hepatectomized rats, which were given postoperatively a zinc-deficient diet, Duncan and Dreosti (1976) found decreased activity of thymidine kinase, compared with that of control animals, before a reduction in DNA and protein synthesis occurred. Whereas the synthesis of protein was reduced after 48 hrs, and that of DNA after 18 hrs, of postoperative zinc deficiency, the thymidine kinase activity was diminished after only 10 hrs. Similarly, there was a markedly reduced activity of thymidine kinase in 12-day-old embryos of zinc-deficient rats compared with those of pair-fed control animals (Dreosti and Hurley, 1975). Reduced activity of thymidine kinase became evident even when pregnant females were given the zinc-deficient diet from the 9th to the 12th days of pregnancy only.

These results show that either the thymidine activity or the biochemical processes necessary for the activity of thymidine kinase are extremely sensitive to zinc depletion. In the metabolic sequence the activity of this enzyme stands before the polymerization of DNA and before cell division. Reduced activity or delayed response of the activity of this enzyme is one of the earliest and most sensitive metabolic alterations and could ultimately prove to be responsible for the rapid decline in growth rate, the anorexia, the delayed wound healing, and so on in zinc-deficient subjects (Oberleas and Prasad, 1974; Prasad, 1976).

3.2. Biochemical Aspects

Differences in the Responses of Individual Zinc
Metalloenzymes to Zinc Depletion

The research results presented in this chapter clearly reveal that among the zinc metalloenzymes only specific ones change their activity, and then only in certain tissues. Therefore the probability of detecting biochemical changes is highest in the tissues that respond rather sensitively to a lack of available zinc, for example, serum, bones, pancreas, and intestinal mucosa (Prasad et al., 1971).

It is not to be expected that zinc-dependent enzymes are affected to the same extent in all tissues of a zinc-deficient animal. Differences in the sensitivity of enzymes evidently result from differences in both the zinc-ligand affinities of the various zinc metalloenzymes and their turnover rates in the cells of the affected tissues (Prasad et al., 1971). Thus the zinc metalloenzymes that bind zinc with a very high affinity should still be fully active even in extreme stages of zinc deficiency.

Swenerton (1971), Swenerton and Hurley (1968), and Swenerton et al. (1972) could not find reduced activity of lactate, glutamate, or alcohol dehydrogenase in the liver of zinc-depleted rats, nor were there reduced activites of malate and lactate dehydrogenases in the testes showing histological lesions. Therefore they did not agree with the hypothesis that reduced enzyme activities are responsible for the severe physiological and morphological changes observed in zinc-deficient animals. The complete lack of responsiveness of the activity of these enzymes, even when symptoms of severe zinc deficiency are apparent, may again be explained on the basis that the affinity of zinc for these enzymes is high, and, consequently, that the turnover rate in these tissues remains unaltered.

In addition, the possibility cannot be excluded that a specific enzyme requiring zinc for its activity is replaced by another enzyme that does not contain zinc but nevertheless catalyzes the same reaction. Furthermore, numerous metalloenzymes are able to remain functionally intact despite a deficiency of a particular metal, because some other metal may substitute (Riordan, 1976). This is particularly true of the zinc metalloenzymes, in which the metal atom is not subjected to an electron transfer with a specific redox potential, but rather functions more generally as a Lewis acid. At least, there is no evidence as yet that zinc is oxidized or reduced in biological reactions (Vallee, 1976).

Cobalt can substitute for zinc in the formation of an active metalloenzyme, and in a few cases this cobalt derivative is more active than the native zinc enzyme (Riordan, 1976). This could be another reason why the activities of a few enzymes do not change with zinc depletion. Consequently, it is difficult to demonstrate the influence of zinc deficiency on a specific enzyme if the activity is the sole criterion tested. The possibility that zinc is replaced by other cations may be excluded, however, in enzymes that show greatly reduced activities in zinc deficiency.

In zinc deficiency the tissue concentrations of other essential trace elements, besides zinc, may also be greatly affected (Burch et al., 1975; Roth and Kirchgessner, 1977; Kirchgessner et al., 1978). Burch et al. (1975) therefore suggest that the altered contents of trace elements in the tissues, as they occur in zinc deficiency, may lead to changes in the activities of metalloenzymes or metalactivated enzymes that are per se independent of zinc. Zinc-deficiency symptoms could thus also be due to changes in the tissue contents of other trace elements, which however, are induced by zinc deficiency.

Relationships between Reduced Enzyme
Activities and Zinc-Deficiency Symptoms

Since many enzymes need zinc for their physiological functions, the manifestation of severe zinc-deficiency symptoms may be associated with reduced activities of a number of zinc-containing enzymes. Since tissues bind zinc with different affinities, dietary zinc depletion may rapidly lead to a deficiency in labile zinc, especially in certain organs, and to a corresponding loss in the activities of specific zinc metalloenzymes. Adequate zinc supplementation rapidly overcomes this deficiency and raises the activity of these zinc metalloenzymes to normal levels. The extent to which a metalloenzyme loses its activity also depends on the functional role of zinc in maintaining the enzyme structure. With some zinc-dependent enzymes, for example, alkaline phosphatase, zinc deficiency may induce structural changes that increase the chance for degradation. The consequence is an increased turnover rate and a lower activity of the enzyme in the tissues (Reinhold and Kfoury, 1969). Mills et al. (1969), Prasad and Oberleas (1970), and Prasad et al. (1971) suggested that the rapidity with which biochemical changes arise in response to zinc depletion and disappear after repletion helps to identify the primary site of metabolic functions of zinc. In studies applying dietary zinc depletion, the early changes in enzyme activities, detectable before a general depletion is evident from tissue zinc levels, indicate that the primary role of zinc must be associated with a tissue component having an extremely high turnover or that zinc is essential at a site where it is freely exchangeable.

According to Prasad et al. (1969a, 1969b), many metabolic processes are regulated by zinc metalloenzymes, which in turn depend on the tissue levels of zinc available for the control of their synthesis and activity. None of the currently known enzymes is sufficiently sensitive or important in metabolism to be solely responsible for the very early and basic effects of zinc depletion on growth and appetite. Zinc metalloenzymes are, however, fundamentally involved in many intracellular biochemical mechanisms, as they, in particular, regulate the various stages of protein synthesis and the early nucleic acid synthesis before cell division (Fell and Burns, 1976).

In the following discussion possible relationships between the known zinc-deficiency symptoms and the described enzyme changes will be considered more closely. Two enzymes that affect protein digestion are the pancreas carboxypeptidases A and B. The activities of these two enzymes are reduced in zinc deficiency; pancreas carboxypeptidase A lost about one quarter of its activity within 2 days (Kirchgessner and Roth, 1975). However, the digestiblity of the dietary dry matter and crude protein was reduced just slightly (1.5 and 3%, respectively) in young zinc-deficient rats compared with pair-fed control animals (Pallauf and Kirchgessner, 1975). On the other hand, a pronounced decrease in feed efficiency and growth rate is generally observed in zinc deficiency. Thus Kirchgessner and Roth (1975) and Weigand and Kirchgessner (1977) recorded the food expenditure for body weight gain to be three to seven times higher in zinc-deficient rats than in control animals. Consequently, the reasons for these drastic effects of zinc deficiency must be sought, not so much in the intestinal tract, as in the biological processes of the cells in the body tissues.

In addition to carboxypeptidases A and B, alkaline phosphatase may be related to the many disorders caused by zinc deficiency. This zinc-containing metalloenzyme is important for a number of processes in the biochemistry of normal bone formation (Ellul-Micallef et al., 1976, Bourne, 1972; Thawley and Willoughby, 1977). Since the alkaline phosphatase in serum is known to stem from various tissues, a reduction of the serum enzyme in response to zinc deficiency is the result of a reduction of the tissue enzyme. In zinc-deficient rats the alkaline phosphatase of the femur bone is reduced by two thirds compared with pair-fed control animals (Roth and Kirchgessner, 1974d), so that, in turn, the alkaline phosphatase in serum is reduced, since bones are the major source of serum alkaline phosphatase (Lin, 1970; Thawley and Willoughby, 1977). Westmoreland (1969, 1971) reported on the possible consequences of reduced alkaline phosphatase for the growth and calcification of bones and for the leg abnormalities observed in zinc-deficient chicks. Zinc deficiency causes hair defects and a reduction of the alkaline phosphatase in rat skin because of a suppressed hair cycle and of a simultaneous decrease in the hair follicles, which are rich in alkaline phosphatase (Lin, 1970). Moreover, a reduced activity of the alkaline phosphatase found in high concentrations in the membranes of the taste buds (Trefz, 1972; Lum and Henkin, 1976) could possibly be responsible for the blunting of the sense of taste in zinc deficiency (Catalanotto and Nanda, 1977). These studies demonstrate that changes in alkaline phosphatase could indeed be partly responsible for many of the defects observed in zinc deficiency. Thus significant growth promotion could be obtained by injecting zinc-deficient rats with alkaline phosphatase (Roth and Kirchgessner, 1976); however, the impaired growth could be only partly reversed.

Although the symptoms of zinc deficiency in animals and human beings are well known, the metabolic and biochemical defects responsible for them are little understood. The best defined biochemical changes in zinc deficiency are the changes in the activities of certain zinc metalloenzymes (Kirchgessner et al., 1976a). Alkaline phosphatase is a widely distributed zinc metalloenzyme in the body; it participates in many biological processes and responds to zinc deficiency with a reduced activity in certain tissues of various animal species and also of human beings (Kirchgessner et al., 1976a). The relationship of this enzyme to zinc may be important in explaining some of the severe defects in zinc deficiency and in understanding the biological functions of zinc. Thus there was reduced activity of the alkaline phosphatase in the duodenum (Williams, 1972) and of the carboxypeptidases A and B in the pancreas (Roth and Kirchgessner, 1974c) before growth rate and food intake were reduced. The level of food intake did not influence the activities of these enzymes. It may be assumed that the decrease in their activities was not caused by lack of sufficient zinc to maintain these activities; rather, the content of the enzymes themselves was reduced because of diminished enzyme synthesis or increased breakdown. Accordingly, no activating effect was observed by adding zinc to preparations of alkaline phosphatase from zinc-deficient versus control animals (Davies and Motzok, 1971; Williams, 1972; Roth and Kirchgessner, 1979c). The relevance of the effects of the reduced activities of alkaline phosphatase and the carboxypeptidases on the functions of the gastrointestinal tract of zinc-deficient rats is, however, uncertain. The activities of these enzymes were already markedly reduced shortly after the start of the zinc-deficient nutrition (Williams, 1972; Roth and Kirchgessner, 1974c), that is, during the early stage of zinc deficiency, when food intake was still high and growth rate normal, and also digestion and absorption were still normal or nearly normal.

After 5 days of zinc-deficient nutrition, the activities of the carboxypeptidases and the intestinal alkaline phosphatase had fallen to half their original values, followed by a direct decrease in food consumption (Mills et al., 1967; Williams, 1972). This suggests that an actual defect had arisen in the absorption or the metabolism of the food. During later stages of zinc deficiency the average daily food intake was adequate for maintaining the energy balance. It seems, therefore, that at the peak of the cyclic feeding pattern typical of zinc deficiency, when the food intake exceeds the energy requirements, digestion and absorption are unaffected, but storage of the excessive energy is reduced because of the growth failure due to zinc deficiency (Williams, 1972). The reduced activities of thymidine kinase and the polymerases during zinc deficiency impede RNA and DNA synthesis and hence cell division and protein synthesis, the prerequisite for normal growth. Consequently, additional food intake is inhibited. The suppression of food intake that occurs soon after the transfer of rats to a zinc-deficient diet would therefore more likely be the consequence of their failure to grow, rather than the converse. Accordingly, Weigand and Kirchgessner (1977) observed that the growth rate of rats submitted to a zinc-deficient diet was depressed before the voluntary food intake was affected. The altered food conversion ratio in zinc deficiency was attributed not so much to utilization of the dietary energy as to the lower protein efficiency ratio, since dietary protein evidently cannot be used in metabolism for body gain.

To better understand the biochemical and physiological apsect of zinc, additional research is needed to elucidate the fundamental mechanisms ultimately responsible for the clinical symptoms of zinc deficiency in human beings and animals.

REFERENCES

Abbasi, A. A., Prasad, A. S., and Ortega, J. (1976). "Gonadal Function Abnormalities in Sickle Cell Anemia: Studies in Adult Male Patients," *Ann. Intern. Med.,* **85**, 601-605.

Apgar, J. (1970). "Effect of Zinc Deficiency on Maintenance of Pregnancy in the Rat," *J. Nutr.,* **100**, 470-476.

Auld, D. S., Kawaguchi, H., Livingston, D. M., and Vallee, B. L. (1974a), "Reverse Transcriptase from Avian Myeloblastosis Virus: A Zinc Metalloenzyme," *Biochem. Biophys. Res. Commun.,* **57**, 967-972.

Auld, D. S., Kawaguchi, H., Livingston, D. M., and Vallee, B. L. (1974b). "RNA Dependent DNA Polymerase (Reverse Transcriptase) from Avian Myeloblastosis Virus: A Zinc Metalloenzyme," *Proc. Natl. Acad. Sci.,* **71**, 2091-2095.

Bischoff, F. (1936). "Histone Combinations of the Protein Hormones," *Am. J. Physiol.,* **117**, 182-187.

Bischoff, F. (1938). "Factors Influencing the Augmentation Effects of Zinc or Copper When Mixed with Gonadotrophic Extract," *Am. J. Physiol.,* **121**, 765-770.

Boquist, L. (1967). "Some Aspects of the Blood Glucose Regulation and the Gluthatione Content of the Non-diabetic Adult Chinese Hamster *Cricetulus griseus,*" *Acta Soc. Med. Upsal.,* **72**, 358-375.

Boquist, L. (1968). "Alloxan Administration in the Chinese Hamster. I: Blood Glucose Variations, Glucose Tolerance, and Light Microscopical Changes in the Pancreatic Islets and Other Tissues," *Virchows Arch.,* **B1**, 157-168.

Boquist, L. and Lernmark, A. (1969). "Effects of the Endocrine Pancreas in Chinese Hamsters Fed Zinc-Deficient Diets," *Acta Pathol. Microbiol. Scand.,* **76**, 215-228.

Bourne, G. H. (1972). In G. H. Bourne, Ed., *The Biochemistry and Physiology of Bone.* Academic Press, New York, p. 177.

Brewer, G. J., Prasad, A. S., Oelshlegel, F. J., Jr., Schoomaker, E. B., Ortega, J. and Oberleas, D. (1976). "Zinc and Sickle Cell Anemia." In A. S. Prasad, Ed., *Trace Elements in Human Health and Disease,* Vol. 1. Academic Press, New York, pp. 283-294.

Briggs, M. H., Briggs, M., and Austin, J. (1971). "Effects of Steroid Pharmaceuticals on Plasma Zinc," *Nature,* **232**, 480-481.

Burch, R. E., Williams, R. V., Hahn, H. K. J., Jetton, M. M., and Sullivan, J. F. (1975). "Serum and Tissue Enzyme Activity and Trace-Element Content in Response to Zinc Deficiency in the Pig," *Clin. Chem.,* **21**, 568-577.

Cash, R. and Berger, C. K. (1969). "Acrodermatitis Enteropathica: Defective Metabolism of Unsaturated Fatty Acids," *J. Pediatr.,* **74**, 717-729.

Catalanotto, F. A. and Nanda, R. (1977). "The Effects of Feeding a Zinc-Deficient Diet on Taste Acuity and Tongue Epithelium in Rats," *J. Oral Pathol.,* **6**, 211-220.

Chesters, J. K. and Will, M. (1978). "Effect of Age, Weight and Adequacy of Zinc Intake on the Balance between Alkaline Ribonuclease and Ribonuclease Inhibitor in Various Tissues of the Rat," *Br. J. Nutr.,* **39**, 375–382.

Coombs, T. L., Grant, P. T., and Frank, B. H. (1971). "Differences in the Binding of Zinc Ions by Insulin and Proinsulin," *Biochem. J.,* **125**, 62P.

Cox, R. P. and Ruckenstein, A. (1971). "Studies on the Mechanism of Hormonal Stimulation of Zinc Uptake in Human Cell Cultures: Hormone-Cell Interactions and Characteristics of Zinc Accumulation," *J. Cell. Physiol.,* **77**, 71–81.

Davies, M. I. and Motzok, I. (1971). "Zinc Deficiency in the Chick: Effect on Tissue Alkaline Phosphatases," *Comp. Biochem. Physiol.,* **40B**, 129–137.

Dreosti, I. E. and Hurley, L. S. (1975). "Depressed Thymidine Kinase Activity in Zinc Deficient Rat Embryos," *Proc. Soc. Exp. Biol. Med.,* **150**, 161–165.

Duncan, J. R. and Dreosti, I. E. (1976). "A Proposed Site of Action for Zinc in DNA Synthesis," *J. Comp. Pathol.,* **86**, 81–86.

Ellul-Micallef, R., Galdes, A., and Fenech, F. F. (1976). "Serum Zinc Levels in Corticosteroid-Treated Asthmatic Patients," *Postgrad. Med. J.,* **52**, 148–150.

Engelbart, K. and Kief, H. (1970). "Über das funktionelle Verhalten von Zink und Insulin in den B-Zellen des Rattenpankreas," *Virchows Arch.,* **B4**, 294–302.

Evans, G. W. and Johnson, P. E. (1977). "Defective Prostaglandin Synthesis in Acrodermatitis Enteropathica," *Lancet,* **2**, 52.

Evans, G. W., Grace, C. I., and Votava, H. J. (1975). "A Proposed Mechanism for Zinc Absorption in the Rat," *Am. J. Physiol.,* **228**, 501–505.

Fasel, J., Hadjikhani, M. D. H., and Felber, J. P. (1970). "The Insulin Secretory Effect of the Human Duodenal Mucosa," *Gastroenterology,* **59**, 109–113.

Fell, G. S. and Burns, R. R. (1976). "Zinc." *Proc. R. Soc. Med.,* **69**, 474–476.

Fernandez-Madrid, F., Prasad, A. S., and Oberleas, D. (1973). "Effect of Zinc Deficiency on Nucleic Acids, Collagen, and Noncollagenous Protein of the Connective Tissue," *J. Lab. Clin. Med.,* **82**, 951–961.

Flynn, A., Strain, W. H., and Pories, W. J. (1972a). "Corticotropin Dependency on Zinc Ions," *Biochem. Biophys. Res. Commun.,* **46**, 1113–1119.

Flynn, A., Pories, W. J., Strain, W. H., and Hill, O. A., Jr. (1972b). "Corticotropin, Corticosteroids and Zinc," *Lancet,* **2**, 235.

Flynn, A., Pories, W. J., Strain, W. H., and Weiland, F. L. (1973). "The Manupulation of Blood Zinc by Progesterone," *Naturwissenschaften,* **60**, 162.

Giroux, E. L. and Henkin, R. I. (1972). "Competition for Zinc among Serum Albumin and Amino Acids," *Biochim. Biophys. Acta,* **273**, 64–72.

Gombe, S., Apgar, J., and Hansel, W. (1973). "Effect of Zinc Deficiency and Restricted Food Intake on Plasma and Pituitary LH and Hypothalmic LRF in Female Rats," *Biol. Reprod.,* **9**, 415–419.

Halsted, J. A. and Smith, J. C., Jr. (1970). "Plasma-Zinc in Health and Disease," *Lancet,* **1**, 322–324.

Halsted, J. A., Hackley, B. M., and Smith, J. C. (1968). "Plasma Zinc and Copper in Pregnancy and after Oral Contraceptives," *Lancet*, **2**, 278–279.

Hambidge, K. M., Walravens, P. A., and Neldner, K. H. (1977). "The Role of Zinc in the Pathogenesis and Treatment of Acrodermatitis Enteropathica." In G. J. Brewer and A. S. Prasad, Eds., *Zinc Metabolism: Current Aspects in Health and Disease*. Alan R. Liss, New York, pp. 329–340.

Hartoma, R. (1977). "Serum Testosterone Compared with Serum Zinc in Man," *Acta Physiol. Scand.*, **101**, 336–341.

Hendricks, D. G. and Mahoney, A. W. (1972). "Glucose Tolerance in Zinc-Deficient Rats," *J. Nutr.*, **102**, 1079–1084.

Henkin, R. I. (1974a). "On the Role of Adrenocorticosteroids in the Control of Zinc and Copper Metabolism." In W. G. Hoekstra, J. W. Suttie, H. E. Ganther, and W. Mertz, Eds., *Trace Element Metabolism in Animals–2*. University Park Press, Baltimore, pp. 647–651.

Henkin, R. I. (1974b). "Growth-Hormone-Dependent Changes in Zinc and Copper Metabolism in Man." In W. G. Hoekstra, J. W. Suttie, H. E. Ganther, and W. Mertz, Eds., *Trace Element Metabolism in Animals–2*. University Park Press, Baltimore, pp. 652–655.

Henkin, R. I. (1974c). Metal-Albumin-Amino Acid Interactions: Chemical and Physiological Interrelationships." In M. Friedman, Ed., *Protein-Metal Interactions*. Plenum Press, New York, pp. 299–328.

Henkin, R. I. (1976). "Trace Metals in Endocrinology," *Med. Clin. N. Am.*, **60**, 779–797.

Henkin, R. I., Meret, S., and Jacobs, J. B. (1969). "Steroid-Dependent Changes in the Copper and Zinc Metabolism," *J. Clin. Invest.*, **48**, 38a.

Henkin, R. I., Schecter, P. J., Friedewald, W. T., Demets, D. L., and Raff, M. (1976). "A Double Blind Study of the Effects of Zinc Sulfate on Taste and Smell Dysfunction," *Am. J. Med. Sci.*, **272**, 285–299.

Homan, J. D. H., Overbeck, G. A., Neutelings, J. P. J., Booiy, L. J., and Van Der Vies, J. (1954). "Corticotropin Zinc Phosphate and Hydroxide Long Acting Aqueous Preparations," *Lancet*, **2**, 541.

Hove, E., Elvehjem, C. A., and Hart, E. B. (1937). "The Physiology of Zinc in the Nutrition of the Rat," *Am. J. Physiol.*, **119**, 768–775.

Hsu, J. M. (1976). "Zinc as Related to Cystine Metabolism." In A. S. Prasad, Ed., *Trace Elements in Human Health and Disease*. Academic Press, New York, pp. 295–309.

Hsu, J. M., Anilane, J. K., and Scanlan, D. E. (1966). "Pancreatic Carboxypeptidases: Activities in Zinc-Deficient Rats," *Science*, **153**, 882–883.

Huber, A. M. and Gershoff, S. N. (1973). "Effect of Zinc Deficiency in Rats on Insulin Release from the Pancreas," *J. Nutr.*, **103**, 1739–1744.

Iqbal, M. (1971). "Activity of Alkaline Phosphatase and Carbonic Anhydrase in Male and Female Zinc-Deficient Rats," *Enzyme*, **12**, 33–40.

Keilin, D. and Mann, T. (1940a). "Carbonic Anhydrase: Purification and Nature of the Enzyme," *Biochem. J.,* **34,** 1163–1176.

Keilin, D. and Mann, T. (1940b). "Zinc in Carbonic Anhydrase," *Nature,* **145,** 304.

Kirchgessner, M. and Roth, H. –P. (1975). "Beziehungen zwischen klinischen Mangelsymptomen und Enzymaktivitäten bei Zinkmangel," *Zbl. Vet. Med.,* **A22,** 14–26.

Kirchgessner, M. and Schneider, U. A. (1978). "Zum Trächtigkeitsanabolismus von Zink," *Arch. Tierernähr.,* **28,** 211–220.

Kirchgessner, M., Schwarz, W. A., and Roth, H. –P. (1975a). "Zur Aktivität der alkalischen Phosphatase in Serum und Knochen von zinkdepletierten und -repletierten Kühen," *Z. Tierphysiol., Tierernähr. Futtermittelkd.,* **35,** 191–200.

Kirchgessner, M., Stadler, A. E., and Roth, H. –P. (1975b). "Carbonic Anhydrase Activity and Erythrocyte Count in the Blood of Zinc-Deficient Rats," *Bioinorg. Chem.,* **5,** 33–38.

Kirchgessner, M., Roth, H. –P., and Weigand, E. (1976a). "Biochemical Changes in Zinc Deficiency." In A. S. Prasad, Ed., *Trace Elements in Human Health and Disease,* Vol. 1. Academic Press, New York, pp. 189–225.

Kirchgessner, M., Schams, D., and Roth, H. –P. (1976b). "Zum Einfluss mangelnder Zinkversorgung von Milchkühen auf FSH- und LH-Gehalte im Serum," *Z. Tierphysiol., Tierernähr. Futtermittelkd.,* **37,** 151–156.

Kirchgessner, M., Roth, H. –P., and Schwarz, W. A. (1976c). "Zur Wirkung von Zinkmangel auf den Serum-Insulinspiegel bei Milchkühen," *Z. Tierphysiol., Tierernähr. Futtermittelkde.,* **36,** 175–179.

Kirchgessner, M., Schwarz, F. J., Roth, H. –P., and Schwarz, W. A. (1978). "Wechselwirkungen zwischen den Spurenelementen Zink, Kupfer und Eisen nach Zinkdepletion und -repletion von Milchkühen," *Arch. Tierernähr.,* **28,** 723–733.

Kroneman, J., Mey, G. J. W., and Helder, A. (1975). "Hereditary Zinc Deficiency in Dutch Friesian Cattle," *Zbl. Vet. Med.,* **A22,** 201–208.

Ku, P. K. (1971). "Nucleic Acid and Protein Metabolism in the Zinc-Deficient Pig," *Diss. Abstr. Int.,* **B32,** 6717.

Lazaris, Y. A., Lopin, V. I., and Palmina, T. V. (1971). "Zinc Content in Insular Tissue of Pancreas in Experimental Diabetes," *Vop. Med. Khim.,* **17,** 621.

Lei, K. Y., Abbasi, A., and Prasad, A. S. (1976). "Function of Pituitary-Gonadal Axis in Zinc-Deficient rats," *Am. J. Physiol.,* **230,** 1730–1732.

Lieberman, I. and Ove, P. (1962). "Deoxyribonucleic Acid Synthesis and Its Inhibition in Mammalian Cells Cultured from the Animal," *J. Biol. Chem.,* **237,** 1634–1642.

Lieberman, I., Abrams, R., Hunt, N., and Ove, P. (1963). "Levels of Enzyme Activity and Deoxyribonucleic Acid Synthesis in Mammalian Cells Cultured from the Animal," *J. Biol. Chem.,* **238,** 3955–3962.

Lifschitz, M. D. and Henkin, R. I. (1971). "Circadian Variations in Copper and Zinc in Man," *J. Appl. Physiol.*, **31**, 88-92.

Lin, C. W. (1970). "Effects of Dietary Zinc on Serum and Tissue Alkaline Phosphatase of Swine and Rats," *Diss. Abstr. Int. Biochem.*, **32**, 515B.

Lombeck, J., Schnippering, H. G., Kasparek, K., Ritzl, F., Kästner, H., Feinendegen, L. E., and Bremer, H. J. (1975). "Akrodermatitis enteropathica—eine Zinkstoffwechselstörung mit Zinkmalabsorption," *Z. Kinderheilkd.*, **120**, 181-189.

Lum, C. K. and Henkin, R. I. (1976). "Characterization of Fractions from Taste Bud- and Non-Taste-Bud-Enriched Filtrates from and around Bovine Circumvallate Papillae," *Biochim. Biophys. Acta*, **421**, 362-379.

Macapinlac, M. P., Pearson, W. N., and Darby, W. J. (1966). "Some Characteristics of Zinc Deficiency in the Albino Rat." In A. S. Prasad, Ed., *Zinc Metabolism*. Charles C Thomas, Springfield, Ill., pp. 142-166.

Maske, H. (1960). "Role of Zinc in Insulin Secretion." In R. H. Williams, Ed., *Diabetes*. Hoeber, New York, p. 46.

Maxwell, L. C. (1934). "The Quantitative and Qualitative Ovarion Response to Distributed Dosage with Gonadotrophic Extracts," *Am. J. Physiol.*, **110**, 458-463.

McBean, L. D., Smith, J. C., Jr., and Halsted, J. A. (1971). "Effect of Oral Contraceptive Hormones on Zinc Metabolism in the Rat," *Proc. Soc. Exp. Biol. Med.*, **137**, 543.

McIntyre, N., Holdsworth, C. D., and Turner, D. S. (1965). "Intestinal Factors in the Control of Insulin Secretion," *J. Clin. Endocrinol. Metab.*, **25**, 1317-1324.

Millar, M. J., Elcoate, P. V., Fischer, M. I., and Mawson, C. A. (1960). "Effect of Testosterone and Gonadotrophin Injections on the Sex Organ Development of Zinc-Deficient Male Rats," *Can. J. Biochem. Physiol.*, **38**, 1457-1466.

Mills, C. F., Quarterman, J., Williams, R. B., Dalgarno, A. S., and Panic, B. (1967). "The Effects of Zinc Deficiency on Pancreatic Carboxypeptidase Activity and Protein Digestion and Absorption in the Rat," *Biochem. J.*, **102**, 712-718.

Mills, C. F., Quarterman, J., Chesters, J. K., Williams, R. B., and Dalgarno, A. C. (1969). "Metabolic Role of Zinc," *Am. J. Clin. Nutr.*, **22**, 1240-1249.

Moynahan, E. J. (1974). "Acrodermatitis Enteropathica: A Lethal Inherited Human Zinc-Deficiency Disorder," *Lancet*, **1**, 399-400.

Neldner, K. H. and Hambidge, K. M. (1975). "Zinc Therapy of Acrodermatitis Enteropathica," *New Engl. J. Med.*, **292**, 879-882.

Neldner, K. H., Hagler, L., Wise, W. R., Stifel, F. B., Lufkin, E. G., and Herman, R. H. (1974). "Acrodermatitis Enteropathica: A Clinical and Biochemical Survey," *Arch. Dermatol.*, **110**, 711-721.

Oberleas, D. and Prasad, A. S. (1974). "Effect of Zinc on Thymidine Kinase

Activity and DNA Metabolism." In W. G. Hoekstra, J. W. Suttie, H. E. Ganther, and W. Mertz, Eds., *Trace Element Metabolism in Animals*—2. University Park Press, Baltimore, pp. 730-732.

O'Dell, B. L., Reynolds, G., and Reeves, P. G. (1977). "Analogous Effects of Zinc Deficiency and Aspirin Toxicity in the Pregnant Rat," *J. Nutr.*, **107**, 1222-1228.

Pallauf, J. and Kirchgessner, M. (1976). "Einfluss mangelnder Zinkversorgung auf Verdaulichkeit und Verwertung von Nährstoffen, *Arch. Tierernähr.*, **26**, 457-473.

Plocke, D. J., Levinthal, C., and Vallee, B. L. (1962). "Alkaline Phosphatase of *Escherichia coli:* A Zinc Metalloenzyme," *Biochemistry*, **1**, 373-378.

Pories, W. J., Mansour, E. G., Plecha, F. R., Flynn, A., and Strain, W. H. (1976). "Metabolic Factors Affecting Zinc Metabolism in the Surgical Patient." In A. S. Prasad, Ed., *Trace Elements in Human Health and Disease*. Academic Press, New York, pp. 115-141.

Prasad, A. S. (1976). "Deficiency of Zinc in Man and Its Toxicity." In A. S. Prasad, Ed., *Trace Elements in Human Health and Disease*, Vol. 1. Academic Press, New York, pp. 1-20.

Prasad, A. S. and Oberleas, D. (1970). "Zinc: Human Nutrition and Metabolic Effects," *Ann. Intern. Med.*, **73**, 631-636.

Prasad, A. S. and Oberleas, D. (1971). Changes in Activities of Zinc-Dependent Enzymes in Zinc-Deficient Tissues of Rats," *J. Appl. Physiol.*, **31**, 842-846.

Prasad, A. S. and Oberleas, D. (1973). "Ribonuclease and Deoxyribonuclease Activities in Zinc-Deficient Tissues," *J. Lab. Clin. Med.*, **82**, 461-466.

Prasad, A. S. and Oberleas, D. (1974). "Thymidine Kinase Activity and Incorporation of Thymidine into DNA in Zinc-Deficient Tissue," *J. Lab. Clin. Med.*, **83**, 634-639.

Prasad, A. S., Halsted, J. A., and Nadimi, M. (1961). "Syndrome of Iron Deficiency Anemia, Hepatosplenomegaly, Dwarfism, Hypogonadism, and Geophagia," *Am. J. Med.*, **31**, 532-546.

Prasad, A. S., Miale, A., Farid, Z., Schulert, A., and Sandstead, H. H. (1963). "Zinc Metabolism in Normals and Patients with Syndrome of Iron Deficiency Anemia, Hypogonadism and Dwarfism," *J. Lab. Clin. Med.*, **61**, 537-549.

Prasad, A. S., Oberleas, D., Wolf. P., Horwitz, J. P., Miller, E. R., and Luecke, R. W. (1969a). "Changes in Trace Elements and Enzyme Activities in Tissues of Zinc-Deficient Pigs," *Am. J. Clin. Nutr.*, **22**, 628-637.

Prasad, A. S., Oberleas, D., Wolf, P., and Horwitz, J. P. (1969b). "Effect of Growth Hormone on Non-hypophysectomized Zinc-Deficient Rats and Zinc on Hypophysectomized Rats," *J. Lab. Clin. Med.*, **73**, 486-494.

Prasad, A. S., Oberleas, D., Miller, E. R., and Luecke, R. W. (1971). "Biochemical Effects of Zinc Deficiency: Changes in Activities of Zinc Dependent

Enzymes and Ribonucleic Acid and Deoxyribonucleic Acid Content of Tissues," *J. Lab. Clin. Med.*, **77**, 144–152.

Prasad, A. S., Schoomaker, E. B., Ortega, J., Brewer, G. J., Oberleas, D., and Oelshlegel, F. J., Jr. (1974). "Deficiency of Zinc in Sickle Cell Disease Patients." In *Abstract of First National Symposium on Sickle Cell Disease,* pp. 33–34.

Prasad, A. S., Schoomaker, E. B., Ortega, J., Brewer, G. J., Oberleas, D., and Oelshlegel, F. J., Jr. (1975a). "Zinc Deficiency in Sickle Cell Disease," *Clin. Chem.*, **21**, 582–587.

Prasad, A. S., Oberleas, D., Lei, K. Y., Moghissi, K. S., and Stryker, J. C. (1975b). "Effect of Oral Contraceptives Agent on Nutrients," *Am. J. Clin. Nutr.*, **28**, 377–384.

Prasad, A. S., Abbasi, A., and Ortega, J. (1977). "Zinc Deficiency in Man: Studies in Sickle Cell Disease." In G. J. Brewer and A. S. Prasad, Eds., *Zinc Metabolism: Current Aspects in Health and Disease.* Alan R. Liss, New York, pp. 211–236.

Prasad, A. S., Rabbani, P., Abbasi, A., Bowersox, E., and Fox, M. R. S. (1978). "Experimental Production of Zinc Deficiency in Men." In M. Kirchgessner, Ed., *Trace Element Metabolism in Man and Animal–3.* ATW, Freising-Weihenstephan, pp. 280–285.

Quarterman, J. and Florence, E. (1972). "Observations on Glucose Tolerance and Plasma Levels of Free Fatty Acids and Insulin in the Zinc-Deficient Rat," *Br. J. Nutr.*, **28**, 75–79.

Quarterman, J., Mills, C. F., and Humphries, W. R. (1966). "The Reduced Secretion of and Sensitivity to Insulin in Zinc-Deficient Rats," *Biochem. Biophys. Res. Commun.*, **25**, 354–358.

Reeves, P. G., Frissell, S. G., and O'Dell, B. L. (1977). "Response of Serum Corticosterone to ACTH and Stress in the Zinc-Deficient Rat," *Proc. Soc. Exp. Biol. Med.*, **156**, 500–504.

Reinhold, J. G. and Kfoury, G. A. (1969). "Zinc-Dependent Enzymes in Zinc-Depleted Rats: Intestinal Alkaline Phosphatase," *Am. J. Clin. Nutr.*, **22**, 1250–1263.

Ribas, B., Lopez-Calderon, A., Culebras, J. M., and Dean, M. (1978). "Manganese and Zinc in Experimental Diabetes of the Rat." In M. Kirchgessner, Ed., *Trace Element Metabolism in Man and Animal–3.* ATW. Freising-Weihenstephan, pp. 378–381.

Riordan, J. F. (1976). "Biochemistry of Zinc," *Med. Clin. N. Am.*, **60**, 661–674.

Roth, H. –P. and Kirchgessner, M. (1974a). "Aktivitätsveränderungen verschiedener Dehydrogenasen und der alkalischen Phosphatase im Serum bei Zink-Depletion und -Repletion," *Z. Tierphysiol., Tierernähr, Futtermittelkd.*, **32**, 289–296.

Roth, H. –P. and Kirchgessner, (1974b). "Zur Aktivität der Blut-Carboanhydrase bei Zinkmangel wachsender Ratten," *Z. Tierphysiol., Tierernähr. Futtermittelkd.*, **32**, 296–300.

Roth H. -P. and Kirchgessner, M. (1974c). "Zur Aktivität der Pankreas-Carboxy-peptidase A und B bei Zink-Depletion und -Repletion," *Z. Tierphysiol., Tierernähr. Futtermittelkd.*, **33**, 62–67.

Roth, H. -P. and Kirchgessner, M. (1974d). "Zum Einfluss unterschiedlicher Diätzinkgehalte auf die Aktivität der alkalischen Phosphatase im Knochen," *Z. Tierphysiol., Tierernähr. Futtermittelkd.*, **33**, 57–61.

Roth, H. -P. and Kirchgessner, M. (1975). "Insulingehalte im Serum bzw. Plasma von Zinkmangelratten vor und nach Glucosestimulierung," *Int. Z. Vit. Ern. Forsch.* **45**, 201–208.

Roth, H. -P., Schneider, U., and Kirchgessner, M. (1975). "Zur Wirkung von Zinkmangel auf die Glucosetoleranz," *Arch. Tierernähr.*, **25**, 545–549.

Roth, H. -P. and Kirchgessner, M. (1976). "Zur Wirkung von alkalischen Phosphatase-Injektionen bei Zinkmangel," *Zbl. Vet. Med.*, **A23**, 578–587.

Roth, H. -P. and Kirchgessner, M. (1977). "Zum Gehalt von Zink, Kupfer, Eisen, Mangan und Calcium in Knochen und Lebern von an Zink depletierten und repletierten Ratten," *Zbl. Vet. Med.*, **A24**, 177–188.

Roth, H. -P. and Kirchgessner, M. (1978a). "Zn Metalloenzyme Activities—Changes and Biochemical Aspects in Zinc Deficiency." In G. H. Bourne, Ed., *World Review of Nutrition and Dietetics*, S. Karger AG, Basel (in press).

Roth, H. -P. and Kirchgessner, M. (1978b). "Experimentelle Untersuchungen zur Diagnose von marginalem Zinkmangel," *Res. Exp. Med.* (in press).

Roth, H. -P. and Kirchgessner, M. (1979a). "Zink- und Chromgehalte in Serum, Pankreas und Leber von Zn-Mangelratten nach Glucosestimulierung," *Z. Tierphysiol., Tierernähr. Futtermittelkd.*, **43**, in press

Roth, H. -P. and Kirchgessner, M. (1979b). "Zinkmangel und Insulinstoffwech-sel," *Z. Tierphysiol., Tierernähr. Futtermittelkd.*, **43**, in press.

Roth, H. -P. and Kirchgessner, M. (1979c). "In vitro Aktivierung der alkalischen Phosphatase im Serum von Zn-Mangel-und Kontrolltieren," *Zbl. Vet. Med.*, A in press.

Sandstead, H. H., Prasad, A. S., Schulert, A. R., Farid, Z., Miale, A., Jr., Bassilly, S., and Darby, W. J. (1967). "Human Zinc Deficiency, Endocrine Manifesta-tions and Response to Treatment," *Am. J. Clin. Nutr.*, **20**, 422–442.

Sandstead, H. H., Gillespie, D. D., and Brady, R. N. (1972). "Zinc Deficiency: Effect on Brain of the Suckling Rat," *Pediatr. Res.*, **6**, 119–125.

Sato, N. and Henkin, R. I. (1973). "Pituitary-Gonadal Regulation of Copper and Zinc Metabolism in the Female Rat," *Am. J. Physiol.*, **225**, 508–512.

Schenker, J. G., Palishuk, W. Z., and Jungreis, E. (1971) "Serum Copper and Zinc Levels in Patients Taking Oral Contraceptives," *Fertil. Steril.*, **22**, 229–234.

Schneider, U. A. and Kirchgessner, M. (1978). "Veränderungen der Retention von Zink im Organismus während der Gravidität," *Nutr. Metab.* (in press).

Schwarz, W. A. and Kirchgessner, M. (1975). "Experimenteller Zinkmangel bei laktierenden Milchkühen," *Vet. -Med. Nachr.*, **1/2**, 19–40.

Scott, D. A. (1934). "Crystalline Insulin," *Biochem. J.*, **28**, 1592–1602.

Scrutton, M. C., Wu, C. W., and Goldthwait, D. A. (1971). "The Presence and Possible Role of Zinc in RNA Polymerase from *Escherichia coli*," *Proc. Natl. Acad. Sci.*, **68**, 2497–2501.

Simkin, P. A. (1976). "Oral Zinc Sulfate in Rheumatoid Arthritis," *Lancet*, **2**, 539–542.

Simpson, R. T. and Vallee, B. L. (1968). "Two Differentiable Classes of Metal Atoms in Alkaline Phosphatase of *Escherichia coli*," *Biochemistry*, **7**, 4343–4350.

Slater, J. P., Mildvan, A. S., and Loeb, L. A. (1971). "Zinc in DNA Polymerases," *Biochem. Biophys. Res. Commun.*, **44**, 37–43.

Somers, M. and Underwood, E. J. (1969). "Ribonuclease Activity and Nucleic Acid and Protein Metabolism in the Testes of Zinc-Deficient Rats," *Austr. J. Biol. Sci.*, **22**, 1277–1282.

Song, M. K. and Adham, N. F. (1976). "A Possible Role for a Prostaglandin-like Substance in Zinc Absorption," *Fed. Proc.*, **35**, 1667.

Song, M. K. and Adham, N. F. (1977). "Prostaglandin Regulation of Zinc Absorption in Rats," *Fed. Proc.*, **36**, 4583.

Song, M. K. and Adham, N. F. (1978). "Role of Prostaglandin E_2 in Zinc Absorption in the Rat," *Am. J. Physiol.*, **234**, E99–105.

Springgate, C. F., Mildvan, A. S., and Loeb, L. A. (1973a). "Studies on the Role of Zinc in DNA Polymerase," *Fed. Proc. Fed. Am. Soc. Exp. Biol.*, **32**, 451.

Springgate, C. F., Mildvan, A. S., and Abramson, R. (1973b). "*Escherichia coli* Deoxyribonucleic Acid Polymerase. I: A Zinc Metalloenzyme," *J. Biol. Chem.*, **248**, 5987–5993.

Stöber, M. (1971). "Parakeratose beim schwarzbunten Niederungskalb. 1: Klinisches Bild und Ätiologie," *Dtsch. Tierärztl. Wochenschr.*, **78**, 257–284.

Swenerton, H. R. (1971). "The Role of Zinc in Mammalian Development," *Diss. Abstr. Int. Biochem.*, **31**, 5443.

Swenerton, H. R. and Hurley, L. S. (1968). "Severe Zinc Deficiency in Male and Female Rats." *J. Nutr.*, **95**, 8–18.

Swenerton, H. R., Shrader, R., and Hurley, L. S. (1972). "Lactic and Malic Dehydrogenases in Testes of Zinc-Deficient Rats," *Proc. Soc. Exp. Med.*, **141**, 283–286.

Terhune, M. W. and Sandstead, H. H. (1972). "Decreased RNA Polymerase Activity in Mammalian Zinc Deficiency," *Science*, **177**, 68–69.

Thawley, D. G. and Willoughby, R. A. (1977). "Electrophoretic Studies on Alkaline Phosphatases in Normal and Zinc Intoxicated Rats," *Can. J. Comp. Med.*, **41**, 84–88.

Trefz, B. (1972). "Histochemical Investigation of the Modal Specifity of Taste," *J. Dent. Res.*, **51**, Suppl. to No. 5, 1203–1211.

Underwood, E. J. (1977). In: E. J. Underwood (Ed.), *Trace Elements in Human and Animal Nutrition*, 4th ed. Academic Press, New York, pp. 196–242.

Vallee, B. L. (1976). "Zinc Biochemistry in the Normal and Neoplastic Growth Processes," *Cancer Enzymol.* **12**, 159–199. (Miami Winter Symposia).

Van Vloten, W. A. and Bos. L. P. (1978). "Skin Lesions in Aquired Zinc Deficiency Due to Parenteral Nutrition," *Dermatologica*, **156**, 175–183.

Wacker, W. E. C. (1962). "Changes in Nucleic Acid and Metal Content as a Consequence of Zinc Deficiency in *Euglena gracilis*," *Biochemistry*, **1**, 859–865.

Wacker, W. E. C. and Vallee, B. L. (1959). "Nucleic Acids and Metals. I: Chronium, Manganese, Nickel, Iron, and Other Metals in Ribonucleic Acid from Diverse Biological Sources," *J. Biol. Chem.*, **234**, 3257–3262.

Weigand, E. and Kirchgessner, M. (1977). "Dietary Zinc Supply and Efficiency of Food Utilization for Growth," *Z. Tierphysiol., Tierernähr. Futtermittelkd.*, **39**, 16–26.

Weismann, K. and Flagstad, T. (1976). "Hereditary Zinc Deficiency (Adema Disease) in Cattle, an Animal Parallel to Acrodermatitis Enteropathica," *Acta Derm. -Venerol.*, **56**, 151–154.

Westmoreland, N. P. (1969). Studies on the Location and Role of Zinc in Animal Tissues," *Diss. Abstr. Int. Biochem.*, **30**, 63B.

Westmoreland, N. P. (1971). "Connective Tissue Alterations in Zinc Deficiency," *Fed. Proc.*, **30**, 1001–1010.

White, H. B. and Montalvo, J. M. (1973). "Serum Fatty Acids before and after Recovery from Acrodermatitis Enteropathica: Comparison of an Infant with her Family," *J. Pediatr.*, **83**, 999–1006.

Williams, R. B. (1972). "Intestinal Alkaline Phosphatase and Inorganic Pyrophosphatase Activities in the Zinc-Deficient Rat," *Br. J. Nutr.*, **27**, 121–130.

5

ZINC METABOLISM IN HUMAN BEINGS

Herta Spencer
Carol Ann Gatza
Lois Kramer
Dace Osis

Metabolic Section, Veterans Administration Hospital, Hines, Illinois

1. Introduction	**105**
2. Materials and Methods	**106**
3. Results	**107**
4. Discussion	**112**
5. Conclusions	**115**
Acknowledgments	**116**
References	**116**

1. INTRODUCTION

In recent years great emphasis has been placed on the importance of zinc in human nutrition (Halsted and Smith, 1970; Prasad and Oberleas, 1970; Reinhold, 1971), and inadequate zinc nutrition has been reported to result in delayed growth and sexual maturation (Prasad and Oberleas, 1970). Also, zinc, has been found to facilitate wound healing (Pories et al., 1967; Husain, 1969; Henzel et al., 1970). Little was known about the zinc content of the human diet up to a decade ago, when reliable methods of atomic absorption spectroscopy for the analysis of zinc in biological samples became available (Willis, 1962; Osis et al., 1969). This fact, in conjunction with the lack of human zinc balance data, delayed accurate assessment of the dietary zinc intake. Before the use of atomic absorption spectrophotometer, several balance studies were performed in which the dietary zinc content was analyzed by other methods. In long-term zinc balance studies carried out on two adult volunteers on an *ad libitum* diet, the dietary zinc intake ranged from 11 to 18 mg/day (Tipton et al., 1969). In a zinc balance study carried out on preadolescent girls, the zinc intake averaged 7.2 mg/day (Price et al., 1970). Analyzed values of the zinc contents of numerous food items, of an institutional diet, and of two hospital diets have been reported (Schroeder et al., 1967). In the past the average zinc content of the human diet has been estimated to range from 10 to 15 mg/day, according to the Food and Nutrition Board of the National Research Council in 1968.

This chapter reports on zinc balance studies in adults, which were carried out under strictly controlled dietary conditions in the Metabolic Research Unit of this Veterans Administration Hospital in the past decade. In these studies the dietary zinc intake was analyzed throughout the relatively long periods of observation, and the pathways of excretion of ingested zinc were delineated during different nutritional states.

2. MATERIALS AND METHODS

Regular hospital diets, constant diets used for metabolic studies, and individual food items were analyzed for zinc. The total daily hospital diet was collected from the general diet kitchen on 9 randomly chosen days, and the constant diet was the diet used in the Metabolic Research Unit.

Zinc balances were determined during different dietary intake levels of zinc, during low and high calcium intakes, during low and high phosphorus intakes, during a low calorie-low protein intake, during total starvation, during the administration of various medications, after alcohol withdrawal, and during zinc supplementation.

All studies were performed under strictly controlled conditions in the Metabolic Research Ward on fully ambulatory male patients judged to be in good physical condition by all clinical and laboratory criteria. The diet was kept constant throughout all studies and was analyzed for zinc by atomic absorption spectroscopy (Willis, 1962; Osis et al., 1969) in each 6-day study period. This diet contained a daily average of 12.5 mg Zn, 286 g carbohydrate, 75 g protein, 87 g fat, 220 mg Ca, and 800 mg P. The daily fluid intake was kept constant. The different zinc intakes, except for the studies carried out during zinc supplementation, were achieved by increasing or decreasing the protein intake of the diet, as food items that contain a substantial amount of protein are also high in zinc. During zinc supplementation the intake of zinc was increased to approximately 160 mg/day by adding zinc sulfate tablets to the constant diet. To observe the effect of medications on zinc metabolism, patients who received tetracycline or isoniazid for therapeutic purposes were included in the study. In the high-calcium studies, calcium gluconate tablets were given in addition to the constant metabolic diet to raise the calcium intake approximately tenfold, from 200 to 2000 mg/day. The phosphorus intake was increased by the oral intake of sodium glycerophosphate in addition to the constant diet, all other dietary constituents remaining unchanged. In the control study the phosphorus intake was 800 mg/day, which the addition of glycerophosphate increased to 2000 mg/day.

Complete urine and stool collections were obtained from the outset of the study in all investigations. Body weight and 24-hr urine volume were determined daily, as well as the urinary excretion of creatinine to assess the completeness of the daily urine collections.

Metabolic balances of Zn, Ca, P, and N were determined in each 6-day metabolic study period on aliquots of the diet and on aliquots of 6-day collection pools of urine and stool. Zinc and calcium in the diet, plasma, urine, and stool were determined by atomic absorption spectroscopy (Willis, 1962; Osis et al., 1969); phosphorus, by the method of Fiske and SubbaRow (1925); and nitrogen, by the micro-Kjeldahl method. The details of performing metabolic balances of trace elements and of minerals have been described by this group (Spencer et al., 1973).

3. RESULTS

The variability in the zinc contents of the meals that constitute a general hospital diet is illustrated in Table 1. Because of the great differences in the daily composition of these meals, the zinc contents of the meals also differ greatly, by a factor of approximately 4.

Table 2 shows the range and averages of the analyzed values of the zinc contents of the constant metabolic diets used in this Research Unit from 1967 to 1976. A total of 449 diets was analyzed in this 10-year time period. The range in

Table 1. Examples of the Variability of the Zinc Contents of Meals

Meal	Zinc (mg) per Meal	
	Lowest[a]	Highest[a]
Breakfast	1.1	3.6
Lunch	1.9	8.3
Supper	1.7	6.2

[a] As determined from nine samples of hospital meals.

Table 2. Zinc Contents of Metabolic Diets

Year	Number of Diets Analyzed	Zinc (mg) per Day	
		Range	Average
1967	46	10.1–13.5	11.7
1968	29	10.9–13.9	12.7
1969	38	11.4–14.0	12.7
1970	25	10.4–13.6	11.6
1971	53	10.3–14.2	12.4
1972	52	11.0–14.5	12.8
1973	52	11.4–15.3	13.9
1974	53	10.2–15.0	13.0
1975	52	11.0–13.4	12.1
1976	51	11.1–13.0	11.9
Total	449		

dietary zinc content was similar from year to year, with the average value ranging from 11.6 to 13.9 mg/day. The average dietary zinc content of all diets analyzed in this 10-year period was 12.5 mg.

Table 3 shows data on the zinc balances determined in four consecutive 6–day study periods. On an average zinc intake of 14.6 mg/day, the urinary zinc excretion was quite constant from one study period to the next. However, the fecal excretions were more variable and reflected the dietary zinc intake. The zinc balances were slightly positive and ranged from +0.8 to +1.6 mg/day. The average zinc retention in the 24–day study was 1.1 mg/day.

The data listed in Table 4 indicate that the urinary zinc excretions varied considerably from one individual to another; however, the urinary zinc excretion was quite constant from day to day in the same person.

Table 5 shows the zinc balance data determined on four patients during different intake levels of zinc, which were due to the intake of different amounts of protein in the diet. Zinc intakes ranged from 11.9 to 20.3 mg/day. On a normal zinc intake of approximately 12 mg/day, the zinc balances of Patients 1 and 2 were in equilibrium, -0.7 and +0.4 mg/day, respectively. On a higher, but still normal, zinc intake of 14.7 mg/day, the zinc balance was also in equilibrium and was similar, -0.1 mg/day. On a high dietary zinc intake of 20 mg/day, the zinc balance remained negative, -1.0 mg/day. These slightly negative balances in the first three studies resulted primarily from the increased fecal zinc excretions that occurred when the dietary zinc intake was increased from 11.9 to 14.7 mg/day whereas the urinary zinc changed only little. However, during

Table 3. Reproducibility of Zinc Balances in Human Beings

Period	Number of Days	Zinc (mg) per Day			
		Intake	Urine	Stool	Balance
1	6	15.1	0.7	13.6	+0.8
2	6	14.0	0.6	11.8	+1.6
3	6	13.8	0.6	12.6	+0.6
4	6	15.6	0.6	13.6	+1.4
Average		14.6	0.6	12.9	+1.1

Table 4. Urinary Zinc Excretion in Human Beings[a]

	Zinc (mg) per Day		
	Patient 1	Patient 2	Patient 3
	0.38	0.58	1.73
	0.33	0.67	1.52
	0.39	0.57	1.58
	0.40	0.54	1.56
	0.29	0.49	1.66
	0.35	0.59	1.40
Average	0.36	0.57	1.58

[a]Values are averages for each 6-day metabolic period.

Table 5. Zinc Balances During Different Dietary Zinc Intakes

Patient	Number of Days	Zinc (mg) per Day			
		Intake	Urine	Stool	Balance
1	36	11.9	0.4	12.2	−0.7
2	42	12.2	0.8	11.0	+0.4
3	36	14.7	0.9	13.9	−0.1
4	42	20.3	2.3	19.0	−1.0

the highest zinc intake of 20.3 mg/day, the urinary zinc excretion increased distinctly and the fecal zinc again reflected the zinc intake.

Table 6 shows data on zinc balances determined during the addition of calcium and of phosphorus. The effect of calcium on the zinc balance was studied in a 24-

Table 6. Zinc Balances During Different Calcium and Phosphorus Intakes

Patient	Study	Number of Days	Zinc (mg) per Day			
			Intake	Urine	Stool	Balance
1	Control 220 mg Ca[a]	26	15.8	0.5	10.2	+5.1
	High calcium 2400 mg Ca[a]	42	17.7	0.3	12.4	+5.0
2	Control 800 mg P[b]	48	16.0	0.6	12.5	+2.9
	High phosphorus 2000 mg P[b]	24	15.5	0.7	13.5	+1.3

[a] Phosphorus intake, 800 mg/day.
[b] Calcium intake, 220 mg/day.

year-old male. In the control study, during a low calcium intake of 200 mg/day, the zinc intake was relatively high—15.8 mg/day—because of the high protein content of this patient's diet, and it was even greater—17.7 mg/day—during the high calcium intake. The urinary zinc was low during both calcium intakes. The fecal zinc excretion was slightly greater during the high calcium intake; however, the dietary zinc intake was also greater. The zinc balances were highly positive during both the low and the high calcium intakes, but these balances were similar, indicating that the high calcium intake did not affect the zinc balance. The study of the effect of phosphorus on the zinc balance showed that, during the intake of a diet with a relatively low phosphorus content of 800 mg/day, the urinary and fecal zinc excretions were in the expected ranges and the zinc balance was positive—+2.9 mg/day. During the higher phosphorus intake of 2000 mg/day, the urinary zinc excretion did not change, the fecal zinc increased slightly (about 1 mg/day), and the zinc balance was less positive (+1.3 mg/day), reflecting an approximate 50% decrease in zinc retention.

Table 7 shows the zinc balance data of a 55-year-old patient who received an approximately eightfold increase in zinc intake, due to the addition of zinc as zinc sulfate to the constant diet. This zinc supplement increased the total zinc intake from approximately 20 to 160 mg/day. The urinary zinc excretion in-

Table 7. Effect of Zinc Supplements on the Zinc Balance in Human Beings

Patient	Study	Number of Days	Zinc (mg) per Day			
			Intake	Urine	Stool	Balance
1	Control	48	19.9	1.0	15.4	+3.5
	Zinc sulfate	96	161.9	3.7	152.1	+6.1

creased from 1 mg/day in the control study to an average of 3.7 mg/day during the high zinc intake. The fecal zinc increased about tenfold, and the zinc balance rose by about 2.6 mg/day.

Table 8 shows a comparison of the zinc excretions during total starvation and during a low calorie intake of 600 cal/day. During starvation, in the absence of any zinc intake, the loss of zinc was considerable—276 mg during the 60 days. During the low calorie intake, with a dietary zinc intake of 2.1 mg/day, the daily loss of zinc in urine and stool was 6.0 mg, resulting in a negative zinc balance— -3.9 mg/day.The total zinc loss of 250 mg in the 64 days of the low calorie intake is similar to that of 276 mg in 60 days of total starvation.

Table 9 shows the effect of the medications isoniazid (INH) and tetracycline on urinary zinc excretion and zinc balance. A patient who received 250 mg

Table 8. Comparative Loss of Zinc During Starvation and During Low Calorie Intake (all values in milligrams)

Starvation	
Zinc intake	0
Urinary zinc excretion	4.6
Total zinc loss (60 days)	276.0
Low Calorie Diet[a]	
Zinc intake	2.1
Urinary zinc excretion	2.0
Fecal zinc excretion	4.0
Total zinc loss (64 days)	250.0

[a] 600 cal/day.

Table 9. Effect of Medications on the Zinc Balance

Patient	Study	Number of Days	Zinc (mg) per Day			
			Intake	Urine	Stool	Balance
1	INH[a]	96	16.5	1.5	12.2	+2.8
2	INH[a]	24	15.3	1.0	12.6	+1.8
	INH[a]	30	15.0	1.0	11.4	+2.6
3	Tetracycline[b]	48	13.1	1.2	12.9	-1.0
4	Control	66	15.6	0.6	11.3	+3.7
	Tetracycline[b]	24	14.9	1.0	14.6	-0.7

[a] Isoniazid (300 mg) was administered daily.
[b] Tetracycline (250 mg) was administered 4 times daily.

Table 10. Zinc Balances after Alcohol Withdrawal

Patient	Number of Days	Zinc (mg) per Day			
		Intake	Urine	Stool	Balance
1	54	13.1	0.6	7.6	+4.9
2	66	14.1	0.6	8.6	+4.9

tetracycline four times daily for a prolonged period was observed to have a relatively high urinary zinc excretion of 1.2 mg/day (Patient 3), the normal zinc excretion being approximately 0.5 to 0.7 mg/day. In another study, where the zinc balance was determined in a control study and in the initial phase of tetra- cycline administration, an increase in the urinary zinc excretion was also ob- served during tetracycline administration (Patient 4). The fecal zinc excretion of this patient increased also and the zinc balance changed from a positive to a slightly negative value. Two of the patients who received INH, an antituber- culosis drug, also showed an elevated urinary zinc excretion (Patients 1 and 2), while the fecal zinc excretion appears to be in the expected range. The zinc balances of these patients were positive.

Table 10 shows studies of zinc balances of patients with chronic alcoholism after the withdrawal of alcohol. The urinary zinc was in the normal range, whereas the fecal zinc excretion was very low in relation to the dietary zinc intake and the retention of zinc was very high, (4.9 mg/day) for each of these patients.

4. DISCUSSION

Previous studies carried out in this Metabolic Research Unit have shown a cor- relation between dietary zinc intake and amount of protein in the diet (Osis et al., 1972). Great variability in the dietary zinc content of the general hos- pital diet was observed from day to day, although the diet was adequate in all nutrients and in calories. Even the constant metabolic diet differed somewhat in zinc content because different types of meat were used. For instance, if chicken meat, which has a lower zinc content than beef, is consumed, the differ- ence in zinc intake may be as great as 3 mg/day. This difference constitutes a considerable amount of zinc in view of the fact that the normal total dietary zinc intake is 15 mg/day. Studies carried out in this research unit have shown that certain dietary conditions which lead to weight loss are associated with a distinct loss of zinc. The low-calorie diets compatible with the trend toward weight reduction and diet consciousness would therefore not provide the rec- ommended dietary allowance of 15 mgZn/day (Harper, 1973).

The zinc deficiency associated with endocrine changes in human beings (Prasad et al., 1963) may be the result of inadequate dietary intake, including the intake of zinc. On the other hand, human zinc deficiency may develop despite an adequate zinc intake if the diet contains substances that complex zinc and make this trace element unavailable for absorption (O'Dell et al, 1958; Oberleas et al., 1966). The beneficial effects of zinc therapy for wound healing in surgical patients (Pories et al., 1967) and in patients with peripheral vascular disease (Henzel et al., 1970) may indicate that these persons were actually zinc deficient. One may assume that they may have had an inadequate dietary zinc intake for a prolonged period of time, as they were able to utilize the supplemental zinc and the main source of zinc intake is the diet. Impaired wound healing has also been demonstrated in zinc deficiency in experimental animals (Sandstead et al., 1970). In diseases such as myocardial infarction, which occur predominantly in the older age group, the plasma and the tissue levels of zinc have been found to be low (Wacker et al., 1956; Volkov, 1963). As peripheral vascular disease also occurs with great frequency in older persons, it is conceivable that changes in zinc metabolism may occur with age. However, this aspect has not been investigated in human beings. The data on Patient 1 (Table 6), a 24-year-old male, show a highly positive zinc balance. It is possible that the retention of zinc is considerably higher in young than in older persons.

In terms of zinc metabolism in human beings, the main pathway of zinc excretion is via the intestine; the urinary zinc excretion is normally low but may vary from patient to patient. However, these excretions are very constant and are reproducible within the same subject. In previous studies, using ^{65}Zn as tracer, it was shown that the urinary excretion of this radioisotope was extremely low after oral or intravenous administration of ^{65}Zn (Spencer et al., 1965a, 1965b, 1966). The reason for the high intestinal excretion of zinc is not clear. The contribution of zinc from the bile to the intestinal excretions is very small (Miller et al., 1964). *In vitro* studies (Sahagian et al., 1967) and animal studies (Stand et al., 1962) have shown that the zinc content of the intestinal epithelium is high, and the "sloughing off" of these cells may contribute to the high fecal zinc content. In a study on human beings, using intravenous ^{65}Zn as tracer, the 30-day fecal ^{65}Zn excretion was 18%, while the cumulative urinary ^{65}Zn was only approximately 1% in the same time period (Spencer et al., 1965a). In a study on rats it was found that ^{65}Zn was secreted into all segments of the small intestine (Methfessel et al., 1973).

The zinc balances determined in the present research were quite constant from one study period to the next, and the zinc balance was in equilibrium in most cases on a zinc intake of about 12.5 mg/day. However, in a large percentage of the subjects studied, the zinc balance was negative on this intake and an intake of 15 mg/day resulted in zinc equilibrium. The data of Patient 4, shown in Table 5, illustrate that some individuals are not in zinc equilibrium on a higher intake of even 20 mg/day. The officially recommended dietary allowance of zinc is 15 mg/day (Harper, 1973). In view of the individual variability in zinc utilization and retention, further investigations are necessary to determine the changes in the zinc balance occurring during different dietary intake levels of zinc in the same individual. This type of study may further our knowledge of the human zinc requirement.

The interaction of zinc and calcium has been investigated extensively in animal studies (Luecke et al., 1957; O'Dell et al., 1958; Forbes, 1960). Animals receiving borderline intakes of zinc and large amounts of calcium have developed symptoms of zinc deficiency (Tucker and Salmon, 1955; Luecke et al., 1957; O'Dell et al., 1958; Forbes, 1960, 1964). The effects of calcium were shown to be due to the presence of phytic acid (O'Dell et al., 1958). However, other studies have shown that in the absence of phytic acid calcium decreased zinc absorption (Heth and Hoekstra, 1965; Heth et al., 1966). In a review on the antagonism between calcium and zinc in animals, it has been demonstrated that calcium interferes with the absorption of zinc from the intestine in rats (Hoekstra, 1964). The lack of change in the zinc balance during a 12-fold increase in calcium intake in human beings presented here (Table 6) is in agreement with previously reported results obtained with ^{65}Zn in persons receiving a high calcium intake (Spencer et al., 1965b). These negative results may be ascribed to the relatively small amount of calcium given to human beings as compared to the considerably larger amounts fed to animals, as well as to the low phytate content of the constant metabolic diet.

With regard to the effect of phosphorus on zinc metabolism, few studies have been carried out on animals (Heth and Hoekstra, 1965; Heth et al., 1966), and to our knowledge only a single study on human beings has been reported (Pecoud et al., 1975). The studies carried out in this research unit on the effect of phosphorus on zinc metabolism showed a slight increase in fecal zinc (Table 6); further studies are necessary to clarify this effect.

Although large amounts of supplemental zinc have been used therapeutically in human beings in the past decade, little information is availabile on the fate of the added zinc in terms of its elimination from the body, on the pathways of excretion of the excess zinc intake, and on its retention (Spencer et al., 1977). Changes in the plasma levels of zinc, however, have been reported during high zinc intakes (Husain, 1969; Serjeant et al., 1970; Czerwinski, 1974). From animal studies the effects of a high zinc intake on the intestinal absorption of ^{65}Zn (Furchner and Richmond, 1962), on the urinary zinc excretion (Drinker et al., 1927), and on the endogenous fecal zinc excretion (Miller et al., 1970) have been reported.

With regard to the influence of medications on zinc metabolism in human beings, a study of the effect of the antibiotic tetracycline on the plasma level of zinc and on urinary zinc excretion, carried out in a patient who had an *Escherichia coli* infection of the urinary tract, has shown interesting results

(Spencer et al., 1974). Shortly after the start of the antibiotic therapy, both the plasma level of zinc and the urinary zinc excretion increased. The mechanism of these changes is not clear; however, they may be a result of abolishing the infectious process. Low serum zinc values during infection have been attributed to a redistribution of zinc in the body (Pekarek et al., 1971). The plasma zinc level, as well as the urinary excretion of zinc, has been reported to be decreased in human beings with acute infections (Pekarek et al., 1970). A high urinary excretion of zinc has been noted in patients receiving the antituberculosis drug INH (McCall, unpublished data); the same observation has been made regarding the patients reported here. Many drugs may affect various aspects of human metabolism, including the metabolism of protein and therefore also of zinc. Studies in this area are much needed in view of the sparsity of information available on this subject.

Patients with chronic alcoholism have been shown to have a high urinary zinc excretion and to be zinc deficient (Sullivan and Heaney, 1965). The data reported here show a very high zinc retention for relatively long periods of time after alcohol withdrawal, primarily because of the low fecal zinc excretion (Spencer et al., 1978). This high zinc retention is probably compensatory for the zinc depletion occurring during alcoholism and expresses a general anabolic response to retain other nutrients, including the essential trace element zinc. Previous studies carried out on human beings have shown that the intestinal absorption of ^{65}Zn is highly variable but that it is very high in subjects in a poor nutritional state (Spencer et al., 1966). Very high zinc retention has also been observed in patients in the repletion state after severe dietary restriction for weight reduction (Spencer et al., 1976). During weight reduction, which is induced either by calorie restriction or by total starvation, excess loss of zinc has been observed in human beings (Spencer et al., 1971, 1972, 1974). Of interest is the observation that this zinc loss, which was reflected mainly in very high urinary zinc excretions, was not associated with low plasma levels of zinc; in fact, these levels were even elevated. The increased plasma level of zinc during weight reduction may be due to dissociation of the zinc-protein complex secondary to tissue breakdown and to circulation of increased amounts of ionic zinc in plasma before its clearance via the kidney. The increase in the plasma zinc level and the high urinary zinc excretion during starvation may also be related to metabolic acidosis, which may contribute to the dissociation of the zinc-protein complex.

5. CONCLUSIONS

The dietary zinc intake in man has been defined, as well as the disposition of the ingested zinc. The interactions of zinc with other minerals in human beings, as well as the changes in zinc metabolism during different nutritional states, have been described. Preliminary results have been reported on the effects of certain medications on the excretion of zinc; further studies are indicated to explore the effects of other drugs on zinc metabolism.

ACKNOWLEDGMENTS

This study was supported by U.S. Department of Energy Contract EY-76-S-02-1231-124.

REFERENCES

Czerwinski, A. W. (1974). "Safety and Efficacy of Zinc Sulphate in Geriatric Patients," *Clin. Pharmacol. Ther.*, **15**, 436–441.

Drinker, K. R., Thompson, P. K., and Marsh, M. (1927). "An Investigation of the Effect of Long-Continued Ingestion of Zinc, in the Form of Zinc Oxide, by Cats and Dogs, together with observations upon the Excretion and Storage of Zinc." *Am. J. Physiol.*, **80**, 31–64.

Fiske, C. H. and SubbaRow, Y. T. (1925). "The Colorimetric Determination of Phosphorus," *J. Biol. Chem.*, **66**, 375–400.

Forbes, R. M. (1960). "Nutritional Interactions in Zinc and Calcium," *Fed. Proc.*, **19**, 643–647.

Forbes, R. M. (1964). "Mineral Utilization in the Rat," *J. Nutr.*, **83**, 225–232.

Furchner, J. E. and Richmond, C. R. (1962). "Effect of Dietary Zinc on Absorption of Orally Administered Zn-65," *Health Phys.*, **8**, 35–40.

Halsted, J. A. and Smith, J. C., Jr. (1970). "Plasma-Zinc in Health and Disease," *Lancet*, **1**, 322–324.

Harper, A. E. (1973). "Recommended Dietary Allowances (Revised–1973)," *Nutr. Rev.*, **31**, 393–396.

Henzel, J. H., DeWeese, M. S., and Lichti, E. L. (1970). "Zinc Concentration within Healing Wounds," *Arch. Surg.*, **100**, 349–357.

Heth, D. A. and Hoekstra, W. G. (1965). "Zinc-65 Absorption and Turnover in Rats. I: A Procedure to Determine Zinc-65 Absorption and the Antagonistic Effect of Calcium in a Practical Diet," *J. Nutr.*, **85**, 367–374.

Heth, D. A., Becker, W. M., and Hoekstra, W. G. (1966). "Effect of Calcium, Phosphorus and Zinc on Zinc-65 Absorption and Turnover in Rats Fed Semipurified Diets," *J. Nutr.*, **88**, 331–337.

Hoekstra, W. G. (1964). "Recent Observations on Mineral Interrelationships," *Fed. Proc.*, **23**, 1068–1076.

Husain, S. L. (1969). "Oral Zinc Sulphate in Leg Ulcers," *Lancet*, **1**, 1069–1071.

Luecke, R. W., Hoefer, J. A., Brammell, W. S., and Schmidt, D. A. (1957). "Calcium and Zinc in Parakeratosis of Swine," *J. Anim. Sci.,* **16**, 3–11.

Methfessel, A. H. and Spencer, H. (1973). "Zinc Metabolism in the Rat. II: Secretion of Zinc into Intestine," *J. Appl. Physiol.,* **34**, 63–67.

Miller, E. B., Sorscher, A., and Spencer, H. (1964). "Intestinal Zn-65 Secretion in Man," *Radiat. Res.,* **22**, 216–217.

Miller, W. J., Blackmon, D. M., Gentry, R. P., and Pate, F. M. (1970). "Effects of High but Nontoxic Levels of Zinc in Practical Diets on [65]Zn and Zinc Metabolism in Holstein Calves," *J. Nutr.,* **100**, 893–902.

Oberleas, D., Muhrer, M. E., and O'Dell, B. L. (1966). "The Availability of Zinc from Foodstuffs." In A. S. Prasad, Ed., *Zinc Metabolism*. Charles C Thomas, Springfield, Ill., pp. 378–394.

O'Dell, B. L., Newberne, P. M., and Savage, J. E. (1958). "Significance of Dietary Zinc for the Growing Chicken," *J. Nutr.,* **65**, 503–518.

Osis, D., Royston, K., Samachson, J., and Spencer, H. (1969). "Atomic-Absorption Spectrophotometry in Mineral and Trace-Element Studies in Man," *Dev. Appl. Spectrosc.* **7A**, 227–235.

Osis, D., Kramer, L., Wiatrowski, E., and Spencer, H. (1972). "Dietary Zinc Intake in Man," Am. J. Clin. Nutr., **25**, 582–588.

Pecoud, A., Donzel, P., and Schelling, J. L. (1975). "Effect of Foodstuffs on the Absorption of Zinc Sulfate," *Clin. Pharmacol. Ther.,* **17**, 469–474.

Pekarek, R. S. and Beisel, W. R. (1971). "Metabolic Losses of Zinc and Other Trace Elements during Acute Infection." In *Book of Abstracts, Western Hemisphere Nutrition Congress* III, p. 43.

Pekarek, R. S., Burghen, G. A., Bartelloni, P. J., Calia, F. M., Bostian, K. A., and Beisel, W. R. (1970). "The Effect of Live Attenuated Venezuelan Equine Encephalomyelitis Virus Vaccine on Serum Iron, Zinc, and Copper Concentrations in Man," *J. Lab. Clin. Med.,* **76**, 293–303.

Pories, W. J., Henzel, J. H., Rob, C. G., and Strain, W. H. (1967). "Acceleration of Wound Healing in Man with Zinc Sulphate Given by Mouth," *Lancet,* **1**, 121–124.

Prasad, A. S. and Oberleas, D. (1970). "Zinc: Human Nutrition and Metabolic Effects," *Ann. Intern. Med.,* **73**, 631–636.

Prasad, A. S., Schulert, A. R., Sandstead, H. H., Miale, A., and Farid, Z. (1963). "Zinc, Iron, and Nitrogen Content of Sweat in Normal and Deficient Subjects," *J. Lab. Clin. Med.,* **62**, 84–89.

Price, N. O., Bruce, G. E., and Engel, R. W. (1970). "Copper, Manganese, and Zinc Balances in Preadolescent Girls," *Am. J. Clin. Nutr.,* **23**, 258–260.

Reinhold, J. G. (1971). "High Phytate Content of Rural Iranian Bread: A Possible Cause of Human Zinc Deficiency," *Am. J. Clin. Nutr.,* **24**, 1204–1206.

Sahagian, B. M., Harding-Barlow, I., and Perry, H. M., Jr. (1967). "Trasmural Movements of Zinc, Manganese, Cadmium, and Mercury by Rat Small Intestine," *J. Nutr.,* **93**, 291–300.

Sandstead, H. H., Lanier, V. C., Jr., Shephard, G. H., and Gillespie, D. D. (1970). "Zinc and Wound Healing: Effects of Zinc Deficiency and Zinc Supplementation," *Am. J. Clin. Nutr.,* **23,** 514–519.

Schroeder, H. A., Nason, A. D., Tipton, I. H., and Balassa, J. J. (1967). "Essential Trace Metals in Man: Zinc. Relation to Environmental Cadmium," *J. Chron. Dis.,* **20,** 179–210.

Serjeant, G. R., Galloway, R. E., and Gueri, M. C. (1970). "Oral Zinc Sulphate in Sickle-Cell Ulcers," *Lancet,* **2,** 891–892.

Spencer, H., Rosoff, B., Feldstein, A., Cohn, S. H., and Gusmano, E. (1965a). "Metabolism of Zinc-65 in Man," *Radiat. Res.,* **24,** 432–445.

Spencer, H., Vankinscott, V., Lewin, I., and Samachson, J. (1965b). "Zinc-65 Metabolism during Low and High Calcium Intake in Man," *J. Nutr.,* **86,** 169–177.

Spencer, H., Rosoff, B., Lewin, I., and Samachson, J. (1966). "Studies of Zinc Metabolism in Man." In A. S. Prasad, Ed., *Zinc Metabolism.* Charles C Thomas, Springfield, Ill., pp. 339–362.

Spencer, H., Osis, D., Wiatrowski, E., and Samachson, J. (1971). "Zinc Metabolism during Starvation in Man," *Fed. Proc.,* **30,** 643.

Spencer, H., Osis, D., Kramer, L., and Norris, C. (1972). "Studies of Zinc Metabolism in Man." In D. D. Hemphill, Ed., *Trace Substances in Environmental Health*–V. University of Missouri, Columbia, pp. 193–204.

Spencer, H., Friedland, J. A., and Ferguson, V. (1973). "Human Balance Studies in Mineral Metabolism." In I. Zipkin, Ed., *Biological Mineralization.* John Wiley, New York, pp. 689–727.

Spencer, H., Osis, D., Kramer, L., and Wiatrowski, E. (1974). "Studies of Zinc Metabolism in Normal Man and in Patients with Neoplasia." In W. J. Pories, W. H. Strain, J. M. Hsu, and R. L. Woosley, Eds., *Clinical Applications of Zinc Metabolism.* Charles C Thomas, Springfield, Ill., pp. 101–112.

Spencer, H., Osis, D., Kramer, L., and Norris, C. (1976). "Intake, Excretion, and Retention of Zinc in Man." In A. S. Prasad, Ed., *Trace Elements in Human Health and Disease,* Vol. I: *Zinc and Copper.* Academic Press, New York, pp. 345–361.

Spencer, H., Osis, D., and Kramer, L. (1977). "Metabolic Effects of Pharmacologic Doses of Zinc in Man," *Am. J. Clin. Nutr.,* **30,** 611.

Spencer, H., Kramer, L., Osis, D., Norris, C., and DeBartolo, M. (1978). "Studies of Zinc Metabolism Following Alcohol Withdrawal in Man," *Fed. Proc.,* **37,** 254.

Stand, F., Rosoff, B., Williams, G. L., and Spencer, H. (1962). "Tissue Distribution Studies of Ionic and Chelated Zn^{65} in Mice," *J. Pharmacol. Exp. Ther.,* **138,** 399–404.

Sullivan, J. F. and Heaney, R. P. (1965). "Zinc-65 Metabolism in Patients with Cirrhosis," *J. Lab. Clin. Med.,* **66,** 1025.

Tipton, I. H., Stewart, P. L., and Dickson, J. (1969). "Patterns of Elemental Excretion in Long Term Balance Studies," *Health Phys.*, **16**, 455–462.

Tucker, H. F. and Salmon, W. D. (1955). "Parakeratosis or Zinc Deficiency Disease in the Pig," *Proc. Soc. Exp. Biol. Med.*, **88**, 613–616.

Volkov, N. F. (1963). "Cobalt, Manganese and Zinc Content in Blood of Atherosclerosis Patients," *Fed. Proc. Trans. Suppl.*, **22**, 897–899.

Wacker, W. E. C., Ulmer, D. D., and Vallee, B. L. (1956). "Metalloenzymes and Myocardial Infarction. II: Malic and Lactic Dehydrogenase Activities and Zinc Concentrations in Serum," *New Engl. J. Med.*, **255**, 449–456.

Willis, J. B. (1962). "Determination of Lead and Other Heavy Metals in Urine by Atomic Absorption Spectroscopy," *Anal. Chem.*, **34**, 614–617.

6

CELLULAR AND MOLECULAR ASPECTS OF MAMMALIAN ZINC METABOLISM AND HOMEOSTASIS

Robert J. Cousins
Mark L. Failla

Department of Nutrition, Rutgers University—The State University of New Jersey, New Brunswick, New Jersey

1. Introduction	121
2. Plasma Zinc	122
3. Intestinal Zinc	124
4. Liver Zinc	126
5. Prostate Zinc	130
Acknowledgments	131
References	131

1. INTRODUCTION

The term "homeostatic control" means that various organs and associated physiological events act together to maintain constant conditions in the internal environment. With regard to zinc, the concept of homeostasis implies that absorption and excretion of the nutrient are regulated by a series of linked metabolic events. Various aspects of the physiology of zinc have been extensively investigated in intact animal models (Underwood, 1977). However, only recently have developments in biochemistry, cell biology, and molecular biology produced techniques that allow us to examine the basic mechanisms responsible for regulating zinc metabolism. Therefore, in view of this newer information, the primary focus of this chapter will be on mammalian zinc metabolism at the cellular and molecular level.

An overwhelming body of evidence has shown that zinc is absorbed nearly exclusively by the small intestine. Once absorbed, zinc is deposited in both soft tissues and bone. Although bone accounts for an appreciable percentage of the total body zinc (Underwood, 1977), this pool of zinc is relatively unavailable, being mobilized only during dietary and/or other conditions that lead to hypocalcemia (Tao and Hurley, 1975). Therefore bone zinc mobilization is dependent upon bone resorption and is thus regulated by 1,25-dihydroxycholecalciferol and parathyroid hormone (De Luca, 1974). Some recent data indicate, however, that a portion of the zinc associated with bone could be metabolically available (Brown et al., 1978).

A considerable portion of the body zinc not associated with bone is found in the liver and intestine. This complement of zinc represents a metabolically active compartment which exhibits significant changes during fluctuations in dietary zinc intake and metabolic perturbations such as infection and trauma. Homeostatic control of zinc metabolism is directly related to events that alter the dynamics of this zinc pool. The need for regulation of certain phases of zinc metabolism and distribution in mammalian systems is clear when one considers the vital functions of zinc. This metal has established roles as a component of DNA and RNA (Eichhorn et al., 1971; Wacker and Vallee, 1959) and zinc metalloenzymes (Failla, 1977; Vallee and Wacker, 1970) and as a stabilizer of ribosomes (Prask and Plocke, 1971; Tal, 1968) and membranes (Chvapil, 1973; Bettger et al., 1978).

On a daily basis it is difficult for animals with a varied diet to maintain a constant zinc supply. To minimize the effects of this variability homeostatic mechanisms evoke changes in absorption, internal redistribution, and excretion which help to ensure that a constant amount of zinc is available for distribution to various tissues. Although in the past zinc metabolism was viewed in terms of changes occurring on a long-term basis, we now know that, at the cellular level, this metal is in a constant state of flux. Consequently, regulatory control that affects the absorption and metabolism of zinc on a short-term basis is required. These control mechanisms also allow the tissue and plasma concentrations of zinc to be modified by internal stimuli. In this way, animals can adapt to environmental and pathological situations that may require the redistribution of body zinc without the need to rely on the dietary supply.

2. PLASMA ZINC

The plasma zinc concentration can be drastically altered by either dietary zinc or physiological status and often provides a reflection of transitions in zinc metabolism. When the zinc content of the diet is relatively constant and roughly satisfies the minimum daily requirement, the plasma zinc content also remains relatively constant, at about 100 to 120 μg/100 ml. The exact value, however, depends on the methods of sample preparation and analysis. When a zinc-deficient diet is fed to rats, the plasma concentration quickly falls to about 50% of the normal range. When zinc is again added to the diet of these animals, the plasma concentration initially increases above normal limits before returning to average values (Wilkins et al., 1972; Richards and Cousins, 1976a, 1976b). Similar but less rapid reductions in dietary zinc have been reflected in decreases in plasma zinc in human experiments (Prasad et al., 1976; Hess et al., 1976). In both animal and human experiments it has been shown that ingestion of readily available zinc produces a transient, dose-dependent elevation in plasma zinc (Pécoud et al., 1975; Methfessel and Spencer, 1973; Oelshlegel and Brewer, 1977; Richards and Cousins, 1976; Giroux and Prakash, 1977; Wilkins et al., 1972). From the body of data available at this time, we can conclude that the plasma zinc concentration responds homeostatically to a zinc load, that is, elevations are transient and return to within normal limits quickly. However, zinc deficiency does not elicit homeostatic mobilization of zinc stores to elevate plasma zinc to within normal limits.

Apart from plasma zinc changes induced by dietary perturbations, the clinical literature is filled with examples of situations wherein plasma zinc is either depressed (e.g., alcoholism, Cushing's syndrome, sickle-cell anemia, liver disease, and myocardial infarction) or elevated (e.g., adrenal cortical insufficiency). These aspects of zinc metabolism have been well reviewed (Underwood, 1977; Prasad and Oberleas, 1976). Recently, examples of atypical zinc metabolism have been described. Acrodermatitis enteropathica, an autosomal recessive condition in human beings, is characterized by decreased zinc absorption and plasma zinc content and severe skin lesions. Both are rapidly restored to normal with supplemental zinc, in amounts sufficient to override the matabolic defect in absorption. (Moynahan, 1974, Neldner and Hambidge, 1975; Lombeck et al., 1975). A similar syndrome (Adema disease) has been described in cattle (Flagstad et al., 1976). In contrast, abnormally high plasma zinc concentrations in several members of one family (familial hyperzincemia) have been reported (Smith et al., 1976). The metabolic basis of this inherited condition is not known.

Chronic disease that affect zinc metabolism, directly or via a secondary route, tend to produce changes in plasma zinc that remain relatively constant and that are observed for the duration of the condition. As mentioned above, zinc metabolism is dynamic and can be altered drastically on a short-term basis. In chronic disease, however, regulatory control processes appear to be altered, since plasma zinc equilibrates atypically. A different situation seems to exist with regard to the depression of plasma zinc observed during the early stages of infection. This change is associated with decreased plasma iron and elevated plasma copper (Beisel et al., 1976; Kampschmidt et al., 1973). The latter is accounted for primarily in terms of the increased hepatic synthesis and secretion of ceruloplasmin, an acute-phase protein. The decrease in plasma zinc has a duration of about 20 hr, whereas the elevation of plasma copper may last until the infectious episode has been controlled. The agent that has been postulated to be responsible for initiating these changes is leukocytic endogenous mediator (LEM). Although the chemical nature of LEM is not clear, this hormone-like substance (or substances) mediates the accumulation of approximately 60% of the plasma zinc by the liver and perhaps other tissues. As will be discussed in section 4, food restriction and physiological stresses also lead to a redistribution of hepatic zinc. However, the plasma zinc content is not altered during such conditions.

3. INTESTINAL ZINC

The plasma zinc concentration is somewhat related to the extent of dietary zinc absorption by the small intestine. For example, if the amount of zinc in the dietary supply is diminished, the plasma zinc level falls and zinc absorption is increased. Conversely, if the dietary zinc level is substantially increased, there is a transient increase in plasma zinc and a reduction in absorption (Richards and Cousins, 1976b). To accomplish homeostatic regulation of absorption, the intestinal cells must be able to monitor zinc status and use the information to coordinate the flux of zinc entering from the lumen with that which enters the cells from the plasma. Although the concept of homeostatic regulation of zinc was advanced some time ago (Cotzias et al., 1962; Cotzias and Papavasilion, 1964), we are only now beginning to understand specific details and regulatory ' aspects of the absorption process at the cellular level. In our initial experiments in this area, we examined the influence of parenterally administered zinc on the absorption of orally administered ^{65}Zn in rats (Richards and Cousins, 1975a). The zinc load decreased ^{65}Zn transfer from the intestine to the carcass; a similar finding was reported by Evans and associates (1973). Similarly, when rats were fed zinc-adequate and zinc-deficient diets on alternating days, the absorption of orally administered ^{65}Zn was decreased and increased, respectively (Richards and Cousins, 1976b). These data strongly support the concept that the absorption of dietary zinc is homeostatically regulated.

Our studies also demonstrated that fluctuations in the zinc content of intestinal cells could be accounted for to a large extent as metallothionein. The characteristics of this protein isolated from rat intestine (Richards and Cousins, 1977a) are similar to those of metallothioneins from rat liver (Bremner and Davies, 1975), human liver (Buhler and Kaegi, 1974), and equine kidney (Kaegi et al., 1974). Rat intestinal metallothionein, as induced by parenteral zinc, has the following characteristics: Zn/protein ratio, 5.0 to 5.6 g-atoms/mol; Zn/SH ratio, approximately 3.0; lack of aromatic acids; abundance of cysteic acid residues (28 to 31%), and low molecular weight (6000 to 7000 daltons). Previous experiments showed that parenteral zinc administration enhanced the incorporation of [^{14}C]-cystine and ^{65}Zn into this protein (Richards and Cousins, 1975b). Administration of actinomycin D blocked the accumulation of parenteral zinc in intestinal cells and the incorporation of [^{14}C]-cystine into metallothionein (Richards and Cousins, 1975b, 1977b). In all these experiments, the binding of zinc (^{65}Zn) to intestinal metallothionein was correlated with a reduction in zinc absorption. A low molecular weight zinc (^{65}Zn) binding ligand was observed in the cytosol fraction of some intestinal preparations (Evans, et al., 1975; Richards and Cousins, 1975b; Song and Adham, 1978); however, it has recently been shown that low molecular weight zinc-binding species from rat intestine can be accounted for on the basis of both protein degradation and high intestinal cell zinc concentrations (Cousins et al., 1978). A relationship between a low molecular weight zinc-binding fraction in human milk (Hurley et al., 1977) and zinc absorption in the neonate has been suggested, but the factor has yet to be characterized and investigated in detail.

The mechanism of the zinc absorption process is very complex and is undoubtedly influenced by many factors. In an effort to study the mechanism in more detail, we have examined zinc absorption and metabolism using an isolated, vascularly perfused rat intestinal system (Smith et al., 1978). Data from these experiments suggest that the homeostatic absorption of zinc is mediated at the intestinal level. For example, intestines from zinc-deficient rats absorb zinc at a greater rate than those from zinc-sufficient rats. Zinc absorption was lowest in intestines from zinc-loaded (parenteral) rats, wherein the intestinal zinc was above normal (Richards and Cousins, 1977b). Furthermore, a reduction in zinc absorption was directly related to the intracellular binding of zinc to metallothionein. Thus it appears that the synthesis of metallothionein provides the intestinal cell with an additional pool of binding sites with which to alter the net flux of zinc ions. These experiments suggest that zinc enters intestinal cells from both the plasma and the lumen and initially interacts with a "zinc pool." Then, depending on the physiological/nutritional status of the animal, zinc may equilibrate with metalloproteins and metallothionein, or it may be transferred to either the plasma (absorption) or the lumen (endogenous secretion/excretion). These concepts are presented in Figure 1. All of these transitions work in concert to ensure that an adequate supply of zinc is available

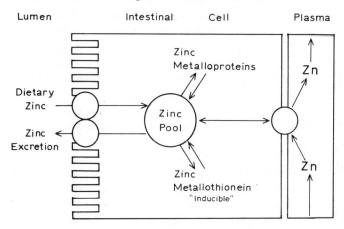

Figure 1. Interaction of dietary and plasma zinc with intracellular zinc in intestinal cells. Regulation of zinc absorption is associated in part with binding of zinc to metalloproteins and the inducible binding component, metallothionein.

to the plasma carrier (viz. albumin) for transport to the tissues. The role of intracellular ligands in zinc absorption has recently been reviewed (Cousins, 1979). Kirchgessner and associates (Schwarz and Kirchgessner, 1978; Weigand and Kirchgessner, 1978) have argued in support of the role that fecal excretion plays in the homeostatic regulatory process. It is likely that the concentrattion of endogenous zinc in the feces, which is a reflection of metabolic events occurring within various tissues, can help to regulate the net direction of zinc flux, particularly on a long-term basis.

4. LIVER ZINC

Liver zinc metabolism undergoes marked changes during fluctuations in the dietary zinc supply, starvation, trauma, and infection. In an effort to understand how the liver processes an influx of zinc, we studied the fate of a parenterally administered zinc load in rats. After the zinc dose, plasma zinc first increased and then returned to a normal level with a concomitant increase in liver zinc. The additional complement of liver zinc was accounted for as metallothionein. Hepatic accumulation of zinc was also accompanied by an increase in the incorporation of labeled cystine into the protein. When actinomycin D was administered before zinc, the metal was not accumulated in the liver and labeled cystine was not incorporated into metallothionein (Richards and Cousins, 1975b, 1977b; Squibb et al., 1977). Subsequent experiments demonstrated that fluctuations in dietary zinc could markedly alter the amount of zinc bound to liver metallothionein in a fashion similar to that observed in the intestine (Richards and Cousins, 1976a, 1976b).

The mechanism of hepatic metallothionein biosynthesis has been extensively studied. Initial experiments demonstrated that labeled amino acids were incorporated into the protein in response to a zinc load, suggesting that metallothionein was synthesized *de novo*. Subsequently, polysomes from livers of zinc-treated rats were shown to direct, in a homologous, cell-free system, the synthesis of more metallothionein than polysomes from control rats (Squibb and Cousins, 1977). These data suggest that more metallothionein mRNA is available for translation in zinc-treated animals. This hypothesis was substantiated using a heterologous cell-free protein-synthesizing system (Shapiro et al., 1978) wherein liver polyadenylated (poly A^+) mRNA was isolated by oligo (dT)-cellulose chromatography and translated in a wheat germ protein-synthesizing system. The amount of metallothionein mRNA in liver polysomes was found to follow the same time course as the rate of synthesis of this hepatic protein observed *in vivo* (Squibb et al., 1977). These data suggest that metallothionein is a gene product whose synthesis is regulated by the zinc status of the cell. Although not as extensively investigated, food restriction (Bremner and Davies, 1975; Richards and Cousins, 1976b), environmental stress (Oh et al., 1978), and infection (Sobocinski et al., 1978) all appear to induce metallothionein via the same general mechanism. It has been proposed that the hepatic zinc content must be in excess of 30 $\mu g/g$ (fresh tissue) for the induction to be initiated (Bremner and Davies, 1975). However, experiments with isolated hepatocytes (Failla and Cousins, 1978b) suggest that only subtle changes in liver zinc are required to initiate the induction process.

Procedures for the preparation of viable and functional adult rat liver parenchymal cells have been developed (Seglen, 1976; Jeejeebhoy and Phillips, 1976). Since hepatic zinc metabolism is responsive to dietary and physiological signals, we have employed monolayers of these nondividing, highly differentiated cells to study more closely the characteristics and regulatory aspects of zinc transport and metabolism. Our results suggest that zinc is accumulated via (a) carrier-mediated system(s) with the following properties (Failla and Cousins, 1978a, 1978b; Failla et al., 1979b):

1. Uptake is time and temperature dependent.
2. Energy poisons (e.g., cyanide and oligomycin) and sulfhydryl alkylating agents (e.g., N-ethylmaleimide) partially inhibit zinc accumulation.
3. The quantity of the metal accumulated is dependent on the zinc concentration of the medium (saturation of the transport system is observed at approximately 11 μM Zn^{2+}).
4. Zinc uptake occurs by means of a high-affinity system (apparent K_m, 8.4 μM) with high specificity. The presence of equimolar amounts of Fe^{3+}, Co^{2+}, Ni^{2+}, and Cu^{2+} does not affect the quantity of zinc accumulated.

However, cadmium, the toxic congener of zinc, reversibly inhibits uptake of the essential metal in a concentration-dependent manner.

5. Finally, zinc accumulation is increased by as much as 100% when glucocorticosteroid hormoes are added to the medium.

The characteristics of zinc uptake by HeLa cells are almost identical (Cox, 1968).

The glucocorticoid-dependent stimulation of zinc accumulation is of particular interest because it is observed when hepatocytes are incubated in medium containing physiological levels of both hormone and metal (Failla and Cousins, 1978b). The ability of various hormones to enhance zinc accumulation was directly related to glucocorticoid potency: dexamethasone $>$ prednisolone $>$ prednisone $>$ corticosterone $>$ cortisone. Other steroids and polypeptide hormones, prostaglandins E_2 and $F_{2\alpha}$, triiodothyronine, and LEM were without effect. Moreover, the glucocorticoid effect did not result from a general increase in the permeability of the plasma membrane, as amino acid uptake and Ca^{2+} uptake were similar in control and dexamethasone-treated cells. These results suggest that the pituitary-adrenal axis is directly involved in mediating the redistribution of zinc from plasma and perhaps other body zinc stores into the liver during episodes of stress (Oh et al., 1978), infection (Sobocinski et al., 1978), and fasting (Bremner and Davies, 1975; Richards and Cousins, 1976b).

The additional complement of zinc accumulated by hepatocytes incubated in medium with dexamethasone was associated with metallothionein (Failla and Cousins, 1978b). The significant increase in $[^3H]$glycine, $[^3H]$lysine, and $[^3H]$serine incorporated into metallothionein (both forms 1 and 2) demonstrated that elevation of the intracellular zinc content was correlated with increased synthesis of the protein. Moreover, in vitro isotopic exchange studies employing ^{109}Cd showed that the level of metallothionein in cells incubated in medium with dexamethasone for 20 hr was twice that in control cells (Failla et al., 1979a).

The above data fail to provide information concerning the specific mechanism(s) by which glucocorticoids alter or regulate hepatocyte zinc metabolism. It is known that steroid hormones act in target tissues by the following sequence of events (O'Malley and Means, 1974). The hormone initially binds to a receptor protein in the cytosol. This steroid-receptor protein complex is next translocated into the nucleus, where, by unknown processes, its presence results in increased mRNA synthesis, from which specific proteins are synthesized. Two possible explanations for the glucocorticoid-dependent enhancement of zinc uptake and metallothionein synthesis are represented schematically in Figure 2. The first hypothesis suggests that zinc uptake is stimulated by the synthesis of one or more glucocorticoid-induced gene products, possibly proteins involved in the transport process. Once the intracellular level of zinc exceeds a threshold con-

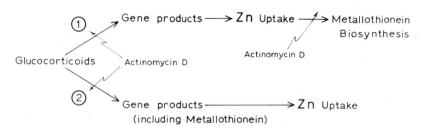

Figure 2. Possible mechanisms for the glucocorticoid-dependent enhancement of zinc uptake and metallothionein synthesis in hepatocytes. Points of possible inhibition by actinomycin D are shown.

centration, a "signal" (perhaps the metal itself) induces or increases the synthesis of metallothionein mRNA. This hypothesis assumes that there are two actinomycin D-sensitive steps and that a certain amount of metal accumulation precedes metallothionein synthesis. The second possibility is that metallothionein itself is one of the glucocorticoid-specific gene products. In either model elevation of the cytoplasmic level of metallothionein affects the binding of zinc as it enters the cell, thereby decreasing efflux and leading to increased accumulation of the metal.

That the sequence outlined in the upper half of Figure 2 best explains the glucocorticoid-mediated stimulation of zinc accumulation by hepatocytes is supported by the following observations. First, incubation of cells in dexamethasone-free medium containing either pharmacological levels of zinc (Failla and Cousins, 1978b) or 0.1 to 3.0 μM cadmium (Failla et al., 1979a) significantly increased *de novo* synthesis of metallothionein. *De novo* synthesis of this protein by suspensions of rat liver parenchymal cells and monolayers of a liver cell line after exposure to cadmium has also been observed (Hildalgo et al., 1978; Rudd and Herschman, 1978). Second, the rate of zinc transport, as well as the quantity of zinc accumulated at later times, was significantly increased in dexamethasone-treated cells. This increase was not due to the presence of an elevated level of metallothionein in the cytoplasm, since cells preincubated in dexamethasone-free medium with either cadmium (2.2 μM) or high zinc (85 μM) accumulated the metals at the same rate as did controls (Failla and Cousins, 1978b; Failla et al., 1979b). These results demonstrate that elevation or alteration of the intracellular metal content is responsible for the induction of metallothionein synthesis. Thus it appears that glucocorticoids enhance the ability of hepatocytes to take up zinc by directly altering one or more components of the transport system (Figure 3). Elucidation of the specific mechanism awaits more detailed studies.

Plasma   Hepatocyte

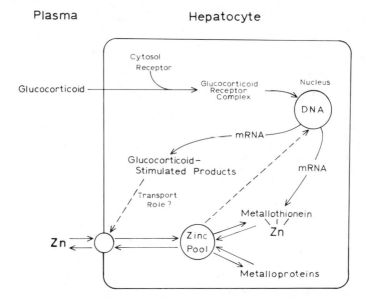

Figure 3. General mechanism for the glucocorticoid-stimulated accumulation of zinc in hepatocytes. Broken lines indicate a possible role of glucocorticoid-induced gene products in zinc transport and the induction of metallothione in mRNA by elevated intracellular zinc.

5. PROSTATE ZINC

The concentration of zinc is greater in human and rat prostate than in any other organ (Underwood, 1977). More than two decades ago Gunn and Gould (1956, 1957) demonstrated that the accumulation of ^{65}Zn by adult rat dorsolateral prostate was significantly decreased after castration and hypophysectomy. Administration of either testosterone to castrated rats (Gunn and Gould, 1956) and dogs (Pront et al., 1959) and/or chorionic gonadotrophin (Gunn and Gould, 1957) and interstitial cell-stimulating hormone (ICSH; Gunn et al.; 1961) to hypophysectomized rats maintained prostatic ^{65}Zn uptake near control levels. Moreover, prolactin and growth hormone enhanced both the testosterone- and ICSH-mediated responses (Gunn et al., 1965). Chandler and associates have recently shown that intracellular distribution of zinc in the prostate is also hormone dependent (Chandler et al., 1974, 1975). In normal tissue the metal is localized with secretory material in the lumen, nucleoli, nucleoplasm, endoplasmic reticulum, and secretory granules. Reduction of either the prolactin or testosterone level resulted in reduction of the total zinc content of the organ and transfer from the above organelles to the chromatin and lysosomes. The concentration of zinc in the lumen remained unchanged. The data clearly demonstrate that the pituitary-gonadal axis has a primary role in regulating zinc transport and metabolism in the prostate and possibly in all male reproductive organs.

ACKNOWLEDGMENTS

Research from the laboratory of R. J. C., discussed in this chapter, is supported by U.S. Public Health Grants AM 18555 and AM 20485 from the National Institute of Arthritis, Metabolism, and Digestive Diseases, the Charles and Johanna Busch Fund of Rutgers University, and the New Jersey Agricultural Experiment Station. M. L. F. was the recipient of a National Research Service Award (AM 05726) from the National Institute of Arthritis, Metabolism, and Digestive Diseases. Present address: Department of Biochemistry and Nutrition, Virginia Polytechnic Institute and State University, Blacksburg, Virgina.

REFERENCES

Beisel, W. R., Pekarek, R. S., and Wannemacker, R. W. (1976). "Homestatic Mechanisms Affecting Plasma Zinc Levels in Acute Stress." In A. S. Prasad and D. Oberleas, Eds., *Trace Elements in Human Health and Disease;* Vol. I: *Zinc and Copper.* Academic Press, New York, pp. 87–106.

Bettger, W. J., Fish, T. J., and O'Dell, B. L. (1978). "Effects of Copper and Zinc Status of Rats on Erythrocyte Stability and Superoxide Dismutase Activity," *Proc. Soc. Exp. Biol. Med.,* **158**, 279–282.

Bremner, I. and Davies, N. T. (1975). "The Induction of Metallothionein in Rat Liver by Zinc Injection and Restriction of Food Intake," *Biochem. J.,* **149**, 733–738.

Brown, E. D., Chan, W., and Smith, J. C., Jr. (1978). "Bone Mineralization during a Developing Zinc Deficiency," *Proc. Soc. Exp. Biol. Med.,* **157**, 211–214.

Buhler, R. H. O. and Kaegi, J. H. R. (1974). "Human Hepatic Metallothioneins," *FEBS Lett.,* **39**, 229–234.

Chandler, J. A., Harper, M. E., and Griffiths, K. (1974). "Studies in Subcellular Zinc Distribution in Relation to Hormone Levels in Rat Prostate Tissue Using the Electron Microanalyzer, EMMA," *J. Endocrinol.,* **61**, 1v–1vi.

Chandler, J. A., Harper, M. E., Blundell, G. K., and Morton, M. (1975). "Examination of the Subcellular Distribution of Zinc in Rat Prostate after Castration Using the Electron Microanalyzer, EMMA," *J. Endocrinol.,* **65**, 34p–35p.

Chvapil, M. (1973). "New Aspects in the Biological Role of Zinc: A Stabilizer of Macromolecules and Biological Membranes," *Life Sci.,* **13**, 1041–1049.

Cotzias, G. C. and Papavasilion, P. S. (1964). "Specificity of Zinc Pathway through the Body: Homeostatic Considerations," *Am. J. Physiol.*, **206**, 787-792.

Cotzias, G. C., Borg, D. C., and Selleck, B. (1962). "Specificity of Zinc Pathways through the Body: Turnover of Zn in the Mouse," *Am. J. Physiol.*, **202**, 359-363.

Cousins, R. J. (1979). "Regulation of Zinc Absorption: Role of Intracellular Ligands," *Am. J. Clin. Nutr.* 32, 339-345.

Cousins, R. J., Smith, K. T., Failla, M. L., and Markowitz, L. A. (1978). "Origin of Low Molecular Weight Zinc-Binding Complexes from Rat Intestine," *Life Sci.*, **23**, 1819-1826.

Cox, R. P. (1968). "Hormonal Stimulation of Zinc Uptake in Mammalian Cell Cultures," *Mol. Pharmacol.*, **4**, 510-521.

De Luca, H. F. (1974). "Vitamin D: The Vitamin and the Hormone," *Fed. Proc.*, **33**, 2211-2219.

Eichhorn, G. L. Berger, N. A., Butzow, J. J., Clark, P. Rifkind, J. M., Shin, Y. A., and Tarien, E. (1971). "The Effect of Metal Ions on the Structure of Nucleic Acids." American Chemical Society, Advances in Chemistry Series 100, pp. 135-154.

Evans, G. W., Grace, C. O., and Hahn, C. (1973). "Homeostatic Regulation of Zinc Absorption in the Rat," *Proc. Soc. Exp. Biol. Med.*, **143**, 723-725.

Evans, G. W. Grace, C. I., and Votava, H. J. (1975). "A Proposed Mechanism for Zinc Absorption in the Rat," *Am. J. Physiol.*, **228**, 501-505.

Failla, M. L. (1977). "Zinc: Functions and Transport in Microorganisms." In E. D. Weinberg, Ed., *Microorganisms and Minerals*. Marcel Dekker, New York, pp. 151-214.

Failla, M. L. and Cousins, R. J. (1978a). "Zinc Uptake by Isolated Rat Liver Parenchymal Cells," *Biochim. Biophys. Acta*, **538**, 435-444.

Failla, M. L. and Cousins, R. J. (1978b). "Zinc Accumulation and Metabolism in Primary Cultures of Adult Rat Liver Cells: Regulation by Glucocorticoids," *Biochim. Biophys. Acta*, **543**, 293-304.

Failla, M. L., Cousins, R. J., and Mascenik, M. J. (1979a). "Cadmium Accumulation and Metabolism by Rat Liver Parenchymal Cells in Primary Monolayer Culture," *Biochim. Biophys. Acta* 583, 63-72.

Failla, M. L., Lauhoff, A. R., and Cousins, R. J. (1980). "Actions of Dexamethosone on Zinc Transport and Metallothionein Synthesis in Primary Cultures of Rat Liver Parenchymal Cells," (in preparation).

Flagstad, T. (1976). "Lethal Train A46 in Cattle: Intestinal Zinc Absorption," *Nord. Vet-Med.*, **28**, 160-169.

Giroux, E. L. and Prakash, N. J. (1977). "Influence of Zinc-Ligand Mixtures on Serum Zinc Levels in Rats," *J. Pharm. Sci.*, **66**, 391-395.

Gunn, S. A. and Gould, T. C. (1956). "The Relative Importance of Androgen and Estrogen in the Selective Uptake of Zn^{65} by the Dorsolateral Prostate of the Rat," *Endocrinology*, **58**, 443-452.

Gunn, S. A. and Gould, T. C. (1957). "Hormone Interrelationships Affecting the Selective Uptake of ^{65}Zn by the Dorsolateral Prostate of the Hypophysectomized Rat," *J. Endocrinol.,* **16,** 18-27.

Gunn, S. A., Gould, T. C., and Anderson, W. A. D. (1961). "Hormonal Control of Zinc in Mature Rat Testis," *J. Endocrinol.,* **23,** 37-45.

Gunn, S. A., Gould, T. C., and Anderson, W. A. D. (1965). "The Effect of Growth Hormone and Prolactin Preparations on the Control of Interstitial Cell-Stimulatory Hormone of Uptake of ^{65}Zn by the Rat," *J. Endocrinol.,* **32,** 205-214.

Hess, F., King, J., Morgen, S. (1976). "The Effect of a Low Zinc Intake on Zinc Excretion in Healthy Young Women," *Fed. Proc.* **35,** 658 (abstr.).

Hildalgo, H. A., Koppa, V., and Bryan, S. E. (1978). "Induction of Cadmium-Thionein in Isolated Rat Liver Cells," *Biochem. J.,* **170,** 219-225.

Hurley, L. S., Duncan, J. R., Sloan, M. V., and Eckhert, C. D. (1977). "Zinc-Binding Ligands in Milk and Intestine: A Role in Neonatal Nutrition? *Proc. Natl. Acad. Sci.,* **74,** 3547-3549.

Jeejeebhoy, K. N. and Phillips, M. J. (1976). "Isolated Mammalian Hepatocytes in Culture," *Gastroenterology,* **71,** 1086-1096.

Kaegi, J. H. R., Himmelhock, R. S. R., Whanger, P. D., Bethune, J. L., and Vallee, B. L. (1974). "Equine Hepatic and Renal Metallothioneins: Purification, Molecular Weight, Amino Acid Composition, and Metal Content, *J. Biol. Chem.,* **249,** 3537-3542.

Kampschmidt, R. F., Upchurch, H. F., Eddington, C. L., and Pulliam, L. A. (1973). "Multiple Biological Activities of a Partially Purified Leukocytic Endogenous Mediator," *Am. J. Physiol.,* **224,** 530-533.

Lombeck, I., Schnippering, H. C., Ritel, H., Feinendegen, L. E., and Bremer, H. J. (1975). "Absorption of Zinc in Acrodermatitis Enteropathica," *Lancet,* **i,** 855.

Methfessel, A. H. and Spencer, H. (1973). "Zinc Metabolism in the Rat. I: Intestinal Absorption of Zinc," *J. Appl. Physiol.,* **34,** 58-62.

Morgan, R. S. and Sattilaro, R. F. (1972). "Zinc in *Entamoeba invaders.*" *Science,* **176,** 929-930.

Moynahan, E. J. (1974). "Acrodermatitis Enteropathica: A Lethal Inherited Human Zinc Deficiency," *Lancet,* **ii,** 399.

Neldner, K. H. and Hambidge, M. (1975). "Zinc Therapy of Acrodermatitis Enteropathica," *New Engl. J. Med.,* **292,** 879-882.

Oelshlegel, F. J. and Brewer, G. J. (1977). "Absorption of Pharmacological Doses of Zinc." In G. J. Brewer and A. S. Prasad, Eds., *Zinc Metabolism—Current Aspects in Health and Disease.* Alan R. Liss, New York, p. 299-311.

Oh, S. H., Deagen, J. T., Whanger, P. D., and Weswig, P. H. (1978). "Biological Function of Metallothionein. V: Its Induction in Rats by Various Stresses," *Am. J. Physiol.,* **234,** E282-E285.

O'Malley, B. W. and Means, A. R. (1974). "Female Steroid Hormones and Target Cell Nuclei," *Science,* **183,** 610-620.

Pécoud, A., Donzel, P. and Schelling, J. L. (1975). "Effect of Foodstuffs on the Absorption of Zinc Sulfate," *Clin. Pharmacol. Ther.*, **17**, 469–474.

Prasad, A. S. and Oberleas, D. (1976). *Trace Elements in Human Health and Disease*, Vol. I: *Zinc and Copper*. Academic Press, New York.

Prasad, A. S., Abbasi, A., Oberleas, D., Rabbani, P., Fernandez-Madrid, F., and Ryan, J. (1976). "Experimental Production of Zinc Deficiency in Man," *Fed. Proc.*, **35**, 658 (abstr.).

Prask, J. A. and Plocke, D. J. (1971). "A Role for Zinc in the Structural Integrity of the Cytoplasmic Ribosomes of *Euglena gracilis*," *Plant Physiol.*, **48**, 150–155.

Pront, G. R., Sierp, M., and Whitmore, W. F. (1959). "Radioactive Zinc in the Prostate: Some Factors Influencing Concentration in Dogs and Man," *J. Am. Med. Assoc.*, **169**, 1703–1710.

Richards, M. P. and Cousins, R. J. (1975a). "Mammalian Zinc Homeostasis: Requirement for RNA and Metallothionein Synthesis," *Biochem. Biophys. Res. Commun.*, **64**, 1215–1222.

Richards, M. P. and Cousins, R. J. (1975b). "Influence of Parenteral Zinc and Actinomycin D on Tissue Zinc Uptake and the Synthesis of a Zinc-Binding Protein," *Bioinorg. Chem.*, **4**, 215–224.

Richards, M. P. and Cousins, R. J. (1976a). "Zinc-Binding Protein: Relationship to Short Term Changes in Zinc Metabolism," *Proc. Soc. Exp. Biol. Med.*, **153**, 52–56.

Richards, M. P. and Cousins, R. J. (1976b). "Metallothionein and Its Relationship to the Metabolism of Dietary Zinc in Rats," *J. Nutr.*, **106**, 1591–1599.

Richards, M. P. and Cousins, R. J. (1977a). "Isolation of an Intestinal Metallothionein Induced by Parenteral Zinc," *Biochem. Biophys. Res. Commun.*, **75**, 286–294.

Richards, M. P. and Cousins, R. J. (1977b). "Influence of Inhibitors of Protein Synthesis on Zinc Metabolism," *Proc. Soc. Exp. Biol. Med.*, **156**, 505–508.

Rudd, C. J. and Herschman, H. R. (1978). "Metallothionein Accumulation in Response to Cadmium in a Clonal Rat Liver Cell Line," *Toxicol., Appl. Pharmacol.*, **44**, 511–522.

Schwarz, F. and Kirchgessner, M. (1978). "Studies on the Regulation of the Intestinal Absorption of Zinc: In M. Kirchgessner, Ed., *Trace Element Metabolism in Man and Animals*–3. ATF, Weilienstephan, West Germany, pp. 110–115.

Seglen, P. O. (1976). "Preparation of Isolated Rat Liver Cells." In D. Prescott, Ed., *Methods in Cell Biology*, Vol. XIII. Academic Press, New York, pp. 29–83.

Shapiro, S. G., Squibb, K. S., Markowitz, L. A., and Cousins, R. J. (1978). "Cell-Free Synthesis of Metallothionein Directed by Rat Liver Polyadenylated Messenger Ribonucleic Acid," *Biochem. J.*, **175**, 833–841.

Smith, J. C., Zeller, J. A., Brown, E. D., and Ong, S. C. (1976). "Elevated Plasma Zinc: A Heritable Anomaly," *Science*, **193**, 496–498.

Smith, K. T., Cousins, R. J., Silbon, B. L., and Failla, M. L. (1978). "Zinc Absorption and Metabolism by Isolated, Vascularly Perfused Rat Intestine," *J. Nutr.*, **108**, 1849–1854.

Sobocinski, P. Z., Canterbury, W. J., Mapes, C. A., and Dinterman, R. E. (1978). "Involvement of Hepatic Metallothioneins in Hypozincemia Associated with Bacterial Infection," *Am. J. Physiol.*, **234**, E399–E406.

Song, M. K. and Adham, N. F. (1978). "Role of Prostaglandin E_2 in Zinc Absorption in the Rat," *Am. J. Physiol.*, **234**, E99–E105.

Squibb, K. S. and Cousins, R. J. (1977). "Synthesis of Metallothionein in a Polysomal Cell-Free Systems," *Biochem. Biophys. Res. Commun.*, **75**, 806–812.

Squibb, K. S., Cousins, R. J., and Feldman, S. L. (1977). "Control of Zinc-Thionein Synthesis in Rat Liver," *Biochem. J.*, **164**, 223–228.

Tal, M. (1968). "On the Role of Zn^{2+} and Ni^{2+} in Ribosome Structure," *Biochim. Biophys. Acta*, **169**, 564–565.

Tao, S. H. and Hurley, L. S. (1975). "Effect of Dietary Calcium Deficiency during Pregnancy on Zinc Mobilization in Intact and Parathyroidectomized Rats," *J. Nutr.*, **105**, 220–225.

Underwood, E. J. (1977). *Trace Elements in Human and Animal Nutrition*, 4th ed. Academic Press, New York, pp. 196–242.

Vallee, B. L. and Wacker, W. E. C. (1970). "Metalloproteins." In H. Neurath, Ed., *The Proteins*, Vol. 5. Academic Press, New York, 192 pp.

Wacker, W. E. C. and Vallee, B. L. (1959). "Nucleic Acids and Metals. I: Chromium, Manganese, Nickel, Iron, and Other Metals in Ribonucleic Acid from Diverse Biological Sources," *J. Biol. Chem.*, **234**, 3257–3262.

Weigand, E. and Kirchgessner, M. (1978). "Homeostatic Adjustments in Zinc Digestion to Widely Varying Dietary Zinc Intake," *Nutr. Metab.*, **22**, 101–112.

Wilkins, P. J., Grey, P. C., and Dreasti, I. E. (1972). "Plasma Zinc as an Indicator of Zinc Status in Rats," *Br. J. Nutr.*, **26**, 113–120.

7

BLOOD ZINC IN HEALTH AND DISEASE

John D. Bogden

Department of Preventive Medicine and Community Health, College of Medicine and Dentistry of New Jersey, New Jersey Medical School, Newark, New Jersey

1. Introduction	138
2. Contamination Control and Analysis	138
3. Diurnal Variation of Plasma Zinc	139
4. Normal Concentrations	141
5. Plasma Protein Binding	143
6. Zinc and Protein Deficiencies	144
7. Pregnancy, Maternal and Cord Blood, and Oral Contraceptive Agents	145
8. Hematological Disorders	146
8.1. Sickle-cell anemia	147
8.2. Zinc toxicity	148
9. Hepatic Disease and Alcoholism	148
10. Renal Disease and Dialysis	149
11. Malignancies	150
12. Infectious Diseases	151
13. Acrodermatitis Enteropathica	153
14. Other Diseases and Conditions	153
15. Mechanisms	155
16. Medications	156
17. Conclusion	158
References	159

137

1. INTRODUCTION

The volume of literature on zinc has increased substantially during the past few years and includes several reviews that contain sections on blood zinc concentrations. Berfenstam (1952) published an extensive study on blood zinc in the early 1950s. Although not strictly a review, since it contained original laboratory data, this publication is still a detailed and valuable reference. More recently, reviews, monographs, or chapters that contain some information on blood zinc have been published or edited by a number of authors, including Prasad (1966a, 1976a, 1977), Spivey Fox (1970), Mikac-Dević (1970), Underwood (1971), Beisel and Pekarek (1972), Chvapil et al. (1972), Halsted et al. (1974), Reinhold (1975), Sandstead (1975), Burch et al. (1975), Fisher (1975), and Brewer and Prasad (1977). These publications serve as a valuable starting point for anyone attempting to assess a specific aspect of the large volume of literature on blood zinc.

2. CONTAMINATION CONTROL AND ANALYSIS

Mikac-Dević (1970) reviewed methods for the determination of zinc in biological material. He discussed colorimetric methods, fluorometry, atomic absorption spectrophotometry, neutron activation analysis, polarography, and X-ray fluorescence. It is not the intent of this chapter to compare the various techniques, except to note that atomic absorption spectrophotometry is the method employed by most laboratories that measure blood zinc. Blood zinc measurements, especially when the fluid to be analyzed is serum or plasma, are not technically difficult. However, many of the values reported in the literature must be considered suspect because of the failure of the investigators to adequately control for the possibility of contamination during sample collection. Anand et al. (1975) reviewed the precautions that must be exercised to obtain uncompromised samples for trace metal analysis. They listed a number of sources of blood contamination, including anticoagulants, collecting tubes and syringes, vacutainers, rubber stoppers, and micropipets, and outlined the uncertainties associated with long-term specimen storage. These authors warned that "unless the complete history of the sample is known with certainty, the analyst is advised not to spend his time analyzing it."

Most authors have reported concentrations for plasma or serum zinc in normal and diseased patients that are less than 150 μg/100 ml unless the patient is being treated with substantial quantities of orally or parenterally administered zinc. If values greater than this are obtained, the possibility of sample contamination or an error in standardization must be considered. There is at least one exception to this guideline, however, since Smith et al. (1976) found very high plasma zinc concentrations in five of seven members of one family and two of three second-generation individuals. This appears to be a heritable, autosomal-dominant trait that thus far has not been associated with clinical symptoms or abnormalities. Erythrocyte zinc and the major zinc-binding proteins were within the normal range in these patients. Smith et al. therefore cautioned that very high plasma zinc concentrations are not necessarily the result of sample contamination.

Moody and Lindstrom (1977) suggested an acid-washing procedure for cleaning plastic ware to be used for the collection and analysis of samples for trace elements. Nackowski et al. (1977) noted that vacutainers produce substantial contamination of blood samples, an observation previously recognized by most laboratories that perfrom zinc analyses. Bogden et al. (1978a) also described procedures for the collection of plasma samples for zinc analysis.

Reimold and Besch (1978) recently outlined precautions to be exercised in avoiding zinc contamination of blood samples. They do not recommend the use of ethylenediamine tetraacetate (EDTA) as an anticoagulant, though Bogden (unpublished data) has found that most batches of dipotassium EDTA are not contaminated with excessive zinc and are a suitable anticoagulant. A permanent anticoagulant, such as EDTA, is preferable to heparin, which is a temporary anticoagulant. Protein precipitation will occur in heparinized plasma after even relatively brief periods of storage.

Most laboratories now appreciate the need to exercise considerable caution to avoid contamination and to check for its presence.

In general, the more recent literature is more likely to contain blood zinc concentrations that are not compromised by failure to minimize the possibility of contamination.

3. DIURNAL VARIATION OF PLASMA ZINC

In addition to changes that may occur as a result of disease, diurnal changes in serum or plasma zinc have been reported by some investigators. Halsted et al. (1968a) and McBean and Halsted (1969) reported no significant difference between fasting and postprandial serum zinc concentrations. Hellwege (1970) reported a substantial daily fluctuation in serum zinc comparable to the fluctuations in serum iron. He recommended the use of only early morning samples.

Lifschitz and Henkin (1971) studied serum zinc in 10 disease-free human volunteers under controlled conditions and found a circadian variation in serum zinc similar to that in serum copper (Figure 1). Serum zinc was as much as 7% above the mean from 10:00 A.M. to 10:00 P.M. Between 2:00 A.M. and 6:00 A.M. it fell to a minimum value.

Walker et al. (1973) found that the mean decrease in plasma zinc between 9:00 A.M. and 1:00 P.M. was similar in normal persons and cirrhotics.

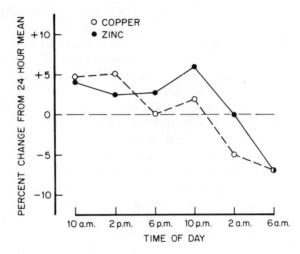

Figure 1. Diurnal variation in plasma copper and zinc. (Adapted from Lifschitz and Henkin, 1971.)

Burr (1973) reported that the diurnal variation in plasma zinc is related to food ingestion, since, when eating was prohibited, the variation disappeared. He recommended drawing blood samples after an overnight fast or at least 3 hr after meals. Hetland and Brubakk (1973) measured plasma zinc in 13 patients at 8:00 A.M., 11:30 A.M., and 3:00 P.M. Mean concentrations found at these times were 104, 96, and 93 μg/100 ml, respectively.

Persigehl et al. (1974) found that serum zinc fell during resting, but increased during "bodily stress." After glucose or protein administration, the serum zinc decreased. They suggested that comparable zinc values can be expected only under comparable conditions of sample collection.

Henry and Elmes (1975) reported that plasma zinc increased significantly during acute starvation for periods of up to 72 hr. Henkin (1976) noted that the variation in serum zinc and copper concentrations was associated with a circadian variation in the blood levels of 17-hydroxycorticosteroids. Beisel et al. (1976) reported that the circadian variation in plasma zinc, ±5 to 10% of mean concentrations, is small in comparison to the effects of inflammatory or infectious processes. Hambidge et al. (1976) collected blood samples for zinc analysis at least 2 hr after meals.

Bogden et al. (1978) suggested that the use of the plasma Cu/Zn ratio can correct for diurnal variation in plasma zinc, since the patterns of variation of these two concentrations are remarkably similar (Figure 1). For example, if, as a result of normal diurnal variation, the plasma copper concentration decreased by 5% over a 12-hr period, the plasma zinc would also be expected to decrease by about 5%. Therefore the plasma Cu/Zn ratio would be unchanged.

In summary, it appears that changes in plasma zinc as a result of diurnal variation are small in comparison to changes occurring as a result of disease. In addition, partial correction for these changes can be made by use of the Cu/Zn ratio. Nevertheless, it is best to collect samples between 8:00 and 9:00 A.M. after an overnight fast if relatively small differences between diseased patients and controls are to be detected.

4. NORMAL CONCENTRATIONS

Enough research has been done on blood zinc concentrations to define broad normal concentration ranges. Berfenstam (1952) reported a mean plasma zinc concentration of 110 µg/100 ml for 84 males and 108 µg/100 ml for 42 females. The mean erythrocyte zinc concentration for these subjects was 1244 µg/100 ml. Berfenstam also reported relatively small variations in plasma zinc, erythrocyte zinc, and whole blood zinc for repeated determinations in the same subject over a period of 14 months.

Vallee (1959) noted that the serum contained 12%, the leukocytes 3%, and the erythrocytes 85% of whole-blood zinc. Thus the leukocytes contain 25 times as much zinc per cell as the erythrocytes. A large percentage of erythrocyte zinc is associated with carbonic anhydrase.

Davies et al. (1968) reported a normal range of 76 to 125 µg/100 ml for plasma zinc. They stated that, in health, plasma zinc is one of the most uniform biochemical characteristics of normal adult blood, and that sex and age differences in adults are not significant. Foley et al. (1968) reported that the zinc content of serum is 16% higher than that of plasma. Kubota et al. (1968) studied 243 blood samples from male residents of 19 U.S. cities; they found some regional differences in whole-blood zinc. The highest mean concentration, for Canandaigua, New York, was five times the lowest. Halsted and Smith (1970) reported mean plasma zinc concentrations of 96 µg/100 ml for males and 97 µg/100 ml for females. They recommend the use of plasma instead of serum because of the release of zinc from platelets during clotting.

Pidduck et al. (1971) indicated that leukocyte zinc was independent of sex, age, weight, and blood glucose, and that plasma and leukocyte zinc levels were not significantly correlated.

Lindeman et al. (1971) found that serial plasma zinc concentrations were remarkably constant over a 6-month period but reported a significant, although small, decrease in plasma zinc with age. Age-adjusted plasma zinc concentrations were significantly lower in females than males, but no attempt to identify females who might be taking oral contraceptives was made.

Burr (1973) reported that serum zinc concentrations were higher than the corresponding plasma concentrations. Willden and Robinson (1975) found that with increasing age the plasma zinc concentration increased in males over 55 years of age and decreased in females. They suggested that these changes were related to the plasma androgen/estrogen ratio. Sunderman (1975), like Halsted and Smith (1970), recommends the use of plasma instead of serum because of the release of zinc from platelets during the clotting process. Douglas and Lines (1976) measured serum zinc in 518 New Zealand children and observed small differences between different racial and age groups. Chooi et al. (1976) reported no significant differences between disease-free males and females. They did find decreasing plasma zinc with increasing age, but all the mean values they reported were somewhat low. Prasad (1977) reviewed the literature on erythrocyte zinc; mean normal values range from 1000 to 1400 μg/100 ml. Beisel et al. (1976) suggest that zinc in the various blood cells is present primarily in metalloenzymes such as carbonic anhydrase and alkaline phosphatase. This zinc does not equilibrate readily with the more labile plasma zinc.

Oelshlegel and Brewer (1977) reported a mean 90-μg/100 ml increase in plasma zinc 2 hr after the ingestion of 110 mg of zinc sulfate under fasting conditions. The plasma zinc returned to the baseline value in about 5 hr. The authors also measured fasting plasma zinc levels in human subjects over a 1-year period and noted variability in these values.

Versieck et al. (1977) reported a mean serum zinc of 113 ± 20 μg/100 ml and a mean packed cell zinc of 1115 ± 183 μg/100 ml in 36 normal subjects. Chvapil (1977) noted that, of the formed elements, platelets contain the highest zinc concentration, mast cells are second, and granulocytes and macrophages are third highest in zinc. Chvapil et al. (1977) also discussed the effect of the serum zinc concentration on leukocyte function. Greger et al. (1978) reported a mean (±s.d.)* serum zinc concentration of 94 ± 15 μg/100 ml for adolescent girls. They found no significant concentration difference between girls who had experienced menarche and those who had not. Reimold and Besch (1978), employing stringent measures to avoid sample contamination, reported a mean (±s.d.) plasma zinc of 88 ± 14 μg/100 ml for 76 disease-free subjects. For a group of 20 disease-free controls, Bogden et al. (1978) found the following mean (±s.e.)† zinc concentrations: plasma, 87 ± 3 μg/100 ml; whole blood, 699 ± 33 μg/100 ml; erythrocytes, 1394 ± 145 μg/100 ml. The last value was estimated from the plasma and whole-blood zinc concentrations and the hematocrit.

Most analyses of blood zinc in disease have been conducted using serum or plasma. In comparing values, it is necessary to be aware that serum zinc is about 16% higher than plasma zinc. The effects of age, sex, and race on serum or plasma zinc appear to be quite small in comparison with the effects of most

*Standard deviation

†Standard Error.

diseases. The protein binding of zinc in plasma and blood zinc concentrations in various diseases will now be discussed.

5. PLASMA PROTEIN BINDING

Vikbladh (1951) demonstrated that plasma zinc is bound to at least two different proteins, a finding supported by more recent research. Berfenstam (1952) reported that plasma is able to bind large quantities of zinc; he could find no upper limit for this binding capacity. Dennes et al. (1962) found that plasma is capable of binding quantities of zinc much higher than normal concentrations. They also demonstrated that loosely bound erythrocyte zinc may form part of the pool of labile zinc, since, *in vitro,* there is a rapid exchange of zinc between erythrocytes and plasma. In addition, they found no transfer from leukocytes to plasma, but a rapid uptake from plasma by white blood cells. Sullivan et al. (1969) reported that the depression of serum zinc in myocardial infarction may be related to simultaneous transient changes in serum protein. Patients with proteinuria have been noted to have increased urinary losses of zinc associated with the urinary protein loss.

Almost all of the plasma zinc is loosely or tightly bound to protein molecules. Parisi and Vallee (1970) reported that the primary zinc-binding protein of the plasma is an $\alpha 2$-macroglobulin that accounts for about 40% of the total zinc. Smaller percentages of zinc may be tightly bound to enzymes or proteins such as transferrin, and about 7% is bound to plasma amino acids, primarily histidine and cysteine (Prasad and Oberleas, 1970).

The remaining plasma zinc exists as a loosely bound albumin complex.

If plasma is ultrafiltered under pressure, 40 to 50% of the zinc can pass through a membrane that excludes molecules with molecular weights greater than 50,000 (Henkin and Smith, 1972).

McBean et al. (1974) found no significant correlation between serum zinc and α_2-macroglobulin. Evans and Winter (1975) reported that the major portion of ^{65}Zn was albumin bound, while the rest of the isotope was associated with the higher molecular weight proteins such as transferrin and α_2-macroglobulins. They also found that zinc is transported from the intestine to the liver in the portal blood as a transferrin complex. The findings of Burns and Fell (1976) are in approximate agreement with the above results, since they report that 60% of zinc is albumin bound, 30% is bound to α_2-macroglobulins, and 10% is bound to low molecular weight complexes. Giroux et al. (1976) reported that albumin zinc is more loosely held than α_2-macroglobulin zinc. Parry (1977) agreed that plasma zinc is bound less tightly to albumin than to α_2-macroglobulin, and that the albumin-zinc bond is more likely to be broken for the maintenance of a pool of unbound and physiologically available plasma zinc.

The data given above are consistent with the findings of Falchuk (1977). He separated two zinc-binding fractions in serum by the use of Sephadex G-100. The mean plasma zinc concentration for fraction I was 38 μg/100 ml and for fraction II was 76 μg/100 ml. Fraction I contained a mixture of proteins, including α_2-macroglobulin and γ-globulin, but no albumin. Fraction II contained albumin and β-globulins. Falchuk found that the decreases in plasma zinc in various diseases or after adrenocorticotropic hormone (ACTH) administration occurred exclusively in fraction II. He speculated that his findings support the hypothesis that ACTH, a hormone released in response to any acute disease, may be responsible, at least in part, for the decreases in serum zinc found in such diseases.

6. ZINC AND PROTEIN DEFICIENCIES

Studies in Egypt first demonstrated that zinc deficiency is responsible for a syndrome of severe anemia, growth retardation, and hypogonadism (Prasad et al., 1963). Sanstead et al. (1967) reported that 22 Egyptian patients with growth retardation and associated zinc deficiency had a mean plasma zinc concentration of 65 μg/100 ml. The mean value for their control patients was 102 μg/100 ml.

Prasad (1966b) demonstrated that plasma zinc and erythrocyte zinc are significantly depressed in zinc-deficient dwarfs. Halsted et al. (1972) found a mean plasma zinc concentration of 48 μg/100 ml in 17 Iranian dwarfs suffering from zinc deficiency. This increased to 77 μg/100 ml after treatment with oral zinc; the normal mean concentration for Iranian medical students was 95 μg/100 ml. Hambidge et al. (1976) reported a mean plasma zinc concentration of 75 μg/100 ml in 74 Denver preschool children—a value significantly different from the mean concentration, 84 μg/100 ml, in 26 controls. The authors stated that this marginal zinc deficiency probably is a result of inadequate zinc nutrition; the 74 children had heights at or below the tenth percentile for their age group. Hambidge and Walravens (1976) summarized previous findings which demonstrate low mean plasma zinc concentrations (40 to 60 μg/100 ml) in kwashiorkor cases; presumably this hypozincemia is partly the result of hypoalbuminemia in these patients. Kutumbale et al. (1976), who found relatively low serum and erythrocyte zinc concentrations in kwashiorkor and marasmus patients, speculated that the low zinc concentrations may be due to altered liver function in advanced malnutrition or to depressed plasma protein concentrations with a resultant reduction in zinc-binding capacity. Golden et al. (1977) also found decreased plasma zinc in marasmus patients.

Thus plasma zinc concentrations, used in conjunction with other evidence, such as taste and smell dysfunction (Henkin et al., 1975), appear to be useful in the evaluation of zinc deficiency. However, they must be interpreted cautiously because of the many other factors that may influence blood zinc levels.

7. PREGNANCY, MATERNAL AND CORD BLOOD, AND ORAL CONTRACEPTIVE AGENTS

It has been appreciated for some time that plasma zinc progressively declines during pregnancy. Berfenstam (1952) reported that most of the decrease was completed by the 30th week of pregnancy. He found no difference in plasma zinc between primapara and multipara women, and no significant difference in erythrocyte zinc between pregnant and nonpregnant women. Berfenstam also reported a considerably higher mean plasma zinc concentration in cord blood (121 μg/100 ml) than in maternal blood collected at delivery (85 μg/100 ml). There was a higher mean plasma zinc concentration in premature infants (187 μg/100 ml) than in term infants, and plasma zinc decreased progressively during the first year of life, whereas erythrocyte zinc gradually increased. Berfenstam also noted that the mean plasma zinc in 10 fetuses with gestational ages of 100 to 150 days was very high (300 μg/100 ml).

Johnson (1961) reported that serum zinc decreases during pregnancy, but Fredricks et al. (1964) found normal postpartum erythrocyte and leukocyte concentrations in seven women. Halsted et al. (1968a) described depressed plasma zinc in pregnant women and in oral contraceptive users. In contrast, O'Leary and Spellacy (1969) found increased plasma zinc in oral contraceptive users; however, their mean value (189 μg/100 ml) is high enough to suggest the possibility of sample contamination. Henkin et al. (1971) reported on the distribution of bound and ultrafilterable zinc in maternal and fetal plasma. They found that two thirds of the zinc in maternal plasma and one third of the fetal plasma zinc were ultrafilterable, and suggested that the "free" metal moved across the placenta by passive transfer. Henkin et al. (1973) also reported that the mean cord plasma zinc concentration in 15 cases was 83 μg/100 ml, but this decreased from birth to 3 months, and then returned to adult values at 4 months. At 12 to 13 months there was another decrease in plasma zinc, with a return to adult levels at 21 months. Hambidge and Droegemueller (1974) found a mean plasma zinc concentration of 68 μg/100 ml in pregnant women at 16 weeks; this decreased to 56 μg/100 ml at 38 weeks. Their mean concentration for oral contraceptive users was 90 μg/100 ml. Heinemann (1974) also found no depression of plasma zinc in women taking oral contraceptives.

Prasad et al. (1974) studied maternal and cord blood collected at delivery; they observed no significant concentration differences between normal neonates and small-for-date neonates for either the maternal or the cord plasma. However, the sex of the neonate had an effect on the maternal plasma level, which was significantly lower for females. Prasad et al. (1975b) found slightly decreased plasma zinc concentrations in oral contraceptive users. Jezerniczky et al. (1976) reported that plasma zinc decreased from birth to 1 month of age, returned to the concentrations measured at birth within 10 months, and soon afterwards attained normal adult values. However, these observations were based on a relatively small number of cases. Jameson (1976) found that serum zinc decreased gradually during the first and second trimesters of pregnancy, but became stabilized at about 25 weeks' gestation. Moreover, women with mature infants and normal deliveries had significantly higher serum zinc concentrations during pregnancy than women with abnormal deliveries or abnormally developed infants. Jameson had speculated that plasma volume expansion during pregnancy is more impotant than hormonal effects in decreasing the plasma zinc concentration. Bogden et al. (1978b) studied maternal and cord plasma zinc concentrations at delivery in low birth weight cases. There was no significant difference, for either maternal or cord plasma, between the low birth weight cases and matched normal birth weight controls. The mean concentrations found were 64 $\mu g/100$ ml in maternal plasma and 105 $\mu g/100$ ml in cord plasma.

It is clear from the above data that plasma zinc decreases during pregnancy, although the reasons for this decline have not been unequivocally delineated. Some investigators have found depressed plasma zinc in oral contraceptive users, whereas others have not. However, in some of these studies not all the potentially relevant variables that might affect plasma zinc were controlled.

8. HEMATOLOGICAL DISORDERS

The zinc content of serum, plasma, whole blood, erythrocytes, or leukocytes has been studied for several hematological disorders. Vallee (1959) reported that the zinc content of leukocytes is decreased to 10% of the normal value in acute or chronic lymphatic and myelogenous leukemia. Dennes et al. (1961) found decreased whole-blood zinc concentrations in chronic lymphocytic leukemia patients, as well as increased erythrocyte zinc in chronic myeloid leukemia patients and decreased leukocyte zinc in both of these types of leukemia. Fredricks et al. (1964) showed that erythrocyte zinc was frequently increased in patients with chronic leukemias, and in acute lymphoblastic and acute monocytic leukemias. They reported that leukocyte zinc was depressed in 46 patients with various types of leukemia, with the lowest values in chronic lymphocytic and granulocytic and acute lymphocytic leukemia. There was an increase in leukocyte zinc in response to therapy in chronic granulocytic leukemia. Auerbach (1965) found significantly depressed whole-blood, erythrocyte, and plasma zinc in Hodgkin's disease patients. He noted that the presence of anemia could by itself produce low whole-blood zinc concentrations because of the high zinc content of red blood cells. Rosner and Gorfien (1968) reported significantly decreased plasma zinc in pernicious anemia, lymphoma, and chronic lymphocytic leukemia. However, the concentrations they reported seem somewhat high, and they provided no details of the precautions exercised to prevent sample contamination. Sullivan et al. (1969) found low serum zinc concentrations in eight lymphoma patients and Milunsky et al. (1970) reported low plasma and leukocyte zinc concentrations in myeloid leukemia. Mikac-Dević (1970) reviewed much of the pre-1970 literature on zinc in hematological diseases and noted that the observations are often contradictory.

Delves et al. (1973) measured plasma zinc and copper in children with acute lymphoblastic leukemia. They found that the plasma Cu/Zn ratio was a more useful measure of disease activity and response to treatment than the individual copper and zinc concentrations. At a Cu/Zn ratio of about 2.00, there was a sharp increase in the percentage of blast cells in the bone marrow. These data suggest that investigation of Cu/Zn ratios in other hematological disorders could also be of value, since increased plasma copper concentrations are associated with many of the conditions that result in decreased plasma zinc.

Kolaric et al. (1975) found no significant difference in serum zinc between treated or untreated malignant lymphoma patients and a control group; he concluded that serum zinc was not of diagnostic or prognostic value in this disease. However, Falchuk (1977) reported a depressed serum zinc concentration in 10 patients with acute leukemias. Although disagreement persists, most investigators have found lowered plasma, serum, or leukocyte zinc concentrations in both acute and chronic leukemia.

8.1. Sickle-Cell Anemia

Zinc deficiency secondary to sickle-cell anemia has been recognized in recent years. Karayalcin et al. (1974) found lower plasma zinc concentrations in sickle-cell disease patients than in controls, but their mean control concentration (177 μg/100 ml) is high enough to suggest the possibility of sample contamination or standardization difficulties. Prasad et al. (1975a) reported lower plasma and erythrocyte zinc concentrations in sickle-cell patients than in controls. However, the mean concentration in the sickle-cell patients (102 μg/100 ml) is quite normal. The authors suggest that the hemolysis occurring in sickle-cell patients tends to falsely elevate the plasma zinc because of the much higher concentration of zinc in the erythrocytes. For this reason, plasma zinc is not a good indicator of deficiency in these patients. Prasad et al. also found that the erythrocyte zinc concentration was strongly correlated (r = .94) with carbonic anhydrase activity in the sickle-cell anemia patients. Additional reports by Prasad et al. (1976) and Brewer et al. (1976) have confirmed the above findings. Prasad et al. (1977) recommended expressing the erythrocyte zinc content as micrograms of zinc per gram of hemoglobin. Expressed in this way, the mean erythrocyte concentration for 84 sickle-cell disease patients was 34 $\mu g/g$ Hb, in comparison to 40 $\mu g/g$ Hb for 61 controls.

8.2. Zinc Toxicity

The most common manifestation of zinc poisoning is severe anemia. Brocks et al. (1977) reported a case of acute fatal zinc intoxication in which the serum zinc concentration was 4184 $\mu g/100$ ml and the hemoglobin concentration was 7.5 g/dl. Brewer et al. (1977) suggested that the anemia of zinc intoxication may be caused by zinc interfering with copper absorption or metabolism. They found that copper deficiency can be induced by relatively large doses of oral zinc (about 150 mg daily) and suggested that sickle-cell disease patients receiving such doses should also receive copper. Porter et al. (1977) also reported the development of severe anemia and low serum copper in a woman receiving 150 mg Zn daily for celiac disease.

These data suggest that it is advisable to measure both plasma copper and plasma zinc, as well as hemoglobin levels, in patients receiving zinc in relatively large doses, for example, 660 mg $ZnSO_4 \cdot 7H_2O$ daily.

9. HEPATIC DISEASE AND ALCOHOLISM

It was reported two decades ago (Vallee, 1956; Vallee et al., 1959) that subjects with alcoholic cirrhosis have depressed serum and liver zinc concentrations. The initial observation that cirrhotics have low serum zinc levels has since been confirmed by numerous investigators. Prasad et al. (1965) found depressed zinc concentrations in both serum and red cells. Kahn et al. (1965), Halsted et al. (1968a), Sullivan et al. (1969), and Sullivan and Heaney (1970) confirmed the presence of low serum zinc in cirrhotic patients. Boyett and Sullivan (1970) noted that the low serum zinc in cirrhotics appears to be related to low serum albumin concentrations. Halsted and Smith (1970) found depressed plasma zinc in alcoholic cirrhosis patients, as well as patients with other liver diseases. Smith et al. (1977) reported a very low mean serum zinc concentration (32 $\mu g/100$ ml) in viral hepatitis patients with active disease; this returned to 84 $\mu g/100$ ml after the illness had subsided. Falchuk (1977) observed a markedly depressed mean serum zinc level (46 ± 26 $\mu g/100$ ml) in 10 liver disease patients with a history of prolonged excessive alcohol intake and physical and laboratory evidence of postalcoholic cirrhosis.

Hartoma et al. (1977) reported elevated serum zinc in alcoholics with normal or fatty liver and low serum zinc in those with alcoholic hepatitis or cirrhosis; the diagnoses were based on studies of liver biopsies. They also determined serum copper concentrations. If the serum Cu/Zn ratios are calculated from their data, it is found that all 10 of the normal or fatty liver patients had ratios below 2.00, while 8 of 12 alcoholic hepatitis and cirrhotic patients had ratios above 2.00. More recently Bogden and Troiano (1978) investigated plasma Cu/Zn ratios in patients undergoing alcohol withdrawal. These ratios were generally below 2.00 for patients with uncomplicated alcohol withdrawal seizures and above 2.00 for patients who eventually developed delirium tremens or a prolonged, severe hallucinatory state. The data suggest that elevated plasma Cu/Zn ratios are to be expected in alcoholic withdrawal seizure patients with the most severe underlying disease.

Furthermore, the Cu/Zn ratio may be of value in predicting which patients are likely to suffer the more serious central nervous system consequences of alcohol withdrawal. It is also possible that a conditioned zinc deficiency, as shown by a depressed plasma zinc concentration, contributes to the development of delirium tremens or prolonged hallucinosis.

10. RENAL DISEASE AND DIALYSIS

Several investigators have noted that serum zinc concentrations are depressed in patients with renal disease. Halsted and Smith (1970) and Condon and Freeman (1970) reported that patients with renal insufficiency had low plasma zinc concentrations. The latter authors found no correlation between plasma zinc and albumin. Halsted and Smith (1970) observed a mean increase of 7 μg/100 ml in the serum zinc concentration as a result of dialysis. Mansouri et al. (1970) found a lower mean plasma zinc concentration (65 μg/100 ml) in nondialyzed patients than in uremic patients on regular dialysis (80 μg/100 ml). Postdialysis plasma concentrations were 7 μg/100 ml higher than the predialysis mean. Mahler et al. (1971) reported lower plasma zinc in dialysis patients than in controls, and found an increase of 6 μg/100 ml as a result of dialysis. There was no significant difference in red cell zinc concentrations between renal failure patients and controls.

Falchuk (1977) reported that renal disease patients had low serum zinc concentrations. He found a mean ($\pm$s.d.) concentration of 50 $\pm$ 20 μg/100 ml for seven renal disease cases.

Lindeman et al. (1977) observed decreased serum zinc concentrations in 23 patients with nephrotic syndrome and 13 patients with renal insufficiency. The mean serum zinc concentration in nephrotic syndrome patients was 66 μg/100 ml; for the controls it was 90 μg/100 ml. Dialysis resulted in an increase of 4 μg/100 ml in the serum zinc. The authors indicated that it is not clear whether low serum zinc in these patients suggests a clinical or subclinical zinc deficiency or is a result of decreased plasma zinc-binding proteins. The correlation found between serum zinc and albumin was statistically significant, but the correlation coefficient was only .34.

Recently, Bogden et al. (unpublished data) have found that dialysis with one type of disposable coil produces increases in plasma zinc averaging almost 100 μg/100 ml. This was due to the release of zinc from the coils. Monitoring of plasma zinc in patients dialyzed with such coils is recommended.

11. MALIGNANCIES

Addink and Frank (1959) reported decreased whole-blood and serum zinc in cancer patients unless the tumor developed in zinc-rich tissues such as bone and lung. A favorable prognosis was associated with a return to normal values. Szmigielski and Litwin (1964) found decreased zinc concentrations in blood granulocytes and suggested that this could have diagnostic value. Davies et al. (1968) reported a striking decrease in plasma zinc in bronchial carcinoma. Sullivan et al. (1969) observed depressed serum zinc in 13 cancer patients. Decreased serum zinc has also been reported in prostatic carcinoma (McBean et al., 1974). Horst-Meyer et al. (1970) found that the zinc content of granulocytes was significantly decreased in patients with tumors in comparison to controls. Since the decrease in plasma zinc appears to precede clinical evidence of neoplasia, measurement of granulocyte zinc concentrations may have predictive value. Versieck et al. (1976) reported that serum zinc concentrations were significantly decreased in patients with liver metastases. Willden and Robinson (1975) studied plasma zinc concentrations in patients with prostatic disease. There were markedly higher concentrations in men with benign prostatic hyperplasia, but the concentration was not related to gland size.

Fisher et al. (1976) showed that patients with primary osteosarcoma had an elevated mean serum zinc concentration (125 μg/100 ml). Patients with metastases had depressed zinc levels, while amputated patients thought to be tumor-free had normal serum zinc concentrations. The authors also measured serum copper levels and suggested that the serum Cu/Zn ratio is useful in differentiating between patients with primary and metastatic osteosarcoma. Falchuk (1977) reported a serum zinc concentration of 66 ± 20 μg/100 ml (mean ± s.d.) in 14 breast, bowel, or lung cancer patients; this was significantly lower than the mean control concentration (96 ± 20 μg/100 ml).

Fong et al. (1977) found decreased serum zinc concentrations in patients with esophogeal cancer. These patients were reported to have "reasonably good" physical and nutritional status at the time of diagnosis and specimen collection.

Most cancers are prolonged illnesses. Additional investigations of blood zinc levels in cancer patients should be undertaken, since these concentrations may be of value in the monitoring of patient response to chemotherapy. It is also possible that zinc deficiency, as indicated by low blood concentrations, may interfere with immune mechanisms against tumor growth.

12. INFECTIOUS DISEASES

Vikbladh (1951) reported decreases in the serum zinc of patients with a variety of acute and chronic infectious conditions. Vallee (1959) also found decreases in serum zinc in acute and chronic infections; concentrations returned to normal with recovery.

Beisel et al. (1970) found decreased serum zinc in volunteers inoculated with a sandfly fever virus. Decreases in serum zinc were also noted in volunteers inoculated with live Venezuelan equine encephalomyelitis virus vaccine (Pekarek et al., 1970). In both of these studies there was an early rapid decline in the serum zinc concentration, even in subjects who remained asymptomatic.

Halsted and Smith (1970) reported decreased plasma zinc in active tuberculosis patients and patients with other types of pulmonary infections. Peterson et al. (1975) provided evidence that it is the bacterial endotoxin, not the associated febrile state, which produces decreases in plasma zinc.

Pekerek et al. (1975) found significantly depressed serum zinc and elevated serum copper concentrations in volunteers infected with *Salmonelli typhi*. After several days of chloramphenicol therapy, the zinc concentrations increased markedly. The authors suggested that the determination of serum trace metals may be useful in monitoring the course of the disease, and may aid in averting drug-induced toxicity due to unnecessarily prolonged or intensive antibiotic treatments. Wannemacher et al. (1975) have found that a substance released by phagocytizing cells, leukocytic endogenous mediator (LEM), is capable of initiating infection-related metabolic responses in laboratory animals, including the depression of serum zinc.

Pekarek and Evans (1976) reported that LEM produces a significant increase in the intestinal absorption of zinc within 7 hr of its administration to rats, and also causes significant decreases in plasma zinc with a simultaneous hepatic uptake of the metal. The authors suggest that LEM may be the key factor in the altering of zinc homeostasis during inflammation. Beisel et al. (1976) have explained the initial depression of serum zinc during infection as being a result of its rapid redistribution among body tissues, primarily as a consequence of movement from the plasma to the liver.

Beisel (1977) noted that serum zinc concentrations increased at certain stages of infectious processes. He therefore suggested that single collections of blood serum samples may be misleading and recommended that multiple samples be collected during the course of an infectious disease. He warned that, if cross-sectional sampling is done without regard to the stage of the infection, increased and decreased values may offset one another, and the mean value may appear to be quite normal. Beisel also noted that the depression of plasma zinc may begin before the onset of clinical symptoms.

Falchuk (1977) reported a significantly depressed serum zinc concentration in hospitalized patients with acute bacterial infections.

Bogden et al. (1977a) found depressed whole-blood and plasma zinc concentrations, but normal erythrocyte zinc, in pulmonary tuberculosis patients. Whole-blood and plasma copper were often increased in these patients. Bogden et al. (1978a) reported that the plasma Cu/Zn ratio was 2.73 ± 0.24 (mean $\pm$ s.e.) for 15 tuberculosis patients and 1.27 ± 0.05 for 20 controls. In addition, there was very little overlap of the ratios, since most tuberculosis patients had ratios above 2.00, and all controls had ratios below this value (Figure 2). The authors pointed out the potential usefulness of the plasma Cu/Zn ratio for monitoring the response of tuberculosis to chemotherapy.

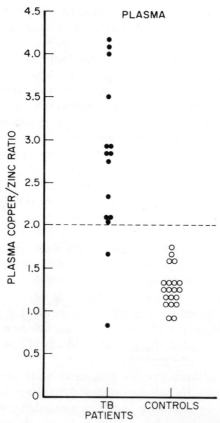

Figure 2. Plasma Cu/Zn ratios in pulmonary tuberculosis patients and controls. (From Bogden et al., 1978a.)

13. ACRODERMATITIS ENTEROPATHICA

Three characteristic symptoms of acrodermatitis enteropathica (AE)—alopecia, dermatitis, and diarrhea—first appear during infancy, usually after weaning if the infant has been breast-fed. If untreated, severe AE results in anorexia, severe failure to thrive, superimposed infection, and a usually fatal outcome. The first use of zinc in the treatment of AE was reported in the mid-1970s (Moynahan and Barnes, 1973; Moynahan, 1974). The dramatic clinical improvement of this condition in response to zinc therapy has since been confirmed by other investigators (Michaelsson, 1974; Portnoy and Molokhia, 1974; Gartside and Allen, 1975). Neldner and Hambidge (1975) reported an adult women in clinical relapse with AE; her plasma zinc concentration was 10 μg/100 ml. Administration of oral zinc sulfate returned the plasma to normal within a few days; this was accompanied by rapid clinical improvement. It is clear that zinc therapy is the treatment of choice for this serious multi-system disease. It is of interest to note that the dose of zinc required to produce dramatic clinical improvement in AE patients is only slightly higher than the recommended daily allowance for adults of 15 mg/day.

Acrodermatitis enteropathica is an autosomal recessive disorder associated with lower plasma zinc concentrations than are found in any other disease investigated to date (Hambidge, 1976). A mean concentration of 25 ± 10 μg/100 ml (laboratory normal, 68 to 100 μg/100 ml) for four cases of AE was reported by Hambidge et al. (1977). This increased to 112 ± 28 μg/100 ml after zinc therapy and concomitant clinical remission. Low normal or depressed erythrocyte zinc is also associated with AE (Hambidge et al., 1977).

Since Brewer et al. (1977) have suggested that copper deficiency and subsequent anemia may be induced by relatively high doses of oral zinc, simultaneous measurement of both plasma zinc and plasma copper, as well as the hemoglobin concentration, might be of value in monitoring AE patients receiving high doses of oral zinc. In one case the development of a hypochromic microcytic anemia in an AE patient receiving 150 mg daily of oral zinc was associated with a very substantial decrease in plasma copper and the plasma Cu/Zn ratio (Oleske and Bogden, 1978, unpublished observations). Oleske (unpublished data) also found that the administration of zinc to an AE patient not only effected clinical improvement, but also restored delayed immune mechanisms, which are profoundly impaired in this disease.

14. OTHER DISEASES AND CONDITIONS

A wide variety of diseases and stresses are associated with a decrease in the plasma zinc concentration. In some of these diseases the low plasma zinc may be an indication of a "conditioned zinc deficiency"—a deficiency occurring as a result of factors other than a lack of dietary zinc. Wacker et al. (1956) reported a significant decrease in serum zinc after myocardial infarction. Sullivan et al. (1969) found decreased serum zinc in myocardial infarction, diabetes mellitus, pneumonia, and chronic lung disease. A number of diseases were studied by Halsted and Smith (1970), who also found decreased plasma zinc in myocardial infarction patients and in subjects with decubitus or leg ulcers. These authors noted that there is a scarcity of conditions in which plasma zinc is elevated. Sinha and Gabrieli (1970) found normal or nearly normal serum zinc concentrations in most diseases they studied, but they noted depressed concentrations in bronchitis and pneumonia. However, some patients with these infections were found to have copper and zinc concentrations within the normal range. Niedermeier and Griggs (1971) found significantly depressed serum zinc in a group of 105 rheumatoid arthritis patients in comparison to 105 controls. Fell et al. (1971) studied plasma zinc and found it to be decreased in a variety of diseases and conditions, including pancreatitis, alcoholic cirrhosis, and iron deficiency, as well as in patients with leg ulcers. Lindeman et al. (1972) reported a rapid fall in plasma zinc after acute tissue injury, regardless of the cause of the injury. They found depressed zinc as a result of surgical procedures or myocardial infarction, and postulated a homeostatic mechanism to return plasma zinc to a relatively narrow normal range after these insults.

Pfeiffer and Iliev (1972) designed an experiment in which eight subjects were stressed by immersion in very cold water for 3 min. This stress significantly increased the plasma zinc concentration; however, interpretation of these data is clouded by the fact that the basal, prestress, and poststress mean plasma zinc levels reported by these authors were rather high (140 to 195 μg/100 ml). Pfeiffer and Iliev also reported that serum zinc was depressed in 11% of their schizophrenia outpatients.

Kumar and Rao (1974) described decreased plasma, erythrocyte, and leukocyte zinc in diabetics. Pangaro et al. (1974) found a decrease in mean erythrocyte zinc in a group of 58 hyperthyroidism cases. McBean et al. (1974) reported a significantly depressed mean plasma zinc concentration (74 ± 22 μg/100 ml) in decubitus ulcer patients. However, the fact that they analyzed serum from their controls and plasma from the decubitus ulcer patients may have contributed to the difference they found because plasma concentrations are generally lower than serum concentrations. Kennedy et al. (1975) reported reduced plasma zinc in rheumatoid arthritics, for both untreated and corticosteroid-treated cases. Versieck et al. (1975) found decreased zinc in myocardial infarction, and Touillon et al. (1975) reported low serum zinc concentrations in burn victims. Pories et al. (1976) noted that patients with low plasma zinc experienced delayed wound healing, and that acutely ill burn patients had persistently low plasma zinc concentrations. Lindeman et al. (1976) reported on the timing of decreases in plasma zinc after myocardial infarction. In a typical patient the plasma zinc was 92 μg/100 ml 2 hr after the onset of chest pain. At 24 hr it was 68 μg/100 ml and decreased further to 50 μg/100 ml at 48 hr. The plasma zinc returned to normal on the fourth postinfarction day.

Beisel et al. (1976) noted that acute stresses, such as surgical trauma, burns, or myocardial infarction, are associated with reduced plasma zinc. According to Sandstead et al. (1976), it is not known whether low plasma zinc is necessarily an indication of zinc deficiency. They discussed mechanisms for the production of "conditioned zinc deficiencies" and reported that low plasma zinc concentrations are found in gastrointestinal disorders such as Crohn's disease, postgastrectomy syndrome, ulcerative colitis, and sprue. Solomons et al. (1976) also found depressed plasma zinc in untreated sprue patients.

Malette and Henkin (1976) reported that serum zinc was normal in 10 patients with primary hyperparathyroidism. In serial pre- and postsurgery measurements of serum zinc Hallböök and Hedelin (1977) found a large mean decrease (60 μg/100 ml) in serum zinc at 2 days postsurgery. The decrease was greatest in the patents undergoing the most extensive surgery.

Bogden et al. (1977b) reported an increased mean plasma Cu/Zn ratio for 21 cerebral infarction patients. Falchuk (1977) studied serum zinc in a number of disease categories. Patients with cardiac, pulmonary, and neurologic illness had depressed mean concentrations. The decreased zinc was associated with the acute phase of the disease and returned to normal in convalescence. Bastek et al. (1977) found significantly higher plasma Cu/Zn ratios in patients with retinitis pigmentosa than in controls from the same family.

Sorenson and DiTommaso (1977) studied the serum zinc concentrations of patients with degenerative connective tissue diseases. Mean zinc concentrations were subnormal for patients with adult rheumatoid arthritis, juvenile rheumatoid arthritis, scleroderma, and ankylosing spondylitis. They were not significantly different from normal for patients with systemic lupus erythematosis, polymyositis, and periarteritis nodosa and for a group categorized as having "mixed connective tissue diseases."

It is obvious that numerous conditions are associated with depressed serum or plasma zinc concentrations. Many authors have failed to take this fact into consideration in categorizing their control and case groups.

15. MECHANISMS

Decreases in plasma zinc in various diseases are often accompanied by increases in plasma copper. However, the mechanisms that produce these changes have not been definitively established.

One possibility is increased loss of zinc in the urine. For example, cirrhosis, a condition in which plasma zinc is decreased, is accompanied by hyperzincuria (Prasad et al., 1965; Halsted et al. 1968a).

It is also possible that part or all of the decrease in plasma zinc may sometimes be due to an altered distribution among body tissues. For example, as discussed in Section 12 on infectious diseases, it has been shown (Pekarek and Evans, 1976) that leukocytic endogenous mediator produces a decrease in plasma zinc which is accompanied by an increased hepatic uptake of the metal.

Another possible explanation is altered plasma protein concentrations, since most plasma zinc is protein bound, with about 60% bound to albumin. Giroux et al. (1976) reported that serum concentrations of zinc and of albumin were highly ($r = .91$) correlated. Falchuk (1977) found that it is the albumin-bound zinc fraction that is decreased during acute diseases.

Hormonal effects may also be a factor. The decreased plasma zinc in pregnancy could be due to hormonal interactions or to hemodilution, the demands of the fetus, or a combination of these factors. Falchuk (1977) demonstrated the ability of ACTH to decrease plasma zinc.

Other factors, such as decreased ingestion in chronic disease due to anorexia, chelating agents in the diet, impaired intestinal absorption, proteinuria, and losses in sweat or feces, are also possible and deserve investigation. The relationship between plasma zinc and tissue concentrations is also worthy of study.

16. MEDICATIONS

The literature contains some reports on the influence of medications on blood zinc concentrations. The effects of oral contraceptive agents were discussed in Section 7 on pregnancy.

Elo and Uksila (1970) reported that serum zinc remained normal during tuberculosis chemotherapy with capreomycin, ethambutol, and rifampicin. According to Flynn et al. (1971), the administration of corticosteroids to burn or surgery patients with low-cardiac-output syndrome resulted in a rapid decrease in the serum zinc concentration. They suggested that this decline was consistent with the hypothesis that the adenohypophyseal-adrenocortical system is involved in maintaining plasma zinc within the normal range.

Flynn et al. (1973a) showed a correlation between delayed wound healing, chronic corticosteroid therapy, and depressed serum zinc concentrations. Flynn et al. (1973b) and Fodor et al. (1973) demonstrated that both anesthesia and surgical injury produce decreases in serum zinc. Auth (1973) found that parenteral administration of the chelating agent calcium-diethylenetriaminepentaacetate (DTPA) resulted in decreases of zinc in blood plasma and other tissues of rats. Henkin et al. (1975) reported that the oral administration of the amino acid histidine to human subjects produced significant decreases in serum zinc.

Klingberg et al. (1976) noted that penicillamine chemotherapy for the treatment of Wilson's disease may produce zinc deficiency. They found a depressed plasma zinc concentration, but a normal erythrocyte zinc level, in a Wilson's disease victim being treated with the drug. Penicillamine is currently used for the treatment of pediatric lead poisoning and rheumatoid arthritis, but its effects on plasma zinc or copper in these patients have not been adequately assessed.

Ellul-Micallef et al. (1976) measured serum zinc in 24 asthmatic patients. Patients not being treated with steroids had normal serum zinc concentrations, with a mean value of 110 $\mu g/100$ ml. Corticosteroid-treated asthmatics had a depressed mean serum zinc concentration (64 $\mu g/100$ ml). Pories et al. (1976) noted that short-term, low-level corticosteroid administration may not produce the depression in plasma zinc occurring in patients receiving chronic corticosteroid therapy.

Sanstead et al. (1976) reported a progressive decline in plasma zinc in patients receiving total parenteral nutrition (TPN), since TPN solutions often contain insufficient zinc to satisfy dietary requirements. Fleming et al. (1976) observed that serum zinc decreased at a rate of 6.6 $\mu g/100$ ml each week of TPN. Greene (1977) has also found that TPN solutions do not contain enough zinc to maintain normal plasma zinc levels.

In rats, plasma zinc was significantly decreased by intravenous histidine infusion (Freeman and Taylor, 1977). Anteby et al. (1978) reported a gradual increase in serum zinc after insertion of an intrauterine device impregnated with copper and zinc.

If a portion of a drug molecule has either of the following general structures, the drug may be able to act as a chelating agent:

$$R_1 - L_1 - X - X - L_2 - R_2$$

or

$$R_1 - L_1 - X - X - X - L_2 - R_2$$

Here R_1 and R_2 are the remaining portions of the drug molecule; L_1 and L_2 are electron-donating atoms—either N, S, O, or a halogen; and X is any other atom. Drugs with the above structures can be expected to function as chelating agents because of the relatively great thermodynamic stability of five- and six-membered rings. An example is the five-membered ring structure of the antitubercular drug ethambutol (Figure 3), where M can be zinc or another metal. If steric factors do not hinder the binding of L_1 and L_2 to a metal atom, drugs with the structures shown above are likely to have chelating ability. Drugs containing one of these moieties include ascorbic acid, aspirin, chloramphenicol, diazepam, diodohydroxyquin, diphenylhydantoin, isoniazid, oxytetracycline,

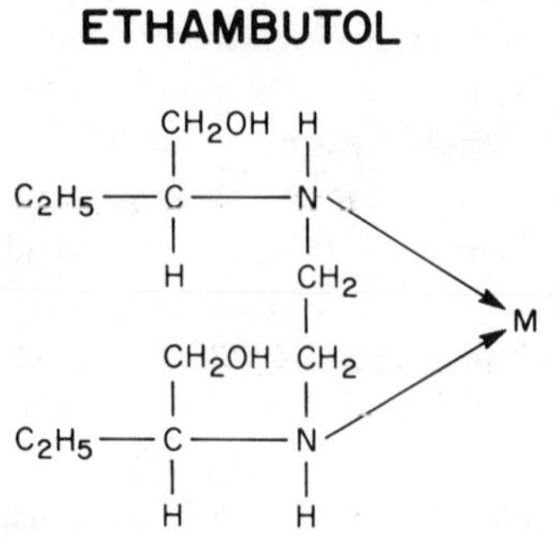

ETHAMBUTOL

Figure 3. Formation of a five-membered chelate ring by ethambutol and a metal ion.

penicillamine, ampicillin, and tetracycline. It remains to be shown whether these drugs produce alterations in blood zinc concentrations.

17. CONCLUSION

It is clear that a wide variety of diseases and conditions, as well as the administration of some drugs, can produce decreases in blood zinc concentrations. Many of the conditions associated with decreased plasma zinc are also associated with increased plasma copper concentrations (Underwood, 1971; Beisel and Pekarek, 1972; Reinhold, 1975; Sandstead, 1975; Burch et al., 1975; Fisher, 1975). As previously discussed, the plasma Cu/Zn ratio may be of greater value than the plasma zinc concentration in monitoring the efficacy of therapeutic regimens in disease states associated with increases in this ratio. Another advantage of the use of the Cu/Zn ratio is that it may partially correct for the diurnal variation in plasma zinc. For laboratories using atomic absorption spectrophotometry, little additional effort is generally required to perform a plasma copper determination in addition to a plasma zinc analysis.

Knowledge of the many conditions that may affect blood zinc concentrations should permit investigators to take these into consideration in the design of future studies on blood zinc. If these conditions have not been considered or corrected for, the investigator cannot have confidence in any differences he or she may see between "control" and "case" patients. It is most important that "cases" and "controls" be matched for relevant variables known to affect plasma zinc concentrations. Increased research on the role of various mechanisms (e.g., altered carrier protein concentrations, altered excretion patterns, hormonal effects, hyperzincuria) in producing changes in blood zinc could also be helpful.

Many studies have reported statistically significant differences in mean blood zinc concentrations between controls and cases with a variety of diseases. However, the clinical utility of these findings is dependent on the range of concentrations found. If there is too much overlap between the case and control concentrations, the means may still be significantly different, but the measurement of plasma zinc or the Cu/Zn ratio will have limited clinical value.

Recent research has provided considerable insight, but clearly much remains to be done for a complete understanding of the factors that affect blood zinc concentrations and the mechanisms by which changes in these concentrations occur.

REFERENCES

Addink, N. W. and Frank, L. J. (1959). "Remarks Apropos of Analysis of Trace Elements in Human Tissues," *Cancer,* 12, 544–551.

Anand, V. D., White, J. M., and Nino, H. V. (1975). "Some Aspects of Specimen Collection and Stability in Trace Element Analysis of Body Fluids," *Clin. Chem.,* 21, 595–602.

Anteby, S. O., Ben Bassat, H. A., Yarkoni, S., Aboulafia, Y., and Sadovsky, E. (1978). "The Effect of Intrauterine Devices Containing Zinc and Copper on Their Levels in Serum," *Fertil. Steril.* 29, 30–34.

Auerbach, S. (1965). "Zinc Content of Plasma, Blood, and Erythrocytes in Normal Subjects and in Patients with Hodgkin's Disease and Various Hematologic Disorders," *J. Lab. Clin. Med.,* 65, 628–637.

Auth, V. (1973). "Metabolism and Toxicity of Therapeutic Chelating Agents," *Strahlentherapie,* 146, 490–497.

Bastek, J., Bogden, J. D., Cinotti, A., Ten Hove, W., Stephens, G., Markopoulos, M., and Charles, J. (1977). "Trace Metals in a Family with Sex-Linked Retinitis Pigmentosa," *Adv. Exp. Med. Biol.,* 77, 43–50.

Beisel, W. R. (1977). "Zinc Metabolism in Infection." In G. J. Brewer and A. S. Prasad, Eds., *Zinc Metabolism: Current Aspects in Health and Disease.* Alan R. Liss, New York, pp. 155–176.

Beisel, W. R. and Pekarek, R. S. (1972). "Acute Stress and Trace Element Metabolism." In C. C. Pfeiffer, Ed., *Neurobiology of the Trace Metals Zinc and Copper.* Academic Press, New York, pp. 53–82.

Beisel, W. R., Wannemacher, R. W., Pekarek, R. S., and Bartelloni, P. J. (1970). "Early Changes in Individual Serum Amino Acids and Trace Metals during a Benign Viral Illness of Man," *Am. J. Clin. Nutr.,* 23, 660.

Beisel, W. R., Pekarek, R. S., and Wannemacher, R. W. (1976). "Homeostatic Mechanisms Affecting Plasma Zinc Levels in Acute Stress." In A. S. Prasad, Ed., *Trace Elements in Human Health and Disease,* Vol. I. Academic Press, New York, pp. 87–106.

Berfenstam, R. (1952). "Studies on Blood Zinc," *Acta Paediatr.,* 41, Suppl. 87, 5–97.

Bogden, J. D. and Troiano, R. A. (1978). "Plasma Calcium, Copper, Magnesium and Zinc Concentrations in Patients With the Alcohol Withdrawal Syndrome," *Clin. Chem.* **24**, 1553–1556.

Bogden, J. D., Lintz, D. I., Joselow, M. M., Charles, J., and Salaki, J. S. (1977a). "Effect of Pulmonary Tuberculosis on Blood Concentrations of Copper and Zinc," *Am. J. Clin. Pathol.,* **67**, 251–256.

Bogden, J. D., Troiano, R. A., and Joselow, M. M. (1977b). "Copper, Zinc, Magnesium, and Calcium in Plasma and Cerebrospinal Fluid of Patients with Neurological Diseases," *Clin. Chem.,* **23**, 485–489.

Bogden, J. D., Lintz, D. I., Joselow, M. M., Charles, J., and Salaki, J. S. (1978a). "Copper/Zinc Ratios in Whole Blood, Plasma, and Erythrocytes in Pulmonary Tuberculosis," *Health Lab. Sci.,* **15**, 38–43.

Bogden, J. D. Thind, I. S. Kemp, F. W., and Caterini, H. (1978b). "Plasma Concentrations of Calcium, Chromium, Copper, Iron, Magnesium and Zinc in Maternal and Cord Blood and Their Relationship to Low Birth Weight," *J. Lab. Clin. Med.* **92**, 455–462.

Boyett, J. D. and Sullivan, J. F. (1970). "Distribution of Protein-Bound Zinc in Normal and Cirrhotic Serum," *Metabolism,* **19**, 148–157.

Brewer, G. J. and Prasad, A. S. (1977). *Zinc Metabolism: Current Aspects in Health and Disease.* Alan R. Liss, New York.

Brewer, G. J., Prasad, A. S., Oelshlegel, F. J., Schoomaker, E. B., Ortega, J., and Oberleas, D. (1976). "Zinc and Sickle Cell Anemia." In A. S. Prasad, Ed., *Trace Elements in Human Health and Disease,* Vol. I. Academic Press, New York, pp. 283–294.

Brewer, G. J., Schoomaker, E. B., Leichtman, D. A., Kruckeberg, W. C., Brewer, L. F., and Meyers, N. (1977). "The Use of Pharmacological Doses of Zinc in the Treatment of Sickle Cell Anemia." In G. J. Brewer and A. S. Prasad, Eds., *Zinc Metabolism: Current Aspects in Health and Disease.* Alan R. Liss, New York, pp. 241–254.

Brocks, A., Reid, H., and Glazer, G. (1977). "Acute Intravenous Zinc Poisoning," *Br. Med. J.,* **1**, 1390–1391.

Burch, R. E., Hahn, H. K., and Sullivan, J. F. (1975). "Newer Aspects of the Roles of Zinc, Manganese, and Copper in Human Nutrition," *Clin. Chem.,* **21**, 501–520.

Burns, R. R. and Fell, G. S. (1976). "Estimation and Interpretation on Plasma Zinc Fractions," *Scot. Med. J.,* **21**, 153–154.

Burr, R. G. (1973). "Blood Zinc in the Spinal Patient," *J. Clin. Pathol.,* **26**, 773–775.

Chooi, M. K., Todd, J. K., and Boyd, N. D. (1976). "Influence of Age and Sex on Plasma Zinc Levels in Normal and Diabetic Individuals," *Nutr. Metab.,* **20**, 135–142.

Chvapil, M. (1977). "Discussion." In G. J. Brewer and A. S. Prasad, Eds., *Zinc Metabolism: Current Aspects in Health and Disease,* Alan R. Liss, New York, p. 179.

Chvapil, M., Elias, S. L., Ryan, J. N., and Zukoski, C. F. (1972). "Pathophysiology of Zinc." In C. C. Pfeiffer, Ed., *Neurobiology of the Trace Metals Zinc and Copper,* Academic Press, New York, pp. 105–124.

Chvapil, M., Stankova, L., Weldy, P., Bernhard, D., Campbell, J., Carlson, E. C., Cox, T., Peacock, J., Bartos, Z., and Zukoski, C. (1977). "The Role of Zinc in the Function of Some Inflammatory Cells." In G. J. Brewer and A. S. Prasad, Eds., *Zinc Metabolism: Current Aspects in Health and Disease,* Alan R. Liss, New York, pp. 103–122.

Condon, C. J. and Freeman, R. M. (1970). "Zinc Metabolism and Renal Failure." *Ann. Intern. Med., 73,* 531–536.

Davies, I. J., Musa, M., and Dormandy, T. L. (1968). "Measurements of Plasma Zinc," *J. Clin. Pathol., 21,* 359–365.

Delves, H. T., Alexander, F. W., and Lay, H. (1973). "Copper and Zinc Concentration in the Plasma of Leukemic Children," *Br. J. Haematol., 24,* 525–531.

Dennes, E., Tupper, R., and Wormall, A. (1961). "The Zinc Content of Erythrocytes and Leukocytes of Blood from Normal and Leukemic Subjects," *Biochem. J., 78,* 578–587.

Dennes, E., Tupper, R., and Wormall, A. (1962). "Studies on Zinc in Blood," *Biochem. J., 82,* 466–476.

Douglas, B. S. and Lines, D. R. (1976). "Serum Zinc Levels in New Zealand Children," *N. Z. Med. J., 83,* 192–194.

Ellul-Micallef, R., Galdes, A., and French, F. F. (1976). "Serum Zinc Levels in Corticosteroid-Treated Asthmatic Patients," *Postgrad. Med. J., 52,* 148–150.

Elo, R. and Uksila, E. (1970). "Serum Iron, Copper, Magnesium, and Zinc Concentration in Chronic Pulmonary Tuberculosis during Chemotherapy with Capreomycin-Ethambutol-Rifampicin Combinations," *Scand. J. Resp. Dis., 51,* 249–255.

Evans, G. W., and Winter, T. W. (1975). "Zinc Transport by Transferrin in Rat Portal Blood Plasma," *Biochem. Biophys. Res. Commun., 66,* 1218–1224.

Falchuk, K. H. (1977). "Effect of Acute Disease and ACTH on Serum Zinc Proteins," *N. Eng. J. Med., 296,* 1129–1134.

Fell, G. S., Canning, E., Husain, S. L., and Scott, R. (1971). "Copper and Zinc in Human Health and Disease." In D. D. Hemphill, Ed., *Trace Substances in Environmental Health,* Vol. V. University of Missouri Press, Columbia, pp. 293–302.

Fleming, C. R., Hodges, R. E., and Hurley, L. S. (1976). "A Prospective Study of Serum Copper and Zinc in Patients Receiving Total Parenteral Nutrition," *Am. J. Clin. Nutr., 29,* 70–77.

Fisher, G. L. (1975). "Function and Homeostasis of Copper and Zinc in Mammals," *Sci. Total Environ., 4,* 373–412.

Fisher, G. L., Byers, V. S., Shifrine, M., and Levin, A. S. (1976). "Copper and

Zinc Levels in Serum from Human Patients with Sarcomas," *Cancer,* **37,** 356–363.

Flynn, A., Pories, W. J., Strain, W. H., Hill, O. A., and Fratianne, R. B. (1971). "Rapid Serum-Zinc Depletion Associated with Corticosteroid Therapy," *Lancet,* **2,** 1169–1172.

Flynn, A., Pories, W. J., Strain, W. H., and Hill, O. A. (1973a). "Zinc Deficiency with Altered Adrenocortical Function and Its Relation to Delayed Healing," *Lancet,* **1,** 789–791.

Flynn, A., Strain, W. H., Pories, W. J., and Hill, O. A. (1973b). "Blood Serum Zinc as an Indicator of Acute Stress." In D. D. Hemphill, Ed., *Trace Substances in Environmental Health,* Vol. VII. University of Missouri Press, Columbia, pp. 271–276.

Fodor, L., Dolp, R., Escher, J., and Ahnefeld, F. W. (1973). "Preliminary Evidence for Zinc Deficiency as a Limiting Factor in Cellular Metabolism," *Anaesthesist,* **22,** 393–399.

Foley, B., Johnson, S. A., Hackley, B., Smith, J. C., and Halsted, J. A. (1968). "Zinc Content of Human Platelets," *Proc. Soc. Exp. Biol. Med.,* **128,** 265–269.

Fong, L. Y., Lin, H. J., Chan, W. C., and Newberne, P. M. (1977). "Zinc and Copper Concentrations in Tissues from Esophageal Cancer Patients and Animals." In D. D. Hemphill, Ed., *Trace Substances in Environmental Health,* Vol. XI. University of Missouri Press, Columbia, pp. 184–192.

Fredricks, R. E., Tanaka, K. R., and Valentine, W. N. (1964). "Variations of Human Blood Cell Zinc in Disease," *J. Clin. Invest.,* **43,** 304–315.

Freeman, R. M. and Taylor, P. R. (1977). "Influence of Histidine Administration on Zinc Metabolism in the Rat," *Am. J. Clin. Nutr.,* **30,** 523–527.

Gartside, J. M. and Allen, B. R. (1975). "Treatment of Acrodermatitis Enteropathica with Zinc Sulfate," *Br. Med. J.,* **3,** 521–522.

Giroux, E. L., Durieux, M., and Schechter, P. J. (1976). "A Study of Zinc Distribution in Human Serum," *Bioinorg. Chem.,* **5,** 211–218.

Golden, M. H., Jackson, A. A., and Golden, B. E. (1977). "Effect of Zinc on Thymus of Recently Malnourished Children," *Lancet,* **2,** 1057–1059.

Greene, H. L. (1977). "Trace Metals in Parenteral Nutrition." In G. J. Brewer and A. S. Prasad, Eds., *Zinc Metabolism: Current Aspects in Health and Disease.* Alan R. Liss, New York, pp. 87–97.

Greger, J. L., Higgins, M. M., Abernathy, R. P., Kirksey, A., DeCorso, M. B., and Baligar, P. (1978). "Nutritional Status of Adolescent Girls in Regard to Zinc, Copper, and Iron," *Am. J. Clin. Nutr.,* **31,** 269–275.

Hallböök, T. and Hedelin, H. (1977). "Zinc Metabolism and Surgical Trauma," *Br. J. Surg.,* **64,** 271–273.

Halsted, J. A. and Smith, J. C. (1970). "Plasma-Zinc in Health and Disease," *Lancet,* **1,** 322–324.

Halsted, J. A., Hackley, B., Rudzki, C., and Smith, J. C. (1968a). "Plasma Zinc Concentrations in Liver Diseases," *Gastroenterology*, **54**, 1098-1105.

Halsted, J. A., Hackley, B. M., and Smith, J. C. (1968b). "Plasma-Zinc and Copper in Pregnancy and after Oral Contraceptives," *Lancet*, **2**, 278-279.

Halsted, J. A., Ronaghy, H. A., Aladi, P., Haghshenass, M., Amirhakemi, G. H., Barakat, R. M., and Reinhold, J. G. (1972). "Zinc Deficiency in Man," *Am. J. Med.*, **53**, 277-284.

Halsted, J. A., Smith, J. C., and Irwin, M. I. (1974). "A Conspectus of Research on Zinc Requirements of Man," *J. Nutr.*, **104**, 345-378.

Hambidge, K. M. (1976). "The Role of Zinc Deficiency in Acrodermatitis Enteropathica," *Int. J. Dermatol.*, **15**, 38-39.

Hambidge, K. M. and Droegemueller, W. (1974). "Changes in Plasma and Hair Concentrations of Zinc, Copper, Chromium, and Manganese during Pregnancy," *Obstet. Gynecol.*, **44**, 666-672.

Hambidge, K. M. and Walravens, P. A. (1976). "Zinc Deficiency in Infants and Preadolescent Children." In A. S. Prasad, Ed., *Trace Elements in Human Health and Disease*, Vol. I. Academic Press, New York, pp. 21-32.

Hambidge, K. M., Walravens, P. A., Brown, R. M., Webster, J., White, S., Anthony, M., and Ruth, M. L. (1976). "Zinc Nutrition of Preschool Children in the Denver Head Start Program," *Am. J. Clin. Nutr.*, **29**, 734-738.

Hambidge, K. M., Walravens, P. A., and Neldner, K. H. (1977). "The Role of Zinc in the Pathogenesis and Treatment of Acrodermatitis Enteropathica." In G. J. Brewer and A. S. Prasad, Eds., *Zinc Metabolism: Current Aspects in Health and Disease*. Alan R. Liss, New York, pp. 329-340.

Hartoma, T. R., Sotaniemi, E. A., Pelkonen, O., and Ahlquist, J. (1977). "Serum Zinc and Serum Copper and Indices of Drug Metabolism in Alcoholics," *Eur. J. Clin. Pharmacol.*, **12**, 147-151.

Heinemann, G. (1974). "Plasma Iron, Serum Copper, and Serum Zinc during Treatment with Ovulation Inhibitors," *Med. Klin.*, **69**, 892-896.

Hellwege, H. H. (1970). "Diurnal Fluctuation of the Serum Zinc Level," *Klin. Wochenschr.*, **48**, 1063-1064.

Henkin, R. I. (1976). "Trace Metals in Endocrinology," *Med. Clin. N. Am.* **60**, 779-797.

Henkin, R. I. and Smith, F. R. (1972). "Zinc and Copper Metabolism in Acute Viral Hepatitis," *Am. J. Med. Sci.*, **264**, 401-409.

Henkin, R. I., Marshall, J. R., and Meret, S. (1971). "Maternal-Fetal Metabolism of Copper and Zinc at Term," *Am J. Obstet. Gynecol.*, **110**, 131-134.

Henkin, R. I., Schulman, J. D., Schulman, C. B., and Bronzert, D. A. (1973). "Changes in Total, Nondiffusible, and Diffusible Plasma Zinc and Copper during Infancy," *J. Pediatr.*, **82**, 831-837.

Henkin, R. I., Patten, B. M., Re, P. K., and Bronzert, D. A. (1975). "A Syndrome of Acute Zinc Loss," *Arch. Neurol.*, **32**, 745-751.

Henry, R. W. and Elmes, M. E. (1975). "Plasma Zinc in Acute Starvation," *Br. Med. J.*, **4**, 625–626.

Hetland, O. and Brubakk, E. (1973). "Diurnal Variation in Serum Zinc Concentration," *Scand. J. Clin. Lab. Invest.*, **32**, 225–226.

Horst-Meyer, H., Borchard, N., and Welzel, P. (1970). "Investigation of the Utility of Zinc Determination in Granulocytes as a Screening Test for Diagnosis of Patients with Malignancies," *Zr. Inn. Med. Jahrg.*, **25**, 437–440.

Jameson, S. (1976). "Variations in Maternal Serum Zinc during Pregnancy and Correlation to Congenital Malformations, Dysmaturity and Abnormal Parturition," *Acta. Med. Scand.* (Suppl.), **593**, 21–37.

Jezerniczky, J., Nagy, Z., Dvoracsek, E., Nagy, B., Ilyes, I., and Csorba, S. (1976). "Trace Elements in the Serum of Mothers and their Children," *Acta Paediatr. Acad. Sci. Hung.*, **17**, 193–197.

Johnson, N. C. (1961). "Study of Copper and Zinc Metabolism during Pregnancy," *Proc. Soc. Exp. Biol. Med.*, **108**, 518–519.

Kahn, A. M., Helwig, H. L., Redecker, A. G., and Reynolds, T. B. (1965). "Urine and Serum Zinc Abnormalities in Disease of the Liver," *Am. J. Clin. Pathol.*, **44**, 426–435.

Karayalcin, G., Rosner, F., Kim, K. Y., and Chandra, P. (1974). "Plasma-Zinc in Sickle Cell Anemia," *Lancet*, **1**, 217.

Kennedy, A. C., Fell, G. S., Rooney, P. J., Stevens, W. H., Dick, W. C., and Buchanan, W. W. (1975). "Zinc: Its Relationship to Osteoporosis in Rheumatoid Arthritis," *Scand. J. Rheumatol.* **4**, 243–245.

Klingberg, W. G., Prasad, A. S., Oberleas, D. (1976). "Zinc Deficiency Following Penicillamine Therapy." In A. S. Prasad, Ed., *Trace Elements in Human Health and Disease*, Vol. I. Academic Press, New York, pp. 51–65.

Kolaric, K., Rouljic, A., Ivankovic, G., and Vukas, D. (1975). "Serum Zinc Levels in Patients with Malignant Lymphomas," *Acta Med. Jugosl.*, **29**, 331–338.

Kubota, J., Lazar, V. A., and Losee, F. (1968). "Copper, Zinc, Cadmium, and Lead in Human Blood from 19 Locations in the United States," *Arch. Environ. Health*, **16**, 788–793.

Kumar, S. and Rao, K. S. (1974). "Blood and Urinary Zinc Levels in Diabetes Mellitus," *Nutr. Metab.* **17**, 231–235.

Kutumbale, A. S., Chhaparwal, B. C., Mehta, S., Vijayvargiya, R., and Mathur, P. S. (1976). "Zinc Levels in Serum and Erythrocytes in Protein Calorie Malnutrition," *Indian Pediatr.* **13**, 837–840.

Lifschitz, M. D. and Henkin, R. I. (1971). "Circadian Variation in Copper and Zinc in Man, *J. Appl. Physiol.*, **31**, 88–92.

Lindeman, R. D., Clark, M. L., and Colmore, J. P. (1971). "Influence of Age and Sex on Plasma and Red-Cell Zinc Concentrations," *J. Gerontol.*, **26**, 358–363.

Lindeman, R. D., Bottomley, R. G., Cornelison, R. L., and Jacobs, L. A. (1972).

"Influence of Acute Tissue Injury on Zinc Metabolism in Man," *J. Lab. Clin. Med.,* **79**, 452–460.

Lindeman, R. D., Yunice, A. A., and Baxter, D. J. (1976). "Zinc Metabolism in Acute Myocardial Infarction." In A. S. Prasad, Ed., *Trace Elements in Human Health and Disease,* Vol I. Academic Press, New York, pp. 143–153.

Lindeman, R. D., Baxter, D. J., Yunice, A., King, R. W., and Kraikit, S. (1977). "Zinc Metabolism in Renal Disease and Renal Control of Zinc Excretion." In G. J. Brewer and A. S. Prasad, Eds., *Zinc Metabolism: Current Aspects in Health and Disease.* Alan R. Liss, New York, pp. 193–208.

Mahler, D. J., Walsh, J. R., and Haynie, G. D. (1971). "Magnesium, Zinc, and Copper in Dialysis Patients," *Am. J. Clin. Pathol.,* **56**, 17–23.

Malette, L. E. and Henkin, R. I. (1976). "Altered Copper and Zinc Metabolism in Primary Hyperparathyroidism," *Am. J. Med. Sci.,* **272**, 167–174.

Mansouri, K., Halsted, J. A., and Gombos, E. A. (1970). "Zinc, Copper, Magnesium, and Calcium in Dialyzed and Nondialyzed Uremic Patients," *Arch. Intern. Med.,* **125**, 88–93.

McBean, L. D. and Halsted, J. A. (1969). "Fasting versus Postprandial Plasma Zinc Levels," *J. Clin. Pathol.,* **22**, 623.

McBean, L. D., Smith, J. C., Berne, B. H., and Halsted, J. A. (1974). "Serum Zinc and α_2-Macroglobulin Concentration in Myocardial Infarction, Decubitus Ulcer, Multiple Myeloma, Prostatic Carcinoma, Down's Syndrome, and Nephrotic Syndrome," *Clin. Chim. Acta,* **50**, 43–51.

Michaelsson, G. (1974). "Zinc Therapy in Acrodermatitis Enteropathica," *Acta Dermatovenerol.,* **54**, 377–381.

Mikac-Dević, D. (1970). "Methodology of Zinc Determinations and the Role of Zinc in Biochemical Processes," *Adv. Clin. Chem.,* **13**, 271–333.

Milunsky, A., Hackley, B. M., and Halsted, J. A. (1970). "Plasma, Erythrocyte and Leucocyte Zinc Levels in Down's Syndrome," *J. Ment. Defic. Res.,* **14**, 99–105.

Moody, J. R. and Lindstrom, R. M. (1977). "Selection and Cleaning of Plastic Containers for Storage of Trace Element Samples," *Anal. Chem.,* **49**, 2264–2267.

Morrison, S. A., Russell, R. M., Carney, E. A., and Oaks, E. V. (1978). "Zinc Deficiency: a Cuase of Abnormal Dark Adaptation in Cirrhotics," *Am. J. Clin. Nutr.,* **31**, 276–281.

Moynahan, E. J. (1974). "Acrodermatitis Enteropathica: A Lethal Inherited Human Zinc-Deficiency Disorder," *Lancet,* **2**, 399–400.

Moynahan, E. J. and Barnes, P. M. (1973). "Zinc Deficiency and a Synthetic Diet for Lactose Intolerance," *Lancet,* **1**, 676–677.

Nackowski, S. B., Putnam, R. D., Robbins, D. A., Varner, M. O., White, L. D., and Nelson, K. W. (1977). "Trace Metal Contamination of Evacuated Blood Collection Tubes," *Am. Ind. Hyg. Assoc. J.,* **38**, 503–508.

Neidermeier, W. and Griggs, J. H. (1971). "Trace Metal Composition of Synovial

Fluid and Blood Serum of Patients with Rheumatoid Arthritis," *J. Chron. Dis.*, **23**, 527–536.

Neldner, K. H. and Hambidge, K. M. (1975). "Zinc Therapy of Acrodermatitis Enteropathica," *N. Engl. J. Med.*, **292**, 879–882.

Oelshlegel, F. J. and Brewer, G. J. (1977). "Absorption of Pharmacological Doses of Zinc." In G. J. Brewer and A. S. Prasad, Eds., *Zinc Metabolism: Current Aspects in Health and Disease*. Alan R. Liss, New York, pp. 299–311.

O'Leary, J. A. and Spellacy, W. N. (1969). "Zinc and Copper Levels in Pregnant Women and Those Taking Oral Contraceptives," *Am. J. Obstet. Gynecol.*, **103**, 131–132.

Pangaro, J. A., Weinstein, M., Devetak, M. C., and Soto, R. J. (1974). "Red Cell Zinc and Red Cell Zinc Metalloenzymes in Hyperthyroidism," *Acta Endocrinol.*, **76**, 645–650.

Parisi, A. F. and Vallee, B. L. (1970). "Isolation of a Zinc-α_2-Macroglobulin from Human Serum," *Biochemistry*, **9**, 2421–2426.

Parry, W. H. (1977). "Distribution of Protein-Bound Zinc in Normal and Zinc-Deficient Lamb Plasma," *Nutr. Med.*, **21** (Suppl. 1), 48–49.

Pekarek, R. S. and Evans, G. W. (1976). "Effect of Leukocytic Endogenus Mediator (LEM) on Zinc Absorption in the Rat," *Proc. Soc. Exp. Biol. Med.*, **152**, 573–575.

Pekarek, R. S. Burghen, G. A., Bartelloni, P. J., Calia, F. N., Bostian, K. A., and Beisel, W. R. (1970). "The Effect of Live Attenuated Venezuelan Equine Encephalomyelitis Virus Vaccine on Serum Iron, Zinc, and Copper Concentrations in Man," *J. Lab. Clin. Med.*, **76**, 293–303.

Pekarek, R. S., Kluge, R. M., DuPont, H. L., Wannemacher, R. W., Hornick, R. B., Bostian, K. A., and Beisel, W. R. (1975). "Serum Zinc, Iron, and Copper Concentrations during Typhoid Fever in Man: Effect of Chloramphenicol Therapy," *Clin. Chem.*, **21**, 528–532.

Persigehl, M., Höck, A., Kasparek, K. Land, E., and Feinendegen, L. E. (1974). "Changes in the Serum Zinc Concentration in Different Metabolic Situations," *Z. Klin. Chem. Klin. Biochem.*, **12**, 171–175.

Peterson, R. A., Voso, D. B., and Smith, J. C. (1975). "The Effect of Heat Exposure and Bacterial Endotoxin on Plasma Zinc Concentration and the Body Temperature in the Broiler," *Poult. Sci.*, **54**, 909–911.

Pfeiffer, C. C. and Iliev, V. (1972). "A Study of Zinc Deficiency and Copper Excess in the Schizophrenias." In C. C. Pfeiffer, Ed., *Neurobiology of the Trace Metals*. Academic Press, New York, pp. 141–165.

Pidduck, H. G., Keenan, J. P., and Price Evans, D. A. (1971). "Leucocyte Zinc in Diabetes Mellitus," *Diabetes*, **20**, 206–213.

Pories, W. J., Mansour, E. G., Plecha, F R., Flynn, A., and Strain, W. H. (1976). "Metabolic Factors Affecting Zinc Metabolism in the Surgical Patient." In A. S. Prasad, Ed., *Trace Elements in Human Health and Disease*, Vol I. Academic Press, New York, pp. 115–141.

Porter, K. G., McMaster, D., Elmes, M. E., and Love, A. H. (1977). "Anemia and Low-Serum Copper during Zinc Therapy," *Lancet,* 2, 774.

Portnoy, B. and Molokhia, M. (1974). "Zinc in Acrodermatitis Enteropathica," *Lancet,* 2, 663–664.

Prasad, A. S. (1966a). *Zinc Metabolism.* Charles C Thomas, Springfield, Ill.

Prasad, A. S. (1966b). *Zinc Metabolism.* Charles C Thomas, Springfield, Ill., p. 270.

Prasad, A. S. (1976a). *Trace Elements in Human Health and Disease,* Vol. I: *Zinc and Copper.* Academic Press, New York.

Prasad, A. S. (1976b). "Deficiency of Zinc in Man and Its Toxicity." In A. S. Prasad, Ed., *Trace Elements in Human Health and Disease,* Vol. I. Academic Press, New York, pp. 1–20.

Prasad, A. S. (1977). "Zinc in Human Nutrition," *CRC Crit. Rev. Clin. Lab. Sci.,* 8, 1–80.

Prasad, A. S. and Oberleas, D. (1970). "Binding of Zinc to Amino Acids and Serum Proteins *in vitro,*" *J. Lab. Clin. Med.,* 76, 416–425.

Prasad, A. S., Miale, A., Farid, Z., Sandstead, H. H., Schulert, A. R., and Darby, W. J. (1963). "Biochemical Studies on Dwarfism, Hypogonadism, and Anemia," *Arch. Intern. Med.,* 111, 407–428.

Prasad, A. S., Oberleas, D., and Halsted, J. A. (1965). "Determination of Zinc in Biological Fluids by Atomic Absorption Spectrophotometry in Normal and Cirrhotic Subjects," *J. Lab. Clin. Med.,* 66, 508–516.

Prasad, A. S., Schoomaker, E. B., Ortega, J., Brewer, G. J., Oberleas, D., and Oelshlegel, F. J. (1975a). "Zinc Deficiency in Sickle Cell Disease," *Clin. Chem.,* 21, 582–587.

Prasad, A. S., Oberleas, D., Lei, K. Y., Moghissi, K. S., and Stryker, J. C. (1975b). "Effect of Oral Contraceptive Agents on Nutrients. I: Minerals," *Am. J. Clin. Nutr.,* 28, 377–384.

Prasad, A. S., Ortega, J. Brewer, G. J., Oberleas, D., and Schoomaker, E. B. (1976). "Trace Elements in Sickle Cell Disease," *J. Am. Med. Assoc.,* 235, 2396–2398.

Prasad, A. S., Abbasi, A., and Ortega, J. (1977). "Zinc Deficiency in Man: Studies in Sickle Cell Disease." In G. J. Brewer and A. S. Prasad, Eds., *Zinc Metabolism: Current Aspects in Health and Disease.* Alan R. Liss, New York, pp. 211–236.

Prasad, L. S., Ganguly, S. K., and Vasuki, K. (1974). "Role of Zinc in Foetal Nutrition," *Indian Pediatr.,* 11, 799–802.

Reimold, E. W. and Besch, D. J. (1978). "Detection and Elimination of Contaminations Interfering with the Determination of Zinc in Plasma," *Clin. Chem.,* 24, 675–680.

Reinhold, J. G. (1975). "Trace Elements—a Selective Survey," *Clin. Chem.,* 21, 476–500.

Rosner, F. and Gorfien, P. C. (1968). "Erythrocyte and Plasma Zinc and Magnesium Levels in Health and Disease," *J. Lab. Clin. Med.*, 72, 213–219.

Sandstead, H. H. (1975). "Some Trace Elements Which Are Essential for Human Nutrition: Zinc, Copper, Manganese, and Chromium," *Prog. Food Nutr. Sci.*, 1, 371–391.

Sandstead, H. H., Prasad, A. S., Schulert, A. R., Faried, A., Miale, A., Bassily, S., and Darby, W. J. (1967). "Human Zinc Deficiency, Endocrine Manifestations and Response to Treatment," *Am. J. Clin. Nutr.*, 20, 422–442.

Sandstead, H. H., Vo-Khactu, K. P., and Solomons, N. (1976). "Conditioned Zinc Deficiencies." In A. S. Prasad, Ed., *Trace Elements in Human Health and Disease*, Vol. I. Academic Press, New York, pp. 33–49.

Sinha, S. N. and Gabrieli, E. R. (1970). "Serum Copper and Zinc Levels in Various Pathologic Conditions," *Am. J. Clin. Pathol.*, 54, 570–577.

Smith, J. C. (1977). "Heritable Hyperzincemia in Humans." In G. J. Brewer and A. S. Prasad, Eds., *Zinc Metabolism: Current Aspects in Health and Disease*. Alan R. Liss, New York, pp. 181–187.

Smith, J. C., Zeller, J. A., Brown, E. D., and Ong, S. C. (1976). "Elevated Plasma Zinc: a Heritable Anomaly," *Science*, 193, 496–498.

Smith, J. C., Brown, E. D., and Cassidy, W. A. (1977). "Zinc and Vitamin A: Interrelationships." In G. J. Brewer and A. S. Prasad, Eds., *Zinc Metabolism: Current Aspects in Health and Disease*. Alan R. Liss, New York, pp. 29–44.

Solomons, N. W., Rosenberg, I. H., and Sandstead, H. H. (1976). "Zinc Nutrition in Celiac Sprue," *Am. J. Clin. Nutr.*, 29, 371–375.

Sorenson, J. R. and DiTommaso, D. J. (1977). "Mean Serum Copper, Magnesium, and Zinc Concentrations in Active Rheumatoid and Other Degenerative Connective Tissue Diseases." In D. D. Hemphill, Ed., *Trace Substances in Environmental Health*, Vol. XI. University of Missouri Press, Columbia, pp. 15–22.

Spivey Fox, M. R. (1970). "Status of Zinc in Human Nutrition," *World Rev. Nutr. Diet.*, 12, 208–226.

Sullivan, J. F. and Heaney, R. P. (1970). "Zinc Metabolism in Alcoholic Liver Disease," *Am. J. Clin. Nutr.*, 23, 170–177.

Sullivan, J. F., Parker, M. M., and Boyett, J. D. (1969). "Incidence of Low Serum Zinc in Noncirrhotic Patients," *Proc. Soc. Exp. Biol. Med.*, 130, 591–594.

Sunderman, F. W. (1975). "Current Status of Zinc Deficiency in the Pathogenesis of Neurolological, Dermatological, and Musculoskeletal Disorders," *Ann. Clin. Lab. Sci.*, 5, 132–145.

Szmigielski, S. and Litwin, J. (1964). "The Histochemical Demonstration of Zinc in Blood Granulocytes—the New Test in Diagnosis of Neoplastic Disease," *Cancer*, 17, 1381–1384.

Touillon, C., Bansillon, V., Vallon, J. J., Badinand, A., and Comtet, J. J. (1975). "Study of the Zinc Levels in Serum and Erythrocytes in Burn Patients," *Clin. Chim. Acta*, 63, 115–120.

Underwood, E. J. (1971). *Trace Elements in Human and Animal Nutrition.* Academic Press, New York, pp. 57–115, 208–252.

Vallee, B. (1959). "Biochemistry, Physiology and Pathology of Zinc," *Physiol. Rev.,* **39**, 443–490.

Vallee, B. L., Wacker, W. E., Bartholomay, A. F., and Robin, E. D. (1956). "Zinc Metabolism in Hepatic Dysfunction," *N. Engl. J. Med.,* **255**, 403–408.

Versieck, J., Barbier, S., Speecke, A., and Hoste, J. (1975). "Influence of Myocardial Infarction on Serum Manganese, Copper, and Zinc Concentrations," *Clin. Chem.,* **21**, 578–581.

Versieck, J., Hoste, J., and Barbier, F. (1976). "Determination of Manganese, Copper, and Zinc in Serum in Normal Controls and Patients with Liver Metastases," *Acta Gastro-Enterol. Belg.,* **39**, 340–349.

Versieck, J., Hoste, J., Barbier, F., Michels, H., and DeRudder, J. (1977). "Simultaneous Determination of Iron, Zinc, Selenium, Rubidium and Cesium in Serum and Packed Blood Cells by Neutron Activation Analysis," *Clin. Chem.,* **23**, 1301–1305.

Vikbladh, I. (1951). "Studies on Zinc in Blood, II, *Scand, J. Clin. Lab. Invest.,* **3** (Suppl. 2), 1–74.

Wacker, W. E., Ulmer, D. D., and Vallee, B. L. (1956). "Metalloenzymes and Myocardial Infarction," *N. Engl. J. Med.,* **255**, 449–456.

Walker, R. E., Dawson, J. B., Kelleher, J., and Losowsky, M. S. (1973). "Plasma and Urinary Zinc in Patients with Malabsorption Syndrome or Hepatic Cirrhosis," *Gut,* **14**, 943–948.

Wannemacher, R. W., Pekarek, R. S., Klainen, A. S., Bartelloni, P. J. Dupont, H. C., Hornick, R. B., and Beisel, W. R. (1975). "Detection of a Leukocytic Endogenous Mediator-like Mediator of Serum Amino Acid and Zinc Depression during Various Infectious Illnesses," *Infect. Immun.,* **11**, 873–875.

Willden, E. G. and Robinson, M. R. (1975). "Plasma Zinc Levels in Prostatic Disease," *Br. J. Urol.,* **47**, 295–299.

8

BLOOD ZINC AND LEAD POISONING

Morris M. Joselow

College of Medicine and Dentistry of New Jersey, New Jersey Medical School, Newark, New Jersey

1. Protective Action of Zinc   171
2. Enzyme-Activating Effect of Zinc   174
3. Zinc Protoporphyrin: A Simple, Sensitive Test for Lead Poisoning   175
References   179

1. PROTECTIVE ACTION OF ZINC

Excessive absorption of lead, that is, lead poisoning, can cause dysfunctions of the hematoporetic, central nervous, and/or gastrointestinal systems (Browder et al., 1973). Somewhat surprisingly, the ingestion of zinc has been found to offer protection against some of the toxic manifestations of lead absorption.

Among the first to note such effects were Willoughby et al. (1972). In young growing horses grazing on pastures contaminated with both lead and zinc from nearby refinery effluents, these authors found that the signs and symptoms of lead poisoning were considerably mitigated. Although the tissue lead contents were nearly doubled, the horses showed less evidence of intoxication than animals exposed to lead alone.

A similar protective effect of zinc against lead intoxication in young rats was noted by Cerklewski and Forbes (1976). As the dietary zinc content was increased, the severity of lead toxicity decreased. Furthermore, lead concentrations were lowered in the blood, liver, kidneys, and bones, with concomitant decreases also in some metabolic indicators of lead poisoning, such as excretion of δ-aminolevulinic acid, accumulation of free erythrocyte protoporphyrins, and inhibition of the enzyme δ-aminolevulinic acid dehydratase (ALAD; EC 4.2.1.24).

Since injected zinc did not afford such protection, the explanation offered was that the increased dietary intake of zinc impaired the intestinal absorption of lead, thereby blocking its action. Such reasoning, however, would not explain the report of Willoughby et al. (1972), in which enhanced absorption of lead, as well as lessened symptoms of lead toxicity, were noted in the zinc-protected animals.

The ability of dietary zinc to exert a protective effect was confirmed by Finelli et al. (1975). Weanling rats, fed lead in their drinking water, showed a dramatic drop in their erythrocyte ALAD, an enzyme involved in the biosynthesis of heme (Figure 1). Since ALAD is a sulfhydryl enzyme, its activity is inhibited by heavy metals, such as lead. Hence, measurement of ALAD activity can serve as an indicator of the effects of lead absorption (Hernberg and Nikkanen, 1972). The addition of high levels of zinc to the diet of rats that had been lead-treated to the point of almost complete inhibition of ALAD resulted in a relatively quick return of this enzyme activity toward normal values.

BIOSYNTHESIS OF HEME

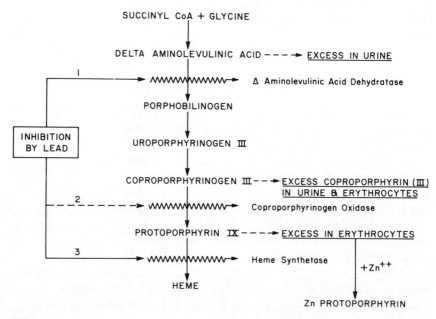

Figure 1.

The *in vitro* effect of zinc in reversing the lead-induced inhibition of ALAD was also demonstrated by Border et al. (1976). When zinc was added to blood obtained from a worker with potentially dangerous levels of lead, the ALAD values obtained did not reflect the true picture of lead absorption. This conceivably could cause a misdiagnosis when exposure to both lead and zinc prevails.

Excess dietary administration of zinc to rats was also shown by Thawley et al. (1977) to be capable of lowering the accumulation of lead in the body, as well as reducing some specific signs of toxicity, for example, a decrease in the δ-aminolevulinic acid levels in the urine.

The antagonistic effects of zinc on both the *in vivo* and *in vitro* lead inhibition of ALAD, similar to those observed in lower animals, were confirmed in baboons by Cantrell et al. (1977). The results also demonstrated the utility of the baboon as a primate model for human exposure to lead.

The overriding actions of zinc suggested to Thomasino et al. (1977) a possible therapeutic role for this element in the treatment of lead poisoning. A lead-intoxicated patient, treated in the usual manner by chelation (intramuscular injection of calcium disodium edetate), was also, during the course of the treatment, given zinc sulfate orally. Without zinc supplementation both lead and zinc were chelated and excreted, but ALAD activity decreased, an undesirable side effect. The adjunctive use of zinc sulfate, however, reversed the enzyme-inhibiting effects of lead and reactivated the ALAD. The authors proposed, subject to corroboration, that zinc supplementation would be valuable during chelation therapy for lead poisoning.

An interesting and somewhat different mechanism for the protective action of zinc against lead poisoning was postulated by Anderson (1978), specifically for zinc sulfide. The ingestion of lead-containing paint chips is undoubtedly the most widespread cause of childhood lead poisoning (Browder et al., 1973). When such paint also contains zinc sulfide, a commonly used ingredient, the absorption of the lead may be significantly hindered. Under simulated stomach conditions, zinc sulfide precipitated lead sulfide and released soluble zinc ions. Since lead sulfide is highly insoluble, lead was presumed to be rendered innocuous by this lead-for-zinc ion exchange. The concept, though attractive, ignores the potential for solubilization, even of highly insoluble compounds, by the physiological dynamics of the body.

In apparent contrast to all the protective actions of zinc salts noted above are the findings of Hsu et al. (1975). When both zinc and lead were fed together to young swine, the lead levels in blood were higher than those found when lead alone was fed, and the zinc tended to aggravate the effects of lead toxicity. These findings, if confirmed, would imply that the young growing pig handles lead differently from other animals, for which a protective compensating action for zinc has been reported.

2. ENZYME-ACTIVATING EFFECT OF ZINC

The ameliorating effects of zinc on lead poisoning have been partially attributed to its ability to activate or reactivate ALAD, one of the key enzymes involved in the biosynthesis of heme. All aerobic cells in animals and plants can synthesize heme (Haeger-Aronsen, 1976). Among the early steps in this biosynthesis (Figure 1) is the condensation of two molecules of δ-aminolevulinic acid to one molecule of porphobilinogen (Muller et al., 1974). δ-Aminolevulinic acid dehydratase from beef liver has a molecular weight of about 250,000 and an estimated number of 28 free sulfhydryl groups per molecule (Wilson et al., 1972). Thus it is vulnerable to combination with and inactivation by heavy metals, such as Pb, Hg, and Ar.

The inhibitory effect of lead has been explored and verified repeatedly (Nakao et al., 1968; Haeger-Aronsen et al., 1971; Hernberg and Nikkanen, 1972). Reversal of this inhibition by zinc, noted in several instances (see above) as an accompaniment to the protective action of this element, was first reported by Abdulla and Haeger-Aronsen (1971). When human blood with significant concentrations of zinc chloride (0.05 to 0.1 g Zn/l) was incubated, increases in enzyme activity were observed. Increased ALAD activity was also noted *in vivo* after oral ingestion of zinc sulfate (ca. 30 to 40 mg Zn/day for 4 weeks).

The functional relationship of zinc to ALAD was explored further by Finelli and co-workers. The activity of rat erythrocyte ALAD was found to be dependent on the dietary zinc intake (Finelli et al., 1974). Low activity was observed with low zinc intake. Since the *in vitro* addition of zinc to blood with low ALAD activity did not substantially increase this activity, it was concluded that zinc may act as a metal ion activator of ALAD, but that the requirement for this metal is at the site of synthesis of the enzyme.

In later studies, however (Finelli et al., 1975), reactivation of enzymatic activity depressed by lead absorption could be achieved by both increased dietary intake of zinc and the *in vitro* addition of zinc to the blood. A relatively high concentration of Zn^{2+} was required *in vitro* to reactivate ALAD inhibited by a much lower amount of Pb^{2+}, implying that the enzyme has a greater affinity for lead than for zinc (Waldron and Stofen, 1974) and raising the question of whether ALAD is a zinc-containing metalloenzyme or a zinc-activated enzyme.

More recent work tends to support the view that the zinc functions as an activator for ALAD activity, rather than as part of the enzyme molecule. The activating effect on ALAD, for example, was found to be directly related to the added zinc concentration (Schlipkoter et al., 1975), while no correlation was found between ALAD activity and zinc content of erythrocytes (Haeger-Aronsen et al., 1976). Nor is the reactivation of the enzyme after lead inhibition specific for zinc ions; aluminum, as well as zinc, can reactivate ALAD (Meredith et al., 1977). When both zinc and aluminum were added, the effect was stoichiometrically additive, indicating an equivalent contribution from each metal ion.

The inhibition of ALAD is not a specific effect of lead absorption but is shown also by other heavy metals capable of combining with sulfhydryl groups (Haeger-Aronsen et al., 1976). Such inhibition has generally been demonstrated *in vitro,* since heavy metal ions do not normally occur in significant concentrations in human blood.

The validity of using ALAD inhibition as a specific measure of increased lead absorption for general populations was challenged by Thompson et al. (1977). Inhibition of the activity of this enzyme was found to be a function of the total metal ion concentrations found in blood, including Fe, Cu, and Zn. Under certain conditions the lead-induced inhibition of ALAD could be enhanced, rather than lessened, by zinc. The explanation offered was that zinc can replace lead from binding sites of compounds resulting from the hemolysis of red cells, thereby freeing the lead for further reaction with ALAD.

That zinc is not unique in its ability to reverse the lead-induced inhibition of ALAD was shown by Davis and Avram (1978). Cadmium, the element below zinc in group IIB of the periodic table, was far more potent than zinc in activating normal erythrocyte ALAD, as well as in reversing *in vitro* lead-induced inhibition of the enzyme. The reason advanced for this similarity in behavior of zinc and cadmium was that they possess common structural features that favor similar reactions with peripheral binding sites of the enzyme.

3. ZINC PROTOPORPHYRIN: A SIMPLE, SENSITIVE TEST FOR LEAD POISONING

Quite aside from the ability of zinc to overcome the lead-induced inactivation of the enzyme ALAD and thereby, to some extent, compensate also for the effects of lead poisoning, a zinc-containing compound has proved to be extraordinarily valuable in providing a simple, sensitive, and rapid screening test for lead poisoning.

Lead interferes with the formation of heme from protoporphyrin IX (Figure 1) by inhibition of the enzyme that catalyzes the insertion of iron into the porphyrin ring (Grinstein et al., 1959; Schwartz et al., 1959). Under such conditions, excess protoporphyrin accumulates in erythrocytes, where it can chelate with available zinc ion to form zinc protoporphyrin (ZP). (Buildup of protoporphyrin and hence of ZP can occur also when there is insufficient iron, as in an iron-deficiency anemia, to combine with protoporphyrin IX to form heme.)

Zinc protoporphyrin, a relatively stable compound, has it own characteristic fluorescence spectrum (Figure 2). When blood containing ZP is diluted considerably (400- to 1000-fold) to minimize self-absorption, and excited with incident light at 424 nm, it shows a characteristic peak at 594 nm (Lamola and Yamane, 1974; Lamola et al., 1975). This can be readily measured by a spectro-

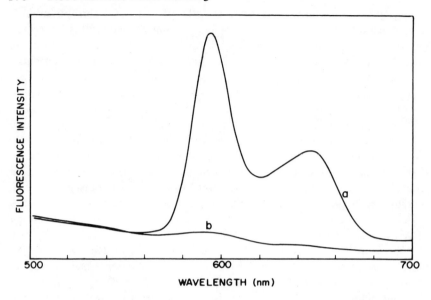

Figure 2. Characteristic emission spectra of whole blood containing ZP. (*a*) whole blood of occupationally exposed man; (*b*) whole blood of male city-dweller, not occupationally exposed to lead.

photofluorometer of sufficient sensitivity and resolving power. Therefore ZP is amenable to rapid, simple determination directly in blood, with no treatment required other than dilution of the blood.

More importantly, the ZP value can provide an indication of the toxic effect of lead, rather than the lead burden, which is commonly obtained by a determination of the lead concentration in blood. A valuable peripheral benefit is the ability of a ZP measurement to provide a clue to a possible iron-deficiency anemia state.

The relationship between lead and ZP concentrations in blood was determined (Joselow et al., 1977; Joselow and Flores, 1977a) for two populations: (1) adults occupationally exposed to inorganic lead; and (2) children living in an old city, who absorbed lead mainly through ingestion of lead-containing paint chips. Characteristic dose-response curves were obtained for both groups (Figures 3 and 4), but with differing "thresholds." For children a blood lead concentration of about 35 μg/100 ml, and for adults a concentration of about 55 μg/100 ml, corresponded to rapid rises in ZP values. The use of ZP as a biological indicator of lead exposure among another group of occupationally exposed adults, secondary lead smelter workers, was reported by Lilis et al. (1977). A similar relationship between blood lead and ZP was found, with a rapid rise in ZP occurring at blood lead levels exceeding about 50 μg/100 ml.

ZINC PROTOPORPHYRIN AND LEAD IN WHOLE BLOOD
(CHILDREN)

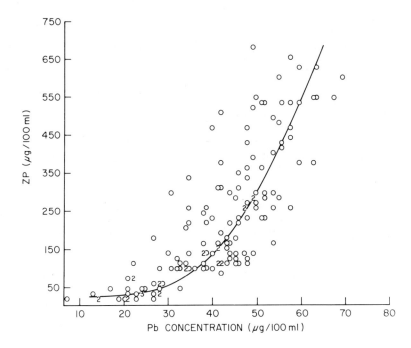

Figure 3.

At such blood lead levels, where the rapid rise in ZP values begins, it could be inferred that compensating physiological mechanisms are overwhelmed, and ZP concentrations reflect the toxic insult of lead. These findings provide support for recommendations that the upper limits for blood lead concentrations in children be below 40 $\mu g/100$ ml (Center for Disease Control, 1978), and in adults, below 60 $\mu g/100$ ml (OSHA, 1978).

The value of ZP as a monitor of lead absorption has prompted the development of practical applications of the procedure. Joselow and Flores (1977b) developed a filter paper disk technique for field collection and storage of a drop of blood with subsequent elution of the blood and measurement of its ZP content in the laboratory.

What is perhaps the simplest and most expeditious method for the determination of ZP in blood utilizes a special portable fluorometer developed by Bell Laboratories, New Jersey (Blumberg et al., 1976). This instrument, termed a

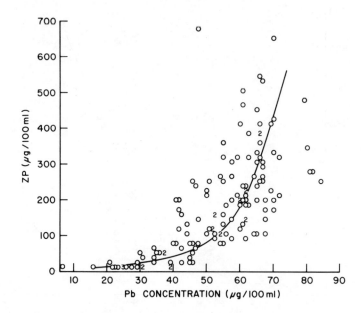

ZINC PROTOPORPHYRIN AND LEAD IN BLOOD
(ADULT WORKERS)

Figure 4.

hematofluorometer, makes use of "front-surface" optics. Essentially all of the exciting light is absorbed on the surface of the blood, regardless of the volume of the specimen. The surface-emitted ZP fluorescence is detected quantitatively by a phototube positioned at an angle away from the exciting source. In practice, a drop of blood is smeared on a cover slip and submitted to the optics of the system, that is, excitation and detection of the emitted fluorescence. After calibration, blood concentrations of ZP can be obtained directly, rapidly, and inexpensively.

The obvious utility of such an instrument has evoked official interest in the United States and has led to the recommendation that this approach be the method of choice in screening for the toxic effects of lead absorption in children (Houk, 1978). Thus the measurement of a zinc-containing compound, occurring in blood as a result of the interference of lead with heme synthesis, has now became the most widely used technique for the detection of lead poisoning in the United States.

REFERENCES

Abdulla, M. and Haeger-Aronsen, B. (1971). "ALA-Dehydratase Activations by Zinc," *Enzyme,* **12,** 708–710.

Anderson, W. S. (1978). "Role of Zinc Sulfide in Lead Poisoning Prevention." Paper presented at 175th Annual Meeting of American Chemical Society, Anaheim, Calif.

Blumberg, E. I., Eisinger, J., Lamola, A. A., and Zuckerman, D. M. (1977). "The Hematofluorometer," *Clin. Chem.,* **23,** 270–274.

Border, E. A., Cantrell, A. C., and Kilroe-Smith, T. A. (1976). "The *in vitro* Effect of Zinc on the Inhibition of Human δ-Aminolevulinic Acid Dehydratase by Lead," *Br. J. Ind. Med.,* **33,** 85–87.

Browder, A. A., Joselow, M. M., and Louria, D. B. (1973). "The Problem of Lead Poisoning," *Medicine,* **52,** 121–129.

Cantrell, A. C., Kilroe-Smith, T. A., Simoes, M. M., and Border, E. A. (1977). "The Effect of Zinc and pH on the Behavior of δ-Aminolevulinic Acid Dehydratase Activity in Baboons Exposed to Lead," *Br. J. Ind. Med.,* **34,** 110–113.

Center for Disease Control (1978). *Preventing Lead Poisoning in Young Children.* U.S. Public Health Service, April.

Cerklewski, F. L. and Forbes, R. M. (1976). "Influence of Dietary Zinc on Lead Toxicity in the Rat," *J. Nutr.,* **106,** 689–96.

Chisolm, J. (1964). "Disturbances in the Biosynthesis of Heme in Lead Intoxication," *J. Pediatr.,* **64,** 174–187.

Davis, J. R. and Avram, M. J. (1978). "A Comparison of the Stimulatory Effects of Cadmium and Zinc in Normal and Lead-inhibited Human Erythrocyte δ-Aminolevulinic Acid Dehydratase Activity *in vitro,*" *Toxicol. Appl. Pharmacol.,* **44,** 181–190.

Finelli, V. N., Murthy, L., Peirano, W. B., and Petering, H. G. (1974). "δ-Aminolevulinate Dehydratase, a Zinc Dependent Enzyme," *Biochem. Biophys. Res. Commun.,* **60,** 1418–1424.

Finelli, V. N., Klauder, D. S., Kanaffa, M. A., and Petering, H. G. (1975). "Interaction of Zinc and Lead on δ-Aminolevulinate Dehydratase," *Biochem. Biophys. Res. Commun.,* **65,** 303–310.

Grinstein, R. M., Bannerman, R. M., and Moore, C. V. (1959). "The Utilization of Protoporphyrin in Heme Synthesis," *Blood,* **14,** 476–485.

Haeger-Aronsen, B., Abdulla, M., and Fristedt, B. I. (1971). "Effect of Lead on δ-Aminolevulinic Acid Dehydrase Activity in Red Blood Cells," *Arch. Environ. Health,* **23,** 440–445.

Haeger-Aronsen, B., Schutz, A., and Abdulla, M. (1976). Antagonistic Effect *in Vivo* of Zinc on Inhibition of δ-Aminolevulinic Acid Dehydratase by Lead," *Arch. Environ. Health,* **31,** 215–220.

Hernberg, S. and Nikkanen, J. (1972). "Effect of Lead on δ-Aminolevulinic Acid Dehydratase: A Selective Review," *Prac. Lek.*, **24**, 77–83.

Houk, V. N. (1978). Letter to childhood lead poisoning programs, Dec. 5, 1978. Center for Disease Control, U.S. Public Health Service, Atlanta, Ga.

Hsu, F. S., Krook, L., Pond, W. G., and Duncan, J. (1975). "Interaction of Dietary Calcium with Toxic Levels of Lead and Zinc in Pigs," *J. Nutr.*, **105**, 112–118.

Joselow, M. M. and Flores, J. (1977a). "Comparison of Zinc Protoporphyrin and Free Erythrocyte Porphyrin in Whole Blood," *Health Lab. Sci.*, **14**, 126–128.

Joselow, M. M. and Flores, J. (1977b). "Application of the Zinc Protoporphyrin (ZP) Test as a Monitor of Occupational Exposure to Lead," *Am. Ind. Hyg. Assoc. J.*, **38**, 63–66.

Joselow, M., Lamola, A., and Flores, J. (1977). "Biochemical Monitoring for Lead Intoxication in Workers and Children by ZP (Zinc Protoporphyrin) Assay." Paper presented at International Symposium in Clinical Chemistry and Chemical Toxicology of Metals, Monte Carlo, Monaco, Mar. 2, 1977.

Lamola, A. A. and Yamane, T. (1974). "Zinc Protoporphyrin in the Erythrocytes of Patients with Lead Intoxication and Iron Deficiency Anemia," *Science*, **186**, 936–938.

Lamola, A. A., Joselow, M., and Yamane, T. (1975). "Zinc Protoporphyrin (ZPP): A Simple, Sensitive Fluorometric Screening Test for Lead Poisoning," *Clin. Chem.*, **21**, 93–97.

Lilis, R., Fischbein, A., Eisinger, J., et al. (1977). "Prevalence of Lead Disease among Secondary Lead Smelter Workers and Biological Indicators of Lead Exposure," *Environ. Res.*, **14**, 255–285.

Meredith, P. A., Moore, M. R., and Goldberg, A. (1977). "Effects of Aluminum, Lead and Zinc on δ-Aminolevulinic Acid Dehydratase," *Enzyme*, **22**, 22–27.

Muller, W., Lassner, R., and Appen, H. (1974). Delta Aminolevulinic Acid Dehydratase: A Review," *Pyrrole Inf.*, **2**, 21–22.

Nakao, K., Wada, O., and Yano, Y. (1968). "δ-Aminolevulinic Acid Dehydratase Activity in Erythrocytes for the Evaluation of Lead Poisoning," *Clin. Chim. Acta*, **19**, 319–325.

OSHA (Occupational Safety and Health Administration) U.S. Department of Labor (1978). Standard for Occupational Exposure to Lead.

Schlipkoter, H. W., Ghelerter, L., and Ost, B. (1975). "Untersuchungen zur Kombinationswirkung von Zink und Blei," *Zbl. Bakteriol. Orig.* B160, **2**, 130–8.

Schwartz, H. C., Cartwright, G. E., Smith, E. L., and Wintrobe, M. W. (1959). "Studies in the Biosynthesis of Heme from Iron and Protoporphyrin," *Blood*, **14**, 486–497.

Thawley, D. G., Willoughby, R. A., McSherry, B. J., MacLeod, G. K., MacKay, K. H., and Mitchell, W. R. (1977). "Toxic Interactions among Pb, Zn, and

Cd with Varying Levels of Dietary Ca and Vitamin D: Hematological System," *Environ. Res.*, **14**, 463–475.

Thomasino, J. A., Zuroweste, E., Brooks, S. N., Petering, H. G., Lerner, S. I., and Finelli, V. N. (1978). "Lead, Zinc, and Erythrocyte δ-Aminolevulinic Acid Dehydratase: Relationships in Lead Toxicity," *Arch. Environ. Health*, **32**, 244–247.

Thompson, J., Jones, D.D., and Beasley, W. H. (1977). "The effect of metal ions on the activity of δ-Aminolevulinic acid dehydratase." *Brit. J. Ind. Med.* **34**, 32–36.

Waldron, H. A. and Stofen, D. (1974). *Subclinical Lead Poisoning.* Academic Press, New York, p. 50.

Willoughby, R. A., McDonald, E., and McSherry, B. J. (1972). "The Interaction of Toxic Amounts of Lead and Zinc Fed to Young Growing Horses," *Vet. Rec.*, **91**, 382–3.

Wilson, E. L., Burger, P. E., and Dowdle, E. B. (1972). "Beef-Liver δ-Aminolevulinic Acid Dehydratase: Purification and Properties," *Environ. J. Biochem.*, **29**, 563–571.

9

ZINC AND PREGNANCY

Sten Jameson

Medical Department, University Hospital, Uppsala, Sweden

1. Introduction   183
2. Zinc Requirements   184
3. Dietary Availability and Absorption   185
4. Zinc Balance in Pregnancy   186
5. Zinc and Malformations   188
6. Zinc Therapy in Pregnancy   188
7. Summary   191
References   192

1. INTRODUCTION

Zinc is essential for the normal growth of plants, animals, and human beings. It is a component of many metalloenzymes and is required for the metabolism of lipids and carbohydrates and the synthesis of protein and nucleic acids.

Human requirements for zinc change throughout life but seem to be high during periods of rapid growth, such as embryonic life, pregnancy, and tissue repair. Zinc-responsive growth failure and sexual infantilism have been observed in both sexes (Prasad, et al., 1976a; Sandstead et al., 1967; Halsted et al., 1972; Ronaghy and Halsted, 1975). Poor growth, anorexia, and hypogeusia have been reported in children, with improvement following zinc treatment (Hambidge et al., 1972).

Acrodermatitis enteropathica is an inherited human zinc-deficiency disorder (Moynahan, 1974). Among the few pregnancies that have been reported in women suffering from this disease, two resulted in malformed infants—one

183

anencephalic fetus and one achondroplastic dwarf (Verburg et al., 1974; Hambidge et al., 1975).

A connection between maternal zinc deficiency and congenital malformations has been postulated (Halsted, 1973; Sever and Emanuel, 1973; Burch et al., 1975; Hambidge et al., 1975). We found that abnormally low serum zinc values in pregnancy were correlated with increased maternal morbidity and risks to the infant. The findings of very low serum zinc values in 6 of 10 mothers who gave birth to infants with congenital malformations support the view that zinc deficiency may act teratogenically in human beings (Jameson, 1976).

A human adult contains a total zinc pool of about 2 g (Vallee, 1959). A weight gain of 12 to 13 kg during pregnancy means that an additional 350 to 375 mg Zn is needed. These requirements—the normal and the extra—have to be met to achieve a positive zinc balance. It may be important to avoid a negative zinc balance because zinc seems to counteract the toxic effects of Hg, Cd (Ferm and Carpenter, 1967; Gale, 1973; Yamane et al., 1977; Parzyck et al., 1978), Pb (Thomasino et al., 1977), and alcohol (Yunice and Lindeman, 1977) in experimental animals.

Obviously more knowledge is needed about the role of zinc in human pregnancy.

2. ZINC REQUIREMENTS

Zinc is lost mainly in the feces (Cotzias et al., 1962) but also via sweat, urine, nails, hair, skin (Schraer and Calloway, 1974; Molin and Wester, 1976), lochia, and mother's milk (Berfenstam, 1952). A positive correlation has been observed between fecal dry matter and excretion of zinc (Schraer and Calloway, 1974). This has been confirmed to be a matter of fiber content (Reinhold et al., 1976); high fiber content may induce a negative mineral balance when the dietary metal content is low.

The calculated daily loss of zinc from normal epidermis via desquamation amounts to 20 to 40 μg/24 hr (Molin and Wester, 1976). A mean daily total dermal loss of 2.8 mg, however, was registered in zinc balance studies in pregnant teenagers (Schraer and Calloway, 1974).

It has been suggested that the loss of zinc in sweat amounts to 2 to 3 mg/day when sweating is excessive (Prasad et al., 1963b). The zinc in urine normally ranges between 0.4 and 0.6 mg/24 hr. In addition, endogenous zinc losses in young women have been calculated at about 1.6 mg/24 hr (Hess et al., 1977a).

Obviously these data will vary according to climatological factors, dietary habits and contents, lactation habits, and other variables. These calculated daily zinc requirements must be added to the 350 to 375 mg Zn needed for the weight gain of 12 to 13 kg during a normal pregnancy.

Fetal requirements for zinc increase rapidly during the first trimester, sevenfold between the 31st and 35th days. Hepatic zinc did not increase per gram of tissue during the second and third trimesters, but owing to the increase in liver size during this interval there was more than a 50-fold increase in total hepatic zinc (Chaube et al., 1973). On the other hand, a decrease in liver zinc with increasing gestational age was reported (Casey and Robinson, 1978), but the liver samples were taken from abnormal gestations and the data may instead indicate a negative zinc balance or deficiency.

A daily retention of 750 μg Zn has been calculated to meet the requirements during pregnancy and to cover the needs of the products of conception (Sandstead, 1973). These figures seem, however, to be too low.

3. DIETARY AVAILABILITY AND ABSORPTION

The major dietary sources of zinc are foods of animal origin (Murphy et al., 1975; Freeland and Cousins, 1976).

Estimations of the zinc contents of several diets have been summarized by Halsted et al. (1974). A study of the daily dietary intakes of zinc in American college women showed a mean value of 13.8 mg Zn (White, 1976). When the daily intakes of zinc in New Zealand women were analyzed using duplicate diets, a mean daily intake of 8.9 mg Zn was found in noninstitutional diets and a daily mean energy intake of 7.7 kJ (Guthrie and Robinson, 1977). A similar study of Swedish pensioners showed a mean daily intake of 8.2 mg Zn (Abdulla and Nordén, 1974).

We studied 20 primigravidae, mean age 22 (range, 18 to 27 years). A diet history was recorded in the 14th week of gestation and was compared to a 3-day dietary recall in the 36th week of gestation (Jameson et al., 1977b). We found a close correlation between data from diet history and dietary recall. The mean energy intake was 7.9 kJ (range, 5.1 to 11.6); the protein intake, 78 g (range, 51 to 104); the zinc intake, 9.4 mg (range, 6.1 to 13.3); the iron intake, 11.2 mg (range, 8.2 to 14.9); and the calcium intake, 1.3 g (range, 0.7 to 1.8). Zinc content parallelled protein and energy intake.

The Food and Nutrition Board of the National Academy of Sciences recommends a daily dietary intake of 20 mg Zn/day in pregnancy and 25 mg/day during lactation (National Research Council, 1974). Many factors affect zinc absorption, and requirements are based on assumed losses and availability (World Health Organization, 1973). Zinc availability from ordinary diets, especially in pregnancy, is an unknown factor, as are the size of body zinc stores, if any, and the eventual adjustment of zinc absorption to meet the increased needs.

Phytate from cereals and vegetables binds zinc and is possibly the most important factor that decreases zinc availability (Reinhold et al., 1973). Phytate in the presence of excess calcium binds zinc in complexes (Fox, 1970).

With increasing fecal dry matter, especially high fiber content, there is increased metal excretion (Schraer and Calloway, 1974; Reinhold et al., 1976). Zinc from plant products is less readily absorbed than that from animal products (O'Dell, 1969).

Although absorption of zinc seems to take place in the proximal part of the small intestine, the exact, controlled(?) mechanism in human beings is not clear (Halsted et al., 1974; Burch et al., 1975). Low serum zinc levels have been registered in patients with malabsorption (MacMahon et al., 1968; Caggiano et al., 1969; Sandstead et al., 1970; Walker et al., 1973).

Low zinc concentrations were found in the hair of patients with celiac disease (Amador et al., 1975), and low serum zinc in celiac disease was observed earlier (Hellwege, 1971; Walker et al., 1973). The signs and symptoms of zinc deficiency in celiac disease were reported by us. Infertility of long standing was reversible after zinc therapy was instituted in one woman, but later she aborted spontaneously. She conceived again and gave birth to a normal baby, with slight signs of dysmaturity (Jameson, 1976). Celiac disease with infertility that is reversible by the institution of a gluten-free diet has been recorded in both sexes (Foss, 1962; Morris et al., 1970; Merianos, 1975).

Reduced intestinal zinc absorption was found in patients treated for obesity by jejunoileostomy (Andersson et al., 1976).

The following groups probably have the greatest risk of developing zinc deficiency in pregnancy: women with low protein intake and high levels of dietary phytates; women with dermatitis herpetiformis and celiac disease or other types of malabsorption; women who have undergone resections of portions of the gut or ventricle; women receiving drugs that interfere with absorption, such as thiol-containing drugs, chelating agents like penicillamine (Ekberg et al., 1974) and EDTA (Spencer et al., 1965), and possibly tetracyclines, antacids, anticonvulsives, tuberculostatics, and drugs that increase excretion (e.g., thiazides); women with multiple pregnancies; multiparae; and teenagers who have not yet completed their own growth.

4. ZINC BALANCE IN PREGNANCY

Few data have been collected on this problem, but much can be learned from newly published studies of nonpregnant women.

In a study of young, nonpregnant university women it was found that, on a diet of usual foods, an average intake of 11.5 mg Zn/day equalled the fecal (11.4 mg) and urinary (0.17 mg) excretions (White and Gynne, 1971). This must mean that the zinc balance was negative, if all other possible losses are taken into account.

Nonpregnant women on an experimental diet containing only 0.17 mg Zn/day had endogenous zinc losses of about 1.6 mg the first day of the study. Their usual zinc intake before the study had been about 10 mg/day. During the study serum zinc dropped 21 to 47% in 1 month. It was suggested that accessible zinc stores were not extensive and that depletion of these stores caused the fall in serum zinc (Hess et al., 1977a). In 7 of 10 participants symptoms of sore throat, dermatologic changes, diarrhea, and taste complaints developed during the study (Hess et al., 1977b).

Healthy young women with an ordinary zinc dietary intake of 11 to 11.3 mg/day were put on an experimental diet with low calcium, magnesium, and phosphorus intakes, supplemented with a trace mineral solution to 10.7 mg Zn/day. Absorption was measured with stable isotopes and averaged 38% of the given zinc, or 4.1 mg Zn daily (King et al., 1978). Such high absorption was achieved during optimal conditions with an experimental, semipurified diet.

Four pregnant teenagers were studied during their last trimester. The controlled diet consisted of natural foods supplemented with mineral salts and vitamins and contained 26, 29, or 32 mg Zn daily. Most of the protein was of animal origin. The mean zinc balance was 3.6 mg/day, intake minus fecal, urinary, and dermal losses. The apparent net absorption was 25% of the mean intake of 29.4 mg Zn/day (Schraer and Calloway, 1974).

Serum or plasma zinc concentrations decrease during human pregnancy (Berfenstam, 1952; Rothe et al., 1960; Halsted et al., 1968; Hähn et al., 1972; Hambidge and Droegemueller, 1974; Jameson, 1976). This decline is to some extent secondary to blood volume expansion, plasma protein changes, and hormonal effects. The decrease is due mostly to diminished albumin binding of zinc, however, as well as to a decreased affinity of albumin for zinc (Giroux et al., 1976).

The lowered concentrations of zinc in hair found during late pregnancy (Hambidge and Droegemueller, 1974) may indicate a negative zinc balance. Such changes seem to be accentuated in women of low socioeconomic status (Sarram et al., 1969).

Urinary zinc excretion gradually decreases in a zinc-deficient situation, judging from the results of Hess et al. (1977a).

We found that women with abnormal deliveries, especially inefficient labor, atonic bleeding, and delivery outside normal term, had lower serum zinc during early pregnancy than women with normal deliveries. Seven gravidae studied before and during oral zinc sulfate therapy showed low serum zinc and very low urinary zinc excretion; both values rose significantly after 1 week's therapy. Three women spontaneously reported an improvement in taste acuity. Normotensive, nonanaemic women who, after normal pregnancies, were delivered normally of normal babies at the normal time, showed only a slight, if any, reduction of serum zinc levels throughout pregnancy, as compared to normal, fertile women not on contraceptive pills (Jameson, 1976).

These and other data seem to indicate the presence of a zinc-deficiency syndrome in pregnancy, which includes increased maternal morbidity as well as higher risks to the fetus, and especially abnormal taste sensations, prolonged gestation, inefficient labor, atonic bleeding, and postmaturity.

5. ZINC AND MALFORMATIONS

A possible correlation between human maternal zinc deficiency and congenital malformations, especially anomalies of the central nervous system, has been postulated. As mentioned in the Introduction, two infants with severe malformations have been born to mothers with acrodermatitis enteropathica, a zinc-deficiency disorder. We found 10 cases of congenital malformations; 6 of the 10 mothers showed the lowest serum zinc values that were recorded in our series (Jameson, 1976).

In laboratory animals zinc deficiency results in reduced fertility, and the incidence of fetal malformations increases (Blamberg et al., 1960; Hurley and Swenerton, 1966; Warkany and Petering, 1973; Adeloye and Warkany, 1976). Preimplantation rat eggs developed abnormally after 3 days of maternal dietary zinc deficiency (Hurley and Shrader, 1975).

The highest serum zinc concentrations in women with normal menstrual cycles were found during the time of ovulation (Hähn et al., 1972). This variation was abolished when using contraceptive pills.

An interaction of zinc with copper has been documented in animal experiments (Suttle and Mills, 1966). The fact that two infants with congenital malformations were born to mothers who had used copper-containing intrauterine contraceptive devices (Barrie, 1976) seems to indicate that tubal zinc concentration is a crucial factor in this respect.

One mother, in a series of 312, gave birth to an infant with myelomeningocele. During pregnancy her serum zinc values dropped dramatically, from 14.7 μmol/l in the 14th week of gestation to 8.0 μmol/l in the 36th week. The infant had one of the lowest serum zinc values (10.0 μmol/l) that we have recorded in cord blood in a series of 105 infants. The records provide no other possible explanation for this severe, lethal congenital malformation (Jameson et al., 1977a).

These data support the suspicion that human zinc deficiency may act teratogenically, directly or perhaps indirectly via other toxic agents, (e.g., metals such as Hg, Cd, and Pb, or alcohol).

6. ZINC THERAPY IN PREGNANCY

It is not known whether zinc absorption in women can adjust to meet the needs of pregnancy, although zinc absorption in rats increases during pregnancy and further during lactation (Davies and Williams, 1975).

Nonpregnant young women seem to have easily depletable zinc stores, when given a zinc-deficient diet (Hess et al., 1977a). Moreover, abnormally low serum zinc levels during pregnancy seem to indicate a state of zinc deficiency with increased risks to mother and fetus (Jameson, 1976).

Taste disturbances with cravings and aversions are common in pregnancy (in our country, between 20 and 30% of all pregnancies), and the taste threshold is raised during pregnancy (Hansen and Langer, 1935). Metallic ions—Zn, Cu, and Ni—seem to be essential to the early steps of taste transmission (Henkin et al., 1969). Taste acuity in children with hypogeusia and poor growth was restored to normal by zinc therapy (Hambidge et al., 1972). In one girl with signs of zinc deficiency a rapid improvement of pica occurred when zinc therapy was instituted (Hambidge and Silverman, 1973).

We treated 7 of 20 women with low serum zinc and low urinary zinc excretion with zinc sulfate (Solvezinc®), 0.2 g twice daily during the third trimester, corresponding to a dose of 90 mg Zn/day. Serum zinc and urinary zinc excretion increased significantly. Three women reported spontaneously an improvement in taste acuity. Whereas 6 of 13 untreated women in the control group suffered severe hemorrhage at delivery, with signs of uterine atony, none of the 7 in the treatment group had abnormal bleeding. One of the 13 mothers in the control group delivered in the 42nd , and 4 in the 43rd, week of gestation or later, and four infants were dysmature. Two women in the treatment group delivered normal infants in the 42nd week; one infant delivered in the 40th week showed slight signs of dysmaturity (Jameson, 1976).

The zinc therapy given to women with lowered serum zinc values seemed to reduce complications at delivery in this selected group. Side effects, except for nausea in one case, were not registered.

Women in underdeveloped countries give birth to small-for-date infants in high frequency. Most of these women have cereal-based diets, which provide only around 6 mg Zn/day. In a pilot study from India supplements of 0.1 g zinc sulfate were given during the third trimester of pregnancy. The study was discontinued, however, when four consecutive births were premature and one infant was stillborn (Kumar, 1976).

From rat studies it was suggested that moderately high levels of zinc in the diet might be associated with harmful effects on the course of pregnancy (Kumar, 1976). We have not been able to confirm this hypothesis in women.

It must be borne in mind that geographical differences exist in trace element concentrations in the human body (Forssén, 1972). Also, a possible interaction between zinc and other metals must always be considered. For instance, there was a report of a woman with nonresponsive celiac disease, who developed severe anemia, neutropenia, and hypocupremia after 14 months of zinc sulfate therapy in a dose of 0.66 g daily. She recovered after the withdrawal of zinc therapy, together with transfusions and copper sulfate substitution (Porter et al., 1977).

In further studies we recorded serum zinc values for 312 randomly selected women in the 14th week of gestation. According to date of birth, every other woman with serum zinc lower than 10 μmol/l was assigned randomly to oral zinc sulfate therapy or a control group. Zinc therapy (Solvezinc®), 0.2 g daily, corresponding to 45 mg Zn/day, was given throughout pregnancy to 64 women; 69 belonged to the control group with initially low serum zinc; and 179 were controls with initial serum zinc values $\geqslant$ 10 μmol/l. The results are listed in Table 1.

Prolonged gestations, inefficient labors, and postmature infants were fewer in the zinc treatment group than in the low-zinc control group. The frequency of normal deliveries of normal infants was higher, also compared to the control group with high initial serum zinc.

One boy with hypospadia glandis was born in the treatment group. Two infants in the low-zinc control group had malformations, one luxatio coxae, one neurological defects with bilateral median nerve palsy and fits. One infant in the high-zinc control group had a myelomeningocele. The serum zinc in cord blood was one of the lowest we registered, and maternal serum zinc values dropped dramatically during pregnancy, as described earlier. Except for these findings there were no other abnormal signs or symptoms during pregnancy.

Table 1. Some Relevant Data about 312 Randomly Selected Women (see text)[a]

| | Group | | |
| | Zinc Therapy | Control Serum Zinc < 10 μmol/1 | Control Serum Zinc $\geqslant$ 10 μmol/1 |
Item			
Number of women	64	69	179
Normal birth and normal infant	40 (63%)	33 (48%)	98 (55%)
Labor > 20 hr	2	7	10
Gestation $\geqslant$ 42 weeks	13	19	37
Postmaturity	1	6	8
Malformations— all diagnoses	2	2	10
Caesarean section	4	7	15
Vacuum extraction	6	7	15
Bleeding at delivery (ml, mean)	307	328	
Duration of labor (hr, mean)	10.0	10.5	

[a] Serum zinc mean 10.3 μmol/1 in the 14th week of gestation.

We found no difference in maternal hemoglobin concentrations or serum copper concentrations in the 36th week of gestation, birth weights, or frequency of immaturity between the zinc treatment group and the low-zinc control group. No side effects, except nausea in a few cases, were registered.

This general study was divided into substudies. A random sample of women came to an extra control situation in the 36th week of gestation. Of 136 women who received no zinc therapy, in 22% the serum zinc level remained unchanged or increased, perhaps indicating zinc equilibrium or positive balance, and 63% of those with a spontaneously positive zinc balance delivered normal infants normally, irrespective of the initial serum zinc level. Untreated women with initial serum zinc levels above the mean, but with decreasing serum zinc, also delivered normally in 61% of the cases. By contrast, women with initial serum zinc below the mean and with subsequently decreasing serum zinc levels, which might indicate zinc disequilibrium or negative balance, delivered normal infants normally in only 26% of the cases, a difference highly significant by the chi-square test ($p < .001$).

These results mean that women in the 14th week of gestation with zinc levels below the mean, which subsequently decreased even further, suffered from complications (all obstetrical and/or fetal diagnoses) in very high frequency, 74%. This figure should be compared to 37% of complications among women in the zinc therapy group, and with 37% of complications among women who spontaneously achieved zinc equilibrium.

There seems to be no doubt that it is beneficial to achieve a positive zinc balance in pregnancy, and that zinc sulfate therapy in a dosage that gives 45 mg Zn/day reduces the frequency of maternal and fetal complications, if initial serum zinc levels indicate zinc deficiency (Jameson et al., 1977a).

In another study we measured taste acuity by an electric stimulation method, electrogustatometry. Women with high taste thresholds ($\geq 10 \ \mu A$) had significantly lower serum zinc values, and also lower serum copper, than women with low taste threshold ($< 10 \ \mu A$). Women with pica and cravings had significantly higher taste thresholds than the other women. The frequency of pica and cravings was reduced in the zinc therapy group, compared to the control group (Jameson et al., 1977c).

These findings may indicate that the abnormal taste acuity and aversions often experienced by pregnant women are a disorder of metal deficiency, and that zinc deficiency plays a central role.

7. SUMMARY

The normal weight gain in pregnancy entails a need for an additional 350 to 375 mg Zn. The zinc content in normal food parallels the protein and energy intake. Zinc availability is an uncertain factor but in any case, is reduced when phytate intake is high.

Zinc absorption is reduced in celiac disease and after bypass operations. It is not known whether absorption adjusts for increasing needs in human beings. Although zinc depots are not well defined, they seem to be easily depleted in young women who are fed a zinc-deficient diet.

Zinc losses increase with increasing fecal mass and fiber content. Serum zinc and urinary zinc excretion decrease when the zinc balance is negative.

Women with initially low and subsequently decreasing serum zinc levels were found to suffer from complications at delivery in high frequency. For women with initially low serum zinc levels zinc therapy reduced the frequency of complications; normal deliveries of normal infants were as frequent as among women who spontaneously achieved zinc equilibrium or increasing levels of zinc.

Congenital malformations were recorded among infants born to mothers with very low or decreasing serum zinc levels.

Diminished taste acuity was found in women with low serum zinc and copper. Women with pica and cravings had higher taste thresholds than the other women under study. Zinc therapy reduced the frequency of pica and cravings. It is suggested that these data are compatible with the presence of a zinc-deficiency syndrome in pregnancy, which includes increased maternal morbidity, abnormal taste sensations, prolonged gestation, inefficient labor, atonic bleeding, and increased risks to the fetus, especially postmaturity.

REFERENCES

Abdulla, M. and Nordén, Å. (1974). "Zink i Kosten." In B. Zederfeldt, Ed., *Symposium om Zink*. AB Tika, Lund, pp. 55–59.

Adeloye, A. and Warkany, J. (1976). "Experimental Congenital Hydrocephalus," *Child's Brain*, 2, 325–360.

Amador, M., Pena, M., Garcia-Miranda, A., Gonzales, A., and Hermelo, M. (1975). "Low Hair-Zinc Concentrations in Acrodermatitis Enteropathica," *Lancet*, 1, 1379.

Andersson, K. -E., Bratt, L., Dencker, H., and Lanner, E. (1976). "Some Aspects of the Intestinal Absorption of Zinc in Man," *Eur. J. Clin. Pharmacol.*, 9, 423–428.

Barrie, H. (1976). "Congenital Malformation Associated with Intrauterine Contraceptive Device," *Br. Med. J.*, 1, 488–490.

Berfenstam, R. (1952). "Studies of Blood Zinc: a Clinical and Experimental Investigation into the Zinc Content of Plasma and Blood Corpuscles with Special Reference to Infancy," *Acta Paediatr.*, 41, Suppl. 87, 3–97.

Blamberg, D. L., Blackwood, U. B., Supplee, W. C., and Combs, G. F. (1960).

"Effect of Zinc Deficiency in Hens on Hatchability and Embryonic Development," *Proc. Soc. Exp. Biol. Med.,* **104**, 217–220.

Burch, R. E., Hahn, K. K. J., and Sullivan, J. F. (1975). "Newer Aspects on the Roles of Zinc, Manganese, and Copper in Human Nutrition," *Clin. Chem.,* **21**, 501–520.

Caggiano, V., Schnitzler, R., Strauss, W., Baker, R. K., Carter, A. C., Josephson, A. S., and Wallack, S. (1969). "Zinc Deficiency in a Patient with Retarded Growth, Hypogonadism, Hypogammaglobulinemia and Chronic Infection," *Am. J. Med. Sci.,* **257**, 305–319.

Casey, C. E. and Robinson, M. F. (1978). "Copper, Manganese, Zinc, Nickel, Cadmium and Lead in Human Foetal Tissues," *Br. J. Nutr.,* **39**, 639–647.

Chaube, S., Nishimura, H., and Swinyard, C. A. (1973). "Zinc and Cadmium in Normal Human Embryos and Fetuses," *Arch. Environ. Health,* **26**, 237–240.

Cotzias, G. C., Borg, D. C., and Selleck, B. (1962). "Specificity of Zinc Pathway through the Body: Turnover of Zn in the Mouse," *Am. J. Physiol.,* **202**, 359–363.

Davies, N. T. and Williams, R. B. (1975). "Zinc Absorption in Pregnancy and Lactation," *Proc. Nutr. Soc.,* **35**, 5A–6A.

Ekberg, M., Jeppsson, J. -O., and Denneberg, T. (1974). "Penicillamine Treatment of Cysteinuria," *Acta Med. Scand.,* **195**, 415–419.

Ferm, V. H. and Carpenter, S. J. (1967). "Teratogenic Effect of Cadmium and Its Inhibition by Zinc," *Nature,* **216**, 1123.

Forssén, A. (1972). "Occurrence of Ba, Br, Ca, Cd, Cs, Cu, K. Mn, Ni, Sn, Sr, Y and Zn in the Human Body," *Ann. Med. Exp. Biol. Fenn.,* **50**, 99–196.

Foss, G. L. (1962). "Infantilism to Fatherhood in the Malabsorption Syndrome," *Br. Med. J.,* **2**, 368–371.

Fox, M. R. S. (1970). "The Status of Zinc in Human Nutrition," *World Rev. Nutr. Diet.,* **12**, 208–226.

Freeland, J. H. and Cousins, R. J. (1976). "Zinc Content of Selected Foods," *J. Am. Diet. Assoc.,* **68**, 526–529.

Gale, T. F. (1973), "The Interaction of Mercury with Cadmium and Zinc in the Mammalian Embryonic Development," *Environ. Res.,* **6**, 95–105.

Giroux, E., Schechter, P. J., and Schoun, J. (1976). "Diminished Albumin Binding of Zinc in Serum of Pregnant Women," *Clin. Sci. Mol. Med.,* **51**, 545–549.

Guthrie, B. E.and Robinson, M. F. (1977). "Daily Intakes of Manganese, Copper, Zinc and Cadmium by New Zealand Women," *Br. J. Nutr.,* **38**, 55–63.

Hähn, N., Paschen, K., and Haller, J. (1972). "Serum Levels of Copper, Iron, Magnesium, Calcium and Zinc in Women with a Normal Menstrual Cycle, under Treatment with Oral Contraceptives and in Pregnancy," *Arch. Gynaekol.,* **213**, 176–186.

Halsted, J. A. (1973). "Zinc Deficiency and Congenital Malformations," *Lancet,* **1**, 1323.

Halsted, J. A., Hackley, B. M., and Smith, J. C. (1968). "Plasma-Zinc and Copper in Pregnancy and after Oral Contraceptives," *Lancet,* 2, 278.

Halsted, J. A., Ronaghy, H. A., Abadi, P., Haghshenass, M., Amirhakemi, G. H., Barakat, R. M., and Reinhold, J. G. (1972). "Zinc Deficiency in Man: The Schiraz Experiment," *Am. J. Med.,* 53, 277-284.

Halsted, J. A., Smith, J. C., and Irwin, M. I. (1974). "A Conspectus of Research on Zinc Requirements of Man," *J. Nutr.,* 104, 345-378.

Hambidge, K. M. and Droegemueller, W. (1974). "Changes in Plasma and Hair Concentrations of Zinc, Copper, Chromium, and Manganese during Pregnancy," *Obstet. Gynecol.,* 44, 666-672.

Hambidge, K. M. and Silverman, A. (1973). "Pica with Rapid Improvement after Dietary Zinc Supplementation." *Arch. Dis. Child.,* 48, 567-568.

Hambidge, K. M., Hambidge, C., Jacobs, M., and Baum, J. D. (1972). "Low Levels of Zinc in Hair, Anorexia, Poor Growth, and Hypogeusia in Children," *Pediatr. Res.,* 6, 868-874.

Hambidge, K. M., Neldner, K. H., and Walravens, P. A. (1975). "Zinc, Acrodermatitis Enteropathica, and Congenital Malformations," *Lancet,* 1, 577-578.

Hansen, R. and Langer, W. (1935). "Über Geschmacksveränderungen in der Schwangerschaft," *Klin. Wochenschr.,* 14, 1173-1176.

Hellwege, H. H. (1971). "Serum Zinc Level and Its Variations in Some Diseases in Infancy," *Monatsschr. Kinderheilkd.,* 119, 37-41.

Henkin, R. I., Graziadei, P. P. G., and Bradley, D. F. (1969). "The Molecular Basis of Taste and Its Disorders," *Ann. Intern. Med.,* 71, 791-821.

Hess, F. M., King, J. C., and Margen, S. (1977a). "Zinc Excretion in Young Women on Low Zinc Intakes and Oral Contraceptive Agents," *J. Nutr.,* 107, 1610-1620.

Hess, F. M., King, J. C., and Margen, S. (1977b). "Effect of Low Zinc Intake and Oral Contraceptive Agents on Nitrogen Utilization and Clinical Findings in Young Women," *J. Nutr.,* 107, 2219-2227.

Hurley, L. S. and Shrader, R. E. (1975). "Abnormal Development of Preimplantation Rat Eggs after Three Days of Maternal Dietary Zinc Deficiency," *Nature,* 254, 427-429.

Hurley, L. S. and Swenerton, H. (1966). "Congenital Malformations Resulting from Zinc Deficiency in Rats," *Proc. Soc. Exp. Biol. Med.,* 123, 692-696.

Jameson, S. (1976). "Effects of Zinc Deficiency in Human Reproduction," *Acta Med. Scand.,* 197, Suppl. 593, 3-89.

Jameson, S., Bárány, S., Ohlsson, A., and Ursing, I. (1977a). "Zinc Deficiency and Pregnancy—Results of Therapy," *Hygiea,* 86(4), 418 (abstr.).

Jameson, S., Svärdström, G., and Ralling, M. (1977b). "Zinc and Iron Content of Food in Pregnancy—a Risk of Deficiency?" *Hygiea,* 86(4), 409 (abstr.).

Jameson, S., Widström, Å, Ursing, I., Bárány, S., and Ohlsson, A. (1977c).

"Copper, Zinc, and Taste Threshold Measurements in Pregnancy," *Hygiea,* 86(4), 482 (abstr.).

King, J. C., Raynolds, W. L., and Margen, S. (1978). "Absorption of Stable Isotopes of Iron, Copper, and Zinc during Oral Contraceptive Use," *Am. J. Clin. Nutr.,* **31,** 1198–1203.

Kumar, S. (1976). "Effect of Zinc Supplementation on Rats during Pregnancy," *Nutr. Rep. Int.,* **13,** 33–36.

MacMahon, R. A., Parker, M. L., and McKinnon, M. -C. (1968). "Zinc Treatment in Malabsorption," *Med. J. Aust.,* **2,** 210–212.

Merianos, P. (1975). "Reversible Infertility in Male Coeliac Patients," *Br. Med. J.,* **1,** 316–317.

Molin, L. and Wester, P. O. (1976). "The Estimated Daily Loss of Trace Elements from Normal Skin by Desquamation," *Scand. J. Clin. Lab. Invest.,* **36,** 679–682.

Morris, J. S., Adjukiewicz, A. B., and Read, A. E. (1970). "Coeliac Infertility? An Indication for Dietary Gluten Restriction?" *Lancet,* **1,** 213–214.

Moynahan, E. J. (1974). "Acrodermatitis Enteropathica: a Lethal Inherited Human Zinc-Deficiency Disorder," *Lancet,* **2,** 399–400.

Murphy, E. W., Willis, B. W., and Watt, B. (1975). "Provisional Tables on the Zinc Content of Foods," *J. Am. Diet. Assoc.,* **66,** 345–355.

National Research Council: Food and Nutrition Board (1974). *Recommended Dietary Allowances,* 8th ed. National Academy of Sciences, Washington, D.C.

O'Dell, B. L. (1969). "Effect of Dietary Components upon Zinc Availability: A Review with Original Data," *Am. J. Clin. Nutr.,* **22,** 1315–1322.

Parzyck, D. C., Shaw, S. M., Kessler, W. V., Vetter, R. J., Van Sickle, D. C., and Mayes, R. A. (1978). "Fetal Effects of Cadmium in Pregnant Rats on Normal and Zinc Deficiency Diets," *Bull. Environ. Contam. Toxicol.,* **19,** 206–214.

Porter, K. G., McMaster, D., Elmes, M. E., and Love, A. H. G. (1977). "Anaemia and Low Serum-Copper during Zinc Therapy," *Lancet,* **2,** 774.

Prasad, A. S., Miale, A., Farid, Z., Sandstead, H. H., and Schulert, A. R. (1963a). "Zinc Metabolism in Patients with the Syndrome of Iron Deficiency Anemia, Hepatosplenomegaly, Dwarfism and Hypogonadism," *J. Lab. Clin. Med.,* **61,** 537–549.

Prasad, A. S., Schulert, A. R., Sandstead, H. H., Miale, A., and Farid, Z. (1963b). "Zinc, Iron and Nitrogen Content of Sweat in Normal and Deficient Subjects," *J. Lab. Clin. Med.,* **62,** 84–89.

Reinhold, J. G., Nasr, K., Lahimgarzadeh, A., and Hedayati, H. (1973). "Effects of Purified Phytate and Phytate-Rich Bread upon Metabolism of Zinc, Calcium, Phosphorus, and Nitrogen in Man," *Lancet,* **1,** 283–288.

Reinhold, J. G., Faradji, B., Abadi, P., and Ismail-Beigi, F. (1976). "Decreased

Absorption of Calcium, Magnesium, Zinc and Phosphorus by Humans Due to increased Fiber and Phosphorus Consumption as Wheat Bread," *J. Nutr.,* **106**, 493–503.

Ronaghy, H. A. and Halsted, J. A. (1975). "Zinc Deficiency Occurring in Females: Report of Two Cases," *Am. J. Clin. Nutr.,* **28**, 831–836.

Rothe, K., Piskazek, K., and Bilek, K. (1960). "Serum Zinc in Normal Pregnancy and in Early and Late Toxicosis," *Arch. Gynaekol.,* **192**, 349–364.

Sandstead, H. H. (1973). "Zinc Nutrition in the United States," *Am. J. Clin. Nutr.,* **26**, 1251–1260.

Sandstead, H. H., Prasad, A. S., Schulert, A. S., Farid, Z., Miale, A., Bassily, S., and Darby, W. J. (1967). "Human Zinc Deficiency, Endocrine Manifestations and Response to Treatment," *Am. J. Clin. Nutr.,* **20**, 422–442.

Sandstead, H. H., Burk, R. F., Booth, G. H., and Darby, W. J. (1970). "Current Concepts on Trace Minerals: Clinical Considerations," *Med. Clin. N. Am.,* **54**, 1509–1531.

Sarram, M., Younessi, M., Khorvash, P., Kfoury, G. A., and Reinhold, J. G. (1969). "Zinc Nutrition in Human Pregnancy in Fars Province, Iran," *Am. J. Clin. Nutr.,* **22**, 726–732.

Schraer, K. K. and Calloway, D. H. (1974). "Zinc Balance in Pregnant Teenagers," *Nutr. Metab.,* **17**, 205–212.

Sever, L. E. and Emanuel, I. (1973). "Is There a Connection between Maternal Zinc-Deficiency and Congenital Malformations of the Central Nervous System in Man?" *Teratology,* **7**, 117–118.

Spencer, H., Vankinscott, V., Lewin, I., and Samachson, J. (1965). "Zinc-65 Metabolism during Low and High Calcium Intake in Man," *J. Nutr.,* **86**, 169–177.

Suttle, N. F. and Mills, C. F. (1966). "Studies on the Toxicity of Copper to Pigs," *Br. J. Nutr.,* **20**, 135–148.

Swenerton, H. and Hurley, L. S. (1971). "Teratogenic Effects of a Chelating Agent and Their Prevention by Zinc," *Science,* **173**, 62–64.

Thomasino, J. A., Zuroweste, E., Brooks, S. M., Petering, H. G., Lerner, S. I., and Finelli, V. N. (1977). "Lead, Zinc, and Erythrocyte Delta-Aminolevulinic Acid Dehydratase: Relationships in Lead Toxicity," *Arch. Environ. Health,* **32**, 244–247.

Vallee, B. L. (1959). "Biochemistry, Physiology and Pathology of Zinc," *Physiol. Rev.,* **39**, 443–490.

Verburg, D. J., Burd, L. I., Hoxtell, E. O., and Merrill, L. K. (1974). "Acrodermatitis Enteropathica and Pregnancy," *Obstet. Gynecol.,* **44**, 233–237.

Walker, B. E., Dawson, J. B., Kelleher, J., and Sosowsky, M. S. (1973). "Plasma and Urinary Zinc in Patients with Malabsorption Syndromes or Hepatic Cirrhosis," *Gut,* **14**, 943–948.

Warkany, J. and Petering, H. G. (1973). "Congenital Malformations of the Brain

Produced by Short Zinc Deficiencies in Rats," *Am. J. Ment. Defic.*, **77**, 645–653.

White, H. S. (1976). "Zinc Content and the Zinc-to-Calorie Ratio of Weighed Diets," *J. Am. Diet. Assoc.*, **68**, 243–245.

White, H. S. and Gynne, T. N. (1971). "Utilization of Inorganic Elements by Young Women Eating Iron-Fortified Foods," *J. Am. Diet. Assoc.*, **59**, 27–33.

World Health Organization. (1973). *Trace Elements in Human Nutrition.* Technical Report Series 532, pp. 9–15.

Yamane, Y., Fukina, H., and Imagawa, M. (1977). "Suppressive Effect of Zinc on the Toxicity of Mercury," *Chem. Pharm. Bull.*, **25**, 1509–1518.

Yunice, A. A. and Lindeman, R. D. (1977). "Effect of Ascorbic Acid and Zinc Sulfate on Ethanol Toxicity and Metabolism," *Proc. Soc. Exp. Biol. Med.*, **154**, 146–150.

10

TUMORGENESIS AND ZINC

Jerry L. Phillips
Mary K. Kindred

University of Texas at San Antonio, Division of Allied Health and Life Sciences, San Antonio, Texas

1. Introduction	199
2. Tumorgenesis by Zinc	200
3. Tissue and Blood Serum Levels of Zinc in Malignancy	201
4. Suppression of Tumorgenesis and Tumor Growth by Zinc Deficiency	204
5. Suppression of Tumorgenesis and Tumor Growth by Administration of Excess Zinc	206
Acknowledgment	207
References	208

1. INTRODUCTION

Zinc is a ubiquitous trace metal that has been shown to be essential for normal cell differentiation and growth in both plants and animals (Mikac-Dević, 1970). In recent years, numerous studies on the physiology and biochemistry of zinc have stressed both its importance to normal and abnormal biological systems and the need to understand the precise molecular mechanisms by which zinc exerts its profound effects. Early attempts to elucidate the biological functions of zinc involved primarily the isolation and characterization of zinc-containing proteins, particularly zinc metalloenzymes. These studies have been fruitful, allowing the identification of zinc as an essential constituent of many key cellular enzymes (Vallee and Wacker, 1970). Included are the enzymes of nucleic acid biosynthesis, the RNA polymerases (Auld and Atsuya, 1976; Lattke and Weser,

1976; Petranyi et al., 1977; Wandzilak and Benson, 1977; Falchuk et al., 1977), the DNA polymerases (Slater et al., 1971; Stavrianopoulos et al., 1972; Levinson et al., 1973; Berger et al., 1975), and terminal deoxynucleotidyl transferase (Sabbioni, 1976). Additionally, some researchers, in an attempt to outline a regulatory function for zinc, have studied its ability to inhibit, rather than stimulate or maintain, the activity of cellular enzymes. From numerous reports to date, it is probable that the zinc status of a cell may control certain key cellular functions by regulating the activities of key cellular enzymes. This concept is particularly important to understanding the possible involvement of zinc in certain disease states, such as malignancy. Consequently, this chapter will attempt to detail the relationship between zinc and malignancy by concentrating on four areas: tumorgenesis by zinc, changes in tissue and body fluid levels accompanying malignancy, suppression of tumorgenesis and tumor growth by zinc deficiency, and suppression of tumorgenesis and tumor growth by administration of excess zinc.

2. TUMORGENESIS BY ZINC

The earliest demonstration of metal carcinogenesis was provided by Michalowsky (1926), who found that intratesticular injection of zinc chloride solution in cockerels produced testicular teratomas, when injections were made during periods of high gonadal activity (i.e., the spring). Similar tumors have been induced by intratesticular injection of other zinc salts, such as zinc sulfate (Falin, 1940) and zinc nitrate (Falin, 1941), as well as by intratesticular injection of other metal salts, for example, copper sulfate (Falin and Anissimowa, 1940) and cadmium chloride (Guthrie, 1964). Two researchers have induced testicular teratomas in fowl during seasons not normally associated with gonadal activity, by stimulating gonadal activity either with gonadotrophins (Bagg, 1936) or by changes of photoperiod (Guthrie, 1971) in addition to intratesticular injection of zinc chloride solution.

Attempts have also been made to induce testicular teratomas in mammals, particularly mice and rats, with inconsistent results. Early reports indicated that intratesticular injection of zinc chloride in Wistar rats did not produce tumors (Willis, 1934; Guthrie, 1956). However, Riviere et al. (1960) claimed that similar injections in rats produced seminomas, interstitial cell tumors, and teratomas.

Additionally, one report in the literature indicated that ingestion of zinc chloride dissolved in drinking water could produce mammary carcinomas in rodents (Halme, 1961). This report, however, was later disputed by Roe and Lancaster (1964). Indeed, it is significant that zinc and zinc compounds have not been demonstrated conclusively to be carcinogenic when administered by any route other than intratesticular injection.

Whether a relationship exists between malignancy and zinc ingestion is still uncertain, although reports have appeared that indicate a statistical correlation between possible excessive intake of dietary zinc and various forms of cancer. Stocks and Davies (1964), working in England, reported that the logarithm of the Zn/Cu ratio in soil samples was significantly higher in gardens at houses where an individual had just died of cancer of the stomach after 10 or more years of residence than at houses where an individual had died of something other than cancer. Hirayama (1962) also found a statistical correlation between excessive zinc intake and stomach cancer in Japan, while McGlashan (1967) reported a similar relationship between zinc intake and esophageal cancer in Africa. Similarly, Chahovitch (1955) showed that injection of zinc sulfate accelerated the growth of experimental sarcomas.

3. TISSUE AND BLOOD SERUM LEVELS OF
ZINC IN MALIGNANCY

It has been observed by numerous investigators that tissue and body fluid levels of essential trace elements, including zinc, are different in malignant as compared to normal systems. Such changes in the levels of biologically important metals may indicate that they play some role in the etiology of or the metabolic alterations that exist in malignancy. It is unfortunate, however, that reports of zinc values in fluids and tissues of cancer patients have been highly variable, perhaps because serum zinc levels exhibit diurnal variation and are dependent on the age and sex of the patient as well as on the time elapsing between eating and sampling (Burr, 1974). Nevertheless, certain trends that exist in serum and tissue zinc levels in malignancy merit discussion.

Wright and Dormandy (1972) found that malignant tissue from liver carcinomas had significantly (on the average, 44%) lower zinc levels than normal tissue. Additionally, they reported finding a steep zinc concentration gradient when samples were analyzed moving from the necrotic core of the tumor (the region of lowest zinc concentration) to the apparently uninvaded surrounding tissue (the region of highest zinc concentration). On the basis of their data, they postulate that the increase in zinc in tissue adjacent to and away from the tumor may be part of a defense mechanism against the invasion by malignant cells. This speculation is based on the known relationship between zinc and wound healing (see Van Rij and Pories, this volume). Kew and Mallett (1974) reported similar data. They found that in normal liver tissue the zinc concentration was 78 ± 41 µg/g (wet weight), while the zinc concentration of the primary liver cancer itself was 18 ± 7 µg/g. Additionally, liver tissue in which metastases exist has been shown to have a higher zinc content than either normal liver tissue or the tumor itself (Kaltenbach and Egen, 1968). Olson et al. (1958) and McBean et al. (1972) both reported increased liver zinc concentrations in patients with a variety of cancers, including bronchiogenic, gastric, and nasopharyngeal cancer.

Cancerous mammary tissue has also been reported to have a higher zinc concentration than does normal tissue (Mulay et al., 1971). In the study of Schwartz et al. (1974), the mean zinc concentration of cancerous mammary tissue was 5.7 times that of normal mammary tissue, although the absolute values for tissue zinc concentrations were extremely variable among individuals. The studies correlated with the earlier work of Tupper et al. (1955), which demonstrated an increase in the uptake of ^{65}Zn (injected as a ^{65}Zn-containing zinc-glycine complex) by tumorous mammary tissue as compared to normal tissue in mice.

A number of investigators have reported significantly reduced levels of zinc in cancerous prostatic tissue (Mawson and Fischer, 1952; Hoare et al., 1956; Voigt, 1958; Kerr et al., 1960). Habib et al. (1976) have confirmed these data and, additionally, have extended their studies to probe the relationship between prostatic levels of zinc and of the steroid hormones testosterone and dihydrotestosterone. This was done since earlier work demonstrated that prostatic zinc is under hormonal control (Hoare et al., 1956; Prout et al., 1959). Habib et al. (1976) concluded that the concentrations of testosterone and dihydrotestosterone in prostatic carcinoma are inversely proportional to the levels of zinc in the tumor tissue. At present, it is not clear what functions are served by zinc in hormone metabolism or by hormones in zinc metabolism, although there appears to be an inverse relationship between zinc and 5 α-reductase activity in human hyperplastic prostate (Wallace and Grant, 1975). Also, zinc has been shown to influence the binding of dihydrotestosterone to its cytosolic receptor (Dube and Tremblay, 1974).

Zinc levels apparently decrease in cancers other than those discussed above. Serum and plasma zinc concentrations are significantly decreased in bronchial carcinoma (Davies et al., 1968b; Beeley et al., 1974), carcinoma of the colon (Davies et al., 1968a), and primary osteosarcoma with metastases (Fisher et al., 1976), as well as other carcinomas (Danielsen and Steinnes, 1970). Interestingly, Szmigielski and Litwin (1964) found a significant decrease in the zinc content of blood granulocytes in each of 50 neoplastic diseases investigated.

Nearly 30 years ago, Vallee and co-workers reported markedly decreased zinc concentrations in leukocytes from patients with chronic leukemias (Vallee and Gibson, 1949; Gibson et al., 1950). This observation has subsequently been confirmed by numerous investigators, who, in addition, have reported that both serum and leukocyte zinc levels are also decreased in patients with lymphoma, acute leukemias, Hodgkin's disease, and multiple myeloma. Moreover, there appears to be a reciprocal relationship between leukocyte and erythrocyte zinc levels (Fredricks et al., 1964; Rosner and Gorfien, 1968; Delves et al., 1973). It is possible that the increased uptake and retention of zinc by erythrocytes may be responsible for decreasing serum and leukocyte zinc in blood disorders. The significance of increased erythrocyte zinc levels and the relationship, if any, to the etiology or progress of the disease remain unknown.

The level of zinc in the tissues and body fluids of cancer patients is related to the levels of several other essential elements. For example, Rosner and Gorfien (1968) found parallel decreases in both serum magnesium and zinc in patients with various leukemias, lymphoma, and multiple myeloma. In the same patients both erythrocyte zinc and magnesium were correspondingly elevated. Pfeilsticher (1965) found that cancer patients excrete up to three times as much zinc in the urine as do apparently healthy persons, and, additionally, in cancer patients molybdenum excretion is decreased. Indeed, Pfeilsticher suggested that a urinary Zn/Mo ratio greater than 300 indicates advanced cancer. Danielsen and Steinnes (1970) studied tissues from a variety of human cancers and discovered that the K/Zn ratio was much greater in cancerous tissue than in corresponding normal tissue. The change in this ratio appears to be due to both a decrease in tissue zinc and an increase in tissue potassium. Finally, there appears to be an inverse relationship between serum or plasma levels of copper and zinc in some malignancies. Delves et al. (1973) reported that plasma copper concentrations were higher and plasma zinc concentrations lower in children with untreated acute lymphoblastic leukemia than in the same children after successful treatment or in healthy children. Fisher et al. (1976) found the serum Cu/Zn ratio in patients with metastatic osteosarcoma to be higher than that in patients with primary osteosarcoma; the latter, in turn, was higher than the ratio in healthy individuals. This group also found that patients with more advanced disease and poorest prognoses had the highest serum Cu/Zn ratios.

It appears, therefore, that tissue and body fluid levels of zinc are indeed altered in malignancy, although the significance of these changes remains unknown. There is presently no indication that changes in zinc levels contribute to the etiology of cancer rather than simply representing the system's response to the abnormal state. It is interesting, however, that changes in tissue or body fluid levels or the ratio of zinc level to the level of some other metal, such as copper, may be of diagnostic and prognostic value as well as of assistance in evaluating the response of a patient to therapy.

With the advent of new techniques and increased interest in elucidating lesions in zinc metabolism at the molecular level, several recent studies have provided insight into the relationship between zinc and certain characteristics of malignant cells. In 1974, Andronikashvili et al. (1974) reported that DNA and RNA fractions isolated from tumor tissues contained two to four times more zinc than similar fractions from liver. It is postulated that the increased zinc levels may be due to excess DNA and RNA polymerases, which, as discussed earlier, are zinc metalloenzymes. Additional studies were reported by this group in 1976 (Andronikashvili et al., 1976). Working with CCRF-CEM human leukemic cells, they found that the DNA of these cells contained one-fourth the zinc of lymphocytes from a normal donor. RNA from the leukemic cells, however, contained nearly four times the zinc of lymphocytes from a control donor. Also, the total histone fraction from the leukemic cells had approximately one-fourth the zinc content of the same fraction from normal lymphocytes. This study is significant because it documents for the first time the presence of trace metals, one of which is zinc, in nuclear proteins. Since the histones are known to be involved in the regulation of gene expression, it is postulated that changes in the trace metal content of the histone fraction may alter the interaction of histones with DNA and as a consequence alter gene activity.

Finally, Yarom et al. (1976) used electron microscopic X-ray microanalysis to perform a comparative study of the elemental contents of normal and chronic lymphocytic leukemic human lymphocytes. The most prominent difference observed was approximately one-third the nuclear zinc and one-third to one-fifth the cytoplasmic zinc in the leukemic lymphocytes as compared to the normal cells. The authors point out that the causes of low intracellular zinc in lymphocytic leukemia are still unknown, although work in our laboratory indicates that leukemic lymphocytes incorporate transferrin-bound zinc more slowly and to a lesser extent than do normal lymphocytes (J. L. Phillips, unpublished observation). In addition, normal lymphocytes respond to increased intracellular zinc by synthesizing a low molecular weight protein, possibly a metallothionein, while lymphocytes from donors with chronic lymphocytic leukemia fail to synthesize this intracellular zinc-binding protein (J. L. Phillips, unpublished observation). The resulting intracellular zinc deficiency can result in alterations in a number of cellular processes, as discussed in detail below.

4. SUPPRESSION OF TUMORGENESIS AND TUMOR GROWTH BY ZINC DEFICIENCY

As discussed earlier, zinc is a requirement for growth in all species studied. Consequently, researchers have employed the rationale that producing a state of zinc deficiency might inhibit the proliferative capacity of tumor cells. Early experiments to test this hypothesis were performed by McQuitty et al. (1970), who studied the effect of different levels of dietary zinc intake on the growth of Walker 256 carcinosarcoma in rats. They found that the mean survival of rats maintained on a zinc-deficient diet was 19 days, compared to a mean survival of 11 days for rats maintained on zinc-adequate or zinc-excess diets. The prolongation of mean survival time was attributed to the effect of zinc deficiency in decreasing tumor growth. Furthermore, this group speculated on the reason for the decreased levels of zinc in tissues and body fluids in cancer patients. They state that serum zinc may decrease because the zinc is sequestered by the tumor at the host's expense. Tumor tissue levels of zinc may then be low because the available zinc is divided among the tumor cells competing for it.

This study was confirmed by DeWys and Pories (1972), who also demonstrated that the growth of a slowly dividing tumor, Lewis lung carcinoma, as well as several ascites tumors in mice and rats, was inhibited by dietary zinc deficiency. They note, however, that zinc deficiency affects only tumor growth and not tumorgenesis. Moreover, there are adverse effects of zinc deficiency on the animal as a whole that are linearly related to the degree of deficiency. These adverse effects may limit the usefulness of dietary zinc deficiency as a means of inhibiting tumor growth.

In similar studies Barr and Harris (1973) found decreased growth of P388 leukemia cells in zinc-deficient mice, and Duncan and co-workers (Duncan et al., 1974; Duncan and Dreosti, 1975, 1976), observed decreased growth of a chemically induced hepatoma in zinc-deficient rats. The work of the latter group is especially interesting, since they also attempted to determine the primary site of action of zinc deficiency in limiting tumor growth. They reported that DNA synthesis was significantly reduced in chemically induced hepatomas as compared to controls. Moreover, the activities of two zinc-dependent enzymes associated with DNA synthesis, thymidine kinase and DNA polymerase, were significantly reduced in zinc-deficient animals as compared to zinc-normal controls. Consequently, they hypothesize that the primary action of zinc deficiency is to reduce thymidine kinase and DNA polymerase activities, resulting in decreased DNA synthesis and thus producing decreased tumor growth.

There are, however, additional possibilities to explain the role of zinc deficiency in decreasing tumor growth. For instance, zinc deficiency may (1) decrease RNA synthesis by decreasing the activity of the RNA polymerases, which, as pointed out earlier, are zinc metalloenzymes; (2) decrease the activities of other zinc metalloenzymes associated with nucleic acid synthesis, such as aspartate transcarbamylase (Nelbach et al., 1972); (3) decrease protein biosynthesis by decreasing the activities of essential zinc metalloenzymes, such as elongation factor 1 (Kotsiopoulos and Mohr, 1975); (4) decrease the activity of RNA-dependent DNA polymerases ("reverse transcriptases") in certain virus-induced cancers, since these enzymes have been shown to be zinc metalloenzymes (Auld et al., 1974a, 1974b, 1975; Poiesz et al., 1974a, 1974b); (5) produce changes in the structure, and so the function, of both DNA and RNA. Zinc has been found associated with DNA, apparently bound to the phosphate groups of the molecule, and causes heated, uncoiled DNA to rewind on cooling (Shin and Eichhorn, 1968a, 1968b). Zinc is also associated with the RNA isolated from a variety of cells, affecting the structure of the RNA and thus having important implications in nucleic acid and protein metabolism (Wacker and Vallee, 1959; Prask and Plocke, 1971). Additionally, recent studies employing laser excitation cytofluorometry to study the cell cycle in *Euglena gracilis* have demonstrated that all the biochemical processes required for cells to pass from the G_1 phase into S, from S to G_2, and from G_2 to mitosis require zinc, and that zinc deficiency can block all phases of the growth cycle in this organism (Falchuk et al., 1975). Later studies with leukemic lymphocytes (Vallee, 1976) have indicated that such cells may be destroyed selectively by treatment with the zinc-chelating agent 1,10-phenanthroline, apparently by exerting some effect on the G_1 and S premitotic phases of the cell cycle. It is not yet known, however, which specific zinc systems are affected in zinc deficiency.

5. SUPPRESSION OF TUMORGENESIS AND TUMOR GROWTH BY ADMINISTRATION OF EXCESS ZINC

Early work by Gunn et al. (1963) demonstrated that in both rats and mice a single subcutaneous injection of cadmium chloride produced interstitial cell tumors of the testes. Moreover, they demonstrated that tumorgenesis could be prevented by simultaneous injection of zinc acetate at 100 times the dose of the cadmium chloride. A subsequent report from this group confirmed the protective effect of zinc on cadmium-induced tumor formation, as long as the molar ratio of zinc to cadmium was 100:1. Gunn et al. (1964) speculated that the cadmium might "injure" some zinc-dependent growth-regulating mechanism in cells, perhaps by successful competition with cellular zinc. Consequently, administration of excess zinc could then prevent or retard the effect of the cadmium.

Poswillo and Cohen (1971) took a different approach to the problem, and demonstrated an inhibition of tumorgenesis in hamster cheek pouch by oral administration of excess zinc. Addition of 100 ppm zinc sulfate to the drinking water inhibited formation of dimethylbenzanthracene-induced carcinomas in eight of nine experimental animals. Similar results were reported by Duncan et al. (1974), who found significantly reduced growth of a chemically induced hepatoma in rats maintained on a diet high in zinc ($\geqslant 500 \ \mu g/g$) as compared to controls maintained on a zinc-normal diet (60 $\mu g/g$).

Woster et al. (1975) reported that subcutaneous injection of 80 μg zinc sulfate daily suppressed initiation of sarcoma 180 growth in up to 60% of the host mice, as long as the zinc injections were begun within 2 days after intraperitoneal inoculation of tumor cells. Additionally, daily administration of zinc sulfate was required for 7 days after tumor cell inoculation to protect the hosts successfully. This study confirmed the earlier one of Cherkasova (1968). A similar study was performed by Phillips and Sheridan (1976), who reported that aqueous zinc acetate (0.1 to 0.5 mg) injected intraperitoneally prevented tumorgenesis in 50 to 70% of mice previously inoculated intraperitoneally with L1210 leukemia cells. However, subcutaneous injection of zinc acetate did not prevent initiation of tumor growth in mice inoculated intramuscularly with BW5147 lymphatic leukemia cells. In this group only a small, although statistically significant, increase in mean survival was noted.

The precise mechanism by which administration of zinc suppresses tumorgenesis and tumor growth is unknown. An understanding of the many cellular and molecular effects of zinc, however, is the key to advancing several theories. For instances, zinc is known to enhance lymphocyte transformation (Kirchner and Ruhl, 1970; Berger and Skinner, 1974; Phillips and Azari, 1974) and hence may be involved in augmentation of the immune response to foreign tumor cells. In the same vein, zinc has been shown to activate the cytotoxic potency of cultured mouse macrophages (Melsom and Seljelid, 1973). Additionally, some researchers, attempting to outline a regulatory function for zinc, have studied its ability to inhibit, rather than stimulate or maintain, the activity of cellular enzymes. Mustafa et al. (1971) and Donaldson et al. (1971) have shown that both Mg^{2+}-dependent and Na^+- and K^+-stimulated ATPases are inhibited by zinc. Zinc has also been shown to be a particularly effective inhibitor of both phospholipase A_2 (Stossel et al., 1970) and guanyl cyclase (Hardman and Sutherland, 1969). Recent studies have demonstrated that zinc inhibits electron transport in rat liver mitochondria by acting at a specific site between ubiquinone and the b cytochromes (Kleiner, 1974). Furthermore, zinc inhibits phosphoprotein phosphatase activity in rat uterus (Vokaer et al., 1974) and produces increased phosphorylation of nucleolar proteins in Novikoff hepatoma ascites cells (Kang et al., 1974) and also of F_1 histone from HTC cells (Tanphaichitr and Chalkley, 1976). Because tumor cells exhibit uncontrolled growth, their requirement for zinc should be greater than that of normal cells. Indeed, tumor cells may sequester zinc from body stores to fulfill their metabolic needs. In that case perhaps the administration of excess zinc selectively overloads and thus poisons tumor cells and at the same time provides necessary zinc for zinc-depleted tissues. Indeed, it is probable that the zinc status of a cell may control certain key cellular functions by regulating the activities of key cellular enzymes. This is of obvious importance in understanding the role of zinc in both initiation and suppression of tumorgenesis.

ACKNOWLEDGMENT

The authors wish to acknowledge the support of Grant HL-19132 from the National Institutes of Health during the preparation of this chapter.

REFERENCES

Andronikashvili, E., Mosulishvile, A., Belokobilski, A., Kharabadze, N., Tevzieva, T., and Efremova, E. (1974). "Content of Some Trace Elements in Sarcoma M-1 DNA in Dynamics of Malignant Growth," *Cancer Res.*, **34**, 271-274.

Andronikashvili, E., Mosulishvili, L. M., Belokobil'skiy, A. I., Kharbadze, N. E., Shonia, N. I., Desai, L. S., and Foley, G. E. (1976). "Human Leukemic Cells: Determination of Trace Elements in Nucleic Acids and Histones by Neutron-Activation Analysis," *Biochem. J.*, **157**, 529-533.

Auld, D. S. and Atsuya, I. (1976). "Yeast RNA Polymerase I: a Eukaryotic Zinc Metalloenzyme," *Biochem. Biophys. Res. Commun.*, **69**, 548-554.

Auld, D. S., Kawaguchi, H., Livingston, D. M., and Vallee, B. L. (1974a). "Reverse Transcriptase from Avian Myeloblastosis Virus: a Zinc Metalloenzyme," *Biochem. Biophys. Res. Commun.*, **57**, 967-972.

Auld, D. S., Kawaguchi, H., Livingston, D. M., and Vallee, B. L. (1974b). "RNA-Dependent DNA Polymerase (Reverse Transcriptase) from Avian Myeloblastosis Virus: a Zinc Metalloenzyme," *Proc. Natl. Acad. Sci.*, **71**, 2091-2095.

Auld, D. S., Kawaguchi, H., Livingston, D. M., and Vallee, B. L. (1975). "Zinc Reverse Transcriptases from Mammalian RNA Type C Viruses," *Biochem. Biophys. Res. Commun.*, **62**, 296-302.

Bagg, H. J. (1936). "Experimental Production of Teratoma Testis in Fowl," *Am. J. Cancer*, **26**, 69-84.

Barr, D. H. and Harris, J. W. (1973). "Growth of the P388 Leukemia as an Ascites Tumor in Zinc-Deficient Mice," *Proc. Soc. Exp. Biol. Med.*, **144**, 284-287.

Beeley, J. M., Darke, C. S., Owen, G., and Cooper, R. D. (1974). "Serum Zinc, Bronchiectasis, and Bronchial Carcinoma," *Thorax*, **29**, 21-25.

Berger, N. A. and Skinner, A. M. (1974). "Characterization of Lymphocyte Transformation Induced by Zinc Ions," *J. Cell Biol.*, **61**, 45-55.

Berger, N. A., Johnson, E. S., and Skinner, A. M. (1975). "Orthophenanthroline Inhibition of DNA Synthesis in Mammalian Cells," *Exp. Cell Res.*, **96**, 145-155.

Burr, R. G. (1974). "Plasma-Zinc Levels," *Lancet*, **1**, 879.

Chahovitch, X. (1955). "The Action of Zinc on the Growth of Experimental Tumors Incited by Carcinogens," *Gl. Srpske Akad. Nauka, Od. Med. Nauka*, **215**, 143-146.

Cherkasova, E. B. (1968). "Effect of Zinc on Oxygen Consumption, Phagocytosis, and the Development of Pliss Lymphosarcoma," *Vopr. Onkol.*, **15**, 81-85.

Danielsen, A. and Steinnes, E. (1970). "A Study of Some Selected Trace Elements in Normal and Cancerous Tissues by Neutron Activation Analysis," *J. Nucl. Med.*, **11**, 260-264.

Davies, I. J. T., Musa, M., and Dormandy, T. L. (1968a). "Measurements of Serum Zinc. I: In Health and Disease," *J. Clin. Pathol.*, **21**, 359–363.

Davies, I. J. T., Musa, M., and Dormandy, T. L. (1968b). "Measurements of Serum Zinc. II: In Malignant Disease," *J. Clin. Pathol.*, **21**, 363–365.

Delves, H. T., Alexander, F. W., and Lay, H. (1973). "Copper and Zinc Concentration in the Plasma of Leukemic Children," *Br. J. Haematol.*, **24**, 525–531.

DeWys, W. and Pories, W. (1972). "Inhibition of a Spectrum of Animal Tumors by Dietary Zinc Deficiency," *J. Natl. Cancer Inst.*, **48**, 375–381.

Donaldson, J., St.-Pierre, T., Minnich, J., and Barbeau, A. (1971). "Seizures in Rats Associated with Divalent Cation Inhibition of Na^+-K^+-ATPase," *Can. J. Biochem.*, **49**, 1217–1224.

Dube, J. Y. and Tremblay, R. E. (1974). "Androgen Binding Protein in Cock's Tissues; Properties of Ear Lobe Protein and Determination of Binding Sites in Head Appendages and Other Tissues," *Endocrinology*, **95**, 1105–1112.

Duncan, J. R. and Dreosti, I. E. (1975). "Zinc Intake, Neoplastic DNA Synthesis, and Chemical Carcinogenesis in Rats and Mice," *J. Natl. Cancer Inst.*, **55**, 195–196.

Duncan, J. R. and Dreosti, I. E. (1976). "A Site of Action for Zinc in Neoplastic Tissue," *S. Afr. Med. J.*, **50**, 711–713.

Duncan, J. R., Dreosti, I. E., and Albrecht, C. F. (1974). "Zinc Intake and Growth of a Transplanted Hepatoma Induced by 3'-Methyl-4-dimethylaminoazobenzene in Rats," *J. Natl. Cancer Inst.*, **53**, 277–278.

Falchuk, K. H., Fawcett, D. W., and Vallee, B. L. (1975). "Role of Zinc in Cell Division of *Euglena gracilis*," *J. Cell Sci.*, **17**, 57–78.

Falchuk, K. H., Ulpino, L., Mazus, B., and Vallee, B. L. (1977). *E. gracilis* RNA Polymerase I: a Zinc Metalloenzyme," *Biochem. Biophys. Res. Commun.*, **74**, 1206–1212.

Falin, L. I. (1940). "Experimental Teratoma Testis in Fowl," *Am. J. Cancer*, **38**, 199–211.

Falin, L. I. (1941). "Morphologie und Differenzcering der Nervemelement in der Experimentellen Teratoiden," *Z. Mikroskop. Anat. Forsch.*, **49**, 193–224.

Falin, L. I. and Anissimowa, W. W. (1940). "Zur Pathogenese der Experimentellen Teratoiden Geschwulste der Geschlechtsdrusen. Teratoide Hodengeschwulst belm Hahn, eryeugt durch Emfuhrung von $CuSO_4$," *Z. Krebsforsch.*, **50**, 339–351.

Fisher, G. L., Byers, V. S., Shifrine, M., and Levin, A. S. (1976). "Copper and Zinc Levels in Serum from Human Patients with Sarcomas," *Cancer*, **37**, 356–363.

Fredricks, R. E., Tanaka, K. R., and Valentine, W. N. (1964). "Variations of Human Blood Cell Zinc in Disease," *J. Clin. Invest.*, **43**, 304–315.

Gibson, J. G., Vallee, B. L., Fluharty, R. G., and Nelson, J. E. (1950). "Studies of the Zinc Content of the Leukocytes in Myelogenous Leukemia," *Union Int. Clin. Cancrum Acta*, 6, 1102-1107.

Gunn, S. A., Gould, T. C., Anderson, W. A. D. (1963). "Cadmium-Induced Interstitial Cell Tumors in Rats and Mice and Their Prevention by Zinc," *J. Natl. Cancer Inst.*, 31, 745-753.

Gunn, S. A., Gould, T. C., and Anderson, W. A. D. (1964). "Effect of Zinc on Carcinogenesis by Cadmium," *Proc. Soc. Exp. Biol. Med.*, 115, 653-657.

Guthrie, J. (1956). "Attempts to Produce Seminomata in the Albino Rat by Inoculation of Hydrocarbons and Other Carcinogens into Normally Situated and Ectopic Testes," *Br. J. Cancer*, 10, 134-144.

Guthrie, J. (1964). "Histological Effects of Intratesticular Injections of Cadmium Chloride in Domestic Fowl," *Br. J. Cancer*, 18, 255-260.

Guthrie, J. (1971). "Zinc Induction of Testicular Teratomas in Japanese Quail (*Coturnix coturnix japonica*) after Photoperiodic Stimulation of Testis," *Br. J. Cancer*, 25, 311-314.

Halme, E. (1961). "The Carcinomatous Effect of Zinc in Drinking Water," Vitalstoffe Zivilisationskrankh., 6, 59-66.

Habib, F. K., Hammond, G. L., Lee, I. R., Dawson, J. B., Mason, M. K., Smith, P. H., and Stritch, S. R. (1976). "Metal-Androgen Interrelationships in Carcinoma and Hyperplasia of the Human Prostate," J. Endocr., 71, 133-141.

Hardman, J. G. and Sutherland, E. W. (1969). "Guanyl Cyclase, an Enzyme Catalyzing the Formation of Guanosine-3'-5'-monophosphate from Guanosine Triphosphate," *J. Biol. Chem.*, 244, 6363-6370.

Hirayama, T. (1962). Quoted in *S. Afr. Cancer Bull.*, 6, 114.

Hoare, R., Delory, G. E., and Penner, D. W. (1956). "Zinc and Acid Phosphatase in Human Prostate," *Cancer*, 9, 721-726.

Kaltenbach, T. and Egen, E. (1968). "Beitrage zum Histochemie Nachweis von Eisen, Kupfer, und Zinc in der Menschlich an Leber unter besonderer Buruchsichtigrung des Selberolfid-Verhavens nach Tumor," *Acta Histochem.*, 25, 329-359.

Kang, Y. J., Olson, M. O. J., and Busch, H. (1974). "Phosphorylation of Acid-Soluble Proteins in Isolated Nucleoli of Novikoff Hepatoma Ascites Cells: Effects of Divalent Cations," *J. Biol. Chem.*, 249, 5580-5585.

Kerr, W. K., Keresteci, A. G., and Mayoh, H. (1960). "Distribution of Zinc within Human Prostate," *Cancer*, 13, 550-554.

Kew, M. C. and Mallett, R. C. (1974). "Hepatic Zinc Concentrations in Primary Cancer of the Liver," *Br. J. Cancer*, 29, 80-83.

Kirchner, H. and Ruhl, H. (1970). "Stimulation of Human Peripheral Lymphocytes by Zn *in vitro*," *Exp. Ce.. Res.*, 61, 229-230.

Kleiner, D. (1974). "The Effect of Zn^{2+} Ions on Mitochondrial Electron Transport," *Arch. Biochem. Biophys.*, 165, 121-125.

Kotsiopoulos, P. S. and Mohr, S. C. (1975). "Protein Synthesis Elongation Factor 1 from Rat Liver: a Zinc Metalloenzyme," *Biochem. Biophys. Res. Commun.*, **67**, 979–987.

Lattke, H. and Weser, U. (1976). "Yeast RNA Polymerase B: a Zinc Protein," *FEBS Lett.*, **65**, 288–292.

Levinson, W., Faras, B., Jackson, W. J., and Bishop, J. M. (1973). "Inhibition of RNA-dependent DNA Polymerase of Rous-Sarcoma Virus by Thiosemicarbazones and Several Cations," *Proc. Natl. Acad. Sci.*, **70**, 164.

Mawson, C. A. and Fischer, M. I. (1952). "Occurrence of Zinc in Human Prostate Gland," *Can. J. Med. Sci.*, **30**, 336–339.

McBean, L. D., Dove, J. T., Halsted, J. A., and Smith, J. C. (1972). "Zinc Concentration in Human Tissues," *Am. J. Clin. Nutr.*, **25**, 672–676.

McGlashan, N. D. (1967). "Zinc and Oesophageal Cancer," *Lancet*, **1**, 578.

McQuitty, J. T., DeWys, W. D., Monaco, L., Strain, W. H., Rob, C. G., Apgar, J., and Pories, W. J. (1970). "Inhibition of Tumor Growth by Dietary Zinc Deficiency," *Cancer Res.*, **30**, 1387–1390.

Melsom, H. and Seljelid, R., (1973). "The Cytotoxic Effect of Mouse Macrophages on Syngeneic and Allogeneic Erythrocytes," *J. Exp. Med.*, **137**, 807–820.

Michalowsky, I. (1926). "Die experimentelle Erzeugung einer Teratoiden Neubildung der hoden beim Hahn Mitteilung," *Zentr. Allgem. Pathol. Anat.*, **38**, 385–587.

Mikac-Dević, D. (1970). "Methodology of Zinc Determinations and the Role of Zinc in the Biochemical Processes," *Adv. Clin. Chem.*, **13**, 271.

Mulay, I. L., Roy, R., Knox, B. E., Suhr, N. H., and Delaney, W. E. (1971). "Trace Metal Analysis of Cancerous and Noncancerous Human Tissues," *J. Natl. Cancer Inst.*, **47**, 1–13.

Mustafa, M. Cross, C., Munn, R., and Ttardie, J. (1971). "Effects of Divalent Metal Ions on Alveolar Macrophage Membrane Adenosine Triphosphatase Activity," *J. Lab. Clin. Med.*, **77**, 563–571.

Nelbach, M. E., Pigiet, W. P., and Gerhart, J. C. (1972). "A Role for Zinc in the Quaternary Structure of Aspartate Transcarbamylase from *Escherichia coli*," *Biochemistry*, **11**, 315–327.

Olson, K. B., Heggen, G. E., and Edwards, C. F. (1958). "Analysis of 5 Trace Elements in Liver of Patients Dying of Cancer and Noncancerous Disease," *Cancer*, **11**, 554–561.

Petranyi, P., Jendrisak, J. J., and Burgess, R. R. (1977). "RNA Polymerase II from Wheat Germ Contains Tightly Bound Zinc," *Biochem. Biophys. Res. Commun.*, **74**, 1031–1038.

Pfeilsticher, K. (1965). "Spurenelemente in Organen und Urin bei Krebs," *Z. Klin. Chem. Klin. Biochem.*, **3**, 145–150.

Phillips, J. L. and Azari, P. (1974). "Zinc Transferrin: Enhancement of Nucleic Acid Synthesis in Phytohemagglutinin-Stimulated Human Lymphocytes," *Cell. Immunol.*, **10**, 31–37.

Phillips, J. L. and Sheridan, P. J. (1976). "Effect of Zinc Administration on the Growth of L1210 and BW5147 Tumors in Mice," *J. Natl. Cancer Inst.*, 57, 361–363.

Poiesz, B. J., Battula, N., and Loeb, L. A. (1974a). "Zinc in Reverse Transcriptase," *Biochem. Biophys. Res. Commun.*, 56, 959–964.

Poiesz, B. J., Seal, G., and Loeb, L. A. (1974b). "Reverse Transcriptase: Correlation of Zinc Content with Activity," *Proc. Natl. Acad. Sci.*, 71, 4892–4896.

Poswillo, D. E. and Cohen, B. (1971). "Inhibition of Carcinogenesis by Dietary Zinc," *Nature*, 231, 447–448.

Prask, J. A. and Plocke, D. J. (1971). "A Role for Zinc in the Structural Integrity of the Cytoplasmic Ribosomes of *E. gracilis*," *Plant Physiol.*, 48, 150–155.

Prout, C. R., Sierp, M., and Whitmore, W. F. (1959). "Radioactive Zinc in the Prostate," *J. Am. Med. Assoc.*, 169, 1703–1710.

Riviere, M., Chouroulinkov, J., and Guerin, M. (1960). "Production de Tumeurs par Injections Intratesticulaires de Chlorure de Zinc chez le Rat," *Bull. Assoc. Fr. Étude Cancer*, 47, 55–87.

Roe, F. J. C. and Lancaster, M. C. (1964). "Natural, Metallic, and Other Substances as Carcinogens," *Br. Med. Bull.*, 20, 127–133.

Rosner, F. and Gorfien, P. C. (1968). "Erythrocyte and Plasma Zinc and Magnesium Levels in Health and Disease," *J. Lab. Clin. Med.*, 72, 213–219.

Sabbioni, E. (1976). "The Metalloenzyme Nature of Calf Thymus Deoxynucleotidyl Transferase," *FEBS Lett.*, 71, 233–235.

Schwartz, A. E., Leddicotlo, G. W., Fink, R. W., and Friedman, E. W. (1974). "Trace Elements in Normal and Malignant Human Breast Tissue," *Surgery*, 76, 325–329.

Shin, Y. A. and Eichhorn, G. L. (1968a). "Reversible Unwinding and Rewinding of Deoxyribonucleic Acid by Zinc. II: Ions through Temperature Manipulation," *Biochemistry*, 7, 1026–1032.

Shin, Y. A. and Eichhorn, G. L. (1968b). "Interaction of Metal Ions with Polynucleotides and Related Compounds," *Biochemistry*, 7, 3003–3006.

Slater, J. P., Mildvan, A. S., and Loeb, L. A. (1971). "Zinc in DNA Polymerases," *Biochem. Biophys. Res. Commun.*, 44, 37–43.

Stavrianopoulos, J. G., Karkas, J. D., and Chargoff, E. (1972). "DNA Polymerase of Chicken Embryo—Purification and Properties," *Proc. Natl. Acad. Sci.*, 69, 1781.

Stocks, P. and Davies, R. I. (1964). "Zinc and Copper Content of Soils Associated with the Incidence of Cancer of the Stomach and Other Organs," *Br. J. Cancer*, 18, 14–24.

Stossel, T. P., Murad, F., Mason, R. J., and Vaughan, M. (1970). "Regulation of Glycogen Metabolism in Polymorphonuclear Leukocytes," *J. Biol. Chem.*, 245, 6228–6234.

Szmigielski, S. and Litwin, J. (1964). "The Histochemical Demonstration of Zinc in Blood Granulocytes–the New Test in Diagnosis of Neoplastic Diseases," *Cancer,* **17**, 1381–1384.

Tanphaichitr, N. and Chalkley, R. (1976). "Production of High Levels of Phosphorylated F_1 Histone by Zinc Chloride," *Biochemistry,* **15**, 1610–1614.

Tupper, R., Watts, R. W. E., and Wormall, A. (1955). "The Incorporation of Zn^{65} in Mammary Tumors and Some Other Tissues of Mice after Injection of the Isotopes," *Biochem. J.,* **59**, 264–268.

Vallee, B. L. (1976). "Zinc Biochemistry: a Perspective," *Trends Biochem. Sci.,* **2**, 88–91.

Vallee, B. L. and Gibson, J. G. (1949). "The Zinc Content of Whole Blood, Plasma, Leukocytes and Erythrocytes in the Anemias," *Blood,* **4**, 455.

Vallee, B. L. and Wacker, W. E. C. (1970). *The Proteins,* Vol. 5. Academic Press, New York.

Voigt, G. E. (1958). "Histochemische Untersuchungen uber das Zinc in der normalen, hypertrophischen, und carcinomatosen Prostata," *Acta Pathol. Microbiol. Scand.,* **42**, 242–246.

Vokaer, A., Iacobelli, S., and Kram, R. (1974). "Phosphorprotein Phosphatase Activity Associated with Estrogen-Induced Protein in Rat Uterus," *Proc. Natl. Acad. Sci.,* **71**, 4482–4486.

Wacker, W. E. C. and Vallee, B. L. (1959). "Nucleic Acids and Metals. I: Chromium, Manganese, Nickel, Iron and Other Metals in Ribonucleic Acid from Diverse Biological Sources," *J. Biol. Chem.,* **234**, 3257–3262.

Wallace, A. M. and Grant, J. K. (1975). "Effect of Zinc on Androgen Metabolism in the Human Hyperplastic Prostate," *Biochem. Soc. Trans.,* **3**, 540–542.

Wandzilak, T. M. and Benson, R. W. (1977). "Yeast RNA Polymerase III: a Zinc Metalloenzyme," *Biochem. Biophys. Res. Commun.,* **76**, 247–252.

Willis, R. A. (1934). "Experimental Study of Possible Influence of Injury in Genesis of Tumors of Gonads," *Br. J. Exp. Pathol.,* **15**, 234–236.

Woster, A. D., Failla, M. L., Taylor, M. W., and Weinberg, E. D. (1975). "Zinc Suppression of Initiation of Sarcoma 180 Growth," *J. Natl. Cancer Inst.,* **54**, 1001–1003.

Wright, E. B. and Dormandy, T. L. (1972). "Liver Zinc in Carcinoma," *Nature,* **237**, 166.

Yarom, R., Hall, T. A., and Polliack, A. (1976). "Electron Microscopic X-ray Microanalysis of Normal and Leukemic Human Lymphocytes," *Proc. Natl. Acad. Sci.,* **73**, 3690–3694.

11

ZINC IN WOUND HEALING

Andre M. van Rij
Walter J. Pories

Department of Surgery, East Carolina University, Greenville, North Carolina

1. Introduction	215
2. The Wound	216
3. Zinc and Enzyme Function	217
4. Zinc and Wound Protein Synthesis	218
5. Zinc and Wound Nucleic Acid Synthesis	220
6. Zinc and Cellular Membrane Effects	223
7. Zinc and the Inflammatory Cells in Wounds	224
8. Zinc Tissue Transport Following Injury	225
9. Zinc Therapy: Animal Studies	226
10. Zinc Therapy: Human Studies	226
References	228

1. INTRODUCTION

The earliest record of medicinal zinc (in calamine) appears in the Ebers Papyrus (Ebers, 1875) dating back to 1500 B.C. Ever since then, a variety of zinc compounds have been used in salves, ointments, and solutions for the promotion of healing. This practice has continued into modern medicine. Despite this well-established use of zinc, the specific site of action of the element and the indications for zinc therapy in wound healing have yet to be clarified.

Strain et al. (1953, 1954) observed that wounded rats on a control diet healed less rapidly than rats fed a zinc-enriched diet. In retrospect, it appears that the control diets in these experiments were probably low in zinc. Savlov et al. (1962)

demonstrated an increased and rapid uptake of zinc into incised wounds. Subsequently other experimental observations demonstrating an essential role for zinc in healing wounds have been made in a variety of species and in tissues of different types (Sandstead and Shepard, 1968; Sandstead et al., 1970; Oberleas et al., 1971; Lavy, 1972; McCray et al., 1972; Battistone a, b, 1972; Miller, 1970). Zinc deficiency results in impaired wound healing.

A number of clinical trials of zinc supplements to improve healing in human beings have been reported (Brewer et al., 1966; Pories et al., 1967; Husain, 1969; Barcia, 1970; Greaves and Skillen, 1970; Myers and Cherry, 1970; Haeger et al., 1972, 1974; Hallbrook and Lanner, 1972; Haeger and Lanner, 1974; Sedlacek et al., 1976). Although some of these trials have not confirmed enhanced healing with zinc, casting some doubt on the benefit of zinc therapy for human beings, in patients with inadequate tissue levels of the element zinc supplementation does promote healing. Zinc is essential for normal wound healing.

This review of the experimental studies and clinical trials of zinc in wound healing recognizes the wide spectrum of events during the healing process in which zinc participates. Discussed are the apparent discrepancies in the ability of zinc supplementation to enhance wound healing. In particular, attention is drawn to the prior requirement of a degree of zinc deficiency, with its deleterious effects on cellular metabolism, before zinc supplements result in the enhancement of wound healing. Zinc levels in excess of normal do not accelerate wound healing.

2. THE WOUND

Healing in a wound is a dynamic process consisting of a well-ordered sequence of complex events. The duration of healing (Peacock and Van Winkle, 1976) from initial injury to the final maturation of the scar is many days or even months. The early inflammatory response results in the outpouring of exudate containing proteins, kinins, and other active substances into the wound. Associated leukocytes are lysed, releasing further proteolytic enzymes, and other macrophages enter to clear up the debris. Subsequently there is a proliferation of additional cellular elements, including endothelial cells and, particularly, fibroblasts. These cells set about establishing the bulk of the scar with the production of collagen and its associated ground substance. Over the ensuing weeks the wound increases in strength as changes in the collagen protein occur with increased intra- and intermolecular cross-linking. Collagen synthesis then decreases, and resorption by collagenase enzymes contributes to scar molding and shrinkage. In specialized tissues there are other cellular responses: in skin proliferation and migration of the covering epithelial cells occur. With such complexity it is no surprise that healing wounds are influenced by numerous factors.

Included among these are the trace metals such as zinc, which is essential in a variety of biological processes, for example, enzyme function, protein and nucleic acid synthesis, cell replication, and maintenance of cellular membrane integrity. These processes are essential to the general phenomena of growth and occur also in the healing wound.

To elucidate further the specific action of zinc in wound healing a review of the action of zinc in some of these processes, and their correlation with the events of wound healing, is necessary.

3. ZINC AND ENZYME FUNCTION

Some 24 metalloenzymes have been described to date (Riordan and Vallee, 1976) in which zinc is bound in a fixed stochiometric amount to the apoenzyme (Vallee, 1955). The zinc may function directly in catalysis, in synthesis and stablization of the protein structure, and or in a regulatory role of the enzyme (Carpenter and Vahl, 1973; Vallee et al., 1959; Drum et al., 1967; Rosenbusch and Weber, 1971). Zinc is also more loosely associated with other proteins to form metal-enzyme complexes that significantly influence the behavior of the enzyme. Although these different zinc-associated enzymes have not as yet been specifically characterized in healing tissues, they are known to affect a wide variety of metabolic processes, including carbohydrate, lipid, protein, and nucleic acid synthesis.

Depletion of zinc in the tissues containing these metalloenzymes results in a reduction in the activity of the enzymes (Prasad et al., 1967; Prasad and Oberleas, 1970) and in some instances may also cause structural changes in the enzymes which increase their degradation (Reinhold and Kfoury, 1969). These effects of zinc depletion vary with the affinity of the enzyme ligands for zinc, the turnover rate of the enzyme, and the sensitivity of the particular tissue and cell to changes in available zinc. The more loosely bound the zinc, or the more rapid the enzyme turnover, or the more readily a tissue gives up zinc, the more rapidly the process dependent on that enzyme will be impaired by zinc deficiency (Prasad et al., 1971). This is illustrated by the lowered activity of the less tightly bound zinc metalloenzyme alkaline phosphatase in zinc-sensitive tissues such as serum, bone, and intestines with zinc depletion, which consequently is considered to lead to some of the defects of zinc deficiency (Prasad et al., 1971; Kirchgessner and Roth, 1975; Kirchgessner et al., 1975; Huber and Gershoff, 1973). In contrast, carbonic anhydrase, a more tightly binding metalloenzyme, remains relatively unchanged during zinc deficiency (Huber and Gershoff, 1973).

More indirect effects of zinc on enzymes in wound healing may also occur. Because some of the zinc metalloenzymes are essential in the synthetic pathways of other proteins, the production of other enzymes also is zinc dependent. For example, the copper-containing enzyme lysl oxidase, which is required for normal collagen formation in a wound, is decreased in animals on zinc-deficient diets (Chvapil, 1974). It is probable that zinc is required in the synthesis of this enzyme. Furthermore, animals fed high-zinc diets also have decreased lysl oxidase (Chvapil, 1974). In this instance the inhibitory effect of zinc provides an example of the antagonism between zinc and other cations, in particular, copper. Such antagonism has been reported (Van Campen, 1966; Magee and Matrone, 1960; Hill et al., 1963; Starcher, 1969; Prasad et al., 1967), with decreased copper tissue and serum levels being observed after the administration of zinc. *In vitro* the addition of zinc also interferes with lysl oxidase (Chvapil and Walsh, 1972), perhaps because of zinc displacement of copper from the enzyme.

In attempting to elucidate where zinc is most critical, it has been suggested by several different workers (Mills et al., 1969; Prasad and Oberleas, 1970; Prasad et al., 1971) that the rapidity of an enzyme response to zinc deficiency and repletion helps to identify the primary site of the metabolic function of zinc. Unfortunately such characteristics with respect to enzymes in healing wounds are not known. Of the enzymes found in healing tissue alkaline phosphatase in the granulocyte inflammatory cells is a zinc metalloenzyme. Thymidine kinase, observed in implanted cellulose sponges, is sensitive to zinc deficiency (Prasad and Oberleas, 1974). Other identified zinc metalloenzymes include DNA and RNA polymerases (Scrutton et al., 1971; Slater et al., 1971), which are important in cell replication and tissue growth and consequently are of significance in healing. There are probably many others yet to be identified.

4. ZINC AND WOUND PROTEIN SYNTHESIS

The essential role of zinc in general protein synthesis is well documented. The stunted growth in young animals deprived of adequate zinc or the marked decrease in body protein in older animals fed a zinc-deficient diet in comparison to pair-fed control groups of animals is readily observed. The total protein content—in particular, collagen protein—of various tissues is decreased in zinc-deficient animals (Macapinlac et al., 1968; Somers and Underwood, 1969; Sandstead et al., 1971). In association with these changes there is a decreased utilization of amino acids (Hsu and Anthony, 1971; Hsu, 1976), an increased urinary amino acid excretion (Hsu, 1976) and nitrogen excretion (Somers and Underwood, 1969), and a change in plasma proteins (Tao and Hurley, 1971; Fox and Harrison, 1965) and amino acids (Sommers and Underwood, 1969). It has been suggested (Macapinlac et al., 1968; Somers and Underwood, 1969) that zinc deficiency stimulates protein catabolism. Other work, in both plants and animals, however, indicates that zinc deficiency impairs protein synthesis (Theuer and Hoekstra, 1966; Mills et al., 1967; Fernandez-Madrid et al., 1973;

Wacker, 1976). From the additional observations of the role of zinc in RNA synthesis it is apparent that zinc is required for normal protein synthesis. Probably the decreased protein synthesis characteristic of zinc deficiency only secondarily induces protein catabolism. It is interesting to note that the catabolism which follows major trauma is associated with the mobilization of zinc (Fell et al., 1973) to areas of privileged protein synthesis, namely, the wound and the liver (Savlov et al., 1962; van Rij et al., unpublished observation).

In the healing wound with scar formation there is a prolific production of protein, the bulk of which is collagen. The period of maximum synthesis of collagen occurs at about 10 to 20 days and corresponds to the time when the effect of zinc deficiency to decrease tensile strength becomes apparent in comparison with pair-fed controls (Pories et al., 1976; Sandstead and Shepherd, 1968). Similarly in zinc-deficient animals after wounding total collagen in the wound is decreased (Fernandez-Madrid et al., 1973). Zinc may appear to have a specific effect on collagen synthesis.

Not only is collagen content decreased in zinc-deficient wounds, but also there is a similar difference in the content of noncollagen proteins and other parameters of protein synthesis such as the RNA/DNA ratio. Zinc supplement given to deficient animals has a prompt enhancing effect on all these parameters, as well as collagen production (Elias and Chvapil, 1973), suggesting that zinc has a generalized effect on protein synthesis and consequently a nonspecific effect on collagen synthesis (Fernandez-Madrid et al., 1973).

Some confusion has resulted from experiments that failed to show an effect of zinc on collagen synthesis. Two studies measuring changes in wound tensile strength were not continued long enought to include the zinc-dependent critical period (O'Riain et al., 1968; Groundwater and McLeod, 1970). The observation that collagen synthesis in fibroblast culture as measured by the conversion of proline to hydroxyproline was not stimulated by the addition of 10^{-4} to 10^{-8} M zinc, but was inhibited (Waters et al., 1971), also seems contrary. However, the total zinc content of the incubation media was not given, and it is possible that the effect of toxic levels of zinc on collagen synthesis was in fact observed. Lack of a stimulatory effect of supplemental zinc on the rate of collagen synthesis and of collagen deposition in skin wounds in normal rats was described by Elias and Chvapil (1973). Added zinc had no effect on the mechanical properties of the skin wound or the collagen content in granulation tissue in normal animals, whereas these were enhanced when supplements were given to animals with presumed depleted zinc levels. Zinc supplement has no observable histological effects on the sequence of healing in normal zinc-sufficient rats (Lee et al., 1976). There appears to be no acceleration of protein synthesis when tissue zinc is replete.

After the period of maximum collagen synthesis the tensile strength of a wound continues to increase, even as some resorption of collagen occurs from a wound. Although tensile strength is a composite measure, the major factor in its later increase is a qualitative change in the collagen. Zinc appears to play no part in the primary and the secondary structure of the protein. McClain et al. (1973) suggested that zinc may have a fundamental role in the more complex structure changes associated with cross-linkage formation. New collagen is soluble in cold neutral salt solution; aging collagen dissolves only in acid solution; and mature collagen becomes insoluble in both salt and acid solutions. Oxidative deamination of lysine and hydroxylysine residues forms aldehyde radicals which condense to form stable chemical cross-links between collagen molecules. In zinc-deficient rats these workers report an increase in the dimer components and aldehyde content of the salt-soluble collagen, indicating an increased formation of covalent intramolecular cross-links in the more recently formed collagen. The reason for this increase was not clear from the study, but it was suggested that antagonism between copper and zinc could produce the cross-linkage abnormality. Tissues in zinc-deficient rats have a high copper content (Prasad et al., 1967), and consequently the activity of lysl oxidase, a copper enzyme essential in the formation of cross-linkages, may be modified. The McClain et al. study included the apparently paradoxical observation that in the acid-soluble collagen, which is biologically older, there was an inhibition of the intermolecular cross-linking mechanism with zinc deficiency, although the yield of this type of collagen was greater. Fernandez-Madrid and his workers (1973) in a detailed analysis of acid-soluble collagen were not able to observe a difference between collagen in normal and collagen in zinc-deficient animals. The workers concluded that the effects of zinc involve general protein synthesis and that zinc probably has no fundamental role in collagen cross-linkage. Furthermore, it seems inappropriate that the allegedly zinc-dependent changes of increased cross-linkage formation in collagen with scar maturation occur simultaneously with an egress of ^{65}Zn from the scar to a level less than that in the surrounding skin (Savlov et al., 1962).

Collagenase enzymes are instrumental in resorption of collagen from the scar. Like some zinc metalloenzymes, these enzymes are active at neutral pH and are inhibited by metal chelators such as ethylenediamine tetraacetate (EDTA) and cysteine (Peacock and Van Winkle, 1976), but the effect of zinc on these enzymes is still unknown.

5. ZINC AND WOUND NUCLEIC ACID SYNTHESIS

Before the deposition of collagen in the cellular response of a wound there is a spurt of DNA synthesis (Kulonen, 1970), and any process interfering with this spurt will markedly influence collagen formation. From the studies of Sandstead and co-workers (1970) and Fernandez-Madrid et al. (1973) it is evident that zinc is required not only for adequate collagen formation but also for tne preceding nucleic acid synthesis. In these studies decreased RNA content and evidence of decreased DNA synthesis in wounds of zinc-deficient animals were observed. Inhibition by EDTA of DNA synthesis in regenerating tissue is reversable by the addition of zinc (Rubin, 1972; Fujioka and Lieberman, 1964). Further studies indicate the importance of zinc in both RNA and DNA synthesis (Rubin and Koide, 1973; Sandstead and Rinaldi, 1969).

The major effect of zinc on the fibroblast, in which DNA synthesis precedes the production of collagen (Woessner and Boucek, 1961), appears to be on nucleic acid synthesis. Muscle collagen formation as measured by the incorpora- tion of L-$[^{14}C]$ leucine into protein per milligram of polysomal RNA is ap- proximately the same in zinc-supplemented and zinc-deficient animals, but the yield from the polysomes was decreased 32% in zinc deficiency. Although normal translation in these polysomes still occurred, the decreased yield of polysomal RNA suggests an alteration in RNA synthesis. Similarly, $[H^3]$ thy- midine incorporation into skin DNA is impaired in zinc deficiency (McClain et al., 1973). The role of zinc in nucleic acid synthesis is further emphasized by the recognition of DNA polymerase (Scrutton et al., 1971), RNA polymerase (Springgate et al., 1973), and viral, RNA-dependent DNA polymerase, reverse transcriptase (Auld et al., 1974), as zinc metalloenzymes. The RNA polymerase in *Escherichia coli* was reported to require zinc for the initiation of RNA synthe- sis, and Rubin and Koide (1973) suggested that zinc liberation for ribosomal RNA synthesis was the controlling event in DNA synthesis. By extrapolation it is evident that the role of zinc in wound healing is intimately associated with nucleic acid synthesis. There is probably no one critical site for which zinc is essential; rather, there is a complex interaction of several zinc-dependent processes in nucleic acid production in a healing wound.

Zinc may influence nucleic acids by means other than involvement in zinc metalloenzymes. In zinc deficiency increased RNA and DNA catabolism occurs with increased ribonuclease levels. Zinc at 10^{-4} M inhibits the activity of yeast ribonuclease (Somers and Underwood, 1969; Ohtaka et al., 1963).

The presence of zinc in highly purified RNA and DNA was observed by Wacker and Vallee in 1959. Possible structural effects of zinc in these nucleic acids have been examined (Tal, 1969; Wacker and Vallee, 1959), and the con- formation of polyribosome polynucleotides may also be zinc dependent. Fernandez-Madrid and co-workers (1973), studying connective tissue from zinc- deficient animals, observed a decrease in polyribosomes and a corresponding increase in inactive monosomes. The administration of zinc appeared to stimu- late polysomal formation (Sandstead et al., 1971).

Nucleic acid synthesis impairment in zinc deficiency results not only in the interruption of cell cycle growth, with decreased protein and RNA content and increased free amino acids, but also in impaired cell division. Falchuk and co-workers (1975a, 1975b) observed in zinc-deficient *Euglena* that most cells are arrested in the S and G_2 phases of mitosis. However, by inducing synchronous growth of these cells it became evident that the whole mitotic process was adversely affected. Zinc is also now thought to influence normal mitotic spindle formation and function.

Attention has been given thus far to fibroblast proliferation and subsequent protein synthesis. However, in the early wound there is also a proliferation of other metabolically active cell types, and these are equally susceptible to the need for zinc in nucleic acid synthesis. Of particular interest are the epithelial cells. Lactic dehydrogenase, a zinc-dependent enzyme in the epithelium, is increased there during wound repair (Imes and Hoopes, 1970). In both human beings and animals zinc deficiency characteristically affects the skin epithelium, with parakeratosis and subsequent ulceration. The early studies of Strain et al. (1953, 1954) with excised wounds in rats demonstrated the zinc supplementation effect to be most evident during the state of epithelialization. Rahmet et al. (1974) also studied epithelialization of excised wounds and found zinc deficiency to impair this aspect of healing. In the oral cavity healing wounds of the gingival mucous membrane show a dominance of epithelial cell proliferation in proportion to the fibroblastic activity that more often dominates in skin wounds. Zinc supplementation of Syrian hamsters with gingival wounds resulted in noticeably better healing after 3 days and a greater amount of epithelialization and tissue continuity (Mesrobian and Shklar, 1968). The tensile strength of the early mucosal wound is predominently the result of epithelial cell proliferation. Anything inhibiting this increased cell replication reduces the normal increase in tensile strength. and such is the case with zinc deficiency. This inhibitory effect tensile strength, and such is the case with zinc deficiency. This inhibitory effect of zinc deficiency occurs in mucosal wounds after 2 to 4 days but in skin wounds critical role of zinc in cell replication.

The effect of zinc on the metabolic processes described thus far requires further qualification in that, although zinc deficiency may have well-recognized consequences in these processes, the effects of zinc supplements are often unpredictable. Some animal experiments and *in vitro* studies have demonstrated enhanced effects with zinc supplements, but more often these do not occur or, on the contrary, inhibitory, possibly toxic effects are observed. However, zinc supplements generally do enhance the processes already suppressed by zinc deficiency. The supplement then appears to have an accelerating effect per se. Consequently the experimental conditions with respect to preceding zinc levels are important in the interpretation of zinc supplementation in wound healing.

6. ZINC AND CELLULAR MEMBRANE EFFECTS

The observation by ctyologists that zinc can be used to isolate intact cell membranes (Warren et al., 1966) and by Chvapil et al. (1976) that zinc occurs in small amounts in various cell membranes gives substance to the suggestion that zinc modifies cell plasma membranes (Chvapil, 1973). Zinc is bound to certain fractions of cell membranes. In red cell ghosts most of the zinc is bound to a lipoprotein fraction and is predominantly in the lipid moiety, whereas in the whole membrane the distribution of zinc between protein and lipid phases is approximately 45 to 55% (Chvapil et al., 1976). As a result of the interaction of zinc with either reactive macromolecular extrinsic components of the membranes (e.g., sialic acid) or with intrisic structures of the plasma membrane, certain modifications of membrane structure occur. These changes may indirectly modify the cell functions dependent on the activity of the plasma membrane. Consequently a regulatory role for zinc on cell function, which is not directly enzymatic, has been suggested (Chvapil, 1973).

In isolated peritoneal macrophages functional changes in migration and phagocytocytic activity are observed with changes in available zinc (Chvapil et al., 1976). Increasing amounts of zinc reduce membrane activity, an effect that is reversible with removal of the zinc. Furthermore, the ATPase system of activated alveolar macrophages is membrane bound and located predominantly in surface membranes (Mustafa et al., 1970) and is completely inhibited by 0.5 mM zinc (Chvapil et al., 1976). This ATPase system has been assumed to be essential for phagocytosis and pinocytosis (Mustafa et al., 1970).

There is some similarity between the zinc membrane-stabilizing effects and the membrane-stabilizing properties of corticosteroids and of some anti-inflammatory drugs. These substances have, however, been associated in large doses with retardation of wound repair (Morton and Malone, 1972). Such a possible inhibitory effect of pharmacological doses of zinc on wound repair has not been observed. However, some therapeutic application has been suggested for the anti-inflammatory effect of zinc, for example, in the suppression of renal graft rejection (Chvapil et al., 1976) and in antiallergic propylaxis (Cho et al., 1977).

Other membrane effects of zinc have also been described. Lymphocyte transformation, which is induced by phytohemagglutanin is also caused by the addition of zinc ions in lymphocyte cultures and has been attributed to intracellular effects of zinc on DNA synthesis. However, a similar effect is produced when zinc is complexed as zinc-8-hydroxyquinolone (Chvapil et al., 1976). Zinc in this complex no longer permeates biological membranes and consequently does not enter the cell (Albert, 1965). This implies a direct membrane effect by the zinc, possibly similar to that of phytohemagglutinin.

In sickle-cell anemia the red cells are abnormal, and treatment of the red cell membranes with zinc *in vitro* produces an improvement in membrane filterability (Brewer and Oelshegel, 1974), Many of these membrane effects are observed *in vitro* with pharmacological doses of zinc or with ionic zinc. In contrast, *in vivo,* at physiological levels zinc is predominantly protein bound in the circulating body fluids (Giroux and Henkin, 1972), with small fractions complexed with amino acids and an undetermined, minute amount occurring in the ionic form. Consequently zinc is less readily available to cells *in vivo.* Furthermore, these *in vitro* conditions may not be reproducible *in vivo* with large doses of zinc, without other, toxic effects of the element resulting. Some reservations as to the significance of zinc effects observed with *in vitro* models is warranted.

7. ZINC AND THE INFLAMMATORY CELLS IN WOUNDS

It is also appropriate to draw attention briefly to the role of zinc in the more specific activities of some of the inflammatory cells as they contribute to the initiation of much of the cellular activity that follows in the wound. In response to added zinc *in vitro,* macrophages, previously alluded to, are inhibited in their phagocytic and bacterial killing ability (Chvapil et al., 1976). Inhibition of oxygen consumption and of the activity of the enzyme NADPH oxidase also occurs. This enzyme is associated with the stimulated metabolism of phagocytosis and the production of hydrogen peroxide needed for bactericidal effects in these cells.

Although the leukocytes are relatively rich in zinc, similar inhibitory effects of added zinc have been demonstrated in activated dog and human peripheral granulocytes (Stankova et al., 1976). Leukocyte inflammatory mediating factor synthesis in these cells and prostaglandin synthesis are also inhibited by zinc (Mapes et al., 1978; Chvapil et al., 1977). The increased stabilization of lysosomal membranes produced by zinc treatment inhibits, *in vitro,* the release of proteolytic enzymes and fibroblast-stimulating factors contained within the lysozomes. This effect is somewhat paradoxical in view of the stimulatory effect of zinc on fibroblast proliferation. Since zinc deficiency is not associated with increased lysosomal fragility, some workers have concluded that lysosomal fragility is not a prime locus of zinc action during cell division (Dreosti and Record, 1978).

Platelets, also present in the initial phase after injury, are influenced by zinc, which inhibits both platelet aggregation and serotonin release (Chvapil et al., 1975). Released serotonin and histamine, as well as, possibly, other platelet factors, stimulate both DNA synthesis in fibroblasts and collagen cross-linkage. Some of these actions of serontonin and histamine are probably mediated by prostaglandins, which have themselves been implicated as mediators of the late phase of the inflammatory response and the early initiation of repair (Peacock and Van Winkle, 1976). More recently zinc has been shown to influence prostaglandin production and action (Mapes et al., 1978).

Lymphocytes, which are particularly important in immune reactions, are also characteristically present in the inflammatory response, especially when a wound is infected or chronic. Both types of lymphocytes, the thymus-derived T-cells and bone-derived B-cells, are present. Administration of zinc within narrow limits into lymphocyte cultures stimulates DNA synthesis within 6 to 7 days and results in increased lymphocyte transformation (Ruhl et al., 1971; Chesters, 1972; Berger and Skinner, 1974). Other evidence is now appearing that supports the importance of zinc in the immune response, in which lymphocytes play a central role. Zinc deficiency per se in mice causes a loss of immunity affecting thymus weight and the antibody-mediated response (Luecke et al., 1978). Small thymus size associated with low zinc blood levels in children responds to zinc supplements with increase in thymus size.

8. ZINC TISSUE TRANSPORT FOLLOWING INJURY

In the past there has been considerable interest in the apparent preferential redistribution of zinc to the healing wound to facilitate repair. Little is known about this process except that it occurs rapidly, within hours of the trauma (Savlov et al., 1963; Lichti et al., 1970; van Rij et al., unpublished observations), and that early changes in wound fluid zinc follow the changes in venous plasma and subsequently remain elevated as the plasma zinc falls (Lichti et al., 1970). These early changes correspond to the period of maximum vascularity and transudation, as demonstrated by injection of ^{131}I-labeled albumin and ^{51}Cr-labeled red cells (Weiber, 1959). Red cells are rich in zinc. Granulocytes have a high zinc content, but it is unlikely that they contribute greatly to the increased zinc content of traumatized tissue (Lindeman et al., 1973). Zinc is bound to various proteins entering the wound, particularly albumin. Parenthetically, not all tissue injury and repair has been associated with increased zinc levels, in particular, in myocardial infarcts (Linderman et al., 1970; Mathew et al., 1978). It seems, however, that zinc enters the wound as part of the general vascular and inflammatory response initiated by trauma. This zinc is taken up into the active proliferating cells in the wound and consequently results in the zinc retention observed there.

Trauma and its associated stress do, however, initiate changes in zinc distribution in tissues remote from the site of injury. Within the first few hours after trauma, reported elevated serum levels have been attributed to progesterone-induced mobilization of zinc from tissues (Pories et al., 1976). After injury decreased serum zinc levels are more frequently observed (Henzel et al., 1970; Lindeman et al., 1972; Flynn et al., 1973; Walker et al., 1978). Zinc uptake in liver (Pedarek and Beisel, 1971) and some other tissues is increased and in bone and muscle is decreased (van Rij et al., unpublished). Increased urinary zinc excretion also occurs after major trauma (Fell et al., 1973). Corticosteroids simulate some of these effects and lower plasma zinc (Flynn et al., 1971, 1973), but a hormone-like substance, leukocyte endogenous mediator (LEM), a product of activated peripheral granulocytes, also appears to have a critical role in initiating some of these changes (Pekarek and Beisel, 1969, 1971). Some of these observations appear to be conflicting, and further elucidation of the cause and timing of these movements of zinc is required, as well as a better understanding of their significance.

9. ZINC THERAPY: ANIMAL STUDIES

The need for zinc in wound healing has been demonstrated in a variety of species, including rats, hamsters, guinea pigs, and domestic animals (Miller 1970), and in a variety of tissues. Skin wounds have been most thoroughly studied, but zinc requirements for healing have also been observed in gingival mucosa, in alveolar bone after tooth extraction (Battistone et al., 1972b), and in fractured long bone (Battistone et al., 1972a). In a number of animal studies supplemental zinc has not resulted in acceleration of healing (Lee et al., 1976). It is now increasingly evident that optimal zinc levels exist for many of the critical processes in a wound but that exceeding these levels provides no advantage. In retrospect it seems likely that some of the early animal studies showing accelerated wound healing with zinc used control animals with low zinc status (Pories et al., 1976). In the 1950's, when these experiments were performed, most animal feeds were woefully low in zinc. Failure to observe effects of zinc in other studies (O'Riain et al., 1968; Groundwater and McLeod, 1970) resulted from the delay period between the critical events of cellular proliferation and protein synthesis and their subsequent effect on tensile strength, the indicator of zinc activity in the wound. Too early measurement of this parameter, as in those studies, would be expected to miss the effect of zinc, which is observed a few days later (Sandstead and Shepherd, 1968).

10. ZINC THERAPY: HUMAN STUDIES

In human beings the therapeutic effect of zinc has been well tested. In 1966 Pories et al. (1966) developed a method of measuring changes in wound volume by alginate impressions and first began testing the effectiveness of oral zinc sulfate therapy. Twenty young men, prior to operation for pilonidal sinus, were randomly distributed into two groups. The zinc-medicated group, despite having larger wounds than the controls, healed 34.3 days faster with zinc therapy (45.8 ± 2.6 days versus 80.1 ± 13.7 days) ($p < .02$). The pattern of healing was similar to that observed in experimental animals, there being no difference in the rate of healing in the first 15 days after wounding; the major effect was apparent during the stage of epithelialization.

Since these observations many other workers have demonstrated the beneficial effects of zinc therapy. These studies include controlled trials on venous leg ulcers (Husain, 1969; Greaves and Skillen, 1970; Haeger et al., 1972, 1974; Hallbook and Lanner, 1972; Haeger and Lanner 1974). Improved graft accep-tance and epithelialization in children with major burns (Larson, 1974), the healing of gastric ulcers (Frommer, 1975) and of indolent ulcers (Halsted and Smith, 1970), and improved healing in surgical patients with abnormally slowly healing wounds (Henzel et al., 1970) and in the wounds of patients on prolonged corticosteroid therapy (Flynn et al., 1973) have all been observed after zinc therapy. Acne has been improved with zinc therapy (Michealsson et al., 1977), reflecting the importance of zinc in normal function and integrity of the skin.

Clinical zinc deficiency causes the marked, unsightly skin changes of parakera-tosis, and wound healing is impaired (Tucker et al., 1976; Kay et al., 1976; van Rij and McKenzie, 1977). These and other abnormalities rapidly respond to zinc therapy. Unfortunately subclinical zinc deficiency is not uncommon in hos-pital patients (Sullivan et al., 1969) and appears to be related to inadequate zinc nutrition and the specific effects of some disorders on zinc metabolism (Sand-stead et al., 1976). Slow healing has been associated with low serum zinc levels (Pories et al., 1971; Sarjeant et al., 1970; Halsted and Smith, 1970; Flynn et al., 1973). Henzel and co-workers (1970, 1974) identified 16 patients with impaired wound healing and low zinc levels in serum, epithelium, and granulation tissue from the wound. Healing and zinc levels in these tissues improved concomitantly with zinc therapy.

Several groups of workers have failed to demonstrate a beneficial effect of zinc therapy on healing (Brewer et al., 1966, 1967; Myers and Cherry, 1970; Barcia, 1970). Some of these studies did not establish the serum zinc levels before therapy, and it is now recognized from laboratory and clinical studies that accelerated healing occurs only when zinc levels are low before supplemen-tation. In the study of Myers and Cherry (1970), statistical interpretations differ and an effect of zinc may exist (Pories et al., 1976).

The role of zinc therapy in tissue repair is now established experimentally and clinically. Routine zinc supplementation is indicated in patients with measured lowered serum zinc levels or those known to be at high risk of developing con-ditioned zinc deficiencies, of which there is a wise spectrum (Sandstead et al., 1976). These patients are at risk of impaired wound healing and other associated complications such as infection, and zinc replacement is effective in minimizing these complications.

228   Zinc in Wound Healing

REFERENCES

Albert, A. (1965). *Selective Toxicity*. John Wiley, New York, p. 222.

Auld, D. S., Kawaguchi, H., Livingston, D. M., and Vallee, B. L. (1974). "Reverse Transcriptase from Ava in Myeloblastosis Virus: a Zinc Metalloenzyme," *Biochem. Biophys. Res. Commun.*, 57, 967–972.

Barcia, P. J. (1970). "Lack of Acceleration of Healing with Zinc Sulfate," *Ann. Surg.*, 172, 1048–1050.

Battistone, G. C., Rubin, M. I., Cutright, D. E., Miller, R. A., and Harmuth-Hoene, A. E. (1972a). "Zinc and Bone Healing: Effect of Zinc Systeamine-*N*-acetic Acid on the Healing of Experimentally Injured Guinea Pig Bone," *Oral Surg., Oral Med., Oral Pathol.*, 34, 542–552.

Battistone, G. C., Posey, W. R., Barone, J. J., and Cutright, D. E. (1972b). "Zinc and Bone Healing: The Effect of Zinc-Cysteamine-*N*-acetic Acid on the Healing of Extraction Wounds in Rats," *Oral Surg., Oral Med., Oral Pathol.*, 34, 704–711.

Berger, N. A. and Skinner, A. M. (1974). "Characterization of Lymphocyte Transformation Induced by Zinc Ions," *J. Cell. Biol.*, 61, 45–55.

Brewer, G. J. and Oelshlegel, F. J., Jr. (1974). "Antisickling Effects of Zinc," *Biochem. Biophys. Rev. Commun.*, 58, 854–861.

Brewer, R. D., Jr., Lea, J. F., and Mihaldzic, N. (1966). "Preliminary Observations on the Effect of Oral Zinc Sulfate on the Healing of Decubitus Ulcers." *Proceedings of the Annual Clinical Spinal Cord Injury Conference*, Vol. 15, pp. 93–96.

Brewer, R. D., Jr., Mihaldzie, N., and Dietz, A. (1967). "The Effect of Oral Zinc Sulfate on the Healing of Decubitus Ulcers in Spinal Cord Injured Patients." In *Proceedings of the Annual Clinical Spinal Cord Injury Conference*, Vol. 17, pp. 70–72.

Carpenter, F. H. and Vahl, M. J. (1973). "Leucine Aminopeptidase (Bovine Lens) Mechanism of Activation by Magnesium $^{2+}$ and Manganese $^{2+}$ of the Zinc Metallo-enzyme, Amino Acid Composition and Sulfhydryl Content," *J. Biol. Chem.*, 248, 294–240.

Chesters, J. K. (1972). "The Role of Zinc Ions in the Transformation of Lymphocytes by Phytohaemaglutinin," *Biochem. J.*, 130, 133–139.

Cho, C. H., Dai, S., and Ogle, C. W. (1977). "The Effect of Zinc on Anaphylaxis *in vivo* in the Guinea-Pig," *Br. J. Pharmacol.*, 60, 607–608.

Chvapil, M. (1973). "New Aspects in the Biological Role of Zinc: A Stabilizer of Macromolecules and Biological Membranes," *Life Sci.*, 13, 1041–1049.

Chvapil, M. (1974). "Zinc and Wound Healing." In B. Zederfeldt, Ed., *Symposium on Zinc*. A. B. Tika, Lund, Sweden.

Chvapil, M. and Walsh, D. (1972). *Excerpta Med. Found. Int. Congr. Ser.* 264, pp. 226–228.

Chvapil, M., Weldy, P. L., Stankova, L., Clark, D. S., and Zukoski, C. F. (1975).

"Inhibitory Effect of Zinc Ions on Platelet Aggregation and Serotonin Release Reaction," *Life Sci.,* **16**, 561–572.

Chvapil, M., Zukoski, C. F., Hattler, B. G., Stankova, L., Montgomery, D., Carlson, E. C., and Ludwig, J. C. (1976). "Zinc and Cells." In A. S. Prasad, Ed., *Trace Elements in Human Health and Disease,* Vol. 1. Academic Press, New York, pp. 269–281.

Chvapil, M., Stankova, L., Zukoski, C., and Zukoski, C. (1977). "Inhibition of Some Functions of Polymorphonuclear Leukocytes by *in vitro* Zinc," *J. Lab. Clin. Med.,* **89**, 135–146.

Dreosti, I. E. and Record, I. R. (1978). "Lysosomal Stability, Superoxide Dismutase and Zinc Deficiency in Regenerating Rat Liver," *Br. J. Nutr.,* **40**, 133–137.

Drum, D. E., Harrison, J. H., Li, T.-K., Bethune, J. L., and Vallee, B. L. (1967). "Structural and Functional Zinc in Horse Liver Alcohol Dehydrogenase," *Proc. Natl. Acad. Sci.,* **57**, 1434–1440.

Ebers, G. (1875). *Papyrus Ebers. Das hermetiches Buch über die Arzneimittel,* Vol. I, p. 36, Tafel I–LXIX; Vol. II, p. 63, Tafel LXX–CS. Engelmann, Leipzig.

Elias, S. and Chvapil, M. (1973). "Zinc and Wound Healing in Normal and Chronically Ill Rats," *J. Surg. Res.,* **15**, 59–66.

Falchuk, K. H., Fawcett, D. W., and Vallee, B. L. (1975). "Role of Zinc in Cell Divison of *Euglena gracilis,*" *J. Cell. Sci.,* **17**, 57–78.

Fell, G. S., Fleck, A., Cuthbertson, D. P., Queen, K., Morrison, C., Bessent, R. G., and Husain, S. L. (1973). "Urinary Zinc Levels as an Indication of Muscle Catabolism," *Lancet,* **1**, 280–282.

Fernandez-Madrid, F., Prasad, A. S., and Oberleas, D. (1973). "Effect of Zinc Deficiency on Nucleic Acids, Collagen, and Noncollagenous Protein of the Connective Tissue," *J. Lab. Clin. Med.,* **82**, 951–961.

Flynn, A., Pories, W. J., Strain, W. H., Hill, O. A., Jr., and Fratianne, R. B. (1971). "Rapid Serum-Zinc Depletion Associated with Corticosteroid Therapy," *Lancet,* **2**, 1169–1172.

Flynn, A., Pories, W. J., Strain, W. H., and Hill, O. A., Jr. (1973). "Zinc Deficiency with Altered Adrenocortical Function and Its Relation to Delayed Healing," *Lancet,* **1**, 780–791.

Fox, M. R. S. and Harrison, B. N. (1965). Effects of Zinc Deficiency on Plasma Proteins of Young Japanese Quail," *J. Nutr.,* **86**, 89–92.

Frommer, Donald J. (1975). "The Healing of Gastric Ulcers by Zinc Sulphate," *Med. J. Aust.,* **2**, 793–796.

Fujioka, M. and Lieberman, I. (1964). "A Zn^{2+} Requirement of Synthesis of Deoxyribonucleic Acid by Rat Liver," *J. Biol. Chem.,* **239**, 1164–1167.

Giroux, E. L. and Henkin, R. I. (1972). "Competition for Zinc among Serum Albumin and Amino Acids," *Biochem. Biophys. Acta,* **273**, 64–72.

Greaves, M. W. and Skillen, A. W. (1970). "Effects of Long-Continued Ingestion of Zinc Sulphate in Patients with Venous Leg Ulceration," *Lancet,* 2, 889–891.

Groundwater, W. and MacLeod, I. B. (1970). "The Effects of Systemic Zinc Supplements on the Strength of Healing Incised Wounds in Normal Rats," *Br. J. Surg.,* 57, 222–225.

Haeger, K. and Lanner, E. (1974). "Oral Zinc Sulphate and Ischaemic Leg Ulcers," *Vasa,* 3, 77–81.

Haeger, K., Lanner, E., and Magnusson, P. -O. (1974). "Oral zinc Sulphate in the Treatment of Venous Leg Ulcers," *Vasa,* 1, 62–69.

Haeger, K., Lanner, E., and Magnusson, P. –O. (1974). "Oral zinc Sulphate in the Treatment of Venous Leg Ulcers." In W. J. Pories et al., Eds., *Clinical Applications of Zinc Metabolism.* Charles C Thomas, Springfield, Ill., pp. 158–167.

Hallbook, T. and Lanner, E. (1972). "Serum-Zinc and Healing of Venous Leg Ulcers," *Lancet,* 2, 780–782.

Halsted, J. A. and Smith, J. C., Jr. (1970). "Plasma-Zinc in Health and Disease," *Lancet,* 1, 322–324.

Henzel, J. H., DeWeese, M. S., and Lichti, E. L. (1970). Zinc Concentrations within Healing Wounds," *Arch. Surg.,* 100, 349–356.

Henzel, J. H., Lichti, E. L., Shepard, W., and Paone, J. (1974). Long-Term Oral Zinc Sulfate in the Treatment of Atherosclerotic Peripheral Vascular Disease: Efficacy of Possible Mechanism of Action." In W. J. Pories et al., Eds., *Clinical Applications of Zinc Metabolism.* Charles C Thomas, Springfield, Ill., pp. 243–259.

Hill, C. H., Matrone, G., Payne, W. L., and Barber, C. W. (1963). "*In vivo* Interactions of Cadmium with Copper, Zinc and Iron," *J. Nutr.,* 80, 227–235.

Hsu, J. M. (1976). "Zinc as Related to Cystine Metabolism." In A. S. Prasad, Ed., *Trace Elements in Human Health and Disease,* Vol. 1. Academic Press, New York, pp. 295–310.

Hsu, J. M. and Anthony, W. L. (1971). "Impairment of Cystine [35]S Incorporation into Skin Protein by Zinc-Deficient Rats," *J. Nutr.,* 101, 445–452.

Huber, A. M. and Gershoff, S. N. (1973). "Effects of Dietary Zinc on the Enzymes in the Rat," *J. Nutr.,* 103, 1175–1181.

Husain, S. L. (1969). "Oral Zinc Sulphate in Leg Ulcers," *Lancet,* 1, 1069–1071.

Imes, M. J. C. and Hoopes, J. E. (1970). "Enzyme Activities in the Repairing Epithelium during Wound Healing," *J. Surg. Res.,* 104, 173.

Kay, R. G., Tasman-Jones, C., Pybus, J., Whiting, R., and Black, H. (1976). "A Syndrome of Acute Zinc Deficiency during Total Parenteral Alimentation in Man," *Ann. Surg.,* 183, 331–340.

Kirchgessner, M. and Roth H. P. (1975). "Zur Bestimmung der Verfugbarkeit von Zink im Stoffwechsel sowie zur Ermittlung des Zinkberdarfs mittels Aktivitatsanderungen von Zn-Metalloenzymen," *Arch. Tierenaehr.,* 25, 83–92.

Kirchgessner, M., Schwarz, W. A., and Roth, H. P. (1975). "Zur Aktivitat der alkalischen Phosphatase in Serum und Knochen von zinkdepletierten und -repletierten Kuhn," *Z. Tierphysiol., Tierernaehr. Futtermittelkd.*, **35**, 191–200.

Kulonen, E. (1970). "Studies on Experimental Granuloma." In E. A. Balazs, Ed., *Chemistry and Molecular Biology of the Intracellular Matrix*, Vol. 3. Academic Press, New York, pp. 1811–1818.

Larson, D. L. (1974). "Oral Zinc Sulfate in the Management of Severely Burned Patients." In W. J. Pories et al., Eds., *Clinical Applications of Zinc Metabolism*. Charles C Thomas, Springfield, Ill., pp. 234–235.

Lavy, U. I. (1972). "The Effect of Oral Supplementation of Zinc Sulphate on Primary Wound Healing in Rats," *Br. J. Surg.*, **59**, 194–196.

Lee, P. W. R., Green, M. A., Long, W. B., and Gill, W. (1976). "Zinc and Wound Healing," *Surg., Gynecol. Obstet.*, **143**, 549–554.

Lichti, E. L., Turner, M., and Henzel, J. H. (1970). "Changes in Serum Zinc Levels Following Periods of Increased Metabolic Activity and Differences in Arterial and Venous Zinc Concentrations Following Surgically Inflicted Wounds." In D. D. Hemphill, Ed., *Trace Substances in Environmental Health*, University of Missouri, Columbia.

Lichti, E. L., Schilling, J. A., and Shirley, H. M. (1972). "Wound Fluid and Plasma Zinc Levels in Rats during Tissue Repair," *Am. J. Surg.*, **123**, 253–256.

Lindeman, R. D., Bottomley, R. G., Cornelison, R. L., Jr., and Jacobs, L. A. (1972). "Influence of Acute Tissue injury on Zinc Metabolism in Man," *J. Lab. Clin. Med.*, **79**, 452–460.

Lindeman, R. D., Yunice, A. A., Baxter, D. J., Miller, L. R., and Nordquist, J. (1973). "Myocardial zinc Metabolism in Experimental Myocardial Infarction," *J. Lab. Clin. Med.*, **81**, 194–204.

Luecke, R. W., Simonel, C. E., and Fraker, P. J. (1978). "The Effect of Restricted Dietary Intake on the Antibody Mediated Response of the Zinc Deficient A/J Mouse," *J. Nutr.*, **108**, 881–887.

McClain, P. E., Wiley, E. R., Beecher, G. R., and Anthony, W. L. (1973). "Influence of Zinc Deficiency on Synthesis and Cross-linking of Rat-Skin Collagen," *Biochim. Biophys. Acta*, **304**, 457–465.

McCray, L. A., Higa, L. H., and Soni, N. N. (1972). "The Effect of Orally Administered Zinc Sulfate on Extraction Wound Healing in Hamsters," *Oral Surg., Oral Med., Oral Pathol.*, **33**, 314–322.

Macapinlac, M. P., Pearson, W. N., Barney, G. H., and Darby, W. J. (1968). "Protein and Nucleic Acid Metabolism in the Testes of Zinc-Deficient Rats," *J. Nutr.*, **95**, pp. 569–577.

Magee, A. C. and Matrone, G. (1960). "Studies on Growth, Copper Metabolism and Iron Metabolism of Rats Fed High Levels of Zinc," *J. Nutr.*, **72**, pp. 233–242.

Mapes, C. A., Bailey, P. T., Matson, C. F., Hauer, E. C., and Sobocinski, P. Z. (1978). "*In vitro* and *vivo* Actions of Zinc Ion Affecting Cellular Substances Which Influence Host Metabolic Responses to Inflammation," *J. Cell. Physiol.*, 95, 115-124.

Mathew, B. M., Kumar, S., Ahmad, M., Salahuddin, Seth, T. D., Hasan, M. Z., and Jamil, S. A. (1978). "A Temporal Profile of Myocardial Zinc Changes after Isoproterenol Induced Cardiac Necrosis," *Jap. Circ. J.*, 42, 353-357.

Mesrobian, A. Z. and Shklar, G. (1968). "The Effect on Gingival Wound Healing Of Dietary Supplements of Zinc Sulfate in the Syrian Hamster," *Periodontics*, 6, 224-229.

Michaelsson, G., Juhlin, L., and Ljunghall, K. (1977). "A Double-Blind Study of the Effect of Zinc and Oxytetracycline in Acne Vulgaris," *Br. J. Dermatol.*, 97, 561-566.

Miller, W. J. (1970). "Zinc Nutrition of Cattle: A Review," *J. Dairy Sci.*, 53, 1123-1135.

Mills, C. F., Quarterman, J., Williams, R. B., Dalgarno, A. C., and Panic, B. (1967). "The Effects of Zinc Deficiency on Pancreatic Carboxypeptidase Activity and Protein Digestion and Absorption in the Rat," *Biochem. J.*, 102, 712-718.

Mills, C. F., Quarterman, J., Chesters, J. K., Williams, R. B., and Dalgarno, A. C. (1969). "Metabolic Role of Zinc," *Am. J. Clin. Nutr.*, 22, 1240-1249.

Morton, J. J. P. and Malone, M. H. (1972). "Evaluation of Vulnerary Activity by an Open Wound Procedure in Rats," *Arch. Int. Pharmacodyn. Ther.*, 196, 117-126.

Mustafa, M. G., Cross, C. E., and Hardie, J. A. (1970). "Localization of Na^+-K^+, Mg^{++} Adenosinetriphosphatase Activity in Pulmonary Alveolar Macrophage Subcellular Fractions," *Life Sci.*, 9, 947-954.

Myers, M. B. and Cherry, G. (1970). "Zinc and the Healing of Chronic Leg Ulcers," *Am. J. Surg.*, 120, 77-81.

Oberleas, D., Seymour, J. K., Lenaghan, R., Hovanesian, J., Wilson, R. F., and Prasad, A. S. (1971). "Effect of Zinc Deficiency on Wound-Healing in Rats," *Am. J. Surg.*, 121, 556-571.

O'Riain, S., Copenhagen, H. J., and Calnan, J. S. (1968). "The Effect of Zinc Sulfate on the Healing of Incised Wounds in Rats," *Br. J. Plast. Surg.*, 21, 240-243.

Peacock, E. E. and Van Winkle, W. (1976). "Wound Repair." In A. S. Prasad, Philadelphia, 1972.

Pekarek, R. S. and Beisel, W. R. (1969). "Effect of Endotoxin on Serum Zinc Concentrations in the Rat," *Appl. Microbiol.*, 18, 482-484.

Pekarek, R. S. and Beisel, W. R. (1971). "Characterization of the Endogenous Mediator(s) of Serum Zinc and Iron Depression during Infection and Other Stresses," *Proc. Soc. Exp. Biol. Med.*, 138, 728-732.

Pories, W. J., Schaer, E. W., Jordan, D. R., Chase, J., Parkinson, G., Whittaker,

R., Strain, W. H., and Rob, C. G. (1966). "The Measurement of Human Wound Healing," *Surgery,* **59**, 821–824.

Pories, W. J., Henzel, J. H., Rob, C. G., and Strain, W. H. (1967a). "Acceleration of Wound Healing in Man with Zinc sulphate Given by Mouth," *Lancet,* **1**, 121–124.

Pories, W. J., Henzel, J. H., Rob, C. G., and Strain, W. H. (1967b). "Acceleration of Healing with Zinc Sulfate," *Ann. Surg.,* **165**, 432–436.

Pories, W. J., Strain, W. H., and Rob, C. G. (1971). "Zinc Deficiency in Delayed Healing and Chronic Disease," *Geol. Soc. Am. Mem.* **123**, pp. 73–95.

Pories, W. J., Mansour, E. G., Plecha, F. R., Flynn, A., and Strain, W. H. (1976). "Metabolic Factors Affecting Zinc Metabolism in the Surgical Patient." In A. S. Prasad, Ed., *Trace Elements in Human Health and Disease,* Vol. 1. Academic Press, New York.

Prasad, A. S., and Oberleas, D. (1970). "Zinc: Human Nutrition and Metabolic Effects," *Ann. Intern. Med.,* **73**, 631–636.

Prasad, A. S. and Oberleas, D. (1974). "Thymidine Kinase Activity and Incorporation of Thymidine into DNA in Zinc Deficient Tissue," *J. Lab. Clin. Med.,* **83**, 634–639.

Prasad, A. S., Oberleas, D., Wolf, P., and Horwitz, J. P. (1967). "Studies on Zinc Deficiency: Changes of Trace Elements and Enzyme Activities in Tissues of Zinc-Deficient Rats," *J. Clin. Invest.,* **46**, 549–557.

Pullen, F. W., II. (1974). "Oral Zinc and Vitamin Therapy for Laryngotracheal Trauma and Surgical Aftercare." In W. J. Pories et al., Eds., *Clinical Applications of Zinc Metabolism,* Charles Thomas, Springfield, Ill., pp. 237–242.

Rahmat, A., Norman, J. N., and Smith, G. (1974). "The Effect of Zinc Deficiency on Wound Healing," *Br. J. Surg.,* **61**, 271–273.

Reinhold, J. G. and Kfoury, G. A. (1969). "Zinc-Dependent Enzymes in Zinc-Depleted Rats: Intestinal Alkaline Phosphatase," *Am. J. Clin. Nutr.,* **22**, 1250–1263.

Riordan, J. F. and Vallee, B. C. (1976). "Structure and Function of Zinc Metalloenzymes in Trace Elements in Human Health and Disease." In A. S. Prasad, Ed., *Trace Elements in Human Health and Disease,* Vol. 1. Academic Press, New York, pp. 227–256.

Rosenbusch, J. P. and Weber, K. (1971). "Localization of the Zinc Binding Site of Asparate Transcarbamoylase in the Regulatory Subunit," *Proc. Natl. Acad. Sci.,* **68**, 1019–1023.

Rubin, H. (1972). "Inhibition of DNA Synthesis in Animal Cells by Ethylene Diamine Tetraacetate and Its Reversal by Zinc," *Proc. Natl. Acad. Sci.,* **69**, pp. 712–716.

Rubin, H. and Koide, T. (1973). "Inhibition of DNA Synthesis in Chick Embryo Cultures by Deprivation of Either Serum or Zinc," *J. Cell Biol.,* **56**, 777–786.

Ruhl, H., Kirchner, H., and Bochert, G. (1971). "Kinetics of the Zn^{2+} Stimulation of Human Peripheral Lymphocytes *in vitro*," *Proc. Soc. Exp. Biol. Med.*, **137**, 1089-1092.

Sandstead, H. H. and Rinaldi, R. A. (1969). "Impairment of Deoxyribonucleic Acid Synthesis by Dietary Zinc Deficiency in the Rat," *J. Cell. Physiol.*, **73**, 81-83.

Sandstead, H. H. and Shepard, G. H. (1968). "The Effect of Zinc Deficiency on the Tensile Strength of Healing Surgical Incisions in the Rat," *Proc. Soc. Exp. Biol. Med.*, **128**, 687-689.

Sandstead, H. H., Lanier, V. C., Jr., Shepard, G. H., and Gillespie, D. D. (1970). "Zinc and Wound Healing: Effects of Zinc Deficiency and Zinc Supplementation," *Am. J. Clin. Nutr.*, **23**, 514-519.

Sandstead, H. H., Hollaway, W. L., and Baum, V. (1971a). "Zinc Deficiency: Effect on Polysomes," *Fed. Proc. Fed. Am. Soc. Exp. Biol.*, **30**, 517.

Sandstead, H. H., Terhune, M., Brady, R. N., Gillespie, D., and Hollaway, W. L. (1971b). "Zinc Deficiency: Brain DNA, Protein and Lipids and Liver Ribosomes and RNA Polymerase," *Clin. Res.*, **19**, 83.

Sandstead, H. H., Vo-Khacty, K. P., and Solomons, N. (1976). "Conditioned Zinc Deficiencies." In A. S. Prasad, Ed., *Trace Elements in Human Health and Disease*, Vol. 1. Academic Press, New York, pp. 33-50.

Savlov, E. D., Strain, W. H., and Huegin, F. (1962). "Radiozinc Studies in Experimental Wound Healing," *J. Surg. Res.*, **2**, 209-212.

Scrutton, M. C., Wu, C. W., and Goldthwait, D. A. (1971). "The Presence and Possible Role of Zinc in RNA Polymerase Obtained from *Escherichia Coli*," *Proc. Natl. Acad. Sci.*, **68**, 2497-2501.

Sedlacek, T. V., Mangan, C. E., Giuntoli, R. L., and Mikuta, J. J. (1976). "Zinc Sulfate: An Adjuvant to Wound Healing in Patients Undergoing Radical Vulvectomy," *Gynecol. Oncology*, **4**, 324-327.

Sarjeant, G. R., Galloway, R. E., and Gueri, M. C. (1970). "Oral Zinc Sulphate in Sickle-Cell Ulcers," *Lancet*, **2**, 891-892.

Slater, J. P., Mildvan, A. S., and Loeb, L. A. (1971). "Zinc in DNA Polymerases," *Biochem. Biophys. Res. Commun.*, **44**, 37-43.

Somers, M. and Underwood, E. J. (1969). "Ribonuclease Activity and Nucleic Acid and Protein Metabolism in the Testes of Zinc-Deficient Rats," *Aust. J. Biol. Sci.*, **22**, 1277-1282.

Springgaate, C. F., Mildvan, A. S., and Loeb, L. A. (1973). "Studies on the Role of Zinc in DNA Polymerase," *Fed. Proc. Fed. Am. Soc. Exp. Biol.*, **32**, p. 451.

Stankova, L., Drach, G. W., Hicks, T., Zukoski, C. F., and Chvapil, M. (1976). "Regulation of Some Functions of Granulocytes by Zinc of the Prostatic Fluid and Prostate Tissue," *J. Lab. Clin. Med.*, **88**, 640-648.

Starcher, B. C. (1969). "Studies on the Mechanism of Copper Absorption in the Chick," *J. Nutr.*, **97**, 321-326.

Strain, W. H., Dutton, A. M., Heyer, H. B., and Ramsey, G. H. (1953). *Experimental Studies on the Acceleration of Burn and Wound Healing.* University of Rochester Reports, Rochester, N.Y., p. 18.

Strain, W. H., Dutton, A. M., Heyer, H. B., Pories, W. J., and Ramsey, G. H. (1954). *Acceleration of Burns and Wound Healing with Methionine Zinc.* University of Rochester Reports, Rochester, N.Y., p. 16.

Sullivan, J. F., Parker, M. M., and Boyett, J. D. (1969). "Incidence of Low Serum Zinc in Noncirrhotic Patients," *Proc. Soc. Exp. Biol. Med.,* **130**, 591–594.

Tal, M. (1969). "Metal Ions and Ribonsomal Conformation," *Biochim. Biophys. Acta,* **195**, 76–86.

Tao, S. and Hurley, L. S. (1971). "Changes in Plasma Proteins in Zinc-Deficient Rats," *Proc. Soc. Exp. Biol. Med.,* **136**, 165–167.

Theuer, R. C. and Hoekstra, W. G. (1966). "Oxidation of ^{14}C-labeled Carbohydrate, Fat, and Amino Acid Substrates by Zinc-Deficient Rats," *J. Nutr.,* **89**, 448–454.

Tucker, S. B., Schroeter, A. L., Brown, P. W., Jr., and McCall, J. T. (1976). "Acquired Zinc Deficiency." *J. Am. Med. Assoc.,* **235**, 2399–2402.

Vallee, B. L. (1955). "Zinc and Metalloenzymes." *Adv. Protein Chem.,* **10**, 317–384.

Vallee, B. L., Stein, E. A., Summerwell, W. N., and Fischer, E. H. (1959). "Metal Content of α-Amylases of Various Origins," *J. Biol. Chem.,* **234**, 2901–2905.

Van Campen, D. R. (1966). "Effects of Zinc, Cadmium, Silver and Mercury on the Absorption and Distribution of Copper-64 in Rats," *J. Nutr.,* **88**, 125–130.

van Rij, A. M. and McKenzie, J. M. (1977). "Hyperzincuria and Zinc Deficiency in Total Parenteral Nutrition." In M. Kirchgessner, Ed., *Trace Element Metabolism in Man and Animals–3,* pp. 288–291.

Wacker W. E. C. (1976). "Role of Zinc in Wound Healing: A Critical Review." In A. S. Prasad, Ed, Trace Elements in Human Health and Disease, Vol. 1 Academic Press New York pp 197–113.

Wacker, W. E. C. and Vallee, B. L. (1959). "Nucleic Acids and Metals. I: Chromium, Maganese, Nickel, Iron, and Other Metals in Ribonucleic Acid from Diverse Biological Sources," *J. Biol. Chem.,* **234**, 3257–3262.

Walker, B. E., Hughes, S., Simmons, A. V., and Chandler, G. N. (1978). "Plasma Zinc After Myocardial Infarction," *Eur. J. Clin. Invest.,* **8**, 193–195.

Warren, L., Glick, M., and Nass, M. (1966). Membranes of Animal cells. I. Methods of Isolation of the Surface Membrane J. Cell. Physiol. 68, 269–288.

Waters, M. D., Moore, R. D., Amato, J. J., and Houck, J. C. (1971). "Zinc Sulfate Failure as an Accelerator of Collagen Biosynthesis and Fibroblast Proliferation," *Proc. Soc. Exp. Biol. Med.,* **138**, 373–377.

Weiber, A. (1959). "Studies in Vascularization of Healing Wounds with Radioactive Isotopes," *Acta Chir. Scand.*, Suppl., p. 237.

Woessner, J. F. Jr. and Boucek, R. J. (1961). "Connective Tissue Development in subcutaneously Implanted Polyvinyl Sponge. I: Biochemical Changes during Development," *Arch. Biochem. Biophys.*, **93**, 85–94.

12

ZINC DENTAL CEMENTS UNDER PHYSIOLOGICAL CONDITIONS

Ralph G. Silvey
George E. Myers

School of Dentistry, University of Michigan, Ann Arbor, Michigan

1. Introduction   238
2. Zinc Phosphate Cement   238
3. Zinc Oxide and Eugenol Cements   239
4. Silicophosphate Cements   241
5. Zinc Polyacrylic Acid Cements   243
6. Tissue Response to Dental Cements   243
7. Clinical Uses of Cements   246
   7.1. Temporary restorations   246
   7.2. Interim luting agent   247
   7.3. Final luting agent   247
   7.4. Luting agent for orthodontic brackets and bands   250
   7.5. Base materials (thermal insulators)   251
   7.6. Periodontal packs   252
   7.7. Impression materials   253
   7.8. Root canal sealers   253
References   253

1. INTRODUCTION

The element zinc appears in dental cements as a compound of oxygen. Zinc oxide powder is the basis for a very diverse group of dental materials whose uses range from dressing materials for both hard and soft oral tissues to temporary and final luting agents and even impression materials.

The use to which the cement will be put determines its composition. Various fillers, plasticizers, accelerators, and reactants are added to the basic zinc oxide, which can in turn be combined with different liquid reactants to produce cements with a wide range of properties. It is this range of physical, chemical, and biophysiological properties that accounts for the wide use of zinc oxide-containing cements.

2. ZINC PHOSPHATE CEMENT

Zinc phosphate cement has been used in the dental profession since the 1870s and is supplied as a powder and a liquid. The powder is principally zinc oxide; however, magnesium oxide is also present. The liquid is a solution of orthophosphoric acid buffered by aluminum salts or occasionally aluminum and zinc salts. The water content is $33 \pm 5\%$ (Paffenbarger et al., 1933).

Dental literature has often referred to zinc phosphate cement as zinc oxyphosphate, but this is not correct since the set cement contains no oxy salts (*Guide to Dental Materials and Devices*, 1976). Compounds formed by the action of the liquid on the powder are noncrystalline phosphates of Zn, Mg, and Al (Sevignac et al., 1965). The greatest portion of the powder in the cement mix does not react, so the set cement consists of powder particles in a matrix of phosphates (*Guide to Dental Materials and Devices*, 1976).

The manipulation of zinc phosphate cement during mixing has a profound effect on the physical properties of the set cement, and individuals working with this material must be trained in its proper use to achieve good results. The manipulation of the cement during mixing determines how much of the powder will react with a given quantity of liquid. The matrix binding the set cement together is quite weak and is also soluble in oral fluids; therefore more desirable properties are obtained in the set cement by increasing the amount of unreacted cement particles present. The mixing process is aimed at wetting the greatest amount of cement powder with the minimum amount of liquid.

Since the reaction is exothermic, it may be retarded by a cool environment and by minimizing heat production. A cool environment is provided clinically by using a thick glass slab that has been chilled. Care is taken to assure that the slab is not cooled below the dew point, however, since premature contact with water is deleterious to the mix. Heat production is minimized by bringing small increments of powder into the mix at a time and by spreading the mix over a wide area on the slab. Mixing time is limited to 90 sec, and the amount of powder incorporated is determined by the consistency of the mix.

In an alternative mixing technique suggested by Jendresen (1973) the slab is cooled to 45°F; any condensation is allowed to stay on the slab. A preweighed amount of powder is brought into the liquid in one increment and mixed for 30 sec. According to Jendresen, the physical properties are essentially unchanged, the working time is increased, and the setting time is decreased. This technique has the added advantage of introducing more consistency into the handling of this material. If subsequent investigations support this technique, it could become widely adopted as the manipulation regemin of choice because of the advantages just stated.

3. ZINC OXIDE AND EUGENOL CEMENTS

Zinc oxide and eugenol cements are available to the dental profession in many forms. Different additives greatly influence the physical properties of the set cement, and these properties determine the use of the material. Applications of these various cements will be discussed more fully in Section 7. At this point the applications will simply be used to classify the cements into groups.

A cement consisting of only zinc oxide and eugenol proved to be unsatisfactory for dental use. The addition of rosin increased the strength and wear properties of the cement to the extent that it made an acceptable interim restoration (Wallace and Hansen, 1939).

A stronger interim resorative material can be formulated by adding polystyrene (Weiss, 1958; Messing, 1961) to zinc oxide and eugenol cements. Up to 10% polystyrene is dissolved in the eugenol, resulting in a cement that, when set, is stronger and has better sealing properties. The polystyrene does not appear to be involved in the setting reaction of the cement (Messing, 1961).

Polymethylmethacrylate (Jendresen et al., 1969; Jendresen and Phillips, 1969) has been added to the powder to increase the strength of zinc oxide and eugenol cements to an even greater degree. This basic type of formulation has been utilized as both an interim restoration and a luting agent.

The highest strengths have been achieved with the zinc oxide and eugenol class of cements by the addition of orthoethoxybenzoic acid (EBA) and alumina. Civjan and Brauer (1964) reported that zinc oxide and eugenol cements with EBA added to the liquid had compressive, tensile, and shear strengths that approached those of zinc phosphate cement. Addition of a heat-treated fused quartz to the powder improved the dimensional stability and strength of the

EBA cements. Further study (Brauer et al., 1968) showed that alumina was superior to fused quartz as a filler, resulting in an improved cement with higher strength, lower solubility, and lower film thickness.

To this point, only additives that increase the strength and/or the hardness of zinc oxide and eugenol cements have been considered. These materials are useful as temporary restorations or luting agents. Zinc oxide and eugenol cement can also be compounded so that it is tough and fracture resistant, making it an acceptable impression material. Anhydrous adeps lanae, or wool fat, has been advocated as a plasticizing filler (Lawrence, 1941). Canada balsam has also been recommended as a plasticizing agent, while balsam of Peru has been used to reduce the burning sensation in soft tissues due to the eugenol (Ross, 1934). These materials may also contain gum rosin, carnauba wax, kauri gum, coumarone resin, olive oil, light mineral oil, or linseed oil as plasticizers and/or fillers, depending on the particular brand of preparation (Peyton and Craig, 1971).

The setting reaction of zinc oxide and eugenol is not clearly understood but appears to involve the methoxy group located ortho to a phenolic hydroxy group. The result of the reaction is the formation of a chelate, zinc eugenolate, which is apparently crystalline in nature (Copeland et al., 1955). Water is necessary for the reaction to proceed. It reacts with the zinc oxide to form zinc hydroxide, which in turn reacts with the eugenol. If the reactants are brought together in the complete absence of water, the material takes an indefinite time to set.

The set cement consists primarily of unreacted zinc oxide and eugenol liquid held in a matrix of zinc eugenolate. Up to 80% of the eugenol used in mixing the cement may be extracted from the set mass. It appears that the eugenol is sorbed by both the zinc oxide and the zinc eugenolate (Copeland et al.. 1955).

The physical properties of the set cement are grossly altered by additives to the reactants. As already pointed out, the inclusion of one or a combination of additives such as rosin, polystyrene, polymethylmethacrylate, fused quartz, and/or alumina can result in a set cement that is harder, stronger, and more resistant to wear and has increased dimensional stability. These improved properties seem to result directly from the properties of the additive, and are not due to participation of the additive in the setting reaction. Other additives can also be included so that the cement mix has a smooth, creamy, fluid, or thyxotropic consistency, and the set cement is tough, somewhat flexible, and thermoplastic, as is needed for impression material. Here again none of these additives enters into the setting reaction, and the change in physical properties is a result of the physical properties of the various constituents of the mix. When EBA is added to zinc oxide and eugenol cements, the situation is qute different. Orthoethoxybenzoic acid is similar to eugenol in that it is a chelating agent, and when it is mixed with zinc oxide powder, the mixture will set to a hardened mass. Mixing zinc oxide powder with a liquid that is 2 parts EBA and 1 part eugenol yields a set cement that has a compressive strength on the order of 9000 to 10,000 psi. It appears that EBA does enter into the setting reaction, although the exact mechanism is unknown (Brauer et al., 1958). There is evidence that chemical species other than zinc eugenolate and zinc 2-ethoxybenzoate result when an EBA-eugenol mixture is reacted with zinc oxide (Wilson and Mesley, 1974). It also appears that the eugenol in set EBA-zinc oxide-eugenol cements is more tightly bound and cannot be extracted from the set cement (Wiedeman et al., 1963).

Many zinc oxide and eugenol preparations contain accelerators, such as metallic salts of acetic, hydrochloric, or nitric acid (Payton and Craig, 1971), to hasten the normally slow setting reaction. Zinc acetate is commonly used for this purpose; large amounts of this compound (5%) encourage the formation of crystalline zinc eugenolate and improve the strength (Peyton and Craig, 1971).

Specific manipulation and handling techniques for zinc oxide and eugenol cements depend somewhat on the function that the particular formulation is intended to perform; however, general statements can be made that apply to the entire class. Moisture and heat accelerate the set of zinc oxide and eugenol cement. The effect is generally quite profound. Since the manufacturer tries to control the setting time with careful formulation, these cements should be mixed in a dry environment at room temperature. The setting reaction is not noticeably exothermic; hence a cool slab is not mandatory, although one can be used to retard the set. Formulations that are used as temporary restorative materials and periodontal tissue packs are mixed to a puttylike consistency. The mix is generally rolled into a rope form for convenient use. Preparations intended for making impressions, those used as temporary or permanent luting agents, and those serving as root canal sealers are mixed to a definite powder/liquid ratio supplied by the manufacturer. The user should not deviate from the manufacturer's recommendations if predictable results are to be obtained.

4. SILICOPHOSPHATE CEMENTS

Zinc silicophosphate cement is supplied as a powder and a liquid, and is a hybrid combination of zinc phosphate and silicate cement (Dennison and Powers, 1974).

The powder is a silicate cement with zinc oxide and magnesium oxide added. The ingredients may be sintered together and then ground, or the powder can be made by simply mixing the silicate and zinc phosphate cement powders together.

Zinc silicophosphate cement liquid is similar to zinc phosphate cement liquid in that it is a buffered phosphoric acid solution. However, it contains more water and less acid (Paffenbarger et al., 1938).

The proportion of zinc phosphate and silicate cement powder is evidently subject to wide variation since chemical analysis of the same brand name product done at different times yielded very different results. The first analysis indicated that the cement powder was 9.4% zinc oxide and 36% silicon dioxide (Paffenbarger et al., 1933). An analysis of the same brand name cement powder done 5 years later indicated that the powder was 53% zinc oxide and 14.6% silicon dioxide (Paffenbarger et al., 1938). The cement liquid was analyzed at the same time and also revealed quite wide variations in content. The first analysis showed the liquid to be 49.6% phosphoric acid, whereas the second analysis showed the acid content to be 57.5%. The degree of difference implies much more than just a variation due to different batches and reflects a change in formulation that was subsequently marketed under the same brand name.

Very few research data have been published on zinc silicophosphate cement, even though it has been used in dentistry since 1878 (Anderson and Paffenbarger, 1962). Therefore only a limited amount of information is available to the profession. The American Dental Association Specification 21 was devised in 1969 to serve as a standard to which this class of cements could be compared. Before this time there was no control at all over the performance of this material.

The specification recognizes three classes of this material, depending on its use. Even though the different uses dictate different consistencies of mixed cement, the mixing technique remains the same. The reaction does not produce as much heat as is evolved with zinc phosphate cement, but a cool slab aids in securing the optimum powder/liquid ratio. Although the material is not as sensitive to manipulative variables as zinc phosphate (Dennison and Powers, 1974), careful and precise mixing procedures will result in the best physical properties. Because of the wide variance of formulation that is allowed by Specification 21, the specification stipulates that the manufacturer must include adequate directions as to temperature, conditions, type of slab and spatula, powder/liquid ratio, rate of incorporating powder, and maximum mixing time for the particular product. In general a measured amount of powder is placed on the slab and divided into three increments (one half of the cement powder is one increment, and one quarter of the powder in each of two others). A stellite or agate spatula is used to prevent discoloration of the cement mix. The largest increment is brought into a measured quantity of liquid and spatulated over a wide area for 15 sec. The remaining two increments are brought into the mix in succession, and each is also spatulated for 15 sec. The entire mix is then spatulated for another 15 sec, making a total mixing time of 1 min. Overmixing tends to cause a deterioration of physical properties.

5. ZINC POLYACRYLIC ACID CEMENTS

Zinc polyacrylic acid cements were introduced to the dental profession in 1968 as luting agents that exhibited true adhesion to the enamel of teeth (Smith, 1968). This type of cement is available as a powder and a liquid.

The cement powder is essentially zinc oxide that has been sintered to reduce its reactivity (Powers et al., 1974). It may contain up to 15% magnesium oxide, which acts to lower the sintering temperature.

The cement liquid is a solution of polyacrylic acid, a solid polymer that is soluble in water. The molecular weight of the polymer varies from 20,000 to 75,000. The viscosity of the liquid may be controlled by varying the molecular weight of the polymer or by adding sodium hydroxide (Dennison and Powers, 1974).

The setting reaction of the polyacrylic acid cements has been subject to many investigations. It was first assumed that the carboxyl groups of the polyacrylic acid formed chelates with both the zinc ions of the cement powder and the calcium ions of the tooth surface. This would account for both the setting of the cement and the adhesion to the tooth surface (Smith, 1968). Later investigations indicated, however, that the bond between the cement and the tooth surface was ionic in nature (Beach, 1972, 1973), as was the bonding between the zinc ions and the cement liquid (Wilson and Crisp, 1974). Hence chelation, which would result in covalent bonds, was ruled out as the setting or adhesion mechanism. The set cement, therefore, consists of unreacted zinc oxide particles within a matrix formed by the ionic cross-linking of polyacrylate chains and zinc ions.

6. TISSUE RESPONSE TO DENTAL CEMENTS

Dental cements come into direct contact with hard and soft oral tissues during clinical use. Generally, it is considered desirable that a dental cement be nonirritating to body tissues. A generally recognized technique to determine the tissue reaction caused by materials is connective tissue implant tests. This is accomplished by taking samples of the material in question and placing them, by surgical means, into the connective tissue of suitable laboratory animals. At various time periods the animals are sacrificed, and the tissue around the implant site is studied histologically for any inflammatory response. Unfortunately, this type of testing has not been done extensively on dental cements. From the literature that is available it is obvious that zinc phosphate, zinc oxide and eugenol, and polyacrylic acid cements all produce a mild inflammatory response when placed into the connective tissue of a rat. (Schaad et al., 1956; Mitchell and Everett,

1957; Boyd and Mitchell, 1961; Truelove et al., 1971; Mitchell, 1959; Phillips and Love, 1961; Guglani and Allen, 1965). Zinc silicophosphate produces a severe inflammatory response when similarly tested (Boyd and Mitchell, 1961).

The response of the gingiva to dental cement is important since even the finest restoration—tooth interface (restoration margin) of a cemented restoration has a thin layer of cement exposed. If the margin is placed beneath the gingival tissue, the gingiva is brought into direct contact with the dental cement over a prolonged period of time. It has been shown that zinc phosphate cement is irritating to the gingival tissue (Waerhaug, 1956; Jensen and Zander, 1955, 1958). This response may be chemical in nature (Waerhaug, 1956), or it may be due to bacterial plaque formation on the cement (Jensen and Zander, 1958). Connective tissue implant studies seem to support the former view.

The effect that a dental cement has on hard structures must also be considered. Unfortunately, research in this area is somewhat limited. It has been shown that the microhardness of dentin within a cavity preparation increases slightly when the dentin comes in contact with zinc oxide and eugenol cement (Mjor, 1962). This does not appear to be simply a response to tooth preparation. When calcium hydroxide, zinc oxide and eugenol, and amalgam are placed in similarly prepared teeth, the observed increases in microhardness of the contiguous dentin are definitely of different magnitudes. Calcium hydroxide produces the greatest increase, and amalgam the least.

Zinc phosphate cement is quite acid during its initial setting stages, and it is well known that acid will decalcify tooth structure. It has been demonstrated that zinc phosphate cement etches the surface enamel of teeth when that material is used to cement orthodontic bands. This can be minimized by using as thick a mix as possible (thereby lowering the pH) and by first treating the tooth with an 8% solution of stannous fluoride (Seniff, 1961).

Zinc phosphate cement has long been regarded as mildly irritating to the dental pulp (Miller, 1891; Manley, 1944; Silberkweit et al., 1955). It has been assumed that the irritation is due to the acidity of this material during setting (Manley, 1950). Some more recent work by Brännström and Nyborg (1974) seems to indicate, however, that bacteria growing in the space between the cement and the cavity walls actually cause the pulpal irritation. Regardless of the mechanism, the use of zinc phosphate produces clinical pulp hypersensitivity in a high number of cases. Further research confirming or refuting Brännström and Nyborg's (1974) work is needed to determine just what effect zinc phosphate has on the pulpal tissue.

Zinc oxide and eugenol cement has been utilized as a sedative treatment dressing in grossly decayed teeth for many years. It is generally assumed to be sedative to the pulpal tissues (Silberkweit et al., 1955) or at least nonirritating (Manley, 1936) and is also known to be bactericidal (Turkheim, 1955a, 1955b). It may well be that the sedative action of zinc oxide and eugenol cements on pulp tissue is based chiefly on its bactericidal properties and not on its tissue tolerance. Connective tissue implant studies have shown that the inflammatory responses to zinc oxide and eugenol and zinc phosphate cements are essentially identical. It has also been demonstrated that when a zinc oxide and eugenol compound is placed directly on the dental pulp it is not sedative in nature but elicits a mild, prolonged irritation (Heys et al., 1975; Zander and Glass, 1949). The major biologic difference between zinc phosphate cement and zinc oxide and eugenol cement appears to be one of bacterial control and not of tissue acceptability. Therefore it appears that a zinc oxide and eugenol cement could give a clinical impression of being sedative to the pulp whereas in actuality it is mildly irritating. It has even been suggested that because of its irritating properties zinc oxide and eugenol cement should not be used on grossly carious teeth (Brännström and Nyborg, 1976), even though many years of clinical use indicate the efficacy of this material.

It is interesting to note that before about 1890 zinc phosphate cement was regarded as sedative to the dental pulp because of its clinical usage. It was common practice at that time to restore grossly carious teeth with this compound. Clinical observation showed that symptomatic teeth so treated often were asymptomatic by the next patient visit (Miller, 1891). In the same way zinc oxide and eugenol cement has come to be regarded as a sedative material, although careful review of the research does not totally support this concept.

Zinc silicophosphate cement produces a severe inflammatory response when used in connective tissue implant tests. By inference it should also produce a severe inflammatory response in the dental pulp. That silicate cement, which is a closely related material, follows this pattern is well documented. There is, however, very little documentation on the pulp response to zinc silicophosphate. Nevertheless this material should be regarded as highly injurious to dental pulp tissue unless proved otherwise by later scientific investigation.

Polyacrylic acid cement, as a class of materials, is the first cement to be introduced to the dental profession with adequate scientific evidence as to its tissue tolerance before it was in use by clinicians. The reason for this is probably twofold. First, the other classes of dental cements discussed here found their way into dentistry in the nineteenth century, when treatment was on an emperical basis and testing of this sort was not done. Second, any new material being introduced to the dental profession today that has medicinal properties must, by law, be adequately tested.

When polyacrylic acid cement was first introduced in 1968 by Smith, evidence of its tolerance by the dental pulp was presented. In tests on dogs, monkeys, and human beings the pulpal response was found to be virtually identical to that elicited by zinc oxide and eugenol (Smith, 1968). These findings were duplicated in a study on monkeys which involved pulpal exposures as well as nonexposures; a moderate inflammatory response persisted for 8 weeks when exposed pulps were treated with either zinc oxide and eugenol or polyacrylic acid cement. In nonexposures the pulpal response was minimal to both materials (Safer et al., 1972). A study on human teeth also indicates that the pulpal response to polyacrylic acid cements is identical to that produced by zinc oxide and eugenol (Barnes and Turner, 1971).

7. CLINICAL USES OF CEMENTS

In the preceding sections dental cements were discussed in terms of the material classes of which they were part. In this section dental cements will be catagorized by clinical usage. The physical properties of a material and the physiologic response to it will determine the best usage for that material in the oral environment. The physical properties are considered only in general terms in this chapter since any dental materials text can be consulted for a detailed discussion of this subject.

Many of the physiologic responses to the various cements were discussed in the preceding section. Additional details on this subject will be presented as the various clinical uses are explored.

7.1. Temporary Restorations

Various types of dental cements have been used as intracoronal interim dressings. The type of cement selected depends on the rationale for placement of the interim restoration.

If the temporary restoration is being used as a means of sealing cut tooth structure and allowing for normal function of a tooth in the interim period between cavity preparation and placement of a cast intracoronal dental restoration, a zinc oxide and eugenol cement containing rosin and/or polystyrene as reinforcement agents is indicated. These preparations are strong enough to withstand the forces of mastication for a period of weeks and yet allow fairly easy removal when it is time to deliver the casting.

If a temporary restoration is to be placed after removal of gross decay, the anticipated time between that procedure and the restoration appointment will be an important consideration in the selection of the proper material. If the restoration is to be done in 2 to 3 weeks, a zinc oxide and eugenol material reinforced with rosin and/or polystyrene is indicated. If a prolonged interval between the initial appointment and the final restoration is anticipated, a zinc oxide and eugenol material reinforced with methylmethacrylate powder is appropriate (Jendresen et al., 1969; Jendresen and Phillips, 1969). This material will last up to a year and must be removed with rotating instruments. Before the development of this material, zinc phosphate and zinc silicophosphate cements were used for this purpose. Pulpal protection of some kind must be utilized under these materials, but they can still serve effectively as long-term interim dressings.

7.2. Interim Luting Agent

In the practice of restorative dentistry it is very often necessary to place temporary restorations of various construction for an interim period of time. It is often desirable to place a permanent restoration for an interim period. For both of these procedures an interim luting agent is indicated. For two very good reasons most of these materials are zinc oxide and eugenol preparations. First, zinc oxide and eugenol cement can be formulated in many different strengths. An interim luting agent must be strong enough to retain the restoration and prevent marginal leakage for the planned period of time, and yet weak enough to allow removal at the desired time. Second, zinc oxide and eugenol cement is well tolerated by the dental pulp as long as the latter is not exposed.

In a clinical study on interim restorations of either acrylic or aluminum shell crowns it was shown that a zinc oxide and eugenol preparation of 2200 to 3500 psi compressive strength was satisfactory (Gilson and Myers, 1968, 1969). A cement of this strength retained the temporary in place for the required time and yet allowed removal with relative ease.

In a similar study zinc oxide and eugenol cements were used to temporarily lute final restorations in place. A 1000-psi cement most frequently fulfilled the above criteria, but cements of 200, 400, 600, 2200, and 3500 psi were used often enough to indicate a need for these materials if all clinical situations were to be met (Gilson and Myers, 1970a).

7.3. Final Luting Agent

Cast restorations are fixed to the abutment teeth with dental cements. At the present time, these materials must be regarded as luting agents and not as adhesives, with the possible exception of polyacrylic acid cements. The cement lute retains the casting in place by forming a mechanical interlock between all the minor irregularities present in the casting and the abutment tooth. It is this area of usage that is currently enjoying the most rapid growth in terms of both new materials and research.

Of the materials discussed in this chapter, zinc phosphate has been used as a luting agent for the longest period of time. This material has become an accepted standard by which other luting agents are evaluated. Except for its alleged pulp-irritating quality there would be little need for any other material. A problem of pulp morbidity, in the form of hot and cold sensitivity, and of pulp mortality does exist, however, when zinc phosphate is used. Whether this problem is due to chemical irritation or bacterial growth under the cement lute is not entirely clear at this time. This has, however, led to a search for other materials that are more compatible with the dental pulp and will retain the casting in place.

The zinc oxide and eugenol class of materials produced the first so-called sedative or nonirritating permanent luting cement. Two basic types were developed at approximately the same time. One type was a cement reinforced with methylmethacrylate; the other type of cement was reinforced with orthoethoxy-benzoic acid and alumina (or, in the case of the earliest ones, fused quartz). Neither of these preparations could match the physical properties of zinc phosphate cement, but they were close enough to warrent a trial. Table 1 shows the range of some reported values for luting cements.

It must be understood that there are no absolute values available that, if met by a material, would qualify it as a permanent luting agent. Zinc phosphate cement has stood the test of time and has proved to be a good luting medium. Research into the physical properties of the cement were instituted only after it had been in clinical use for some time. The properties chosen for investigation were those that, in the view of the investigators, were important in a luting agent. They are by no means the only physical properties that may be important in these materials. Also, it must be noted that, when the specification for zinc phosphate cement was formulated by the American Dental Association, commercially available cements of good quality were tested, and the resultant values were reviewed. With this background, values were selected that both were realistic and could serve as standards by which to judge other zinc phosphate cements. These values must not be considered to be the standard that all luting agents must meet. At the present time only clinical trial can determine whether a material will be an efficient luting agent because all the parameters that a cement must meet to serve in this capacity have not been defined. Richter et al. (1970) support this view in that, when they examined the ability of a zinc phosphate, a hydrophosphate, a polyacrylic acid, and a zinc oxide-eugenol-EBA cement to retain castings in place, they found that "laboratory tests of compressive and tensile strength are not predictive of retentive ability."

In a clinical trial of a zinc oxide and eugenol cement reinforced with polymethylmethacrylate it was found that after 2 years 93.6% of bridge retainers remained successfully luted in place (Gilson and Myers, 1970b). In a follow-up study on the same group of patients and restorations, a success rate of 90.9% was found after 6 to 7 years. The same study reported the success rate of bridge retainers cemented with zinc phosphate cement after a 4-year interval to be 96.06% (Silvey and Myers, 1976).

Table 1. 24-Hour Compressive Strengths (psi) of Luting Cements[a]

Zinc Phosphate		Reinforced Zinc Oxide Eugenol				Polyacrylic Acid	
		Polymethylmethacrylate		EBA and Alumina			
Brand	Compressive Strength	Brand	Compressive Strength	Brand	Compressive Strength	Brand	Compressive Strength
Tenacin[b]	10,000	Fynal[b]	7000	Buffalo[c]	8000	Durelon[d]	8500
				Opotow[e]	8500	Poly C[f]	9000
				Zebacem[b]	9500	PCA[g]	9500

[a] Unpublished data by the authors.
[b] L. D. Caulk Co., Milford, DE 19963.
[c] Buffalo Dental Mfg. Co., 2911–23 Atlantic Ave., Brooklyn, NY 11207.
[d] Premier Dental Products Co., Philadelphia, PA 19107.
[e] Teledyne Dental, Getz-Opotow Division, 1550 Greenleaf Ave., Elk Grove Village, IL 60007.
[f] de Trey Freres S. A., Zürich, Switzerland.
[g] S. S. White Co., Dental Products Division, Philadelphia, PA 19102.

249

In a similar study three cements—a zinc phosphate, a polyacrylic acid, and a zinc oxide and eugenol reinforced with EBA and alumina—were tried. After an observation period that varied from as little as 6 months to a maximum of 30 months (depending on when the restoration was placed), the success rates of the luting agents in keeping retainers sealed to abutment teeth were 98.7% for zinc phosphate, 96.4% for polyacrylic acid, and 96.1% for zinc oxide-eugenol-EBA (Silvey and Myers, 1977). The differences were not thought to be statistically significant.

Several different types of bridge retainers were used in the study mentioned above (Silvey and Myers, 1977). When the cement and retainer type were correlated, a different pattern emerged: there was a definite tendency for pinledge retainers cemented with zinc oxide-eugenol-EBA to loosen (Silvey and Myers, 1978). For other retainers the different cements appeared to be very similar in their rates of success.

The problems of increasing the retentiveness of a cemented restoration and of limiting the exposed cement line have been closely investigated. An increase of 19 to 32% in cemented retention of a cast restoration can be achieved by venting at the time of cementation (Kaufman et al., 1961), and an increase of 25% can be realized by providing a space of 25 μ for the cement lute (Eames et al., 1978). Either technique will reduce the thickness of the cement lute, thereby narrowing the exposed cement line and resulting in better occlusion through more nearly complete seating.

The width of the cement line at the cervical margin of a full crown has been reported to be 0.011 in. (279.4 μ). This can be reduced to 0.003 in. (76.2 μ) by venting the restoration at the time of seating (Kaufman et al., 1961). Another investigation showed that by both venting the restoration and providing room for the cement lute this dimension could be reduced to as little as 10 μ (Bassitt, 1966). It is obvious that the complete seating of cast restorations is highly desirable from many aspects. Retention is increased, occlusal fit of the restoration is improved, and the amount of exposed cement is minimized; therefore dissolution of the cement and gingival irritation due to the cement are reduced.

7.4. Luting Agent for Orthodontic Brackets and Bands

Orthodontic bands have been used for many years to secure brackets to teeth for orthodontic purposes. Wires ligated to these brackets put forces on the teeth that, in turn, reposition them in the alveolar bone. These bands are regarded as a necessary evil since they are unhygienic and, because of their bulk, act as mechanical irritants to the gingiva. It has been suggested that polyacrylic acid cement be used to lute orthodontic brackets on the teeth without using bands (Smith, 1968; Mizrahi and Smith, 1969). This type of treatment seems to be unpredictable, however, and is not recommended as routine clinical practice (Bruce, 1970).

If elimination of the band is not possible, the next best thing from a clinical point of view is to eliminate the problem of loose bands. Fracture of the cement lute under an orthodontic band often results in decalcification of tooth enamel. The band cannot fall off the tooth since it is held in place by the arch wire. Dental plaque grows between the band and the tooth and establishes a highly cariogenic environment. Zinc silicophosphate cement has proved superior to both zinc phosphate and zinc polyacrylic acid cement in retaining orthodontic bands (Clark et al., 1977). Even when bands came loose, no enamel decalcification was detected.

7.5. Base Materials (Thermal Insulators)

Dentin and enamel are poor thermal conductors. When there are sudden changes in mouth temperature due to the ingestion of fluid or to other physiologic functions, the dental pulp is protected from thermal shock by the surrounding tooth structure. Similarly, it is desirable for restorative materials to be poor thermal conductors so that the pulp of a restored tooth is also protected from thermal shock. Clinical experience has shown that moderately deep metallic restorations, at least, cause patient discomfort due to thermal shock. This thermal shock can be minimized by placing an appropriate base material.

The thermal conductivity of dentin has been variously reported to be 0.257×10^{-3} (Phillips et al., 1956), 1.07×10^{-3} (Soyenhoff and Okun, 1958), and 2.29×10^{-3} cal/sec cm^2 (C°/cm.) (Lisanti and Zander, 1950). Inspection of these articles indicates that the thermal conductivity reported depends greatly on the method used for its determination. It is not possible, therefore, to directly compare the thermal conductivity of a restorative or base material reported by one author with the value reported by another. It appears that a determination of the conductivity of tooth structure must be reported as a baseline value when such comparisons are to be made. This makes interpretation of the literature somewhat difficult since not all authors have made such determinations. Jarby (1958) published a very complete discussion of this problem, and the reader is referred to his paper if a more detailed account is desired.

General statements about the thermal conductivities of basing and restorative materials based on research can be made. According to Peyton and Craig (1971), calcium hydroxide and silicate cement have a thermal conductivity of 0.002 cal/sec cm^2 (°C/cm). Since they accept a value for tooth structure of 0.0015 to $0.0022K$, the thermal conductivity of these materials is approximately equal to that of tooth structure. The K values for zinc phosphate and zinc oxide-eugenol are 0.003 and 0.001, respectively. This means that zinc oxide and eugenol is three times as effective as zinc phosphate as an insulator and is superior in this regard to the natural tooth tissue.

The thermal conductivities of silver amalgam and gold are $0.060K$ and $0.710K$, respectively (Peyton and Craig, 1971). This means that amalgam conducts heat about 40 times better than tooth structure, and gold conducts heat almost 500 times as well. (The latter value may be too high since pure gold is rarely used in dentistry. The thermal conductivity of a gold alloy has been reported to be $0.1643K$, which would bring this factor down to about 70 times that of tooth structure. It is interesting to note that pure silver conducts heat even more readily than pure gold.) To prevent thermal shock to the tooth clinical dentists must often place an insulating base under metallic restorations when the cavity preparation goes beyond the ideal. Zinc oxide and eugenol is superior in its insulating ability and appears to be superior in its pulpal response as well. It could be concluded that a reinforced zinc oxide and eugenol cement should be superior to zinc phosphate as a base for physiologic reasons.

It has been shown that the modulus of elasticity of zinc phosphate cement is higher than that of a polymer-modified zinc oxide and eugenol cement (Powers et al., 1976). When an occlusal amalgam restoration is supported by a cement base 2 mm thick, the tensile stresses induced in the restoration are three times greater if it is supported by a zinc oxide and eugenol cement base than if it is supported by zinc phosphate (Farah et al., 1975). Since amalgam is very weak in tension, a ZOE base could result in failure of the restoration. From the mechanical point of view, therefore, zinc phosphate is superior to zinc oxide and eugenol as a base material. There is at present insufficient evidence to determine whether a polyacrylic acid cement would be an acceptable compromise in strength and thermal conductivity between zinc phosphate and zinc oxide-eugenol base materials.

7.6. Periodontal Packs

The periodontal pack was introduced by W. F. Dunlap about 1913 (Miller, 1950). It has been used as an adjunct in the elimination of gingival hyperplasia and shallow periodontal pockets, and also as a postoperative dressing to promote healing although Glickman (1972) reports this may be a misconception. Of the many types of periodontal packs in use only the zinc-containing ones will be mentioned here.

Zinc oxide and eugenol cement reinforced with resin and asbestos fiber is commonly used as a periodontal pack (Glickman, 1972). This type of pack may be in place for as long as 30 days (Goldman and Cohen, 1968) and may be against connective tissue or even bone for some portion of that time. Such a pack has definite antimicrobial properties (Persson and Thilander, 1968a) and is slightly more effective than noneugenol packs as a wound dressing, as far as postoperative sequelae are concerned (Oliver and Heaney, 1970). Degenerative changes in the new epithelium occur under both noneugenol and eugenol-containing packs; there is, however, less inflamation under eugenol-containing packs than those of the noneugenol type (Persson and Thilander, 1968b). This difference was not considered to be of clinical importance.

7.7. Impression Materials

Zinc oxide and eugenol cements have been used for many years to obtain detailed and accurate impressions of edentulous and partially edentulous mouths. The success of these materials for this purpose is largely determined by their physical properties, a subject outside the scope of this chapter. A few patients exhibit allergic responses to this type of preparation and this would contraindicate its use.

7.8. Root Canal Sealers

A root canal sealer is a cement used to seal minute irregularities, accessory canals, and foremen when endodontic therapy is performed on a tooth. The main canals are cleaned, enlarged to a funnel-like shape, and obturated with gutta percha, silver, or some other relatively inert filling material. The sealer is simply an adjunct to fill in minor irregularities.

Zinc oxide and eugenol cement is the basis for many of these sealers. Tissue implant studies have shown that this cement per se causes mild inflammation of connective tissue. Additives to zinc oxide and eugenol cement that are supposed to increase its effectiveness as a sealer can produce severe tissue responses (Browne and Friend, 1969; Friend and Browne, 1968). It appears that root canal sealers are not inert and may elicit a marked degree of tissue irritation for a prolonged period of time. For this reason clinical techniques aim at confining all of the sealer within the canal.

REFERENCES

Anderson, J. N. and Paffenbarger, G. C. (1962). "Properties of Silicophosphate Cements," *Dent. Prog.*, 2, 72–75.
Barnes, D. S. and Turner, E. P. (1971). "Initial Response of Human Pulp to Zinc Polycarboxylate Cement," *J. Can. Dent. Assoc.*, 37, 265–266.

Bassitt, R. W. (1966). "Solving the Problems of Cementing the Full Veneer Cast Gold Crown," *J. Prosthet. Dent.*, 16, 740–747.

Beech, D. R. (1972). "A Spectroscopic Study of the Interaction between Human Tooth Enamel and Polyacrylic Acid (Polycarboxylate Cement)," *Arch. Oral Biol.*, 17, 907–911.

Beech, D. R. (1973). "Improvement in the Adhesion of Polyacrylate Cements to Human Dentine," *Br. Dent. J.*, 135, 442–445.

Boyd, J. B. and Mitchell, D. F. (1961). "Reaction of subcutaneous Connective Tissue of Rats to Implanted Dental Cements," *J. Prosthet. Dent.*, 11, 174–183.

Brännström, M. and Nyborg, H. (1974). "Bacterial Growth and Pulpal Changes under Inlays Cemented with Zinc Phosphate Cement and Epoxylite CBA 9080," *J. Prosthet. Dent.*, 31, 556–565.

Brännström, M. and Nyborg, H. (1976). "Pulp Reaction to a Temporary Zinc Oxide Eugenol Cement," *J. Prosthet. Dent.*, 35, 185–191.

Brauer, G. M., White, E. E., and Mashonas, M. G. (1958). "The Reaction of Metal Oxides with *o*-Ethoxybenzoic Acid and Other Chelating Agents," *J. Dent. Res.*, 37, 547.

Brauer, G. M., McLaughlin, R., and Huget, E. (1968). "Aluminum Oxide as a Reinforcing Agent for Zinc Oxide-Eugenol *o*-Ethoxybenzoic Acid," *J. Dent. Res.*, 47, 622.

Browne, R. M. and Friend, L. A. (1969). "An investigation into the Irritant Properties of Some Root Filling Materials," *Arch. Oral Biol.*, 13, 1355–1369.

Bruce, L. D. (1970). "An Evaluation of Direct Placement of Orthodontic Brackets." Thesis, University of Michigan School of Dentistry, 36 pp.

Civjan, S. and Brauer, G. M. (1964). "Physical Properties of Cements Based on Zinc Oxide, Hydrogenated Rosin, *o*-Ethoxybenzoic Acid and Eugonol," *J. Dent. Res.*, 43, 281.

Clark, J. R. et al. (1977). "An Evaluation of Silicophosphate as an Orthodontic Cement," *Am. J. Orthod.*, 71, 190–196.

Copeland, H. I., Jr., et al. (1955). "Setting Reaction of Zinc Oxide and Eugenol," *J. Res. Natl. Bur. Stand.*, 55, 133.

Dennison, J. D. and Powers, J. M. (1974). "A Review of Dental Cements Used for Permanent Retention of Restorations. I: Composition and Manipulation," *J. Mich. Dent. Assoc.*, 56, 116–121.

Eames, W. B. et al. (1978). "Techniques to Improve the Seating of Castings," *J. Am. Dent. Assoc.*, 96, 432–437.

Farah, J. W. et al. (1975). "Effects of Cement Bases on the Stresses in Amalgam Restorations," *J. Dent. Res.*, 54, 10–15.

Friend, L. A. and Browne, R. M. (1968). "Tissue Reactions to Some Root Filling Materials," *Br. Dent. J.*, 125, 291–298.

Gilson, T. D. and Myers, G. E. (1968). "Clinical Studies of Dental Cements. I: Five Zinc Oxide-Eugenol Cements," *J. Dent. Res.*, 47, 737–741.

Gilson, T. D. and Myers, G. E. (1969). "Clinical Studies of Dental Cements. II: Further Investigation of Two Zinc Oxide-Eugenol Cements for Temporary Restorations," *J. Dent. Res., 48,* 366–367.

Gilson, T. D. and Myers, G. E. (1970a). "Clinical Studies of Dental Cements: III: Seven Zinc Oxide-Eugenol Cements Used for Temporarily Cementing Completed Restorations," *J. Dent. Res., 49,* 14–20.

Gilson, T. D. and Myers, G. E. (1970b). "Clinical Studies of Dental Cements. IV: A Preliminary Study of a Zinc Oxide-Eugenol Cement for Final Cementation," *J. Dent. Res., 49,* 75–78.

Glickman, I. (1972). *Clinical Periodontology,* 4th ed. W. B. Saunders, Philadelphia, p. 820.

Goldman, H. W. and Cohen, W. D. (1968). *Periodontal Therapy,* 4th ed. C. V. Mosby, St. Louis, p. 707.

Guglani, L. M. and Allen, E. F. (1965). "Connective Tissue Reaction to Implants of Periodontal Packs," *J. Periodontol., 36,* 279–282.

Guide to Dental Materials and Devices, 8th ed. (1976). Council of Dental Materials and Devices, American Dental Association, Chicago, p. 137.

Heys, D. R., Heys, R. J., and Racette, W. A. (1975). "A Histological Evaluation of Ten Restorative Compounds." Thesis, University of Michigan School of Dentistry.

Jarby, S. (1958). "On Temperature Measurements in Teeth," *Odont. T., 66,* 423–471.

Jendresen, M. D. (1973). "New Dental Cements and Fixed Prosthodontics. II," *J. Prosthet. Dent., 30,* 684–688.

Jendresen, M. D. and Phillips, R. W. (1969). "A Comparative Study of Four Zinc Oxide and Eugenol Formulations as Restorative Materials. II," *J. Prosthet. Dent., 21,* 300–309.

Jendresen, M. D., Phillips, R. W., Swartz, M. L., and Morman, R. D. (1969). "A Comparative Study of Four Zinc Oxide and Eugenol Formulations as Restorative Materials. I," *J. Prosthet. Dent., 21,* 176–186.

Jensen, J. R. and Zander, H. A. (1955). "Effect of Two Cementing Materials on the Gingival Tissues," *J. Dent. Res., 34,* 699. (Abstr. 77).

Jensen, J. R. and Zander, H. A. (1958). "Effect of Two Cementing Materials on the Gingival Tissues," *North West Dent., 37,* 210–212.

Kaufman, E. G. et al. (1961). "Factors Influencing the Retention of Cemented Gold Castings," *J. Prosthet. Dent., 11,* 487–502.

Lawrence, G. O. (1941). "Impression Pastes," *Aust. Dent. J., 45,* 207.

Lisanti, V. F. and Zander, H. A. (1950). "Thermal Conductivity of Dentin," *J. Dent. Res., 29,* 493–497.

Manley, E. B. (1936). "A Preliminary Investigation into the Reaction of the Pulp to Various Filling Materials," *Br. Dent. J., 60,* 231–331.

Manley, E. B. (1944). "Pulp Reactions to Dental Cements," *Dent. Rec., 64,* 75–86.

256 Zinc Dental Cements under Physiological Conditions

Manley, E. B. (1950). "A Review of Pulp Reactions to Chemical Irritation," *Int. Dent. J.*, **1**, 36-48.

Messing, J. J. (1961). "Polystyrene-Fortified Zinc Oxide-Eugenol Cement," *Br. Dent. J.*, **110**, 95.

Miller, S. C. (1950). *Textbook of Periodontia*, 3rd ed. Blakiston, Philadelphia, p. 219.

Miller, W. D. (1891). "Concerning the Oxyphosphate Cement," *Br. Dent. Assoc. J.*, **12**, 18-24.

Mitchell, D. F. (1959). "The Irritational Qualities of Dental Materials," *J. Am. Dent. Assoc.*, **59**, 954-966.

Mitchell, D. F. and Everett, A. R. (1957). "Reaction of Connective Tissues of Rats to Implanted Dental Materials," *IADR*, **35**, 59-60.

Mizrahi, E. and Smith, D. C. (1969). "Direct Cementation of Orthodontic Brackets to Dental Enamel," *Br. Dent. J.*, **127**, 371-375.

Mjor, I. A. (1962). "The Effect of Zinc Oxide and Eugenol on Dentine Evaluated by Microhardness Testing," *Arch. Oral Biol.*, **7**, 333-336.

Oliver, W. M. and Heaney, T. G. (1970). "Sequelae Following the Use of Eugenol or Non-eugenol Dressings after Gingivectomy and Subgingival Curettage," *Dent. Pract.*, **21**, 49-52.

Paffenbarger, G. C., Sweeney, W. T., and Isaacs, A. (1933). "A Preliminary Report on the Zinc Phosphate Cements," *Am. Dent. Assoc. J.*, **20**, 1960.

Paffenbarger, G. C., Schoonover, I. C., and Souder, W. (1938). "Dental Silicate Cements: Physical and Chemical Properties and a Specification," *J. Am. Dent. Assoc.*, **25**, 32.

Persson, G. and Thilander, H. (1968a). "Experimental Studies of Surgical Packs. I: *In vitro* Experiments on Antimicrobial Effect," *Odont. T.*, **76**, 147-155.

Persson, G. and Thilander, H. (1968b). "Experimental Studies of Surgical Pacts. II: Tissue Reaction to Various Packs," *Odont. T.*, **76**, 157-162.

Peyton, F. A. and Craig, R. G. (1971). *Restorative Dental Materials*, 4th ed. C. V. Mosby, St. Louis, p. 175.

Phillips, R. W. and Love, D. R. (1961). "The Effect of Certain Additive Agents on the Physical Properties of Zinc Oxide-Eugenol Mixtures," *J. Dent. Res.*, **40**, 294-303.

Phillips, R. W. et al. (1956). "An Improved Method for Measuring the Coefficient of Thermal Conductivity of Dental Cement," *J. Am. Dent. Assoc.*, **53**, 577-583.

Powers, J. M., Johnson, Z. G., and Craig, R. G. (1974). "Physical and Mechanical Properties of Zinc Polyacrylate Dental Cements," *J. Am. Dent. Assoc.*, **88**, 380-383.

Powers, J. M. et al. (1976). "Modulus of Elasticity and Strength Properties of Dental Cements," *J. Am. Dent. Assoc.*, **92**, 588-591.

Richter, W. A. et al. (1970). "Predictability of Retentive Values of Dental Cements," *J. Prosthet. Dent.*, **24**, 298-303.

Ross, R. A. (1934). "Zinc Oxide Impression Pastes," *J. Am. Dent. Assoc.,* **21**, 2029.

Safer, D. S., Avery, J. K, and Cox, C. F. (1972). "Histopathologic Evaluation of the Effects of New Polycarboxylate Cements on Monkey Pulps," *Oral Surg., Oral Med., Oral Pathol.,* **33**, 966–975.

Savignac, J. R., Fairhurst, C. W., and Ryge, G. (1965). "Strength, Solubility and Disintegration of Zinc Phosphate Cement with Clinically Determined Powder-Liquid Ratio," *Angle Orthod.,* **35**, 126.

Schaad, T. D., Carter, W. J., and Myers, H. I. (1956). "Reaction of Abdominal Connective Tissue in Rats to Dental Base Materials," *IADR,* **34**, 11 (Abstr. 27).

Seniff, R. W. (1961). "Change in Enamel Surface Caused by Oxyphosphate Cement," *J. Dent. Res.,* **40**, 642 (Abstr. 4).

Silberkweit, M. et al. (1955). "Effects of Filling Materials on the Pulp of the Rat Incisor," *J. Dent. Res.,* **34**, 854–869.

Silvey, R. G. and Myers, G. E. (1976). "Clinical Studies of Dental Cements. V: Recall Evaluation of Restorations Cemented with a Zinc Oxide-Eugenol Cement and a Zinc Phosphate Cement," *J. Dent. Res.,* **55**, 289–291.

Silvey, R. G. and Myers, G. E. (1977). "Clinical Study of Dental Cements. VI: A Study of Zinc Phosphate, EBA-Reinforced Zinc Oxide Eugenol and Poly-acrylic Acid Cements as Luting Agents in Fixed Prosthondontics," *J. Dent. Res.,* **56**, 1215–1218.

Silvey, R. G. and Myers, G. E. (1978). "Clinical Study of Dental Cements. VII: A Study of Bridge Retainers Luted with Three Different Dental Cements," *J. Dent. Res.,* **57**, 703–707.

Smith, D. C. (1968). "A New Dental Cement," *Br. Dent. J.,* **124**, 381–384.

Soyenhoff, B. C. and Okun, J. H. (1958). "Thermal Conductivity Measurement of Dental Tissues with the Aid of Thermistors," *J. Am. Dent. Assoc.,* **57**, 23–30.

Truelove, E. L., Mitchell, D. F., and Phillips, R. W. (1971). "Biological Evaluation of a Carboxylate Cement," *J. Dent. Res.,* **50**, 166.

Turkheim, H. J. (1955a). "*In vitro* Experiments on the Bactericidal Effect of Zinc Oxide-Eugenol Cement on Bacteria-Containing Dentin," *J. Dent. Res.,* **34**, 295–301.

Turkheim, H. J. (1955b). "A Study of the Bactericidal Effect of Zinc Oxide-Eugenol Cement," *Dent. Rec.,* **75**, 29–35.

Waerhaug, J. (1956). "Effect of Zinc Phosphate Cement Fillings on Gingival Tissues," *J. Periodontol.,* **27**, 284–290.

Wallace, D. A. and Hansen, H. L. (1939). "Zinc Oxide-Eugenol Cements," *J. Am. Dent. Assoc.,* **26**, 1536.

Weiss, M. B. (1958). "Improved Zinc Oxide-Eugenol Cement. III," *Dent. J.,* **27**, 261.

Wiedeman, W. H., Civjan, S., and Brauer, G. M. (1963). "Chemical Characteristics of o-Ethoxybenzoic acid-Eugenol-Zinc Oxide Cements," *IADR,* **41,** 143 (abstr.).

Wilson, A. D. and Crisp, S. (1974). "Infra-red Spectroscopic Studies of the Nature of Bonding in Cements Formed from PAA (Polyacrylic Acid)," *J. Dent. Res.,* **53** (Suppl.), 1074.

Wilson, A. D. and Mesley, R. J. (1974). "Chemical Nature of Cementing Matrixes of Cements Formed from Zinc Oxide and 2-Ethoxybenzoic Acid-Eugenol Liquids," *J. Dent. Res.,* **53,** 146.

Zander, H. A. and Glass, R. L. (1949). "Healing of Phenolized Pulp Exposures," *Oral Surg., Oral Med., Oral Pathol.,* **2,** 803.

13

ACCUMULATION OF ZINC BY MARINE BIOTA

Ronald Eisler

U.S. Environmental Protection Agency, Narragansett, Rhode Island

1. Introduction	259
2. Zinc in Field Collections of Marine Biota	260
3. Concentration Factors	261
4. Accumulations by Laboratory Populations	319
5. Factors Modifying Zinc Accumulations	327
5.1. Algae	327
5.2. Protista	328
5.3. Mollusca	328
5.4. Crustacea	330
5.5. Annelida	332
5.6. Echinoderms	333
5.7. Teleosts	333
6. Summary	335
Acknowledgments	335
References	335

1. INTRODUCTION

The biological and toxicological effects of zinc and other metals in marine and coastal ecosystems have become the focus of an increasing number of review articles and bibliographies (Vallee, 1959; Rice, 1963; Bryan, 1971, 1976; Eisler, 1973; Waldichuk, 1974; Bernhard and Zattera, 1975; Eisler and Wapner, 1975; Eisler et al., 1978b). Early interest in the environmental behavior of zinc was

259

stimulated by the observation of Japanese scientists that ^{65}Zn from fallout was concentrated strongly in tunas. Subsequently, the nuclide was reported in large amounts in many species of marine organisms when compared to ambient seawater (Saeki et al., 1955; Yoshii, 1956; Lowman et al., 1957; Seymour, 1959). More recently, analysis of the technical literature published in 1974, 1975, and 1976 on the effects of hazardous metals to marine biota established that zinc had been studied more intensively during this period than any other metal (Eisler, 1979). This continuing and growing interest in zinc, especially stable zinc, is linked to the development and availability of newer analytical technology such as atomic absorption spectrophotometry and neutron activation analysis, and to the research efforts generated by nutritionists, ecologists, and environmentalists. As one consequence, data on stable zinc accumulations in marine and estuarine biota have become especially abundant in recent decades, although dispersed widely throughout the scientific literature.

This abundance and dispersion, and the fact that the last comprehensive listing of zinc levels in marine animals and plants appeared about 25 years ago (Vinogradov, 1953), strongly indicated a need to update this topic. The present account summarizes recent available data on stable zinc in four subject areas: residues in field populations of marine organisms; bioconcentration by field populations; accumulation under controlled laboratory conditions; and factors known to modify uptake, retention, and translocation by marine flora and fauna.

2. ZINC IN FIELD COLLECTIONS OF MARINE BIOTA

Concentrations of total zinc from approximately 650 different species of marine, estuarine, and oceanic plants, invertebrates, and vertebrates collected from numerous geographic locations worldwide are listed in Table 1. The great majority of these data are from molluscs, teleosts, plants, and crustaceans, in that order. Most of this information was published during the past decade; however, many technical reports, including almost all of those summarized by Vinogradov (1953), appeared in print before 1948. It is emphasized that the analytical methods used in a number of these earlier papers were of limited sensitivity, and their results of marginal worth by contemporary standards.

It was apparent from Table 1 that the greatest concentrations of zinc were in filter-feeding bivalve molluscs, especially oysters. Crassostreid oysters, for example, frequently contained more than 4.0 g Zn/kg soft parts on a dry weight basis. Several species of algae also contained substantial zinc, that is, more than 1.0 g/kg weight. Barnacles (*Balanus balanoides*) also contained high zinc concentrations (i.e. > 3.3 g/kg in soft tissues on a dry weight basis); however, barnacle zinc values were considerably higher than similar data for other species of crustaceans. In general, the lowest zinc residue ranges of < 0.05 to 0.4 g/kg dry weight of various tissues were reported among the vertebrates: elasmobranches, teleosts, and mammals.

According to Pequegnot et al. (1969) zinc is not limiting to normal life processes in the marine environment and is accumulated in excess of the organism's immediate needs, at least in respect to enzymatically bound zinc. The information in Table 1 tends to support the observations that zinc is not limiting, and in several cases probably accumulated far in excess of biological requirements. It has also been stated that intraspecies zinc concentrations in a wide range of marine flora and fauna were relatively constant, suggesting definite physiological needs (Robertson et al., 1972). However, much of the information in Table 1 contradicts these findings; for example, marked variations in zinc content were evident both within and among closely related species such as crassostreid and ostreid oysters, limpets of the genus *Patella,* and the alga *Ascophyllum nodosum.* In actuality, numerous factors, of which proximity to anthropogenic point sources of zinc is probably one of the more important, are known to be responsible for variations observed in the zinc contents of marine biota. The influence of these biotic and abiotic zinc modifiers will be reported in detail later. At this time it is sufficient to note from Table 1 that zinc accumulations in various marine species were markedly influenced by the following: age and sex of organism; season, year, and geographic location of collection; and zinc-specific sites of accumulation, such as frond in algae, kidney in molluscs, hepatopancreas in crustaceans, viscera and gonad in teleosts, and liver and kidney in mammals.

3. CONCENTRATION FACTORS

Establishment of accurate zinc concentration factors (CFs) requires continuous determinations of the biologically available zinc in the medium during the entire period of bioindicator organism residence. With continuous monitoring, any surges in zinc levels in seawater that are produced as a result of natural or anthropogenic perturbations will be recorded, and these, in conjunction with duration of exposure, may partially account for the large variability commonly recorded in CF values. However, most CF data, including those for zinc and marine biota (Table 2), have not met these criteria. For example, zinc CF values recorded for the alga *Laminaria digitata* were inversely correlated with ambient seawater zinc concentrations (Bryan, 1969). After exposure for 32 days in seawater containing 2.2 to 22.4 μg Zn/l, *Laminaria* exhibited a CF of 2455; but in 50 to 500 μg Zn/l seawater this was reduced to 700 to 1400. In all instances the relationship of zinc in seawater to that in seaweed was logarithmic (Bryan, 1969). To complicate matters, there is substantial variability in the retention time of ac-

Table 1. Zinc Concentrations in Field Collections of Marine and Estuarine Flora and Fauna.
Values are in milligrams zinc per kilogram fresh wegith (FW), dry weight (DW) or ash weight (AW).

Species	Zinc Concentration	Reference
ALGAE		
Algae		
Whole, 50 spp., Japan	110.0–3950.0 FW	Ishibashi et al. (1965)
Whole, 11 spp.	10.0–80.0 DW	Pillai (1956)
Whole, west Norway coast	2370.0 DW max.	Stenner and Nickless (1974b)
Whole, Rio Tinto estuary, Spain	160.0 DW max.	Stenner and Nickless (1975)
Whole, Newport River estuary, N.C.	33.0–120.0 DW	Wolfe (1974)
Whole, 11 spp.	35.0–97.0 DW	Young and Langille (1958)
Ascophyllum nodosum		
Whole	60.0–103.0 DW	Black and Mitchell (1952)
Whole		
Lofoten, Norway	60.0–110.0 DW	Haug et al. (1974)
Trondheimsfjord	66.0–640.0 DW	Haug et al. (1974)
Hardangerfjord	240.0–3700.0 DW	Haug et al. (1974)
Whole		
Menai Straits	149.0 DW	Foster (1976)
Dulas Bay	199.0 DW	Foster (1976)
Asterionella japonica		
Whole	115.0 DW	Riley and Roth (1971)
Chlamydomonas sp.		
Whole	116.0 DW	Riley and Roth (1971)
Chlorella salina		
Whole	301.0 DW	Riley and Roth (1971)
Cymodocea sp.		

Whole	3.6–8.7 FW; 45.0–92.0 DW; 125.0–250.0 AW	Lowman et al. (1966)
Dunaliella primolecta		
Whole	405.0 DW	Riley and Roth (1971)
Dunaliella tertiolecta		
Whole	285.0 DW	Riley and Roth (1971)
Ecklonia maxima		
Whole	1.7 FW	Van As et al. (1973)
Eisenia bicyclis		
Whole	112.0–127.0 DW	Ishibashi et al. (1964)
Fucus serratus		
Whole	99.0–600.0 DW	Leatherland and Burton (1974)
Whole	70.0–79.0 DW	Black and Mitchell (1952)
Fucus vesiculosus		
Whole	60.0 DW	Black and Mitchell (1952)
Thallus	199.0–1240.0 DW	Bryan and Hummerstone (1973a)
Whole	171.0 DW	Preston et al. (1972)
Whole	216.0–507.0 DW	Ireland (1973)
Whole	800.0 DW max.	Butterworth et al. (1972)
Menai Straits	116.0 DW	Foster (1976)
Dulas Bay	306.0 DW	Foster (1976)
Gigartina rachula		
Whole	6.9 FW	Van As et al. (1973)
Hemiselmis brunescens		
Whole	480.0 DW	Riley and Roth (1971)
Hemiselmis virescens		
Whole	259.0 DW	Riley and Roth (1971)

Table 1. Continued

Species	Zinc Concentration	Reference
Heteromastix longifillis		
Whole	325.0 DW	Riley and Roth (1971)
Laminaria cloustini		
Frond, sterile	117.0 DW (winter)	Black and Mitchell (1952)
Frond, sporing	136.0 DW (winter)	Black and Mitchell (1952)
Frond	76.0 DW (spring)	Black and Mitchell (1952)
Laminaria digitata		
Front	64.0–99.0 DW	Black and Mitchell (1952)
Stipe	N.D. (winter)–	Black and Mitchell (1952)
	62.0 (spring) DW	
Whole	4.0–10.0 FW	Bryan (1969)
Macrocystis integrifolia		
Fronds	97.0 DW	Wort (1955)
	(range 14.0–335.0)	
Stipes	21.0 DW	Wort (1955)
	(range 10.0–34.0)	
Micromonas squamata		
Whole	105.0 DW	Riley and Roth (1971)
Monochrysis lutheri		
Whole	160.0 DW	Riley and Roth (1971)
Nereocystis leutkeana		
Fronds	424.0 DW	Wort (1955)
	(range 90.0–2800.0)	
Stipes	74.0 DW	Wort (1955)
	(range 7.0–310.0)	

Olisthodiscus luteus		
Whole	75.0 DW	Riley and Roth (1971)
Pelvetia canaliculata		
Whole	40.0–47.0 DW	Black and Mitchell (1952)
Phaeodactylum tricornutum		
Whole	325.0 DW	Riley and Roth (1971)
Phytoplankton		
Whole	590.0 DW	Fujita (1972)
	(range 104.0–1757.0)	
Whole	750.0 DW	Knauer and Martin (1973)
Organic fractions		
No-Ti group	19.0 DW	Martin and Knauer (1973)
Ti group	122.0 DW	Martin and Knauer (1973)
Sr-concentrated group	24.0 DW	Martin and Knauer (1973)
Siliceous frustules		
No-Ti group	5.0 DW	Martin and Knauer (1973)
Ti group	10.0 DW	Martin and Knauer (1973)
Sr-concentrated group	N.D.–108.0 DW	Martin and Knauer (1973)
Porphyra capensis		
Whole	11.0 FW	Van As et al. (1973)
Porphrya umbilicalis		
Whole	66.0 DW	Preston et al. (1971)
Pseudopedinella pyriformis		
Whole	243.0 DW	Riley and Roth (1971)
Seaweeds		
Whole, Norway	53.0–520.0 DW	Lunde (1970)
Stichococcus bacillaris		
Whole	251.0 DW	Riley and Roth (1971)
Suhria vittata		
Whole	9.6 FW	Van As et al. (1973)

Table 1. Continued

Species	Zinc Concentration	Reference
Tetraselmis tetrathele		
Whole	410.0 DW	Riley and Roth (1971)
Udotea flabellum		
Whole	4.9–800.0 FW; 8.2– 830.0 DW; 16.0–1900.0 AW	Lowman et al. (1966)
Ulva spp.		
Whole	5.9 FW	Van As et al. (1973)
MACROPHYTES		
Macrophytes		
Whole, Harbour I., Tex.	3.4–50.0 DW	Lytle et al. (1973)
Spartina alterniflora		
Sprout	11.0–20.0 DW	Williams and Murdock (1969)
Mature plant	7.0–12.0 DW	Williams and Murdock (1969)
Dead plant	14.0–37.0 DW	Williams and Murdock (1969)
Thalassia sp.		
Whole	12.0–19.0 FW; 79.0–110.0 DW; 160.0–270.0 AW	Lowman et al. (1966)
Thalassia testudinum		
Whole	5.5–12.0 FW; 35.0–82.0 DW; 75.0–200.0 AW	Lowman et al. (1966)
Zostera sp.		
Whole		
North Carolina	49.0 DW	Wolfe (1974)
Spain	1480.0 DW max.	Stenner and Nickless (1975)

Taxon/Part	Value	Reference
PROTOZOA		
Radiolarians		
Whole	63.0–279.0 DW	Martin and Knauer (1973)
PORIFERA		
Halichondria panicea		
Whole	89.0–152.0 DW	Ireland (1973)
Halichondria sp.		
Whole	150.0 DW	Vinogradov (1953)
Sponges, 2 spp.		
Whole	4.7–30.0 FW; 63.0–180.0 DW; 93.0–360.0 AW	Lowman et al. (1966)
COELENTERATA		
Actinia equina		
Whole	167.0–603.0 DW	Ireland (1973)
Alcyonium digitatum		
Whole	46.0 DW	Riley and Segar (1970)
Aurelia sp.		
Whole	3.4 FW	Matsumoto et al. (1964)
Corals		
Deep open ocean		
Whole, 7 spp.	<2.0–9.0 DW	Livingston and Thompson (1971)
Shallow open ocean		
Whole, 8 spp.	2.0–70.0 DW	Livingston and Thompson (1971)
Shallow coastal		
Whole, 8 spp.	<2.0–7.0 DW	Livingston and Thompson (1971)
Eusmilia fastigiata		
Whole	74.0–80.0 FW; 75.0–80.0 DW; 77.0–83.0 AW	Lowman et al. (1966)

Table 1. Continued

Species	Zinc Concentration	Reference
Manicina aerolata		
Whole	59.0–63.0 FW; 60.0–68.0 DW; 65.0–69.0 AW	Lowman et al. (1966)
Meandrina meandrites		
Whole	54.0–120.0 FW; 55.0–120.0 DW; 56.0–120.0 AW	Lowman et al. (1966)
Pelagia sp.		
Whole	28.0 DW	Leatherland et al. (1973)
Tealia felina		
Whole	200.0 DW	Leatherland and Burton (1974)
Whole	280.0 DW	Riley and Segar (1970)
Whole, 12 spp.	11.5–104.0 DW	Vinogradov (1953)
CTENOPHORA		
Pleurobrachia pileus		
Whole	900.0 AW	Vinogradova and Koual'skiy (1962)
MOLLUSCA		
Abra alba		
Soft parts	1110.0 FW (max.)	Rosenberg (1977)
Acmaea digitalis		
Shell	8.4–29.9 DW	Graham (1972)
Soft parts	429.0–763.0 DW	Graham (1972)
Anodonta cygnea		
Soft parts	141.0–650.0 DW	Leatherland and Burton (1974)
Anodonta sp.		
Shell	6.4 DW	Segar et al. (1971)
Soft parts	120.0 DW	Segar et al. (1971)

Arctica islandica		
Soft parts	2.4–25.8 FW	Palmer and Rand (1977)
Buccinum undatum		
Shell	13.0 DW	Segar et al. (1971)
Soft parts	620.0 DW	Segar et al. (1971)
Busycon canaliculatum		
Muscle	25.0 DW	Greig (1975)
Digestive diverticula	2600.0 DW	Greig (1975)
Cardium edule		
Soft tissues	52.0–81.0 DW	Leatherland and Burton (1974)
Mantle fluids	95.0 DW	Leatherland and Burton (1974)
Shell	6.8 DW	Segar et al. (1971)
Soft parts	130.0 DW	Segar et al. (1971)
Cerastoderma edule		
Soft parts	63.0–208.0 DW	Romeril (1974)
Chenopus pespelicani		
Soft parts	2890.0 FW (max.)	Rosenberg (1977)
Chlamys opercularis		
Soft parts	462.0 DW	Bryan (1973)
Body fluid	45.1 DW	Bryan (1973)
Mantle	152.0 DW	Bryan (1973)
Gill	384.0 DW	Bryan (1973)
Digestive gland	132.0 DW	Bryan (1973)
Striated adductor muscle	61.7 DW	Bryan (1973)
Unstriated adductor muscle	111.0 DW	Bryan (1973)
Gonad and foot	117.0 DW	Bryan (1973)
Kidney	40,800.0 DW	Bryan (1973)
Shell	7.3–11.0 DW	Segar et al. (1971)
Choromytilus meridionalis		
Soft parts	73.0–113.0 DW	Watling and Watling (1976b)

269

Table 1. Continued

Species	Zinc Concentration	Reference
Soft parts	16.0 FW	Van As et al. (1973)
Clams		
Flesh	16.0–58.0 FW; 295.0– 590.0 DW; 2825.0–5650.0 AW	Lowman et al. (1970)
Shell	8.7–28.0 FW; 9.1–44.1 DW; 43.0–88.0 AW	Lowman et al. (1970)
Crassostrea angulata		
Muscle	312.0 FW	Vinogradov (1953)
Mantle	1074.0 FW	Vinogradov (1953)
Gill	652.0 FW	Vinogradov (1953)
Remainder	3223.0 FW	Vinogradov (1953)
Crassostrea commercialis		
Soft parts	318.0 FW	Ratkowsky et al. (1974)
Soft parts	80.0–665.0 FW	Mackay et al. (1975)
Crassostrea cucullata		
Soft parts	254.5 FW	Sastry and Bhatt (1965)
Crassostrea gigas		
Soft parts	86.0–344.0 FW	Pringle et al. (1968)
Soft parts	333.0–10,019.0 FW	Ratkowsky et al. (1974)
Soft parts	99,220.0 DW	Boyden and Romeril (1974)
Soft parts	396.0 DW	Watling and Watling (1976a)
Soft parts	424.0 DW	Watling and Watling (1976b)
Crassostrea madrasensis		
Soft parts	203.4 FW	Sastry and Bhatt (1965)
Crassostrea margaritacea		

Soft parts	886.0 DW	Watling and Watling (1976a)

Crassostrea virginica

Soft parts	Range of means: 200.0–4441.0 FW	Vinogradov (1953)
Muscle	160.0 FW	Vinogradov (1953)
Intestinal gland	237.0 FW	Vinogradov (1953)
Mantle	267.0 FW	Vinogradov (1953)
Soft parts	500.0–13,700.0 DW	Galtsoff (1953)
Soft parts		
Lower Chesapeake Bay	310.0 FW	McFarren et al. (1962)
Near Baltimore, Md.	4000.0 FW	McFarren et al. (1962)
Soft parts	Mean 230.0 FW (range 24.0–820.0 FW)	Kopfler and Mayer (1967)
Soft parts	1428.0 FW (range 180.0–4120.0 FW)	Pringle et al. (1968)
Soft parts	180.0–4120.0 FW	Shuster and Pringle (1968)
Soft parts	Means 1018.0–1641.0 FW (range 204.4–4120.0 FW)	Shuster and Pringle (1969)
Newport River, N.C.		
February, 1966		
Shell	24.7 FW	Wolfe (1970)
Soft parts	159.4 FW	Wolfe (1970)
January, 1969		
Shell	245.1 FW	Wolfe (1970)
Soft parts	210.5 FW	Wolfe (1970)
April, 1965		
Mantle	135.0 FW	Wolfe (1970)
Gill	182.0 FW	Wolfe (1970)
Labial palps	123.0 FW	Wolfe (1970)
Muscle	79.0 FW	Wolfe (1970)

Table 1. Continued

Species	Zinc Concentration	Reference
Digestive gland	260.0 FW	Wolfe (1970)
Remainder	175.0 FW	Wolfe (1970)
North River, N.C.		
November, 1966		
Mantle	450.6 FW	Wolfe (1970)
Gill	508.1 FW	Wolfe (1970)
Labial palp	501.2 FW	Wolfe (1970)
Muscle	126.1 FW	Wolfe (1970)
Digestive gland	491.8 FW	Wolfe (1970)
Rectum	531.0 FW	Wolfe (1970)
Pericardial sac	223.9 FW	Wolfe (1970)
Gonad	435.9 FW	Wolfe (1970)
Shell	1.5–8.1 DW	Windom and Smith (1972)
Soft parts	1200.0–5700.0 DW	Windom and Smith (1972)
Soft parts	12,675.0 DW max.	Valiela et al. (1974)
Soft parts	Mean 3915.0 DW; 10,000.0 DW max.	Drifmeyer (1974)
Soft parts		
North Carolina	125.0–325.0 FW	Huggett et al. (1975)
Virginia	275.0–425.0 FW	Huggett et al. (1975)
Soft parts		
August	5000.0 DW	Frazier (1975)
November	2100.0 DW	Frazier (1975)
Shell	2.5 DW	Frazier (1975)
Soft parts		

From metal-contaminated area	4100.0 DW	Frazier (1976)
From noncontaminated areas	1700.0 DW	Frazier (1976)
Crepidula fornicata		
Soft parts	116.0 DW	Leatherland and Burton (1974)
Shell	1.1 DW	Segar et al. (1971)
Soft parts	940.0 DW	Segar et al. (1971)
Cryptochiton stelleri		
Muscle	18.7 FW	Vinogradov (1953)
Intestinal gland	3.0 FW	Vinogradov (1953)
Mantle	40.0 FW	Vinogradov (1953)
Donax serra		
Soft parts	16.0 FW	Van As et al. (1973)
Ensis arcuata		
Shell	1.4 DW	Bertine and Goldberg (1972)
Ensis siliqua		
Shell	0.8 DW	Bertine and Goldberg (1972)
Gastropods		
Soft parts, 12 spp.	12.7–65.3 FW	Vinogradov (1953)
Soft parts, 13 spp.	152.0–269.0 DW	Vinogradov (1953)
Soft parts, 9 spp.	580.7–1481.6 DW	Vinogradov (1953)
Liver, 2 spp.	188.0–316.0 DW	Vinogradov (1953)
Glycymeris glycymeris		
Shell	160.0 DW	Segar et al. (1971)
Soft parts	120.0 DW	Segar et al. (1971)
Haliotis midea		
Soft parts	12.0 FW	Van As et al. (1973)
Haliotis rufescens		
Gill	28.0–54.0 DW	Anderlini (1974)
Mantle	42.0–74.0 DW	Anderlini (1974)
Digestive gland	536.0–980.0 DW	Anderlini (1974)

Table 1. Continued

Species	Zinc Concentration	Reference
Foot	38.0–46.0 DW	Anderlini (1974)
Haliotis tuberculata		
Soft parts	98.0–103.0 DW	Bryan et al. (1977)
Foot	28.0–33.1 DW	Bryan et al. (1977)
Viscera	238.0–288.0 DW	Bryan et al. (1977)
Digestive gland	624.0–656.0 DW	Bryan et al. (1977)
Blood	19.9 DW	Bryan et al. (1977)
Muscle	38.0 DW	Bryan et al. (1977)
Gill	63.4 DW	Bryan et al. (1977)
Left kidney	124.0 DW	Bryan et al. (1977)
Right kidney	298.0 DW	Bryan et al. (1977)
Hinnites multirugosus		
Kidney	37.2 FW	Vattuone et al. (1976)
Digestive gland	68.8 FW	Vattuone et al. (1976)
Katelysia marmorata		
Soft parts	11.0 FW	Sastry and Bhatt (1965)
Lamellibranchs		
Soft parts, 29 spp.	11.6–174.0 FW	Vinogradov (1953)
Soft parts, 26 spp.	66.0–4966.0 DW	Vinogradov (1953)
Soft parts, 23 spp.	258.9–14,862.5 AW	Vinogradov (1953)
Littorina littoralis		
Soft parts	312.0 DW	Leatherland and Burton (1974)
Littorina littorea		
Soft parts	38.0 DW	Leatherland and Burton (1974)
Soft parts	6.6–20.2 FW	Topping (1973)

Soft parts	520.0 DW max.	Butterworth et al. (1972)
Soft parts	28.0–274.0 DW	Ireland (1973)
Soft parts	88.0 DW	Bryan et al. (1977)
Soft parts	15.5–24.3 FW	Wharfe and Van Den Broek (1977)
Loligo opalescens		
Liver	247.0–449.0 DW	Martin and Flegal, 1975
Loligo vulgaris		
Whole	28.3 FW; 117.2 DW	Vinogradov (1953)
Mercenaria mercenaria		
Soft parts	1359.0 DW	McHargue (1924, 1927)
Soft parts	1680.0 AW	Eisler and Weinstein (1967)
Mantle	810.0 AW	Eisler and Weinstein (1967)
Gill	3070.0 AW	Eisler and Weinstein (1967)
Muscle	1500.0 AW	Eisler and Weinstein (1967)
Soft parts	11.5–40.2 FW	Eisler and Weinstein (1967)
Soft parts	20.6 FW	Shuster and Pringle (1968)
	(range 11.5–40.2)	Pringle et al. (1968)
Shell	3.4 DW	Segar et al. (1971)
Soft parts	94.0 DW	Segar et al. (1971)
Soft parts	12.4–27.6 DW	Romeril (1974)
Soft parts	298.0 DW max.	Valiela et al. (1974)
Meretrix meretrix		
Soft parts	14.8 FW	Sastry and Bhatt (1965)
Modiolus demissus		
Soft parts	52.0 DW max.	Valiela et al. (1974)
Modiolus modiolus		
Shell	6.2 DW	Segar et al. (1971)
Soft parts	320.0–530.0 DW	Segar et al. (1971)
Mantle and gills	180.0 DW	Segar et al. (1971)
Muscle	68.0 DW	Segar et al. (1971)

Table 1. Continued

Species	Zinc Concentration	Reference
Gonad	140.0 DW	Segar et al. (1971)
Gut and digestive gland	190.0 DW	Segar et al. (1971)
Molluscs		
Soft parts		
West Norway coast	2900.0 DW max.	Stenner and Nickless (1974b)
Rio Tinto estuary, Spain	370.0 DW max.	Stenner and Nickless (1975)
Mya arenaria		
Soft parts	17.0 FW	Pringle et al. (1968)
	(range 9.0–28.0)	
Soft parts	9.0–28.0 FW	Shuster and Pringle (1968)
Soft parts	9.5–11.2 FW	Eisler (1977)
Mytilus californianus		
Shell	8.6–26.5 DW	Graham (1972)
Soft parts	164.0–310.0 DW	Graham (1972)
Digestive gland	46.0–110.0 DW	Alexander and Young (1976)
Mytilus edulis		
Shell	0.04 DW	Segar et al. (1971)
Soft parts	91.0 DW	Segar et al. (1971)
Shell	0.6 DW	Bertine and Goldberg (1972)
Soft parts	40.0 DW	Bertine and Goldberg (1972)
Shell	6.9–14.2 DW	Graham (1972)
Soft parts	204.0–341.0 DW	Graham (1972)
Soft parts	253.0–779.0 DW	Ireland (1973)
Soft parts	12.5–82.5 FW	Topping (1973)
Soft parts	45.8 FW	Eustace (1974)

Soft parts	269.0 DW	Leatherland and Burton (1974)
Digestive gland	185.0 FW max.	Young and McDermott (1975)
Gonad	183.0 FW max.	Young and McDermott (1975)
Muscle	172.0 FW max.	Young and McDermott (1975)
Remainder	244.0 FW max.	Young and McDermott (1975)
Soft parts	19.3–60.1 FW	Phillips (1976a)
Soft parts	16.4–97.1 FW	Phillips (1976b)
Soft parts	29.0–41.8 FW	Wharfe and Van Den Broek (1977)
Soft parts	14.0–460.0 DW	Phillips (1977b)
Mytilus edulis aoteanus		
Soft parts	91.0 DW	Brooks and Rumsby (1965)
Mantle	192.0 DW	Brooks and Rumsby (1965)
Gill	336.0 DW	Brooks and Rumsby (1965)
Visceral mass	525.0 DW	Brooks and Rumsby (1965)
Gonad	260.0 DW	Brooks and Rumsby (1965)
Mytilus galloprovincialis		
Soft parts	209.0 DW	Fowler and Oregioni (1976)
Mytilus viridis		
Soft parts	11.0 FW	Sastry and Bhatt (1965)
Notodarius gouldi		
Whole	18.5 FW	Eustace (1974)
Nucella lapillus		
Shell	0.59 DW	Segar et al. (1971)
Soft parts	860.0 DW	Segar et al. (1971)
Soft parts		
Beer, Dorset	345.0 DW	Stenner and Nickless (1974a)
Transplants from Beer to Bristol		
Channel after 5 months	2530.0 DW	Stenner and Nickless (1974a)
Soft parts	415.0 DW	Leatherland and Burton (1974)

Table 1. Continued

Species	Zinc Concentration	Reference
Octopus sp.		
Whole	106.0 FW	Matsumoto et al. (1964)
Whole	18.5 FW	Eustace (1974)
Octopus vulgaris		
Liver	53.2 FW; 144.3 DW	Vinogradov (1953)
Ommastrephes bartrami		
Liver	163.0 DW	Martin and Flegal (1975)
Ostrea angasi		
Soft parts	371.0–8865.0 FW	Ratkowsky et al. (1974)
Soft parts	5657.0 FW	Eustace (1974)
Ostrea edulis		
Soft parts	660.0 DW	Watling and Watling (1976a)
Ostrea gigas		
Soft parts		
Transplanted from control area to metal-contaminated area— After 20 days		
Controls	200.0 FW	Ikuta (1968a)
Transplants	580.0 FW	Ikuta (1968a)
Soft parts		
Transplants to metal-contaminated area after 120 days		
Adductor muscle	250.0 FW	Ikuta (1968b)
Labial palp	800.0–900.0 FW	Ikuta (1968b)
Mantle	800.0–900.0 FW	Ikuta (1968b)
Gill	800.0–900.0 FW	Ikuta (1968b)
Remainder	300.0 FW	Ikuta (1968b)

Soft parts

Transplants from zinc-contaminated area

Mantle	1072.0 FW	Ikuta (1968c)
Gill	4270.0 FW	Ikuta (1968c)
Labial palp	2178.0 FW	Ikuta (1968c)
Muscle	221.0 FW	Ikuta (1968c)
Remainder	1013.0 FW	Ikuta (1968c)
Whole	953.0 FW	Ikuta (1968c)

After 116 days in "clean" water

Mantle	95.0 FW	Ikuta (1968c)
Gill	171.0 FW	Ikuta (1968c)
Labial palp	112.0 FW	Ikuta (1968c)
Muscle	37.0 FW	Ikuta (1968c)
Remainder	122.0 FW	Ikuta (1968c)
Whole	108.0 FW	Ikuta (1968c)

Ostrea laperousii

Whole	60.0 FW; 410.0 DW	Vinogradov (1953)
Hepatopancreas	65.7 FW; 245.0 DW	Vinogradov (1953)
Mantle	70.0 FW	Vinogradov (1953)
Gill	134.0 FW; 1002.0 DW	Vinogradov (1953)
Muscle	43.0 FW; 233.0 DW	Vinogradov (1953)
Remainder	115.0 FW; 661.0 DW	Vinogradov (1953)

Ostrea sinuata

Soft parts	1103.0 DW	Brooks and Rumsby (1965)
Mantle	4760.0 DW	Brooks and Rumsby (1965)
Gill	3300.0 DW	Brooks and Rumsby (1965)
Muscle	369.0–400.0 DW	Brooks and Rumsby (1965)
Visceral mass	1122.0 DW	Brooks and Rumsby (1965)
Kidney	1248.0 DW	Brooks and Rumsby (1965)
Heart	2090.0 DW	Brooks and Rumsby (1965)

279

Table 1. Continued

Species	Zinc Concentration	Reference
Oysters		
Soft parts	21,000.0 FW max.	Thrower and Eustace (1963a, 1973b)
Patella sp.		
Soft parts	91.0 DW	Stenner and Nickless (1974a)
Beer, Dorset		
Transplants from Beer to Bristol		
Channel—after 8 months	340.0 DW	Stenner and Nickless (1974a)
Shell	51.0 DW	Segar et al. (1971)
Soft parts	84.0 DW	Segar et al. (1971)
Soft parts	580.0 DW max.	Butterworth et al. (1972)
Soft parts	158.0 DW	Preston et al. (1972)
Soft parts	95.0–229.0 DW	Leatherland and Burton (1974)
Soft parts	101.0 DW	Bryan et al. (1977)
In desalinization		
Plant effluent	255.0 DW	Romeril (1977)
Controls	175.0 DW	Romeril (1977)
Pecten jacobaeus		
Muscle	50.0 FW	Vinogradov (1953)
Mantle	43.0 FW	Vinogradov (1953)
Gill	86.0 FW	Vinogradov (1953)
Remainder	82.0 FW	Vinogradov (1953)
Pecten maximus		
Upper valve	4.1 DW	Segar et al. (1971)
Lower valve	6.1 DW	Segar et al. (1971)
Soft parts	230.0 DW	Segar et al. (1971)

Muscle	70.0 DW	Segar et al. (1971)
Gut and digestive gland	1100.0 DW	Segar et al. (1971)
Mantle and gills	160.0–420.0 DW	Segar et al. (1971)
Gonad	180.0–360.0 DW	Segar et al. (1971)
Soft parts	273.0 DW	Bryan (1973)
Body fluid	30.1 DW	Bryan (1973)
Mantle	59.4 DW	Bryan (1973)
Gill	100.0 DW	Bryan (1973)
Digestive gland	407.0 DW	Bryan (1973)
Striated adductor muscle	69.9 DW	Bryan (1973)
Unstriated adductor muscle	94.2 DW	Bryan (1973)
Gonad and foot	196.0 DW	Bryan (1973)
Kidney	19,300.0 DW	Bryan (1973)
Edible tissues	19.1–45.5 FW	Topping (1973)
Pecten novae-zelandiae		
Soft parts	283.0 DW	Brooks and Rumsby (1965)
Muscle	108.0 DW	Brooks and Rumsby (1965)
Visceral mass	400.0 DW	Brooks and Rumsby (1965)
Intestine	392.0 DW	Brooks and Rumsby (1965)
Kidney	2630.0 DW	Brooks and Rumsby (1965)
Foot	210.0 DW	Brooks and Rumsby (1965)
Gonads	256.0 DW	Brooks and Rumsby (1965)
Pitar morrhuana		
Soft parts	145.0–276.0 DW	Eisler et al. (1978a)
Placopecten magellanicus		
Soft parts	105.0 DW	Pesch et al. (1977)
Soft parts	9.3–24.0 FW; 109.0 FW max.	Palmer and Rand (1977)
Protothaca staminea		
Shell	9.2 DW	Graham (1972)

Table 1. Continued

Species	Zinc Concentration	Reference
Soft parts	67.7 DW	Graham (1972)
Pteropods		
Shell	Mean 36.0 DW; 312.0 DW max.	Pyle and Tieh (1970)
Scrobicularia plana		
Soft parts	974.0 DW	Bryan and Hummerstone (1977)
Sepia officinalis		
Whole	13.0 FW; 77.0 DW	Vinogradov (1953)
Eggs	63.6 FW; 772.7 DW	Vinogradov (1953)
Liver	69.6 FW; 175.3 DW	Vinogradov (1953)
Gill	99.0 DW	Leatherland and Burton (1974)
Mantle	52.0 DW	Leatherland and Burton (1974)
Sepia sp.		
Whole	50.0 FW	Matsumoto et al. (1964)
Snails		
Flesh	18.0–20.0 FW; 54.0–66.0 DW; 760.0–860.0 AW	Lowman et al. (1970)
Shell	22.0–68.0 FW; 23.0–70.0 DW; 71.0–74.0 AW	Lowman et al. (1970)
Viscera	3.6 FW; 12.3 DW; 93.0 AW	Lowman et al. (1970)
Squid		
Whole	710.0 AW	Robertson (1967)
Strombus pugilis		
Shell	110.0–210.0 FW; 120.0–210.0 DW; 120.0–220.0 AW	Lowman et al. (1966)

Foot	12.0–25.0 FW; 55.0–110.0 DW; 510.0–1100.0 AW	Lowman et al. (1966)
Soft parts	39.0–250.0 FW; 250.0–1500.0 DW; 1500.0–3500.0 AW	Lowman et al. (1966)
Internal organs	23.0–93.0 FW; 96.0–560.0 DW; 280.0–2300.0 AW	Lowman et al. (1966)
Symplectoteuthis oualaniensis		
Liver	513.0 DW	Martin and Flegal (1975)
Tapes semidecussata		
Shell	11.2 DW	Graham (1972)
Soft parts	11.2 DW	Graham (1972)
Tegula funebralis		
Shell	4.8–23.8 DW	Graham (1972)
Soft parts	25.4–261.0 DW	Graham (1972)
Thais emarginata		
Shell	1.4–23.5 DW	Graham (1972)
Soft parts	1701.0 DW	Graham (1972)
Thais lapillus		
Soft parts	3100.0 DW max.	Butterworth et al. (1972)
Soft parts	916.0–1980.0 DW	Ireland (1973)
ARACHNOIDEA		
Limulus polyphemus		
Liver and gonads	670.0 DW	Phillips (1917)
Blood	70.0 DW	Phillips (1917)
Liver, sex organs	532.0 DW	Vinogradov (1953)
Blood	56.0 DW	Vinogradov (1953)
CRUSTACEA		
Acanthephyra eximia		
Whole	65.0 DW	Leatherland et al. (1973)

Table 1. Continued

Species	Zinc Concentration	Reference
Acartia clausi		
Whole	1270.0 DW	Zafiropoulos and Grimanis (1977)
Amphipods		
Whole	43.0 DW	Bohn and McElroy (1976)
Atelecyclus septemdentatus		
Whole	32.0 FW	Bryan (1968)
Blood	9.5 FW	Bryan (1968)
Muscle	61.0 FW	Bryan (1968)
Stomach fluid	6.0 FW	Bryan (1968)
Hepatopancreas	88.0 FW	Bryan (1968)
Urine	1.4 FW	Bryan (1968)
Excretory organ	29.0 FW	Bryan (1968)
Gill	27.0 FW	Bryan (1968)
Shell	7.0 FW	Bryan (1968)
Vas deferens	30.0 FW	Bryan (1968)
Austropotamobius pallipes		
Whole	24.0 FW	Bryan (1968)
Blood	0.9 FW	Bryan (1968)
Muscle	12.0 FW	Bryan (1968)
Stomach fluid	41.0 FW	Bryan (1968)
Hepatopancreas	109.0 FW	Bryan (1968)
Urine	0.02 FW	Bryan (1968)
Excretory organ	7.0 FW	Bryan (1968)
Gill	8.0 FW	Bryan (1968)
Shell	8.0 FW	Bryan (1968)
Vas deferens	14.0 FW	Bryan (1968)

Ovary	26.0 FW	Bryan (1968)
Balanus amphitrite communis		
Soft parts	71.5 FW	Sastry and Bhatt (1965)
Balanus balanoides		
Soft parts	4500.0–23,100.0 DW	Ireland (1973)
Barnacles		
Soft parts	3300.0 DW max.	Stenner and Nickless (1975)
Whole, 3 spp.	138.0–3438.0 FW	Walker et al. (1975)
Soft parts, 3 spp.	110.0–2980.0 FW	Walker et al. (1975)
Gut and parenchyma, 3 spp.	2220.0 FW max.	Walker et al. (1975)
Cancer irroratus		
Digestive diverticula	35.0–43.0 DW	Greig (1975)
Cancer pagurus		
Hepatopancreas	162.0 DW	Vinogradov (1953)
Whole	27.0 FW	Bryan (1968)
Muscle	64.0 FW	Bryan (1968)
Stomach fluid	15.0 FW	Bryan (1968)
Hepatopancreas	45.0 FW	Bryan (1968)
Urine	0.6 FW	Bryan (1968)
Excretory organ	29.0 FW	Bryan (1968)
Gill	42.0 FW	Bryan (1968)
Shell	3.0 FW	Bryan (1968)
Vas deferens	16.0 FW	Bryan (1968)
External eggs	94.0 FW	Bryan (1968)
Blood	49.0 FW	Bryan (1968)
Meat	21.3–34.5 FW	Topping (1973)
Carcinus maenus		
Whole	24.5 FW	Bryan (1966)
Hepatopancreas	56.0 FW	Bryan (1966)
Leg muscle	44.0 FW	Bryan (1966)

285

Table 1. Continued

Species	Zinc Concentration	Reference
Whole	22.0 FW	Bryan (1968)
Blood	36.0 FW	Bryan (1968)
Muscle	44.0 FW	Bryan (1968)
Stomach fluid	18.0 FW	Bryan (1968)
Hepatopancreas	56.0 FW	Bryan (1968)
Urine	0.4 FW	Bryan (1968)
Excretory organ	19.0 FW	Bryan (1968)
Gill	26.0 FW	Bryan (1968)
Shell	3.0 FW	Bryan (1968)
Vas deferens	23.0 FW	Bryan (1968)
Whole	25.1–38.8 FW	Wharfe and Van Den Broek (1977)
Clibanarius strigimanus		
Whole	74.3 FW	Eustace (1974)
Copepods		
Whole	60.0 DW	Bohn and McElroy (1976)
Whole	62.0–170.0 DW	Martin and Knauer (1973)
Corystes cassivelaunus		
Whole	39.0 FW	Bryan (1968)
Blood	11.0 FW	Bryan (1968)
Muscle	52.0 FW	Bryan (1968)
Stomach fluid	12.0 FW	Bryan (1968)
Hepatopancreas	50.0 FW	Bryan (1968)
Excretory organ	16.0 FW	Bryan (1968)
Gill	45.0 FW	Bryan (1968)
Shell	5.0 FW	Bryan (1968)
Vas deferens	27.0 FW	Bryan (1968)

Crab		
Flesh	24.0–51.0 FW; 120.0–250.0 DW; 75.0–1900.0 AW	Lowman et al. (1970)
Carapace	7.0–46.0 FW; 15.0–120.0 DW; 23.0–110.0 AW	Lowman et al. (1970)
Whole	1216.0 DW	McHargue (1924, 1927)
Crangon vulgaris		
Whole	34.0 FW	Bryan (1968)
Blood	23.0 FW	Bryan (1968)
Muscle	14.0 FW	Bryan (1968)
Hepatopancreas	78.0 FW	Bryan (1968)
Whole	26.4 FW	Wharfe and Van Den Broek (1977)
Crustaceans		
Whole	23.7–28.9 FW	Wright (1976)
Soft parts, 12 spp.	55.0–290.0 DW	Stickney et al. (1975)
Whole	275.0 DW max.	Stenner and Nickless (1974)
Whole, 5 spp.	160.0–1216.0 DW	Vinogradov (1953)
Euphausia pacifica		
Whole	13.0 FW	Cutshall et al. (1977b)
Euphausids		
Whole	53.0–83.0 DW	Martin and Knauer (1973)
Whole	430.0 AW	Robertson (1967)
Eupagurus bernhardus		
Whole	28.0 FW	Bryan (1968)
Blood	11.6 FW	Bryan (1968)
Leg muscle	36.0 FW	Bryan (1968)
Abdominal muscle	24.0 FW	Bryan (1968)
Stomach fluid	26.0 FW	Bryan (1968)
Hepatopancreas	69.0 FW	Bryan (1968)
Excretory organ	23.0 FW	Bryan (1968)

Table 1. Continued

Species	Zinc Concentration	Reference
Gill	69.0 FW	Bryan (1968)
Vas deferens	27.0 FW	Bryan (1968)
Ovary	56.0 FW	Bryan (1968)
Galathea squamifera		
Whole	18.0 FW	Bryan (1968)
Blood	2.5 FW	Bryan (1968)
Muscle	10.0 FW	Bryan (1968)
Stomach fluid	47.0 FW	Bryan (1968)
Hepatopancreas	49.0 FW	Bryan (1968)
Gill	27.0 FW	Bryan (1968)
Shell	9.0 FW	Bryan (1968)
Homarus vulgaris		
Slow-contracting muscle	93.0–105.0 DW	Bryan (1967)
Fast-contracting muscle	13.0–20.0 DW	Bryan (1967)
Meat	13.8–16.7 FW	Topping (1973)
Whole	23.0 FW	Bryan (1968)
Blood	7.4 FW	Bryan (1968)
Leg muscle	60.0 FW	Bryan (1968)
Abdominal muscle	15.0 FW	Bryan (1968)
Stomach fluid	0.7 FW	Bryan (1968)
Hepatopancreas	34.0 FW	Bryan (1968)
Urine	2.2 FW	Bryan (1968)
Excretory organ	20.0 FW	Bryan (1968)
Gill	15.0 FW	Bryan (1968)
Shell	5.0 FW	Bryan (1968)
Vas deferens	13.0 FW	Bryan (1968)

Ovary	50.0 FW	Bryan (1968)
Whole	22.0 FW	Bryan (1966)
Hepatopancreas	34.0 FW	Bryan (1966)
Leg muscle	64.0 FW	Bryan (1966)
Blood	4.7–11.8 FW	Bryan (1964)
Urine	0.1–6.1 FW	Bryan (1964)
Excretory organs	17.3–21.9 FW	Bryan (1964)
Abdominal muscle	14.3–17.7 FW	Bryan (1964)
Hepatopancreas	23.3–59.5 FW	Bryan (1964)
Stomach fluid	0.5–1.3 FW	Bryan (1964)
Gill	12.8–19.9 FW	Bryan (1964)
Shell	0.9–8.9 FW	Bryan (1964)
Vas deferens	13.3 FW	Bryan (1964)
Ovary	50.2 FW	Bryan (1964)
Jasus lalandii		
Meat	17.0 FW	Van As et al. (1973)
Lepas anatifera		
Whole	4.0–27.0 FW; 122.0–393.0 DW	Vinogradov (1953)
Lobsters		
Meat	160.0 DW	McHargue (1924, 1927)
Maia squinado		
Whole	21.0 FW	Bryan (1968)
Blood	2.4 FW	Bryan (1968)
Muscle	63.0 FW	Bryan (1968)
Stomach fluid	31.0 FW	Bryan (1968)
Hepatopancreas	71.0 FW	Bryan (1968)
Urine	0.3 FW	Bryan (1968)
Excretory organ	15.0 FW	Bryan (1968)
Gill	10.0 FW	Bryan (1968)

Table 1. Continued

Species	Zinc Concentration	Reference
Shell	5.0 FW	Bryan (1968)
Vas deferens	16.0 FW	Bryan (1968)
Ovary	45.0 FW	Bryan (1968)
Meganyctiphanes norvegica		
Whole	104.0 DW	Leatherland et al. (1973)
Whole	62.0 DW	Fowler (1977)
Moults	146.0 DW	Fowler (1977)
Eggs	318.0 DW	Fowler (1977)
Fecal pellets	950.0 DW	Fowler (1977)
Nephrops norvegicus		
Meat	8.5–12.2 FW	Topping (1973)
Oplophorus sp.		
Whole	98.0 DW	Leatherland et al. (1973)
Orchestoidea corniculata		
Whole	150.0–300.0 DW	Bender (1975)
Paguristes sericeus		
Whole	29.0–37.0 FW; 92.0–120.0 DW; 230.0–360.0 AW	Lowman et al. (1966)
Palaemon serratus		
Whole	21.0 FW	Bryan (1968)
Blood	38.0 FW	Bryan (1968)
Muscle	10.0 FW	Bryan (1968)
Hepatopancreas	64.0 FW	Bryan (1968)
Gill	35.0 FW	Bryan (1968)
External eggs	24.0 FW	Bryan (1968)

Species / Tissue	Value	Reference
Palaemon squilla		
Whole	30.0 FW	Bryan (1968)
Palaemonetes varians		
Whole	20.0 FW	Bryan (1968)
Blood	87.0 FW	Bryan (1968)
Muscle	14.0 FW	Bryan (1968)
Hepatopancreas	65.0 FW	Bryan (1968)
Palinurus sp.		
Blood	0.0 DW	Phillips (1917)
Liver	200.0 DW	Phillips (1917)
Palinurus vulgaris		
Blood	160.0 DW	Vinogradov (1953)
Liver	152.0 DW	Vinogradov (1953)
Blood	3.1 FW	Bryan (1968)
Leg muscle	66.0 FW	Bryan (1968)
Abdominal muscle	20.0 FW	Bryan (1968)
Stomach fluid	48.0 FW	Bryan (1968)
Hepatopancreas	97.0 FW	Bryan (1968)
Urine	0.6 FW	Bryan (1968)
Excretory organ	20.0 FW	Bryan (1968)
Gill	20.0 FW	Bryan (1968)
Shell	16.0 FW	Bryan (1968)
Ovary	82.0 FW	Bryan (1968)
External eggs	107.0 FW	Bryan (1968)
Pandalus jordani		
Whole	58.0 FW	Cutshall et al. (1977b)
Muscle	11.0 FW	Cutshall et al. (1977b)
Penaeus spp.		
Whole	62.0 DW	Knauer (1970)

Table 1. Continued

Species	Zinc Concentration	Reference
Pilumnus hirtellus		
Whole	49.0 FW	Bryan (1968)
Blood	12.3 FW	Bryan (1968)
Stomach fluid	92.0 FW	Bryan (1968)
Hepatopancreas	169.0 FW	Bryan (1968)
Gill	60.0 FW	Bryan (1968)
Shell	17.0 FW	Bryan (1968)
Platicarcinus pagurus		
Gill	35.0 FW; 248.0 DW	Vinogradov (1953)
Muscle	67.0 FW; 284.0 DW	Vinogradov (1953)
Intestine	17.0 FW; 81.0 DW	Vinogradov (1953)
Porcellana platycheles		
Whole	54.0 FW	Bryan (1968)
Portunus puber		
Whole	27.0 FW	Bryan (1968)
Blood	7.0 FW	Bryan (1968)
Muscle	28.0 FW	Bryan (1968)
Stomach fluid	15.0 FW	Bryan (1968)
Hepatopancreas	42.0 FW	Bryan (1968)
Urine	0.2 FW	Bryan (1968)
Excretory organ	10.0 FW	Bryan (1968)
Gill	23.0 FW	Bryan (1968)
Shell	13.0 FW	Bryan (1968)
Vas deferens	20.0 FW	Bryan (1968)
Ovary	87.0 FW	Bryan (1968)

Portunus depurator		
Whole	21.0 FW	Bryan (1968)
Blood	1.8 FW	Bryan (1968)
Muscle	15.0 FW	Bryan (1968)
Stomach fluid	13.0 FW	Bryan (1968)
Hepatopancreas	24.0 FW	Bryan (1968)
Urine	0.3 FW	Bryan (1968)
Excretory organ	8.0 FW	Bryan (1968)
Gill	25.0 FW	Bryan (1968)
Shell	3.0 FW	Bryan (1968)
Potamobius fluviatis		
Muscle	355.0 DW	Vinogradov (1953)
Shrimp		
Flesh	12.0–36.0 FW; 44.0–130.0 DW; 610.0–1100.0 AW	Lowman et al. (1970)
Carapace	4.0–39.0 FW; 17.0–110.0 DW; 190.0–710.0 AW	Lowman et al. (1970)
Systellaspis debilis		
Whole	50.0 DW	Leatherland et al. (1973)
Xantho incisus		
Whole	26.0 FW	Bryan (1968)
SESTON		
Seston	130.0–1400.0 DW; 270.0–3900.0 AW	Lowman et al. (1966)
ZOOPLANKTON		
Microplankton		
Whole	483.0 DW	Fowler (1977)
Plankton		
Whole	2.6–57.0 FW; 43.0–680.0 DW; 75.0–1700.0 AW	Lowman et al. (1966)

293

Table 1. Continued

Species	Zinc Concentration	Reference
Whole	7.1–136.0 DW	Greig et al. (1977)
Zooplankton		
Whole	9.9–625.0 DW	Greig et al. (1977)
Whole	20.0–1069.0 DW	Fujita (1972)
INSECTA		
Halobates sericeus		
Whole	176.0 DW	Cheng et al. (1976)
Halobates sobrinus		
Whole	176.0 DW	Cheng et al. (1976)
Rheumatobates aestuarius		
Whole	197.0 DW	Cheng et al. (1976)
CHAETOGNATHA		
Chaetognaths		
Whole	76.0–90.0 DW	Bohn and McElroy (1976)
Sagitta sp.		
Whole	20,000.0 AW	Vinogradova and Koual'skiy (1962)
ANNELIDA		
Amphitrite ornata		
Whole	78.0 DW	Cross et al. (1970)
Arabella iricolor		
Whole	145.0 DW	Cross et al. (1970)
Chaetopterus variopedatus		
Whole	100.0 DW	Cross et al. (1970)
Diopatra cuprea		
Whole	85.0 DW	Cross et al. (1970)

Glycera americana		
Whole	165.0 DW	Cross et al. (1970)
Limnodrilus sp.		
Whole	3.2 DW	Vinogradov (1953)
Lineus longissimus		
Whole	10,000.0 AW	Vinogradov (1953)
Nephthys sp.		
Whole	10,000.0 AW	Vinogradov (1953)
Nereis diversicolor		
Whole	130.0–315.0 DW	Bryan and Hummerstone (1973b)
Whole	22.0–37.0 FW	Wharfe and Van Den Broek (1977)
Nereis japonicus		
Whole	8.7 DW	Vinogradov (1953)
Nereis sp.		
Whole	80.0 DW	Cross et al. (1970)
ECHINODERMATA		
Asterias rubens		
Whole	190.0 DW	Riley and Segar (1970)
Oral skin	110.0 DW	Riley and Segar (1970)
Aboral skin	110.0 DW	Riley and Segar (1970)
Pyloric caeca	160.0 DW	Riley and Segar (1970)
Gonad	360.0 DW	Riley and Segar (1970)
Whole	160.0 DW	Vinogradov (1953)
Whole	220.0 DW	Leatherland and Burton (1974)
Brissopsis lyrifera		
Exoskeleton	65.0 DW	Vinogradov (1953)
Callorhynchus millii		
Whole	5.0 FW; 6.7 FW max.	Eustace (1974)

295

Table 1. Continued

Species	Zinc Concentration	Reference
Coscinasterias calamaria		
Whole	40.6 FW	Eustace (1974)
Echinoderms		
Whole	1500.0 DW max.	Stenner and Nickless (1974)
Whole, 11 spp.	3.0–46.0 DW	Vinogradov (1953)
Echinus esculentus		
Oral shell	110.0 DW	Riley and Segar (1970)
Aboral shell	12.0 DW	Riley and Segar (1970)
Aristotle's lantern	22.0 DW	Riley and Segar (1970)
Spines	28.0 DW	Riley and Segar (1970)
Intestines	550.0 DW	Riley and Segar (1970)
Gonad	110.0 DW	Riley and Segar (1970)
Henricia sanguinolenta		
Whole	240.0 DW	Riley and Segar (1970)
Holothuria bermudiana		
Viscera	32.0–180.0 DW	Vinogradov (1953)
Muscle	220.0 DW	Phillips (1917)
Intestines	65.0 DW	Phillips (1917)
Holothuria sp.		
Whole	8.7 FW	Matsumoto et al. (1964)
Mellita lata		
Whole	31.0–130.0 FW;	Lowman et al. (1966)
	54.0–220.0 DW;	
	57.0–240.0 AW	
Mellita sexiesperforata		
Whole	24.0–46.0 FW;	Lowman et al. (1966)
	43.0–83.0 DW;	

Organism / tissue	Concentration	Reference
Patiriella regularis		
Whole	46.0–88.0 AW	Eustace (1974)
Porania pulvillus		
Whole	245.0 FW	
Pseudocentrotus depressus	100.0 DW	Riley and Segar (1970)
Spermatozoa		
Homogenate	66.0–79.0 AW	Morisawa and Mohri (1972)
Head	34.0–41.0 AW	Morisawa and Mohri (1972)
Midpiece	281.0–337.0 AW	Morisawa and Mohri (1972)
Tail	172.0–186.0 AW	Morisawa and Mohri (1972)
Microtubules	118.0–142.0 AW	Morisawa and Mohri (1972)
Solaster papposus		
Whole	120.0–130.0 DW	Riley and Segar (1970)
Spatangus purpureus		
Test and spines	35.0 DW	Riley and Segar (1970)
Gonad	170.0 DW	Riley and Segar (1970)
Stichopus tremulus		
Whole	140.0 DW	Vinogradov (1953)
TUNICATA		
Ascidiacea sp.		
Whole	64.0 FW	Eustace (1974)
Botryllus schlosseri		
Whole	179.0–251.0 DW	Ireland (1973)
Whole	135.0 DW	Leatherland and Burton (1974)
Ciona atra		
Whole	30.0 DW	Phillips (1917)
Ciona intestinalis		
Whole	330.0 DW	Vinogradov (1953)

Table 1. Continued

Species	Zinc Concentration	Reference
Pyrosoma sp.		
Whole	105.0 DW	Leatherland et al. (1973)
ELASMOBRANCHII		
Carcharhinus falciformis		
Muscle	10.0 DW	Windom et al. (1973)
Liver	19.0 DW	Windom et al. (1973)
Kidney	25.0 DW	Windom et al. (1973)
Brain	10.0 DW	Windom et al. (1973)
Gonad	10.0 DW	Windom et al. (1973)
Gill	24.0 DW	Windom et al. (1973)
Spleen	28.0 DW	Windom et al. (1973)
Flesh	4.3–61.0 FW; 16.0–170.0 DW; 300.0–580.0 AW	Lowman et al. (1966)
Carcharhinus longimanus		
Flesh	3.4 FW; 16.0 DW; 290.0 AW	Lowman et al. (1966)
Stomach	11.0 FW; 80.0 DW; 1200.0 AW	Lowman et al. (1966)
Skin	32.0 FW; 81.0 DW; 380.0 AW	Lowman et al. (1966)
Liver	2.7 FW; 4.9 DW; 500.0 AW	Lowman et al. (1966)
Vertebrae	11.0 FW; 35.0 DW; 120.0 AW	Lowman et al. (1966)
Carcharhinus milberti		
Liver	9.0 DW	Windom et al. (1973)
Vertebrae		
Small sharks, 64 cm TL	130.0 AW	Eisler (1967b)
Medium sharks, 130 cm TL	80.0 AW	Eisler (1967b)

Large sharks, 209 cm TL	50.0 AW	Eisler (1967b)
Carcharhinus obscurus		
Muscle	19.0 DW	Windom et al. (1973)
Liver	16.0 DW	Windom et al. (1973)
Brain	27.0 DW	Windom et al. (1973)
Pup (total)	12.0–14.0 DW	Windom et al. (1973)
Dasyatis sabina		
Whole	3.8 FW	Vinogradov (1953)
Elasmobranchs		
Whole, 7 spp.	4.6–5.0 FW; 9.6 FW max.	Eustace (1974)
Musteleus canis		
Muscle	3.7–4.7 FW	Greig and Wenzloff (1977)
Liver	3.2–3.6 FW	Greig and Wenzloff (1977)
Raja clavata		
Whole	9.6 FW	Pentreath (1973a)
Blood	4.7 FW	Pentreath (1973a)
Heart	9.4 FW	Pentreath (1973a)
Spleen	15.9 FW	Pentreath (1973a)
Liver	17.1 FW	Pentreath (1973a)
Kidney	11.3 FW	Pentreath (1973a)
Gonad	10.9 FW	Pentreath (1973a)
Gut	14.8 FW	Pentreath (1973a)
Gill filament	11.1 FW	Pentreath (1973a)
Skin	12.3 FW	Pentreath (1973a)
Muscle	4.8 FW	Pentreath (1973a)
Cartilage	19.3 FW	Pentreath (1973a)
Rectal gland	8.5 FW	Pentreath (1973a)
Brain and nerve cord	10.1 FW	Pentreath (1973a)
Raja eglanteria		
Muscle	20.0 DW	Windom et al. (1973)

Table 1. Continued

Species	Zinc Concentration	Reference
Liver	44.0 DW	Windom et al. (1973)
Yolk-sac	31.0 DW	Windom et al. (1973)
Rhinobatis lentiginous		
Muscle	11.0 DW	Windom et al. (1973)
Liver	29.0 DW	Windom et al. (1973)
Stomach	38.0 DW	Windom et al. (1973)
Yolk-sac	19.0 DW	Windom et al. (1973)
Rhinoptera bonasus		
Muscle	8.0 DW	Windom et al. (1973)
Liver	9.0 DW	Windom et al. (1973)
Brain	36.0 DW	Windom et al. (1973)
Stomach	34.0 DW	Windom et al. (1973)
Spiral valve	29.0 DW	Windom et al. (1973)
Spleen	32.0 DW	Windom et al. (1973)
Uterus	30.0 DW	Windom et al. (1973)
Sphyrna lewini		
Muscle	15.0 DW	Windom et al. (1973)
Liver	16.0 DW	Windom et al. (1973)
Stomach	40.0 DW	Windom et al. (1973)
Sphyrna tiburo		
Muscle	8.0 DW	Windom et al. (1973)
Liver	13.0 DW	Windom et al. (1973)
Stomach	28.0 DW	Windom et al. (1973)
Spleen	25.0 DW	Windom et al. (1973)
Ovary	36.0 DW	Windom et al. (1973)

Species / Tissue	Concentration	Reference
Squalus acanthius		
Muscle	12.0 DW	Windom et al. (1973)
Liver	15.0 DW	Windom et al. (1973)
Stomach	70.0 DW	Windom et al. (1973)
Spleen	30.0 DW	Windom et al. (1973)
Yolk-sac	16.0 DW	Windom et al. (1973)
Embryo	37.0 DW	Windom et al. (1973)
Whole	155.0 FW	Vinogradov (1953)
Squatina squatina		
Whole	7.8 FW	Vinogradov (1953)
Torpedo torpedo		
Whole	3.3 FW	Vinogradov (1953)
TELEOSTEI		
Acanthocybium petus		
Flesh	0.11–8.0 FW; 0.39–360.0 DW; 390.0–5 200.0 AW	Lowman et al. (1966)
Ailurichthys marinus		
Muscle	8.1 FW	Vinogradov (1953)
Swim bladder	12.2 FW	Vinogradov (1953)
Gill	102.5 FW	Vinogradov (1953)
Skin	12.2 FW	Vinogradov (1953)
Bone	93.0–102.5 FW	Vinogradov (1953)
Liver	31.0 FW	Vinogradov (1953)
Albacore tuna		
Flesh	13.4 FW	Orvini et al. (1974)
Anchoa mitchelli		
Whole, less head	397.0 DW	Windom et al. (1973)
Anguilla anguilla		
Liver	73.0 FW	Vinogradov (1953)

Table 1. Continued

Species	Zinc Concentration	Reference
Muscle	21.0 FW	Vinogradov (1953)
Muscle	23.8–27.2 FW	Wharfe and Van Den Broek (1977)
Liver	51.9–71.9 FW	Wharfe and Van Den Broek (1977)
Anguilla rostrata		
Muscle	25.0 DW	Windom et al. (1973)
Anoplopoma fimbria		
Muscle	4.0 FW	Cutshall et al. (1977b)
Argyrozona argyrozona		
Flesh	2.9 FW	Van As et al. (1973)
Arripis trutta		
Muscle	9.0–46.0 FW	Brooks and Rumsey (1974)
Liver	21.0–42.0 FW	Brooks and Rumsey (1974)
Kidney	19.0–140.0 FW	Brooks and Rumsey (1974)
Heart	10.0–24.0 FW	Brooks and Rumsey (1974)
Gonad	11.0–120.0 FW	Brooks and Rumsey (1974)
Spleen	10.0–54.0 FW	Brooks and Rumsey (1974)
Gill	14.0–24.0 FW	Brooks and Rumsey (1974)
Vertebrae	15.0–22.0 FW	Brooks and Rumsey (1974)
Bagre marinus		
Muscle	12.0 DW	Windom et al. (1973)
Bairdiella chrysura		
Muscle	30.0 DW	Windom et al. (1973)
Boreogadus saida		
Fillets	34.0 DW	Bohn and McElroy (1976)
Liver	19.0 DW	Bohn and McElroy (1976)

Bottomfish			
Flesh	3.0–93.0 FW; 13.0–420.0 DW; 190.0–7600.0 AW	Lowman et al. (1970)
Skin	13.0–160.0 FW; 42.0–380.0 DW; 130.0–1600.0 AW	Lowman et al. (1970)
Bone	12.0–150.0 FW; 36.0–380.0 DW; 100.0–900.0 AW	Lowman et al. (1970)
Bryothamanion frigultum		
Flesh	33.0 FW; 125.0 DW; 400.0 AW	Lowman et al. (1966)
Calamus bajonado		
GI tract	23.0 FW; 100.0 DW; 730.0 AW	Lowman et al. (1966)
Vertebrae	130.0 FW; 230.0 DW; 420.0 AW	Lowman et al. (1966)
Flesh	8.7 FW; 36.0 DW; 400.0 AW	Lowman et al. (1966)
Eyes	54.0 FW; 280.0 DW; 2200.0 AW	Lowman et al. (1966)
Scales	166.0 FW; 230.0 DW; 430.0 AW	Lowman et al. (1966)
Gill	35.0 FW; 110.0 DW; 390.0 AW	Lowman et al. (1966)
Caranx latus		
Flesh	29.0 FW; 36.0 DW; 190.0 AW	Lowman et al. (1966)
Caranx lutescens		
Muscle	1.0–35.0 FW	Brooks and Rumsey (1974)
Liver	6.0–90.0 FW	Brooks and Rumsey (1974)
Kidney	40.0–354.0 FW	Brooks and Rumsey (1974)
Heart	15.0–30.0 FW	Brooks and Rumsey (1974)
Gonad	21.0–156.0 FW	Brooks and Rumsey (1974)
Spleen	48.0–624.0 FW	Brooks and Rumsey (1974)
Gill	16.0–29.0 FW	Brooks and Rumsey (1974)
Vertebrae	8.0–24.0 FW	Brooks and Rumsey (1974)
Centropristes striatus		
Muscle	7.0 DW	Windom et al. (1973)

Table 1. Continued

Species	Zinc Concentration	Reference
Cephalopholis fulvus		
Flesh	28.0–100.0 FW; 98.0–370.0 DW; 410.0–630.0 AW	Lowman et al. (1966)
Cetengraulis edentulus		
Flesh	6.7–20.0 FW; 31.0–80.0 DW; 290.0–340.0 AW	Lowman et al. (1966)
Chauliodus macouni		
Whole	7.0 FW	Cutshall et al. (1977b)
Cheilodactylis macropterus		
Muscle	0.9–7.0 FW	Brooks and Rumsey (1974)
Liver	90.0–1200.0 FW	Brooks and Rumsey (1974)
Kidney	12.0–20.0 FW	Brooks and Rumsey (1974)
Heart	16.0–19.0 FW	Brooks and Rumsey (1974)
Gonad	52.0–100.0 FW	Brooks and Rumsey (1974)
Spleen	25.0–84.0 FW	Brooks and Rumsey (1974)
Gill	12.0–24.0 FW	Brooks and Rumsey (1974)
Vertebrae	9.0–12.0 FW	Brooks and Rumsey (1974)
Chrysophrys auratus		
Muscle	2.0–10.0 FW	Brooks and Rumsey (1974)
Liver	16.0–146.0 FW	Brooks and Rumsey (1974)
Kidney	18.0–137.0 FW	Brooks and Rumsey (1974)
Heart	6.0–112.0 FW	Brooks and Rumsey (1974)
Gonad	8.0–124.0 FW	Brooks and Rumsey (1974)
Spleen	36.0–62.0 FW	Brooks and Rumsey (1974)
Gill	18.0–26.0 FW	Brooks and Rumsey (1974)
Vertebrae	10.0–34.0 FW	Brooks and Rumsey (1974)

Ciliata mustela		
Whole	180.2 DW	Hardisty et al. (1974)
Clupea harengus		
Whole	119.0 DW	Andersen et al. (1973)
Males		
Muscle	18.0–47.0 FW;	Vinogradov (1953)
	74.0–162.0 DW	Vinogradov (1953)
Testes and milt	33.0–64.0 FW;	Vinogradov (1953)
	200.0–345.0 DW	Vinogradov (1953)
Intestines and gills	13.0 FW; 69.0 DW	Vinogradov (1953)
Females		
Muscle	24.0 FW; 127.0 DW	Vinogradov (1953)
Ovaries and roe	28.0 FW; 124.0 DW	Vinogradov (1953)
Codfish		
Muscle		
Inshore	4.3 FW	Portmann (1972)
North Sea	5.2 FW	Portmann (1972)
Offshore	4.7 FW	Portmann (1972)
Conger sp.		
Muscle	41.0 DW	Windom et al. (1973)
Cynoscion nebulosus		
Muscle	21.0 DW	Windom et al. (1973)
Cynoscion reticulatus		
Muscle	238.0–600.0 AW	Ting (1971)
Viscera	940.0–2400.0 AW	Ting (1971)
Skin and scales	600.0–1750.0 AW	Ting (1971)
Bone	310.0–420.0 AW	Ting (1971)
Decapterus macarellus		
Flesh	27.0–58.0 FW;	Lowman et al. (1966)

305

Table 1. Continued

Species	Zinc Concentration	Reference
Decapterus punctatus		
Muscle	100.0–220.0 DW; 680.0–1500.0 AW	Windom et al. (1973)
Diplectrum euryplectrum		
Muscle	18.0 DW	Ting (1971)
Viscera	350.0–420.0 AW	Ting (1971)
Skin and scales	940.0 AW	Ting (1971)
Bone	260.0–490.0 AW	Ting (1971)
	218.0–350.0 AW	
Engraulis mordax		
Whole	400.0 AW	Goldberg (1962)
Epinephalus striatus		
Muscle	4.0 FW	Taylor and Bright (1973)
Euthynnus alletteratus		
Muscle	8.0 DW	Windom et al. (1973)
Euthynnus pelamis		
Whole	490.0 AW	Goldberg (1962)
White meat	43.0–390.0 AW	Ting (1971)
Dark meat	191.0–880.0 AW	Ting (1971)
Viscera	1150.0–8400.0 AW	Ting (1971)
Skin and scales	520.0–3300.0 AW	Ting (1971)
Bone	340.0–650.0 AW	Ting (1971)
Flesh	17.0 FW; 57.0 DW; 970.0 AW	Lowman et al. (1966)
Fish		

Flesh		
Georgia coast, 11 spp.	21.0–51.0 DW	Stickney et al. (1975)
Northumberland Coast, England	1.9–119.0 FW	Wright (1976)
Rio Tinto estuary, Spain	132.0 DW max.	Stenner and Nickless (1975)
West Norway coast	45.0 DW max.	Stenner and Nickless (1974)
Flesh, 7 spp.	1.7–14.7 FW	Holden and Topping (1972)
Flesh, 23 spp.	Means 5.0–13.5 FW; 146.7 FW max.	Eustace (1974)
Whole, 18 spp.	1.6–124.4 FW; median 8.5 FW	Vinogradov (1953)
Fishmeal		
Presscake	86.0 DW	Lunde (1968)
N liquor	67.0 FW	Lunde (1968)
Meal	180.0 DW	Lunde (1968)
Industrial meal	38.0–639.0 DW	Lunde (1973)
Laboratory-produced meal	62.0 DW max.	Lunde (1973)
Lipid phase	1.1–54.0 FW	Lunde (1973)
Mixture, 5 spp.	74.0–142.0 DW	Jangaard et al. (1974)
Flatfish		
Whole	4.0 FW	Cutshall et al. (1977b)
Fundulus heteroclitus		
Whole fish		
46 mm TL	1180.0 AW	Eisler and LaRoche (1972)
61 mm TL	1010.0 AW	Eisler and LaRoche (1972)
77 mm TL	945.0 AW	Eisler and LaRoche (1972)
88 mm TL	867.0 AW	Eisler and LaRoche (1972)
106 mm TL	867.0 AW	Eisler and LaRoche (1972)
119 mm TL	788.0 AW	Eisler and LaRoche (1972)
Gadus spp.		
Whole	92.0–125.0 DW	Vinogradov (1953)

307

Table 1. Continued

Species	Zinc Concentration	Reference
Gobius minutus		
Whole	75.6 DW	Hardisty et al. (1974)
Holocentrus rufus		
Flesh	13.0–74.0 FW; 33.0–	Lowman et al. (1966)
	150.0 DW; 130.0–480.0 AW	
Johnius hololepidotus		
Flesh	3.5 FW	Van As et al. (1973)
Latridopsis ciliaris		
Muscle	1.8–20.0 FW	Brooks and Rumsey (1974)
Liver	34.0–90.0 FW	Brooks and Rumsey (1974)
Kidney	10.0–16.0 FW	Brooks and Rumsey (1974)
Heart	22.0–52.0 FW	Brooks and Rumsey (1974)
Gonad	70.0–180.0 FW	Brooks and Rumsey (1974)
Spleen	15.0–36.0 FW	Brooks and Rumsey (1974)
Gill	26.0–33.0 FW	Brooks and Rumsey (1974)
Vertebrae	20.0–24.0 FW	Brooks and Rumsey (1974)
Leiostomus xanthurus		
Muscle	20.0 DW	Windom et al. (1973)
Limanda ferruginea		
Muscle	4.2–4.4 DW	Greig (1975)
Muscle	4.2 FW	Greig and Wenzloff (1977)
Liparis liparis		
Whole	86.5 DW	Hardisty et al. (1974)
Liza ramada		
Whole	120.0 DW	Hardisty et al. (1974)

Lophius piscatorius		
Flesh	3.7 FW	Van As et al. (1973)
Lopholatilus chamaeleonticips		
Liver	1800.0 AW	Mears and Eisler (1977)
Lutjanus aya		
Muscle	2.3 FW	Vinogradov (1953)
Swim bladder	3.6 FW	Vinogradov (1953)
Gill	5.6 FW	Vinogradov (1953)
Fins	10.0 FW	Vinogradov (1953)
Skin	10.6 FW	Vinogradov (1953)
Bone	16.5–18.4 FW	Vinogradov (1953)
Stomach	19.1 FW	Vinogradov (1953)
Spleen	43.5 FW	Vinogradov (1953)
Liver	55.5 FW	Vinogradov (1953)
Macrobrachium carcinus		
Flesh	11.0–84.0 FW;	Lowman et al. (1966)
	139.0–300.0 DW;	
	410.0–920.0 AW	
Makaira nigricans		
Flesh	4.1–70.0 FW;	Lowman et al. (1966)
	72.0–210.0 DW;	
	260.0–2100.0 AW	
Marone labrax		
Flesh	15.0 DW	Leatherland and Burton (1974)
Merlangus merlangus		
Whole	124.8 DW	Hardisty et al. (1974)
Whole	72.5–102.4 DW	Badsha and Sainsbury (1977)
Muscle	9.1–9.2 FW	Wharfe and Van Den Broek (1977)
Liver	28.3 FW	Wharfe and Van Den Broek (1977)
Gut wall	23.1 FW	Wharfe and Van Den Broek (1977)

Table 1. Continued

Species	Zinc Concentration	Reference
Merluccius capensis		
Flesh	3.9 FW	Van As et al. (1973)
Merluccius productus		
Muscle	4.0 FW	Cutshall et al. (1977b)
Whole	12.0 FW	Cutshall et al. (1977b)
Microstomus pacificus		
Flesh		
Wastewater outfall site	39.4 DW	McDermott et al. (1976)
Control site	38.0 DW	McDermott et al. (1976)
Muscle	15.0–26.0 DW	Sherwood and Mearns (1977)
Morone saxatilis		
Muscle	12.0 DW	Windom et al. (1973)
Mugil cephalus		
Muscle	17.0 DW	Windom et al. (1973)
Mugil curema		
Muscle	98.0–560.0 AW	Ting (1971)
Viscera	398.0–1430.0 AW	Ting (1971)
Skin and scales	210.0–750.0 AW	Ting (1971)
Bone	98.0–290.0 AW	Ting (1971)
Mycteroperca phenax		
Muscle	3.2 FW	Taylor and Bright (1973)
Mycteroperca tigris		
Muscle	3.7 FW	Taylor and Bright (1973)
Myctophids		
Whole, less head, 8 spp.	15.0–81.0 DW	Windom et al. (1973)

Whole	210.0 AW	Robertson (1967)
Whole	10.0 FW	Cutshall et al. (1977b)
Nematonurus armatus		
Liver	50.0 FW	Greig et al. (1976)
Neothunnus macropterus		
Whole	730.0 AW	Goldberg (1962)
Ophichthus spp.		
Muscle	21.0–26.0 DW	Windom et al. (1973)
Ophiocephalus argus		
Muscle	6.8 FW; 38.6 DW	Vinogradov (1953)
Liver	33.2 FW; 133.5 DW	Vinogradov (1953)
Intestine	22.2 FW; 108.5 DW	Vinogradov (1953)
Opisthonema libertate		
Muscle	770.0 AW	Ting (1971)
Viscera	2550.0 AW	Ting (1971)
Skin and scales	590.0 AW	Ting (1971)
Bone	400.0 AW	Ting (1971)
Opisthonema oglinum		
Muscle	55.0–377.0 AW	Ting (1971)
Viscera	378.0–2454.0 AW	Ting (1971)
Skin and scales	227.0–987.0 AW	Ting (1971)
Bone	43.0–252.0 AW	Ting (1971)
Flesh	21.0–110.0 FW; 82.0–390.0 DW; 400.0–5700.0 AW	Lowman et al. (1966)
Pachymetopon grande		
Flesh	2.5 FW	Van As et al. (1973)
Paralichthys lethostigma		
Muscle	39.0 DW	Windom et al. (1973)

Table 1. Continued

Species	Zinc Concentration	Reference
Pelagic fish		
White flesh	3.0–15.0 FW;	Lowman et al. (1970)
	3.0–65.0 DW;	
	210.0–880.0 AW	
Dark flesh	6.0–60.0 FW;	Lowman et al. (1970)
	6.0–65.0 DW;	
	120.0–1200.0 AW	
Viscera	210.0 FW; 830.0 DW;	Lowman et al. (1970)
	10,000.0 AW	
Bone	21.0–65.0 FW;	Lowman et al. (1970)
	62.0–150.0 DW;	
	140.0–420.0 AW	
Skin	7.0–120.0 FW;	Lowman et al. (1970)
	110.0–320.0 DW;	
	860.0–9600.0 AW	
Plaice		
Muscle		
Inshore	5.4 FW	Portmann (1972)
North Sea	5.7 FW	Portmann (1972)
Offshore	6.6 FW	Portmann (1972)
Platichthys flesus		
Whole		
Barnstaple Bay		
Age 2+	224.5 DW	Hardisty et al. (1974)
Age 3+	209.4 DW	Hardisty et al. (1974)

Age 4+	200.2 DW	Hardisty et al. (1974)
Age 5+	195.2 DW	Hardisty et al. (1974)
Oldbury-on-Severn		
Age 2+	125.2 DW	Hardisty et al. (1974)
Age 3+	128.0 DW	Hardisty et al. (1974)
Age 4+	128.6 DW	Hardisty et al. (1974)
Age 5+	140.0 DW	Hardisty et al. (1974)
Muscle	12.3–18.2 FW	Wharfe and Van Den Broek (1977)
Liver	53.8–68.9 FW	Wharfe and Van Den Broek (1977)
Gut wall	36.2 FW	Wharfe and Van Den Broek (1977)
Ovary	146.8 FW	Wharfe and Van Den Broek (1977)
Pleuronectes platessa		
Muscle	10.0–11.9 FW	Wharfe and Van Den Broek (1977)
Liver	38.9 FW	Wharfe and Van Den Broek (1977)
Gut wall	27.5 FW	Wharfe and Van Den Broek (1977)
Polydactylus virginicus		
Flesh	9.3–15.0 FW;	Lowman et al. (1966)
	39.0–63.0 DW;	
	280.0–410.0 AW	
Polyprion oxygeneios		
Muscle	2.1–12.0 FW	Brooks and Rumsey (1974)
Liver	33.0–90.0 FW	Brooks and Rumsey (1974)
Kidney	15.0–29.0 FW	Brooks and Rumsey (1974)
Heart	8.0–17.0 FW	Brooks and Rumsey (1974)
Gonad	11.0–200.0 FW	Brooks and Rumsey (1974)
Spleen	6.0–26.0 FW	Brooks and Rumsey (1974)
Gill	24.0–38.0 FW	Brooks and Rumsey (1974)
Vertebrae	40.0–104.0 FW	Brooks and Rumsey (1974)
Pomatomus saltatrix		
White muscle	4.5–5.0 FW	Cross et al. (1973)

Table 1. Continued

Species	Zinc Concentration	Reference
Liver, males	1700.0 AW	Mears and Eisler (1977)
Liver, females	3700.0 AW	Mears and Eisler (1977)
Pomatoshistus minutus		
Whole	28.7 FW	Wharfe and Van Den Broek (1977)
Pseudopleuronectes americanus		
Muscle	5.2–6.0 FW	Greig and Wenzloff (1977)
Liver	15.0–45.0 FW	Greig and Wenzloff (1977)
Sardinop ocellata		
Flesh	19.0 FW	Van As et al. (1973)
Scomber japonicus		
Flesh	9.0 FW	Van As et al. (1973)
Scomberomerus cavalla		
Flesh	5.0–94.0 FW; 26.0–260.0 DW; 260.0–3000.0 AW	Lowman et al. (1966)
Scomberomerus maculatus		
Muscle	15.0 DW	Windom et al. (1973)
Scombersox saurus		
Whole	410.0 AW	Robertson (1967)
Scopthalmus aquosus		
Muscle	23.0–28.0 DW	Greig (1975)
Seriola grandis		
Muscle	2.8–56.0 FW	Brooks and Rumsey (1974)
Liver	15.0–37.0 FW	Brooks and Rumsey (1974)
Kidney	75.0–390.0 FW	Brooks and Rumsey (1974)

Vertebrae	15.0–22.0 FW	Brooks and Rumsey (1974)
Seriola pappei		
Flesh	5.5 FW	Van As et al. (1973)
Sphaeroides maculatus		
Gill	4400.0 AW	Eisler and Edmunds (1966)
Liver	37,500.0 AW	Eisler and Edmunds (1966)
Liver	9700.0 AW	Eisler (1967a)
Gill arch	1700.0 AW	Eisler (1967a)
Gill filament	15,300.0 AW	Eisler (1967a)
Muscle	600.0 AW	Eisler (1967a)
Ovary	5700.0 AW	Eisler (1967a)
Testes	2400.0 AW	Eisler (1967a)
Serum	219.0 FW	Eisler (1967a)
Sphaeroides spengleri		
Flesh	30.0 FW; 130.0 DW; 700.0 AW	Lowman et al. (1966)
Sprattus sprattus		
Whole	133.0 DW	Andersen et al. (1973)
Whole	38.5 FW	Wharfe and Van Den Broek (1977)
Stellifer stellifer		
Flesh	9.7–10.0 FW; 37.0–40.0 DW; 200.0–210.0 AW	Lowman et al. (1966)
Synaptura marginata		
Flesh	5.5 FW	Van As et al. (1973)
Tautoga onitis		
Liver, males	1500.0 AW	Mears and Eisler (1977)
Liver, females	1300.0 AW	Mears and Eisler (1977)
Thunnus albacares		
White meat	40.0–140.0 AW	Ting (1971)

Table 1. Continued

Species	Zinc Concentration	Reference
Dark meat	450.0–785.0 AW	Ting (1971)
Viscera	219.0–3020.0 AW	Ting (1971)
Skin and scales	650.0–3800.0 AW	Ting (1971)
Bone	290.0–550.0 AW	Ting (1971)
Trachurus trachurus		
Flesh	3.0 FW	Van As et al. (1973)
Trichiurus lepturus		
Flesh	5.7–47.0 FW;	Lowman et al. (1966)
	26.0–220.0 DW;	
	230.0–1800.0 AW	
Trigla kumu		
Muscle	2.5–16.2 FW	Brooks and Rumsey (1974)
Liver	12.0–102.0 FW	Brooks and Rumsey (1974)
Kidney	8.0–26.0 FW	Brooks and Rumsey (1974)
Heart	8.0–25.0 FW	Brooks and Rumsey (1974)
Gonad	80.0–162.0 FW	Brooks and Rumsey (1974)
Spleen	26.0–36.0 FW	Brooks and Rumsey (1974)
Gill	7.0–13.0 FW	Brooks and Rumsey (1974)
Vertebrae	12.0–24.0 FW	Brooks and Rumsey (1974)
Triglia capensis		
Flesh	3.4 FW	Van As et al. (1973)
Trisopterus minutus		
Whole	158.1 DW	Hardisty et al. (1974)
Urophycis chuss		
Liver	28.0–41.0 FW	Greig and Wenzloff (1977)

Muscle	3.3 FW	Greig and Wenzloff (1977)
Urophycis tenuis		
Muscle	2.9 FW	Greig and Wenzloff (1977)
Muscle	2.9–3.8 DW	Greig (1975)
Xiphiurus capensis		
Flesh	6.9 FW	Van As et al. (1973)
MAMMALIA		
Halichoerus grypus		
Liver	64.0–94.0 FW	Holden (1975)
Phoca vitulina		
Liver	54.0 FW (43.0–84.0 FW)	Holden (1975)
Muscle	34.0 FW (33.0–35.0 FW)	Holden (1975)
Heart	31.0 FW (28.0–32.0 FW)	Holden (1975)
Kidney	37.0 FW (28.0–51.0 FW)	Holden (1975)
Spleen	32.0 FW (28.0–35.0 FW)	Holden (1975)
Brain	26.0 FW (19.0–36.0 FW)	Holden (1975)
Blubber	9.0 FW (4.0–13.0 FW)	Holden (1975)
Liver	27.0–56.0 FW	Drescher et al. (1977)
Kidney	16.3–32.5 FW	Drescher et al. (1977)
Brain	10.8–15.0 FW	Drescher et al. (1977)
Zalophus californianus		
Mothers with premature pups		
Liver	201.0 DW	Martin et al. (1976)
Kidney	149.0 DW	Martin et al. (1976)
Mothers with normal pups		
Liver	220.0 DW	Martin et al. (1976)
Kidney	173.0 DW	Martin et al. (1976)
Premature pups		
Liver	425.0 DW	Martin et al. (1976)

Table 1. Continued

Species	Zinc Concentration	Reference
Kidney	19.0 DW	Martin et al. (1976)
Normal pups		
Liver	505.0 DW	Martin et al. (1976)
Kidney	28.4 DW	Martin et al. (1976)

cumulated zinc between members of different phylogenetic groupings. Thus zinc once accumulated by the alga *L. digitata* has little tendency to be lost (Bryan, 1969); on the other hand, up to 25% of the zinc accumulated by the euphausiid *Meganyctiphanes norvegica* was lost within 12 hr (Antonini-Kane et al., 1972).

It should also be emphasized that CF values differed considerably for different organisms, as well as for different tissues (Kameda et al., 1968); this was particularly evident for molluscan kidney and for barnacle soft parts (Table 2). In the case of barnacles (*Balanus balanoides*) the high CF recorded was attributed to inorganic granules that contained up to 38% Zn and accumulated in tissues surrounding the midgut (Walker et al., 1975). These inorganic granules were localized within the prosomal parenchyma cells and appeared to closely reflect environmental zinc levels (Walker et al., 1975; Walker, 1977).

Concentration factors based on ^{65}Zn accumulation data should be viewed with reservation. It has been reported that the addition of stable zinc reduced ^{65}Zn accumulation in algae, and in viscera from molluscs, crustaceans, echinoderms, and teleosts, sometimes by an order of magnitude (Hiyama and Shimizu, 1964). In the scallop *Chlamys opercularis,* as one case, the CF values for kidney and digestive gland were substantially lower when derived from ^{65}Zn than from stable zinc (Bryan, 1973), suggesting an ability by scallop kidney to discriminate among chemical species of zinc.

Within the above constraints it appears from Table 2 that the more effective bioaccumulators of zinc include red and brown algae, ostreid and crassostreid oysters, and scallops.

4. ACCUMULATIONS BY LABORATORY POPULATIONS

Results of selected studies on zinc accumulation by marine algae, invertebrates, and teleosts under controlled conditions are summarized in Table 3. It is clear from this table that at least several factors affect uptake, retention, and translocation of zinc by marine biota. For example, algae and fish both exhibited significant accumulations after death (Skipnes et al., 1975; Eisler and Gardner, 1973), suggesting that zinc residue data from these and other organisms found dead on collection are of limited worth. Other modifying factors included the temperature (Eisler, 1977b) and salinity (Bryan and Hummerstone, 1973b) of the medium, initial zinc concentrations (Shuster and Pringle, 1968, 1969; Bryan and Hummerstone, 1973b; Eisler, 1977b), and duration of exposure (Shuster and Pringle, 1969, Skipnes et al., 1975; Eisler and Gardner, 1973; Eisler, 1977b); the significance of these and other modifiers will be discussed later.

Intra- and interspecies zinc accumulation rates varied widely. On the basis of

Table 2. Concentration Factors Reported for Zinc in Field Collections of Marine and Estuarine Biota
Values are in milligrams zinc per kilogram wet weight living organism ÷ milligrams zinc per liter seawater, unless stated otherwise.

Species	Concentration Factor	Reference
ALGAE		
Ascophyllum nodosum	1300 (dry weight basis)	Foster (1976)
Ascophyllum nodosum	3600–9700	Young (1975)
Ecklonia maxima	1300	Van As et al. (1973)
Fucus serratus	5200–12,300	Young (1975)
Fucus vesiculosus	1000–6400 (dry weight basis)	Foster (1976)
Fucus vesiculosus	4700–11,300	Young (1975)
Fucus vesiculosus	11,000–64,000	Bryan and Hummerstone (1973a)
Gigartina rachula	5300	Van As et al. (1973)
Laminaria digitata	2455	Bryan (1969)
Porphyra capensis	8500	Van As et al. (1973)
Spartina alterniflora		
Sprout	530	Williams and Murdock (1969)
Mature plant	370	Williams and Murdock (1969)
Dead plant	750	Williams and Murdock (1969)
Suhria vittata	7400	Van As et al. (1973)
Ulva sp.	4500	Van As et al. (1973)
CRUSTACEA		
Balanus balanoides	710,000–1,500,000 (dry weight basis)	Walker et al. (1975)
Jasus lalandi	13,000	Van As et al. (1973)
Meganyctiphanes norvegica	85	Antonini-Kane et al. (1972)
ECHINODERMATA		
Echinoderm digestive tract	> 200	Hiyama and Shimizu (1964)
Sea urchin	15–500	Kameda et al. (1968)
MOLLUSCA		
Chlamys opercularis		
Whole, less shell	32,000	Bryan (1973)

320

Table 2. Continued

Species	Concentration Factor	References
Kidney	4,000,000	Bryan (1973)
Digestive gland	18,000	Bryan (1973)
Choromytilus meridionalis	12,000	Van As et al. (1973)
Crassostrea gryphoides	5500–13,000	Bhatt et al. (1968)
Crassostrea virginica	148,000	Pringle et al. (1968)
Donax serra	12,000	Van As et al. (1973)
Haliotus midea	9200	Van As et al. (1973)
Mercenaria mercenaria	2100	Pringle et al. (1968)
Mussel	2200	Pringle et al. (1968)
Mya arenaria	1700	Pringle et al. (1968)
Mytilus edulis aoteanus	9100	Brooks and Rumsby (1965)
Ommastrephes sp.	20,000	Ketchum and Bowen (1958)
Ostrea sinuata	110,000	Brooks and Rumsby (1965)
Pecten maximus		
Whole, less shell	17,000	Bryan (1973)
Kidney	1,700,000	Bryan (1973)
Digestive gland	58,000	Bryan (1973)
Pecten novae-zelandiae	28,000	Brooks and Rumsby (1965)
Shellfish	50–400	Kameda et al. (1968)
Surf clam	1525	Pringle et al. (1968)
Whelk	8200	Pringle et al. (1968)
TELEOSTEI		
Argyrozona argyrozona	2200	Van As et al. (1973)
Fish	3–20	Kameda et al. (1968)
Johnius holoepidotus	2700	Van As et al. (1973)
Lophius piscatorius	2800	Van As et al. (1963)
Merluccius capensis	3000	Van As et al. (1973)
Pachymetopan grande	1900	Van As et al. (1973)
Pomatomus saltatrix		
White muscle	2100	Cross et al. (1973)
Sardinop ocellata	15,000	Van As et al. (1973)
Scomber japonicus	6900	Van As et al. (1973)
Synaptura marginata	4200	Van As et al. (1973)
Trachurus trachurus	2300	Van As et al. (1973)
Trigla capensis	2600	Van As et al. (1973)
Xiphiurus capensis	5300	Van As et al. (1973)

Table 3. Selected Laboratory Studies on Zinc Accumulation by Marine Biota

ALGAE
Ascophyllum nodosum tips (Skipnes et al., 1975)

Stage	Zinc (mg/1) in Medium	Exposure Period (days)	Concentration (mg Zn/kg FW)	Accumulation Rate (mg Zn/kg FW day)
living	0.1	4	5	0.8
living	0.3	7	27.0	1.3
dead	0.1	7.5	–	95.0

ANNELIDA
Nereis diversicolor, exposure for 816 hr at 17.5% salinity, pH 7.05 to 7.8; worms collected from sediments with low or high zinc levels (Bryan and Hummerstone, 1973b)

Zinc (mg/1) in Medium	Residues in Survivors (mg Zn/kg DW)	Rate of Net Absorption (mg Zn/kg DW day)	
		Low-Zinc Sediments	High-Zinc Sediments
250	2410 (some deaths)	2230	1680
100	2500 (some deaths)	1150	672
25	3630	216	89
10	2040	55	19
0	180	0	0

Nereis diversicolor, explosure for 408 hr at 3.5% or 0.35% salinity; worms collected from sediments with low or high zinc levels (Bryan and Hummerstone, 1973b)

Zinc (mg/1) in Medium	Residues in Survivors (mg Zn/kg DW)	Rate of Net Absorption (mg Zn/kg DW day)	
		Low-Zinc Sediments	High-Zinc Sediments
Salinity 3.5%			
25	1150 (some deaths)	439	442
10	1140 (some deaths)	178	173
5	1010	55	48
0	198	0	0

Table 3. Continued

Salinity 0.35%			
5	812	310	319
2.5	638	144	113
1.0	673	60	31
0.0	216	0	0

CRUSTACEA

Homarus vulgaris, injected with 6600 μg Zn and then transferred to seawater containing 100 μg Zn/1 (Bryan, 1964)

	Residues (mg Zn/kg FW) 3, 5, or 19 Days Postinjection		
Tissue	3 Days	5 Days	19 Days
Blood	21	5	9
Excretory organs	83	28	25
Abdominal muscle	16	14	12
Hepatopancreas	170	87	117
Gill	30	17	29
Exoskeleton	10	13	13
Stomach fluid	0.7	0.3	1.4

MOLLUSCA

Crassostrea virginica, exposure for 20 weeks in raw seawater containing 0.1 or 0.2 mg Zn/1 added (Shuster and Pringle, 1968, 1969)

Exposure Time (weeks)	Residues (mg Zn/kg FW soft parts)		
	0.0 mg Zn/1	0.0 mg Zn/1	0.2 mg Zn/1
Controls	1036–1708	—	—
4	—	1831–2095	1474–1761
8	—	1732–2229	1869–2307
12	—	2189–2251	2642–2660
16	—	2118–2475	2667–3233
20	—	2560–2708	3185–3813

Mercenaria mercenaria, 84-day depletion study at 4 to 12°C (Pringle et al., 1968)

Table 3. Continued

Zinc Level in *Mercenaria* (mg/kg FW)		Depletion Rate (mg Zn/kg day)
Initial	Final	
38	26	0.12

Mya arenaria, experiment conducted during winter at 0 to 10°C in raw seawater with added zinc (Eisler, 1977b)

Zinc Concentration (mg/1)	Exposure Time (weeks)	Residues (mg Zn/kg FW soft parts)	Concentration Factor
0.010 (controls)	16	9.5	> 957
0.500	6	48.3	78
0.500	16	31.2	43
2.500	6	140.4	52

Mya arenaria, experiment conducted during spring at 16 to 22°C in row seawater with added zinc (Eisler, 1977b)

Zinc Concentration (mg/l)	Exposure Time (days)	Residues (mg Zn/kg FW soft parts)	Concentration Factor
0.010 (controls)	14	11	> 1122
0.500	1	13	4
0.500	7	59	96
0.500	14	82	143
2.500	1	43	13
2.500	7	113	41

Mya arenaria, exposure for 50 days to 0.2 mg Zn/1 at 20°C (Pringle et al., 1968)

Zinc Level in *Mya* (mg/kg FW)		Total Accumulation (mg Zn/kg FW)	Accumulation Rate (mg Zn/kg day)
Initial	Final		
10.0	27.0	17.0	0.35

Table 3. Continued

Mytilus edulis, exposure for 13 days at 35% salinity (Phillips, 1977a)

Zinc (mg/1) in Medium	Residue (mg Zn/kg DW) at Day 13
0.0	233–258
0.4	1167
0.8	1215
1.0	1241

TELEOSTEI
Fundulus heteroclitus, exposure for 96 hr (Eisler and Gardner, 1973)

Initial Zinc Concentration (mg/1)	Whole-Body Zinc (mg/kg AW)
0.0	1240
36.0	7650
60.0	7510 (some deaths)

Fundulus heteroclitus, exposure to 40 0 mg/1 Zn (Eisler and Gardner, 1973)

		Zinc (mg/kg whole-body ash)	
Exposure Time (hr)	Initial Condition	Immediately after Exposure	24 hr Postexposure
24	Live	753	824
24	Dead	2660	1380
48	Live	878	743
48	Dead	4730	3470

data from Table 3 and suitable wet/dry correction factors, the maximum ac-cumulation rates (in mg Zn/kg fresh weight/day) for different species were as follows: 1.3 for the alga *Ascophyllum nodosum;* 32.0 for the softshell clam *Mya arenaria;* 19.8 for the oyster *Crassostrea virginica;* 223.0 for the polychaete annelid *Nereis diversicolor;* 32.0 for the killifish *Fundulus heteroclitus;* and 7.7 for the mussel *Mytilus edulis.* The variability in daily zinc accumulation rates by the softshell clam *Mya arenaria* is one example of what is probably typical for marine species. During immersion in 0.50 mg Zn/l in a flow-through system at elevated (16 to 22°C) temperatures, *Mya* flesh accumulated 2.0 mg Zn/kg fresh weight/day during the first 24 hr; between 24 hr and 7 days the accumula-tion was 7.7; and between 7 and 14 days, 3.3. Under the same conditions, except that ambient zinc levels were maintained at 2.5 mg/l, the daily accumula-tion rate during the first 24 hr was 32.0 mg Zn/kg fresh weight; between 24 hr and 7 days this dropped to 11.7 mg Zn/kg. At lower temperatures (0 to 10°C), and 0.5 mg Zn/l, *Mya* accumulated zinc at the rate of 0.9 mg/kg fresh weight/ day during the first 42 days; however, between days 42 and 112, *Mya* lost zinc daily at the rate of 0.24 mg/kg fresh weight. Similar observations have been made for *Nereis, Crassostrea,* and *Fundulus.* It is apparent that changes in accumulation rates of zinc by *Mya* reflect, at least partially, complex interaction effects between water temperature, ambient zinc concentrations, duration and season of exposure, physiological saturation, and detoxification mechanisms.

Results of most laboratory studies on zinc uptake and retention for residues are expressed in whole organisms. It is well to remember, however, that not all tissues concentrated and retained zinc at the same rate, lobster hepatopancreas being a prime example (Bryan, 1964). Marked interspecies differences were evident in the ability to regulate zinc accumulation, bivalve molluscs being a case in point. Thus, at comparatively low ambient zinc levels of 0.1 or 0.2 mg/l, crassostreid oysters continued to accumulate zinc over a 20-week period with little evidence of regulation (Shuster and Pringle, 1968, 1969). But softshell clams (*Mya arenaria*) at 0.5 mg Zn/l exhibited some regulatory ability at com-paratively low water temperatures during exposure for 16 weeks, as shown by a drop in concentration factor (Eisler, 1977b); this was not evident at higher temperatures during exposure for only 2 weeks (Eisler, 1977b), although ac-cumulation was much more rapid. On the other hand, mussels (*Mytilus edulis*) appeared to plateau (regulate) at concentrations above 0.4 mg Zn/l (Phillips, 1977a).

Physiological adaptation of the annelid worm *Nereis diversicolor* collected from zinc-impacted sediments was proposed by Bryan and Hummerstone (1973b). Under laboratory conditions they found that *Nereis* specimens from high-zinc sediments accumulated less zinc per unit of body weight per day and showed better survival than those collected from low-zinc sediments; this pattern appeared to hold at several salinities. Additional studies on physiological adaptation to zinc and other metal stress appear warranted at this time.

5. FACTORS MODIFYING ZINC ACCUMULATIONS

The concentrations of zinc in marine organisms vary widely, even among closely related taxonomic groups. Apart from inherent species differences, most of this variation is attributable to changes in the physicochemical environment and to intrinsic biological factors. According to numerous well-documented reports in the technical literature, these modifiers markedly alter the rate and extent of zinc accumulations in saltwater flora and fauna, sometimes by an order of magnitude or greater.

Appreciation of the importance of modifiers and their interaction effects is essential to an understanding of zinc kinetics in coastal and oceanic environments. To illustrate this point, selected examples of abiotic and biotic modifiers for major indicator organism assemblages are given below.

5.1. Algae

Increasing accumulations of zinc were observed in various species of marine algae with decreasing light intensity (Bachmann and Odum, 1960; Gutknecht, 1961); decreasing pH (Gutknecht, 1961; Parry and Hayward, 1973); increasing temperature (Gutknecht, 1961; Parry and Hayward, 1973; Styron et al., 1976; Baudin, 1974); decreasing levels of the pesticide DDT (Andryushchenko and Polikarpov, 1974); and increasing dissolved oxygen in the medium (Bachmann and Odum, 1960). Other abiotic factors known to modify zinc residues in algae include ambient levels of copper and magnesium (Braek et al., 1976); salinity of the medium (Styron et al., 1976); and season of collection and location of weed in the intertidal zone (Bryan and Hummerstone, 1973a). The physical processes of adsorption or cation exchange were primarily responsible for zinc uptake in algae, while the relationship between photosynthesis and zinc absorbtion was a secondary effect related to surface/volume ratio and pH (Gutknecht, 1961). Zinc residues from algae already dead on collection were of limited worth. For example, dead *Ulva lactosa* accumulated zinc more rapidly than live *Ulva* during a 6-hr period (Gutknecht, 1961); a similar pattern was observed in *Ascophyllum nodosum* (Skipnes et al., 1975). It was suggested that algal zinc-binding substances, presumably proteinaceous compounds (Parry and Hayward, 1973), were not directly accessible to ambient zinc ions in seawater before death (Skipnes et al., 1975).

For reasons that are still unclear, zinc accumulations varied widely among algal species. Of 16 species of seaweeds analyzed, at least 3 did not markedly concentrate zinc over ambient seawater levels, while 7 species exhibited significant biomagnification (Saenko et al., 1976). Some species, such as *Skeletonema costatum,* contained significant accumulations of zinc from ambient concentrations, which did not seem to influence growth or development (Jensen et al.,

1974). The fact that the growth stage of *Phaeodactylum tricornutum* was important in determining the final zinc residue may be linked to the number of zinc-binding sites at different growth cycles, despite the availability of further zinc for uptake (Davies, 1973). In another study with *P. tricornutum,* it was shown that ionic zinc was taken up more rapidly than other physicochemical states, and this may account for differences observed in the uptake rates of stable and radioactive ^{65}Zn (Bernhard and Zattera, 1969). In general, increasing exposure to increasing ambient zinc concentrations was associated with higher algal zinc residues, but this was not always the case (Mehran and Tremblay, 1965; Bryan and Hummerstone, 1973a; Baudin, 1974).

5.2. Protista

Bacteria and their metabolic by-products play a significant role in the removal of zinc and other metals from seawater and their subsequent deposition in marine sediments. In one study, mixed cultures of bacteria removed up to 85% of labeled ^{65}Zn from seawater in 120 hr (McLerran and Holmes, 1974). Interaction effects of Zn^{2+} with salts of other metals on protozoan growth are also documented. *Cristigera,* a ciliate protozoan, experienced growth reduction and presumably altered zinc uptake when held in media containing salts of Zn, Pb, and Hg (Gray, 1974).

5.3. Mollusca

Zinc accumulations in molluscs are mediated by many factors, not the least being interaction effects with salts of Ca, Co, Fe, Cd, and various organic substances. In one study the total amount of zinc found in oysters (*Ostrea edulis*) was far in excess of the amount of the element contributed by zinc-dependent enzymes such as carbonic anhydrase, alkaline phosphatase, carboxypeptidase A, malic dehydrogenase, and α-D-mannoside; however, the amount of nondialyzable zinc was of the same order of magnitude (Coombs, 1972). This apparent excess of dialyzable zinc was considered to be a consequence of the high levels of calcium found in tissues, demonstrating a competition between calcium and zinc in their uptake (Coombs, 1972). Mussels (*Mytilus edulis*), during exposure to 200 μg Cd/l for 20 days, showed only a slight increase in the zinc content of soft parts (George and Coombs, 1977). On the other hand, zinc at 0.5 mg/l substantially decreased cadmium uptake in mussel *M. edulis* and clam *Mulinia lateralis* (Jackim et al., 1977). Also reported were interaction effects of Zn, Pb, and Cd salts on zinc accumulation in mussels (Phillips, 1976a); depressed zinc uptake in oysters by iron or cobalt (Romeril, 1971); increased zinc accumulation in mantle tissue of the quahaug clam *Mercenaria mercenaira* during exposure to organochlorine and organophosphorous insecticides (Eisler and Weinstein, 1967); and influence of dissolved carbonates (Tabata, 1969a) and chelators on zinc uptake and retention in mussels (Keckes et al., 1968, 1969). In the latter case, chelators such as EDTA increased the loss rate of zinc from mussels at 50 mg/ℓ, but uptake rates were depressed, with the effect dependent on EDTA concentration, in the 0.01 to 50.0 mg/l range (Keckes et al., 1968, 1969). To confound matters, the loss rate of zinc from mussels is not constant, suggesting multicompartmental zinc metabolism dependent, perhaps, on the solubility of various chemical species (Keckes et al. 1968; Tabata, 1969a; Van Weers, 1973).

Although the salinity of the medium affects the final zinc concentrations in mussel soft parts (Tabata, 1969b; Phillips, 1977a, 1977b), it was not an important factor governing zinc uptake in mussels within the 15 to 35% range (Philips, 1976a). But oysters (*Crassostrea virginica*) from lower salinity waters generally contained more zinc than oysters collected from higher salinities (Huggett et al., 1973, 1975). Similarly, water temperature had a marked effect on sensitivity and zinc accumulation of the softshell clam *Mya arenaria* (Eisler, 1977a, 1977b), but none on *Mytilus edulis*, within the 10 to 18°C range (Phillips, 1976a). The season of collection affected the zinc content of mussels (Phillips, 1976a, 1976b) and other species of bivalve molluscs (Romeril, 1974).

It has been reported that the zinc content of marine molluscs tended to correlate positively with the zinc content of food items ingested (Young, 1975), with the zinc content of surrounding sediments (Romeril, 1974), with the proximity to heavily carbonized and industrial areas (Ratkowsky et al, 1974; Thrower and Eustace, 1973a, 1973b; Phillips, 1977b), and with the particulate matter produced as a result of dredging operations (Rosenberg, 1977), and negatively with depth and distance from shore (Seymour and Nelson, 1973; Young, 1975). Some of the high body burdens reported may have been influenced by incomplete removal of zinc-contaminated sediments, food, and other materials. For example, contamination of biological samples by ingested sediments has been reported for various species of bivalve molluscs and copepods; in some locations elemental zinc concentrations in organisms correlated significantly with elemental zinc values in the inorganic residues (Flegal and Martin, 1977). In another study, corrosion of galvanized zinc suspension trays was the cause of elevated (99,000 mg/kg dry weight) zinc recorded in soft tissues of *Crassostrea gigas* oysters (Boyden and Romeril, 1974).

The loss rate of zinc from mussels (*Mytilus galloprovincialis*) was greater with shorter contact time during immersion in zinc-contaminated seawater; the loss rate was greater from shell than soft tissues (Keckes et al., 1968). Uptake was rapid in mussels, especially in soft tissues, but equilibrium was reached only after a long period (Keckes et al., 1969); a similar pattern was observed in *Littorina obtusata* (Mehran and Tremblay, 1965). Stomach and digestive gland of mussels had the highest accumulations of zinc, but the loss rate of the element from these tissues was also highest (Pentreath, 1973b). As far as the uptake of ^{65}Zn from seawater is concerned, both the relatively slow accumulation and the subsequent loss reduce the risk of contamination of mussels after incidental releases of ^{65}Zn in coastal waters (Van Weers, 1973). It is emphasized at this juncture that at least one investigator (Romeril, 1971) has suggested that loss rates of zinc in laboratory populations of bivalves were lower than those from natural populations. For the marine gastropod *Littorina irrorata*, respiration rate, although linked to water temperature and body size, appeared to be the deciding factor in excretion of zinc (Mishima and Odum, 1963).

Other biological factors capable of influencing zinc accumulation patterns in marine molluscs include the age of the organism and inherent differences among closely related species. Regarding the age of the organism, increasing zinc per unit weight has been recorded with increasing body weight in the clams *Mercenaria mercenaria* and *Venerupis decussata* (Boyden, 1974); the reverse has been recorded for *Patella vulgata* (Boyden, 1974, 1977). Excretion of zinc by the gastropod *Littorina irrorata* was linked to body size (Mishima and Odum, 1963). Similar trends have been recorded in other species of molluscs (Romeril, 1974; Eisler et al., 1978a). At comparatively high ambient zinc concentrations, clam embryos exhibited reduced survival, and presumably increased uptake (Calabrese et al., 1977), when compared to adults (Eisler and Hennekey, 1977). Although oysters as a whole appear to have the greatest capacity for zinc biomagnification, crassostreid oysters accumulated zinc to a greater extent than ostreid oysters; tissue distributional patterns were similar in both groups (Romeril, 1971). In Chesapeake Bay, Maryland, oysters contained 30 to 40 times more zinc than clams and up to 90 times more zinc than mussels (McFarren, et al., 1962). At present there is no evidence that high body burdens of zinc are translocated to sex products. Unlike the values for other metals, zinc levels of adults and eggs of the oyster *Crassostrea virginica* were essentially the same; adults had dissimilar copper and cadmium concentrations (Greig et al., 1976).

There is no conclusive evidence at present that ingestion of commercial seafood products containing extremely high concentrations of zinc presents any real or potential threat to human health. However, high concentrations of zinc, as well as other metals, in oysters from Tasmania reportedly caused nausea and vomiting in some people after ingestion; these shellfish exceeded the current Australian food regulation of 40 mg Zn/kg fresh weight by a factor of about 500 (Thrower and Eustace, 1973a, 1973b).

5.4. Crustacea

The toxic action of zinc, and presumably the accumulation rates, are most pronounced for crustaceans, polychaete annelids, and oyster embryos, in that order (Eisler and Hennekey, 1977). In general, larval stages of crustaceans were more sensitive than adults (Brown and Ahsannullah, 1971).

Seasonal fluctuations in the zinc body burdens of copepods, barnacles, and decapod crustacenas are documented. Summer maxima in *Acartia* were associated with termination of active vertical transport (Pearcy and Osterberg, 1967; Zafiropoulos and Grimanis 1977) and proximity to anthropogenic discharges (Pearcy and Osterberg, 1967). In barnacles (*Balanus balanoides*) and some decapods, seasonal variations in zinc content were correlated with river flow rates, tidal flow, and primary productivity (Tennant and Forster, 1969; Ireland, 1974; Bender, 1975). Among euphausiids there were no seasonal variations in zinc content from collections made at 550 m; however, some variation was evident from collections of the same species at 150 to 500 m depth (Pearcy and Osterberg, 1967).

Zinc content of euphausiids and pelagic shrimp was inversely correlated with body weight. Depending on species, smaller individuals contained 1.2 to 4.1 times more zinc than larger individuals of the same species (Fowler, 1974). A similar pattern was observed in *Orchestoidea* sp. (Bender, 1975). Other factors known to influence uptake, retention, and translocation of zinc by crustaceans include water temperature (Fowler et al., 1971; Jones, 1975) and interaction with lead salts (Benijts-Claus and Benijts, 1975).

Zinc exchange with the environment is reportedly very slow. Zinc pools within adult aquatic organisms were observed to exchange only slowly, if at all, with zinc atoms in the organism's food or surrounding water (Renfro et al., 1975). but this point is still in contention. For example, among euphausiids there is a major loss of ^{65}Zn due to isotopic exchange with seawater, with 96% of the body burden eliminated over a period of 5 months (Fowler et al., 1971). In any event, food is considered more important than the medium as a pathway for zinc accumulation among decapod crustaceans. (Renfro et al., 1975; Bender, 1975). For the shrimp *Palaemon serratus*, different zinc excretion rates were reported, depending on the size of the animal. In general, ionic zinc is excreted first, then complexed zinc (Small et al., 1974). There is also a possibility that ionic-particulate zinc and dissolved organic compounds may be excreted separately with subsequent combination in water to yield zinc complex (Small et al., 1974). Surface-adsorbed zinc is turned over faster than internally adsorbed zinc in crabs (*Cancer magister*) (Tennant and Forster, 1969). Zooplankters, including copepods, chaetognaths, euphausiids, and pteropods, all eliminate accumulated zinc, presumably surface adsorbed, at rates of 1 to 5%/hr (Kuenzler, 1969). In the well-studied euphausiid *Meganyctiphanes norvegica*, about half the whole-body zinc is shed through molting when accumulated via the medium, but only 33% when accumulated by grazing on brine shrimp (Fowler et al., 1971, 1972); molts constituted one important vehicle of zinc transfer into marine ecosystems (Fowler et al., 1972). Finally, the participatory turnover time, that is, the time required to cycle an element in a system through a given material in that system, for ionic zinc in the Ligurian Sea by adult *M. norvegica* is lengthy, estimated at 498 to 1234 years, depending on available food and considering the food chain as the only route for zinc accumulation (Small and Fowler, 1973); fecal pellet deposition represented over 90% of the zinc flux in this model (Small et al., 1973).

5.5. Annelida

Among the polychaetes the composition of the feeding community in respect to feeding types is an important factor in the partitioning of stable zinc; the less dependent an organism is on the sediment as a source of food, the lower its concentration of zinc (Phelps, 1967). Polychaetes that feed primarily from the sediment-water interface, that is, selective deposit feeders, concentrate iron to the apparent exclusion of zinc, while polychaetes that feed primarily below the sediment-water interface, that is, nonselective deposit feeders, omnivores, and carnivores, preferentially concentrate zinc over iron (Phelps, 1967). *Nereis diversicolor,* for example, could remove up to 4% of the [65] Zn in the upper 2 cm of a hypothetical radioactive estuary (Renfro, 1973). The qualitative and quantitative structure of the benthic faunal community has a direct effect on zinc distribution: a shift in preponderance from polychaete annelids to pelecypod molluscs would result in a decrease in the amount of zinc incorporated into the living community (Phelps, 1967).

Body zinc levels of nereid worms tended to reflect sediment zinc levels (Bryan, 1974). In some cases, however, the trace metal contents of polychaetes did not differ markedly when collected from sediments that contained substantially different quantities of Fe, Mn, or Zn. In another study, the zinc levels in benthic communities, primarily annelids, showed no relation to proximity of the land mass or to changes in substrate, although there was a tendency for higher zinc values to occur in organisms found in the offshore silts and clays (Phelps et al., 1969). The time needed to reach equilibrium with the sediment must also be considered. The polychaete *Hermione hystrix* required 60 days to approach a steady state with zinc in sediments (Renfro, 1973). As stated earlier, nereid worms from sediments of high metal content were more tolerant to the toxic action of zinc than nereids collected from low-metal environments (Bryan, 1974). This tolerance or physiological adaptation to zinc by nereid worms from different environments was attributed in part to decreased body surface permeability (Bryan, 1974).

Life stage of the organism, possible discrimination against chemical species of zinc, and normal interspecies differences also should be considered when interpreting zinc accumulations in annelids. In regard to life stage, juveniles of *Neanthes* and *Capitella* polychaete annelids were more sensitive to zinc effects than adults (Reish et al., 1976). The polychaete *Hermione hystrix* is apparently capable of distinguishing radiozinc from stable zinc: the biological half-life of ^{65}Zn in that species was 52 to 197 days, but for stable zinc the value was 14 to 17 days (Renfro, 1973). Variability in zinc content among closely related species of polychaetes was considerable; among six species zinc contents differed by an eightfold factor (Cross et al., 1970).

5.6. Echinoderms

Diet and water depth seemed to influence zinc accumulations among echinoderms. Carey (1970) stated that detrital feeders contained more zinc than carnivores, and that species which fed near the surface contained more zinc per unit weight than sediment feeders. Furthermore, he found lower zinc levels in echinoderms collected offshore in deeper waters than in the same species collected inshore.

5.7. Teleosts

Direct accumulation from water plays only a minor role relative to food in the metabolism of zinc by the flatfish *Pleuronectes platessa* (Pentreath, 1973c, 1976) and other species (Renfro et al., 1975). However, no relation between diet and whole-fish zinc levels was observed by Hardistry et al. (1974), and Ting (1971) found no significant differences in the zinc contents of muscle, skin, viscera, or bone of seven species of fishes representing herbivorous, planktonic, or benthic and pelagic carnivorous feeding habitats.

Different life stages of plaice (*Pleuronectes platessa*) accumulated zinc from the medium at different rates, ranging from negligible for embryos, to 7.2 µg/kg/day for larvae, to 39.2 µg/kg/day in bone for 8-month-old fish (Pentreath, 1976). Dead fish (*Fundulus heteroclitus*) accumulated zinc at a substantially higher rate than living *Fundulus*, suggesting that zinc residue data from teleosts dead on collection are of limited worth (Eisler, 1967c; Eisler and Gardner, 1973).

Small hake (*Merluccius productus*) reflected zinc contamination in whole body and in muscle tissue earlier than larger hake (Cutshall et al., 1977a), with increasing zinc per unit weight of muscle observed with increasing body size (Cutshall et al., 1977b). However, the closely related species *Merlangus merlangus,* from the 0+ age group, contained zinc levels that were negatively correlated with length and weight, with upper threshold limits reached quickly when the fish entered zinc-contaminated waters (Badsha and Sainsbury, 1977). In the case of bluefish (*Pomatomus saltatrix*) body length had little relation to muscle zinc concentrations (Cross et al., 1973).

Sex or sexual condition of *Fundulus heteroclitus* did not affect whole-body zinc levels (Eisler and LaRoche, 1972). Liver zinc residues in female tilefish (*Lopholatilus chamaeleonticeps*) were not significantly different from those in males, but females, unlike males, contained more zinc with increasing age (Mears and Eisler, 1977).

Liver and viscera were major storage sites of zinc. Viscera from *Fundulus heteroclitus,* which accounted for 4% of the whole-fish ash weight, contained 29% of the whole-fish zinc; heads, which represented 45% of the whole-fish ash weight, contained 33% of the total-fish zinc; the remainder contained 38% of the total-body zinc but comprised 51% of the fish ash weight (Eisler and La Roche, 1972). Liver cells of *Tetraodon hispidus* accumulated zinc against an apparent 7X concentration gradient via a passive mechanism not directly coupled to metabolic energy (Saltman and Boroughs, 1960). Of seven species of marine fishes collected from a deepwater disposal site in the New York Bight, liver of *Nematonurus* contained the highest level (50.0 mg Zn/kg fresh weight) of all tissues and species examined (Greig et al., 1976).

Zinc accumulation from the medium by teleosts was markedly affected by the presence of other constituents in solution. Various organochlorine and organophosphorous pesticides affected the zinc contents of gill, liver, and serum in the puffer *Sphaeroides maculatus* (Eisler and Edmunds, 1966; Eisler, 1967a). Zinc and copper acted more than additively in toxicity to killifish (*Fundulus heteroclitus*), but zinc residues in whole fish were unaffected by the presence of copper (Eisler and Gardner, 1973). The zinc content of seawater can modify the uptake of lead and other heavy metals by fishes; lead and cadmium, for example, are accumulated 10 times more rapidly at elevated ambient zinc levels (Havre et al., 1972). In *Fundulus heteroclitus,* zinc and cadmium act additively in biocidal properties; but zinc residues in whole fish decreased significantly with increasing ambient cadmium levels, suggesting competition between these two elements for the same physiologically active sites (Eisler and Gardner, 1973). Other factors reported to influence zinc uptake by teleosts include respiration rate (Edwards, 1967), temperature of the medium (Saltman and Boroughs, 1960; Shulman et al., 1961; Eisler and LaRoche, 1972; Negilski, 1976), duration of exposure (Edwards, 1967), salinity of the medium (Shulman et al., 1961; Eisler and LaRoche, 1972), distance from point source (Cutshall et al., 1977a), sediment lithology (Sherwood and Mearns, 1977), metabolic transformations of zinc into various chemical species with different retention times (Baptist et al., 1970), and migratory patterns. In regard to the last-named factor, for example, plasma zinc levels in sockeye salmon (*Oncorhynchus nerka*) dropped from 0.23 to 0.05 mg/l during upstream migration from salt to fresh water (Fletcher et al., 1975).

6. SUMMARY

1. Zinc concentrations in tissues and organs from approximately 650 different species of representative marine plants, invertebrates, and vertebrates captured at numerous coastal and offshore locations were listed. Filter-feeding bivalve molluscs contained the greatest concentrations of zinc; recorded values in excess of 4000 mg Zn/kg dry weight soft parts were relatively common among crassostreid oysters. Several species of algae and one species of barnacle also contained substantial zinc, that is, > 1000 mg/kg on a dry weight basis. The lowest zinc residues, < 50 to about 400 mg/kg dry weight, were found in the highest marine trophic levels, namely, elasmobranchs, teleosts, and mammals.

2. Concentration factors reported for zinc in field collections of marine biota were presented. The most effective bioaccumulators of zinc included red and brown algae, ostreid and crassostreid oysters, and scallops.

3. Laboratory studies on the uptake and retention of zinc by marine organisms were summarized. Considerable variability was observed in these parameters, even among closely related species. For different organisms it appears that changes in zinc accumulation rates were governed significantly by water temperature, ambient zinc concentration, duration of exposure, and internal factors such as physiological saturation and detoxification.

4. Biological and abiotic factors reportedly capable of modifying zinc accumulations in saltwater flora and fauna were listed and illustrated with appropriate examples.

ACKNOWLEDGMENTS

I thank Drs. John H. Gentile, Frank G. Lowman, and Donald K. Phelps of the U.S. Environmental Protection Agency, Narragansett, Rhode Island, for editorial comments, and Ms. Gloria Gaboury for technical assistance.

REFERENCES

Alexander, G. V. and Young, D. R. (1976). "Trace Metals in Southern Californian Mussels," *Mar. Pollut. Bull.*, 7(1), 7–9.

Aderlini, V. (1974). "The Distribution of Heavy Metals in the Red Abalone, *Haliotis rufescens,* on the California Coast," *Arch. Environ. Contam.,* 2(3), 253–265.

Anderson, A. T., Dommasnes, A., and Hesthagen, I. H. (1973). "Some Heavy Metals in Sprat (*Sprattus sprattus*) and Herring (*Clupea harengus*) from the Inner Oslofjord," *Aquaculture,* 2, 17–22.

Andryushchenko, V. V. and Polikarpov, G. G. (1974). "An Experimental Study of Uptake of Zn[65] and DDT by *Ulva rigida* from Seawater Polluted with Both Agents," *Hydrobiol. J.*, 10(4), 41–46.

Antonini-Kane, J., Fowler, S. W., Heyraud, M., Keckes, S., Small, L. F. and Veglia, A. (1972). "Accumulation and Loss of Selected Radionuclides by *Meganyctiphanes norvegica* M. Sars," *Rapp. Comm. Int. Mer Medit.*, 21(6), 289–290.

Bachmann, R. W. and Odum, E. P. (1960). "Uptake of Zn[65] and Primary Productivity in Marine Benthic Algae," *Limnol. Oceanogr.*, 5(4), 349–355.

Badsha, K. S. and Sainsbury, M. (1977). "Uptake of Zinc, Lead and Cadmium by Yound Whiting in the Severn Estuary," *Mar. Pollut. Bull.*, 8(7), 164–166.

Baptist, J. P., Hoss, D. E., and Lewis, C. W. (1970). "Retention of Chromium-51, Iron-59, Cobalt-60, Zinc-65, Strontium-85, Niobium-95, Indium-114m, and Iodine-131 by the Atlantic Croaker (*Micropogon undulatus*)," *Health Phys.*, 18, 141–148.

Baudin, J. -P. (1974). "Premieres Donnees sur l'Étude Experimentale du Cycle du Zinc dans l'Etang de l'Olivier," *Vie Milieu*, 24(1), Ser. B, 59–80.

Bender, J. A. (1975). "Trace Metal Levels in Beach Dipterans and Amphipods," *Bull. Environ. Contam. Toxicol.*, 14(2), 187–192.

Benijts-Claus, C. and Benijts, F. (1975). "The Effect of Low Lead and Zinc Concentrations on the Larval Development of the Mudcrab, *Rhithropanopeus harissi* Gould." In J. H. Koeman and J. J. T. W. A. Strik, Eds., *Sublethal Effects of Toxic Chemicals on Aquatic Animals*. Elsevier, Amsterdam, pp. 43–52.

Bernhard, M. and Zattera, A. (1969). "A Comparison between the Uptake of Radioactive and Stable Zinc by a Marine Unicellular Alga." In *Proceedings of the 2nd National Symposium on Radioecology, Ann Arbor, Mich., 1967*, pp. 389–398.

Bernhard, M. and Zattera, A. (1975). "Major Pollutants in the Marine Environment." In Pearson and Frangipane, Eds., *Marine Pollution and Marine Waste Disposal*. Pergamon Press, New York, pp. 195–300.

Bertine, K. K. and Goldberg, E. D. (1972). "Trace Elements in Clams, Mussels, and Shrimp," *Limnol. Oceanogr.*, 17, 877–884.

Bhatt, Y. M., Sastry, V. N., Shah, S. M., and Krishnamoorthy, T. M. (1968). "Zinc, Manganese, and Cobalt Contents of Some Marine Bivalves from Bombay," *Proc. Natl. Inst. Sci. India*, Part B, Biol. Sec., 34(B,6), 283–287.

Black, W. A. P. and Mitchell, R. L. (1952). "Trace Elements in the Common Brown Algae and in Seawater," *J. Mar. Biol. Assoc. U.K.*, 30(3), 575–584.

Bohn, A. and McElroy, R. O. (1976). "Trace Metals (As, Cd, Cu, Fe, and Zn) in Arctic Cod, *Boreogadus saida*, and Selected Zooplankton from Strathcona Sound, Northern Baffin Island," *J. Fish. Res. Board Can.*, 33, 2836–2840.

Boyden, C. R. (1974). "Trace Element Content and Body Size in Molluscs," *Nature*, 251, 311–314.

Boyden, C. R. (1977), "Effect of Size upon Metal Content of Shellfish," *J. Mar. Biol. Assoc. U.K.*, **57**, 675–714.

Boyden, C. R. and Romeril, M. G. (1974). "A Trace Metal Problem in Pond Oyster Culture," *Mar. Pollut. Bull.*, **5**, 74–78.

Braek, G. S., Jensen, A., and Mohus, A. (1976). "Heavy Metal Tolerance of Marine Phytoplankton. III: Combined Effects of Copper and Zinc Ions on Cultures of Four Common Species," *J. Exp. Mar. Biol. Ecol.*, **25**, 37–50.

Brooks, R. R., and Rumsby, M. G. (1965). "The Biogeochemistry of the Element Uptake by Some New Zealand Bivalves," *Limnol. Oceanogr.*, **10**, 521–527.

Brooks, R. R., and Rumsey, D. (1974). "Heavy Metals in Some New Zealand Commercial Sea Fishes," *N. Z. J. Mar. Freshwater Res.*, 8(1), 155–166.

Brown, B. and Ahsanullah, M. (1971). "Effect of Heavy Metals on Mortality and Growth," *Mar. Pollut. Bull.*, **2**, 182–187.

Bryan, G. W. (1964). "Zinc Regulation in the Lobster *Homarus americanus*. I: Tissue Zinc and Copper Concentrations," *J. Mar. Biol. Assoc. U.K.*, **44**, 549–563.

Bryan, G. W. (1966). "The Metabolism of Zn and [65] Zn in Crabs, Lobsters and Freshwater Crayfish." In *Radioecological Concerntration Processes*. Proceedings of an International Symposium, Stockholm, Apr. 25–29. Pergamon Press, New York, pp. 1005–1016.

Bryan, G. W. (1967). "Zinc Concentrations of Fast and Slow Contracting Muscles in the Lobster," *Nature*, **213**, 1043–1044.

Bryan, G. W. (1968). "Concentrations of Zinc and Copper in the Tissues of Decapod Crustaceans," *J. Mar. Biol. Assoc. U.K.*, **48**, 303–321.

Bryan, G. W. (1969). "The Absorption of Zinc and Other Metals by the Brown Seaweed *Laminaria digitata*," *J. Mar. Biol. Assoc. U.K.*, **49**, 225–243.

Bryan, G. W. (1971). "The Effects of Heavy Metals (Other than Mercury) on Marine and Estuarine Organisms," *Proc. R. Soc. London*, **B177**, 389–410.

Bryan, G. W. (1973). "The Occurrence and Seasonal Variation of Trace Metals in the Scallops *Pecten maximus* (L.) and *Chlamys opercularis* (L.)," *J. Mar. Biol. Assoc. U.K.*, **53**, 145–166.

Bryan, G. W. (1974). "Adaptation of an Estuarine Polychaete to Sediments Containing High Concentrations of Heavy Metals." In F. J. Vernberg and W. B. Vernberg, Eds., *Pollution and Physiology of Marine Organisms*. Academic Press, New York, pp. 123–135.

Bryan, G. W. (1976). "Heavy Metal Contamination in the Sea." In R. Johnston, Ed., *Marine Pollution*. Academic Press, London, pp. 185–302.

Bryan, G. W. and Hummerstone, L. G. (1973a). "Brown Seaweed as an Indicator of Heavy Metals in Estuaries in South-west England," *J. Mar. Biol. Assoc. U.K.*, **53**, 705–720.

Bryan, G. W. and Hummerstone, L. G. (1973b). "Adaptation of the Polychaete

Nereis diversicolor to Estuarine Sediments Containing High Concentrations of Zinc and Cadmium," *J. Mar. Biol. Assoc. U.K.*, **53**, 839–857.

Bryan, G. W. and Hummerstone, L. G. (1977). "Indicators of Heavy-Metal Contamination in the Looe Estuary (Cornwall) with Particular Regard to Silver and Lead," *J. Mar. Biol. Assoc. U.K.*, **57**, 75–92.

Bryan, G. W., Potts, G. W., and Forster, G. R. (1977). "Heavy Metals in the Gastropod Mollusc *Haliotis tuberculata* (L.)." *J. Mar. Biol. Assoc. U.K.*, **57**, 379–390.

Butterworth, J., Lester, P., and Nickless, G. (1972). "Distribution of Heavy Metals in the Severn Estuary," *Mar. Pollut. Bull.*, **3**(5), 72–74.

Calabrese, A., MacInnes, J. R., Nelson, D. A., and Miller, J. E. (1977). "Survival and Growth of Bivalve Larvae under Heavy-Metal Stress," *Mar. Biol.*, **41**, 179–184.

Carey, A. G., Jr. (1970). Zn-65 in Benthic Invertebrates off the Oregon Coast. Available as RLO–1750–55, from National Technical Information Service, Springfield, VA 22161, pp. 1–27.

Cheng, L., Alexander, G. V., and Franco, P. J. (1976). "Cadmium and Other Heavy Metals in Seaskaters (Gerridae: *Halobates, Rheumobates*)," *Water, Air Soil Pollut.*, **6**, 33–38.

Chipman, W. A., Rice, T. R. and Price, T. J. (1958). "Uptake and Accumulation of Radioactive Zinc by Marine Plankton, Fish, and Shellfish," *U.S. Fish. Wildl. Serv. Fish. Bull. 135*, **58**, 279–292.

Coombs, T. L. (1972). "The Distribution of Zinc in the Oyster *Ostrea edulis* and Its Relation to Enzymic Activity and to Other Metals," *Mar. Biol.*, **12**, 170–178.

Cross, F. A., Duke, T. W., and Willis, J. N. (1970). "Biogeochemistry of Trace Elements in a Coastal Plain Estuary: Distribution of Manganese, Iron, and Zinc in Sediments, Water and Polychaetous Worms," *Chesapeake Sci.*, **11**(4), 221–234.

Cross, F. A., Hardy, L. H., Jones, N. Y., and Barber, R. T. (1973). "Relation between Total Body Weight and Concentrations of Manganese, Iron, Copper, Zinc and Mercury in White Muscle of Bluefish (*Pomatomus saltatrix*) and a Bathyldemersal Fish (*Antimora rostrata*)," *J. Fish. Res. Board Can.*, **30**, 1287–1291.

Cutshall, N. H., Naidu, J. R. and Pearcy, W. G. (1977a). "Zinc-65 Specific Activities in the Migratory Pacific Hake *Merluccius productus*," *Mar. Biol.*, **40**, 75–80.

Cutshall, N. H., Naidu, J. R., and Pearcy, W. G. (1977b). "Zinc and Cadmium in the Pacific Hake *Mericcius productus* off the Wester U.S. Coast.," *Mar. Biol.*, **44**, 195–202.

Davies, A. G. (1973). "The Kinetics of and a Preliminary Model for the Uptake of Radio-Zinc by *Phaeodactylum tricornutum* in Culture." In *Radioactive Contamination of the Marine Environment*. International Atomic Energy Agency, Vienna, Austria, pp. 403–420.

Drescher, H. E., Harms, U., and Huschenbeth, E. (1977). "Organochlorines and Heavy Metals in the Harbour Seal *Phoca vitulina* from the German North Sea Coast," *Mar. Biol.*, **41**, 99–106.

Drifmeyer, J. E. (1974). "Zn and Cu Levels in the Eastern Oyster, *Crassostrea virginica*, from the Lower James River," *J. Wash. Acad. Sci.*, **64**(4), 292–294.

Edwards, R. R. C. (1967). "Estimation of the Respiratory Rate of Young Plaice (*Pleuronectes platessa* L.) in Natural Conditions Using Zinc-65," *Nature*, **216**, 1335–1337.

Eisler, R. (1967a). *Tissue Changes in Puffers Exposed to Methoxychlor and Methyl Parathion.* U.S. Bureau of Sport Fisheries and Wildlife, Tech. Pap. 17, pp. 1–15.

Eisler, R. (1967b). "Variations in Mineral Content of Sandbar Shark Vertebrae (*Carcharhinus milberti*)," *Nat. Can.*, **94**, 321–326.

Eisler, R. (1967c). "Acute Toxicity of Zinc to the Killifish, *Fundulus heteroclitus*," *Chesapeake Sci.*, **8**, 262–264.

Eisler, R. (1973). *Annotated Bibliography on Biological Effects of Metals in Aquatic Environments* (No. 1–567). U.S. Environmental Protection Agency, Rep. EPA-R3-73-007, 287 pp. Available from National Technical Information Service, Springfield, VA 22161.

Eisler, R. (1977a). "Acute Toxicities of Selected Heavy Metals to the Softshell Clam, *Mya arenaria*," *Bull. Environ. Contam. Toxicol.*, **17**(2), 137–145.

Eisler, R. (1977b). "Toxicity Evaluation of a Complex Metal Mixture to the Softshell Clam. *Mya arenaria*," *Mar. Biol.*, **43**, 265–276.

Eisler, R. (1979). "Toxic Cations and Marine Biota: Analysis of Research Effort during the Three-Year Period 1974–1976." In W. Vernberg, A. Calabrese, F. P. Thurberg, and F. J. Vernberg, Eds., *Marine Pollution: Functional Responses.* Academic Press, New York, pp. 111–146.

Eisler, R. and Edmunds. P. H. (1966). "Effects of Endrin on Blood and Tissue Chemistry of a Marine Fish," *Trans. Am. Fish. Soc.*, **95**, 153–159.

Eisler, R. and Gardner, G. R. (1973). "Acute Toxicology to an Estuarine Teleost of Mixtures of Cadmium, Copper and Zinc Salts," *J. Fish Biol.*, **5**, 131–142.

Eisler, R. and Hennekey, R. J. (1977). "Acute Toxicities of Cd^{2+}, Cr^{+6}, Hg^{2+}, Ni^{2+} and Zn^{2+} to Estuarine Macrofauna," *Arch. Environ. Contam. Toxicol.*, **6**, 315–323.

Eisler, R. and LaRoche, G. (1972). "Elemental Composition of the Estuarine Teleost *Fundulus heteroclitus* (L.)," *J. Exp. Mar. Biol. Ecol.*, **9**(1), 29–42.

Eisler, R. and Wapner, M. (1975). *Second Annotated Bibliography on Biological Effects of Metals in Aquatic Environments* (No. 568–1292). U.S. Environmental Protection Agency Rep. EPA-600/3-75-008, 400 pp. Available from National Technical Information Service, Springfield, VA 21161.

Eisler, R. and Weinstein, M. P. (1967). "Changes in Metal Composition of the Quahaug Clam, *Mercenaria mercenaria*, after Exposure to Insecticides," *Chesapeake Sci.*, **8**, 253–258.

Eisler, R., Barry, M. M., Lapan, R. L., Jr., Telek, G., Davey, E. W., and Soper, A. E. (1978a). "Metal Survey of the Marine Clam *Pitar morrhuana* Collected near a Rhode Island (USA) Electroplating Plant," *Mar. Biol.*, 45, 311–317.

Eisler, R., O'Neill, D. J. Jr., and Thompson, G. W. (1978b). *Third Annotated Bibliography on Biological Effects of Metals in Aquatic Environments* (No. 1293-2246), U.S. Environmental Protection Agency, Rep. 600/3-78-005, 487 pp. Available from National Technical Information Service, Springfield, VA 21161.

Eustace, I. J. (1974). "Zinc, Cadmium, Copper and Manganese in Species of Finfish and Shellfish Caught in the Derwent Estuarine, Tasmania," *Aust. J. Mar. Freshwater Res.*, 25, 209–220.

Flegal, A. R. and Martin, J. H. (1977). "Contamination of Biological Samples by Ingested Sediment," *Mar. Pollut. Bull.*, 8(4), 90–92.

Fletcher, G. L., Watts, E. G., and King, M. J. (1975). "Copper, Zinc, and Total Protein Levels in the Plasma of Sockeye Salmon (*Oncorhynchus nerka*) during Their Spawning Migration," *J. Fish. Res. Board Can.*, 31, 78–82.

Foster, P. (1976). "Concentrations and Concentration Factors of Heavy Metals in Brown Algae," *Environ. Pollut.*, 10, 45–53.

Fowler, S. W. (1974). "The Effect of Organism Size on the Content of Certain Trace Metals in Marine Zooplankton," *Rapp. Comm. Int. Mer Medit.*, 22 (9), 145–146.

Fowler, S. W. (1977). "Trace Elements in Zooplankton Particulate Products," *Nature*, 269(5623), 51–53.

Fowler, S. W. and Oregioni, B. (1976). "Trace Metals in Mussels from the N.W. Mediterranean," *Mar. Pollut. Bull.*, 7, 26–29.

Fowler, S. W., Small, L. F., and Dean, J. M. (1971). "Experimental Studies on Elimination of Zinc-65, Cesium-137 and Cerium-144 by Euphausiids," *Mar. Biol.*, 8, 224–231.

Fowler, S. W., Small, L. F., and LaRosa, J. (1972). "The Role of Euphausiid Molts in the Transport of Radionuclides in the Sea," *Rapp. Comm. Int. Mer Medit.*, 21(6), 291–292.

Frazier, J. M. (1975). "The Dynamics of Metals in the American Oyster, *Crassostrea virginica*. I: Seasonal Effects," *Cheasapeake Sci.*, 16, 162–171.

Frazier, J. M. (1976). "The Dynamics of Metals in the American Oyster, *Crassostrea virginica*. II: Environmental Effects," *Chesapeake Sci.*, 17, 188–197.

Fujita, T. (1972). "The Zinc Content in Marine Plankton," *Rec. Oceanogr. Works Jap.*, 11(2), 73–79.

Galtsoff, P. S. (1953). "Accumulation of Manganese, Iron, Copper, and Zinc in the Body of American Oyster, *Crassortaea (Ostrea) virginica*," *Anat. Rec.*, 117, 601–602.

George, S. G. and Coombs, R. L. (1977). "The Effects of Chelating Agents on the Uptake and Accumulation of Cadmium by *Mytilus edulis*," *Mar. Biol.*, 39, 261–268.

Goldberg, E. D. (1962). "Elemental Composition of Some Pelagic Fishes," *Limnol. Oceanogr., 7* (Suppl.), 72–75.

Graham, D. L. (1972). "Trace Metal Levels in Intertidal Mollusks of California," *Veliger,* **14**(4), 365–372.

Gray, J. S. (1974). "Synergistic Effects of Three Heavy Metals on Growth Rates of a Marine Ciliate Protozoan." In F. J. Vernberg, and W. B. Vernberg, Eds., *Pollution and Physiology of Marine Organisms.* Academic Press, New York, pp. 465–485.

Greig, R. A. (1975). "Comparison of Atomic Absorption and Neutron Activation Analyses for Determination of Silver, Chromium and Zinc in Various Marine Organisms," *Anal. Chem.,* **47**, 1682–1684.

Greig, R. A. and Wenzloff, D. R. (1977). "Trace Metals In Finfish from the New York Bight and Long Island Sound," *Mar. Pollut. Bull.,* **8**(9), 198–200.

Greig, R. A., Nelson, B. A., and Nelson, D. A. (1975). "Trace Metal Content in the American Oyster," *Mar. Pollut. Bull.,* **6**, 72–73.

Greig, R. A., Wenzloff, D. R., and Pearce, J. B. (1976). "Distribution and Abundance of Heavy Metals in Finfish, Invertebrates and Sediments Collected at a Deepwater Disposal Site," *Mar. Pollut. Bull.,* **7**, 185–187.

Greig, R. A., Adams, A., and Wenzloff, D. R. (1977). "Trace Metal Content of Plankton and Zooplankton Collected from the New York Bight and Long Island Sound," *Bull. Environ. Contam. Toxicol.,* **18**(1), 3–8.

Gutknecht, J. (1961). "Mechanism of Radioactive Zinc Uptake by *Ulva lactuca,*" *Limnol. Oceanogr.,* **6**, 426–431.

Hardisty, M. W., Kartar, S., and Sainsbury, M. (1974). "Dietary Habits and Heavy Metal Concentrations in Fish from the Severn Estuary and Bristol Channel," *Mar. Pollut. Bull.,* **5**(4), 61–63.

Haug, A., Melsom, S., and Omang, S. (1974). "Estimation of Heavy Metal Pollution in two Norwegian Fjord Areas by Analysis of the Brown Alga *Ascophyllum nodosum,*" *Environ. Pollut.,* **7**, 179–192.

Havre, G. N., Underal, B., and Christiansen, C. (1972). "The Content of Lead and Some Other Heavy Elements in Different Fish Species from a Fjord in Western Norway." In *Proceedings of the International Symposium on Environmental Health Aspects of Lead, Amsterdam Oct. 2–6,* pp. 99–111.

Hiyama, Y. and Shimizu, M. (1964). "On the Concentration Factors of Radioactive Cs, Sr, Cd, Zn, and Ce in Marine Organisms," *Rec. Oceanogr. Works Jap.,* **7**(2), 43–77.

Holden, A. V. (1975). "The Accumulation of Oceanic Contaminants in Marine Mammals," *Rap. P.-v. Reun. Const. Int. Explor. Mer,* **169**, 353–361.

Holden, A. V. and Topping, G. (1972). "Occurrence of Specific Pollutants in Fish in the Forth and Tay Estuaries," *Proc. R. Soc. Edinb.,* **B71**(14), 189–194.

Huggett, R. J., Bender, M. E., and Slone, H. D. (1973). "Utilizing Metal Concentration Relationships in the Eastern Oyster (*Crassostrea virginica*) to Detect Heavy Metal Pollution," *Water Res., 7*, 451–460.

Huggett, R. J., Cross, F. A., and Bender, M. E. (1975). "Distribution of Copper and Zinc in Oysters and Sediments from Three Coastal-Plain Estuaries." In F. G. Howell, J. B. Gentry, and M. H. Smith, Eds., *Mineral Cycling in Southeastern Ecosystems.* U.S. Energy, pp. 224–238. Available as CONF–740513 from National Technical Information Service, Springfield, VA 22161.

Ikuta, K. (1968a). Studies on Accumulation of Heavy Metals in Aquatic organisms. II: On Accumulation of Copper and Zinc in Oysters," *Bull. Jap. Soc. Sci. Fish., 34*(2), 112–116.

Ikuta, K. (1968b). "Studies on Accumulation of Heavy Metals in Aquatic Organisms. III: On Accumulation of Copper and Zinc in the Parts of Oysters," *Bull. Jap. Soc. Sci. Fish., 34*(2), 117–122.

Ikuta, K. (1968c). "Studies on Accumulation of Heavy Metals in Aquatic Organisms. IV: On Disappearance of Abnormally Accumulated Copper and Zinc in Oysters," *Bull. Jap. Soc. Sci. Fish., 34*(6), 482–487.

Ireland, M. P. (1973). "Result of Fluvial Zinc Pollution on the Zinc Content of Littoral and Sublittoral Organisms in Cardigan Bay, Wales," *Environ. Pollut., 4*, 27–35.

Ireland, M. P. (1974). "Variations in the Zinc, Copper, Manganese and Lead Content of *Balanus balanoides* in Cardigan Bay, Wales," *Environ. Pollut., 7*, 65–75.

Ishibashi, M., Fujinaga, T., Morii, F., Kanchiku, Y., and Kamiyama, F. (1964). "Chemical Studies on the Ocean (Part 94); Chemical Studies on the Seaweeds (Part 19). Determination of Zinc, Copper, Lead, Cadmium, and Nickel in Seaweeds Using Dithizone Extraction and Polarographic Method," *Rec. Oceanogr. Works Jap., 7*(2), 33–36.

Ishibashi, M., Fujinaga, T., Yamamoto, T., Fujita, T., and Watanabe, K. (1965). "Zinc and Iron in Seaweeds," *J. Chem. Soc. Jap., 86*, 728–733.

Jackim, E., Morrison, G., and Steele, R. (1977). "Effects of Environmental Factors on Radiocadmium Uptake by Four Species of Marine Bivalves," *Mar. Biol., 40*, 303–308.

Jangaard, P. M., Regier, L. W., Claggett, F. G., March, B. E., and Biely, J. (1974). "Nutrient Composition of Experimentally Produced Meals from Whole Argentine, Capeline, Sand Lance, and from Flounder and Redfish Filleting Scrap," *J. Fish. Res. Board Can., 31*, 141–146.

Jensen, A., Rystad, B., and Melsom, S. (1974). "Heavy Metal Tolerance of Marine Phytoplankton. I: The Tolerance of Three Algal Species to Zinc in Coastal Seawater," *J. Exp. Mar. Biol. Ecol., 15*, 145–147.

Jones, M. B. (1975). "Synergistic Effects of Salinity, Temperature and Heavy Metals on Mortality and Osmoregulation in Marine and Estuarine Isopods (Crustacea)," *Mar. Biol., 30*, 13–20.

Kameda, K., Shimizu, M., and Hiyama, Y. (1968). "On the Uptake of Zinc-65 and the Concentration Factor of Zinc in Marine Organisms. I: Uptake of zinc-65 in Marine Organisms," *J. Radiat. Res.*, **9**, 50–62.

Keckes, S., Ozretic, B., and Krajnovic, M. (1968). "Loss of Zn-65 in the Mussel *Mytilus galloprovincialis*," *Malacologia*, **7**(1), 1–6.

Keckes, S., Ozretic, B., and Krajnovic, M. (1969). "Metabolism of Zn-65 in Mussles (*Mytilus galloprovincialis* Lam.): Uptake of Zn-65," *Rapp. Comm. Int. Mer Medit.*, **19**(5), 949–952.

Ketchum, B. M. and Bowen, V. T. (1958). "Biological Factors Determining the Distribution of Radioisotopes in the Sea." In *Proceedings of the Second Annual U.N. International Conference on Peaceful Uses of Atomic Energy*, Vol. 18: *Waste Treatment and Environmental Aspects of Atomic Energy*, Geneva, pp. 429–433.

Knauer, G. A. (1970). "The Determination of Magnesium, Manganese, Iron, Copper, and Zinc in marine shrimp." *Analyst*, **93**, 476–480.

Knauer, G. A. and Martin, J. H. (1973). "Seasonal Variations of Cadmium, Copper, Manganese, Lead, and Zinc in Water and Phytoplankton in Monterey Bay, California," *Limnol. Oceanogr.*, **18**, 597–604.

Kopfler, F. C. and Mayer, J. (1967). "Studies of Trace Metals in Shellfish." In *Proceedings of the Gulf and South Atlantic States Shellfish Sanitation Research Conference*, pp. 67–80.

Kuenzler, E. J. (1969). "Elimination of Iodine, Cobalt, Iron, and Zinc by Marine Zooplankton." In *Proceedings of the 2nd National Symposium on Radioecology*. U.S. Atomic Energy Commission (Conf. 670503), pp. 462–473.

Leatherland, T. M. and Burton, J. D. (1974). "The Occurrence of Some Trace Metals in Coastal Organisms with Particular Reference to the Solent Region," *J. Mar. Biol. Assoc. U.K.*, **54**, 457–468.

Leatherland, T. M., Burton, J. B., Culkin, F., McCartney, M. J., and Morris, R. J. (1973). "Concentrations of Some Trace Metals in Pelagic Organisms and of Mercury in Northeast Atlantic Ocean Water," *Deep Sea Res.*, **20**, 679–685.

Livingston, H. D. and Thompson, G. (1971). "Trace Element Concentrations in Some Modern Corals," *Limnol. Oceanogr.*, **16**(5), 786–795.

Lowman, F. G., Martin, J. H., Ting, R. Y., Barnes, S. S., Swift, D. J. P., Seiglie, G. A., Pirie, R. G., Davis, R., Santiago, R. J., Escalera, R. M., Gordon, A. G., Telek, G., Besselievre, H. L., and McCanless, J. B., (1970). *Bioenvironmental and Radiological-Safety Feasibility Studies, Atlantic-Pacific Interoceanic Canal. Estuarine and Marine Ecology*, Vols. I–IV. Prepared for Battelle Memorial Institute, Columbus, Ohio, under Contract AT (26-1)-171.

Lowman, F. G., Palambo, R. F., and South, D. J. (1957). *The Occurrence and Distribution of Radioactive Non-fission Products in Plants and Animals of the Pacific Proving Ground*. Applied Fisheries Laboratory, University of Washington, Seattle; U.S. Atomic Energy Commission, Rept. UWFL-51, 61 pp.

Lowman, F. G., Phelps, D. K., Ting, R. Y., and Escalera, R. M. (1966). *Marine Biology Program June 1965-June 1966*. Progress Summary Rep. 4; Puerto Rico Nuclear Center, Rep. PRNC 85, 57 pp. and Appendices A–F.

Lunde, G. (1968). "Activation Analysis of Trace Elements in Fishmeal," *J. Sci. Food Agric.*, **19**, 432–434.

Lunde, G. (1970). "Analysis of Trace Elements in Seaweed," *J. Sci. Food Agric.*, **21**, 416–418.

Lunde, G. (1973). "Trace Metal Contents of Fish Meal and of the Lipid Phase Extracted from Fish Meal," *J. Sci. Food Agric.*, **24**, 413–419.

Lytle, T. F., Lytle, J. S., and Parker, P. L. (1973). "A Geochemical Study of a Marsh Environment," *Gulf Res. Rep.*, **4**(2), 214–232.

Mackay, N. J., Williams, R. J., Kacprzac, J. L., Kazacos, M. N., Collins, A. J., and Auty, E. H. (1975). "Heavy Metals in Cultivated Oysters (*Crassostrea commercialis = Saccostrea cucullata*) from the Estuaries of New South Wales," *Austr. J. Mar. Freshwater Res.*, **26**, 31–46.

Martin, J. H. and Flegal, A. R. (1975). "High Copper Concentrations in Squid Livers in Association with Elevated Levels of Silver, Cadmium, and Zinc," *Mar Biol.*, **30**, 51–55.

Martin, J. H. and Knauer, G. A. (1973). "The Elemental Composition of Plankton," *Geochim. Cosmochim. Acta*, **37**, 1639–1653.

Martin, J. H., Elliot, P. D., Anderlini, V. C., Girvin, D., Jacobs, S. A., Risebrough, R. W., Delong, R. L. and Gilmartin, W. G. (1976). "Mercury-Selenium-Bromine Imbalance in Premature Parturient California Sea Lions," *Mar. Biol.*, **35**, 91–104.

Matsumoto, T., Satake, M., Yamamoto, J., and Haruna, S. (1964). "On the Microconstituent Elements in Marine Invertebrates," *J. Oceanogr. Soc. Jap.*, **20**(3), 15–19.

McDermott, D. J., Alexander, G. V., Young, D. R., and Mearns, A. J. (1976). "Metal Contamination of Flatfish around a Large Submarine Outfall," *J. Water Pollut. Contr. Fed.*, **48**(8), 1913–1918.

McFarren, E. F., Campbell, J. E., and Engle, J. B. (1962). "The Occurrence of Copper and Zinc in Shellfish." In E. G. Jensen, Ed., *Proceedings of the 1961 Shellfish Sanitation Workshop*. U.S. Public Health Service, pp. 229–234.

McHargue, J. S. (1924). "The Significance of the Occurrence of Copper, Manganese, and Zinc in Shell-fish," *Science*, **60**, 530.

McHargue, J. S. (1927). "The Proportion and Significance of Copper, Iron, Manganese, and Zinc in some Mollusks and Crustaceans," *Trans. Ky. Acad. Sci.*, **2**, 46–52.

McLerran, C. J. and Holmes, C. W. (1974). "Deposition of Zinc and Cadmium by Marine Bacteria in Estuarine Sediments," *Limnol. Oceanogr.*, **19**, 998–1001.

Mears, H. C. and Eisler, R. (1977). "Trace Metals in Liver from Bluefish, Tautog and Tilefish in Relation to Body Length," *Chesapeake Sci.,* 18(3), 315–318.

Mehran, A. R. and Tremblay, J. L. (1965). "An Aspect of Zinc Metabolism in *Littorina obtusata* L. and *Fucus edentatus,*" *Rev. Can. Biol.,* 24, 157–161.

Mishima, J. and Odum, E. P. (1963). "Excretion Rate of Zn[65] by *Littorina irrorata* in Relation to Temperature and Body Size," *Limnol. Oceanogr.,* 8, 39–44.

Morisawa, M. and Mohri, H. (1972). "Heavy Metals and Spermatozoan Motility. I: Distribution of Iron, Zinc, and Copper in Sea Urchin Spermatozoa," *Contrib. Tamano Mar. Lab.* 187, pp. 1–6.

Negilski, D. S. (1976). "Acute Toxicity of Zinc, Cadmium, and Chromium to the Marine Fishes, Yellow-Eye Mullet (*Aldrichetta forsteri* C. & V.) and Small-Mouthed Hardyhead (*Antherinasoma microstoma* Whitley)," *Austr. J. Mar. Freshwater Res.,* 27, 137–149.

Orvini, E., Gills, T. E., and LaFleur, P. D. (1974). "Method for Determination of Selenium, Arsenic, Zinc, Cadmium, and Mercury in Environmental Matrices by Neutron Activation Analysis," *Anal. Chem.,* 46, 1294–1297.

Palmer, J. B. and Rand, G. M. (1977). "Trace Metal Concentrations in Two Shellfish Species of Commercial Importance," *Bull. Environ. Contam. Toxicol.,* 18(4), 512–520.

Parry, G. D. R. and Hayward, J. (1973). "The Uptake of [65]Zn by *Dunaliella tertiolecta* Butcher," *J. Mar. Biol. Assoc. U.K.,* 53, 915–922.

Pearcy, W. G. and Osterberg, C. L. (1967). "Depth, Diel, Seasonal, and Geographic Variation in Zinc-65 of Midwater Animals off Oregon," *Int. J. Oceanol. Limnol.,* 1, 103–116.

Pentreath, R. J. (1973a). "The Accumulation from Seawater of [65]Zn, [54]Mn, [58]Co, and ^{59}Fe by the Thronback Ray, *Raja clavata* L.," *J. Exp. Mar. Biol. Ecol.,* 12, 327–334.

Pentreath, R. J. (1973b). "The Accumulation from Water of [65]Zn, [54]Mn, [58]Co, and ^{59}Fe by the Mussel, *Mytilus edulis,*" *J. Mar. Biol. Assoc. U.K.,* 53, 127–143.

Pentreath, R. J. (1973c). "The Accumulation and Retention of [65]Zn and [54]Mn by the Plaice, *Pleuronectes platessa* L.," *J. Exp. Mar. Biol. Ecol.,* 12, 1–18.

Pentreath, R. J. (1976). "Some Further Studies on the Accumulation and Retention of Zn-65 and Mn-54 by the Plaice, *Pleuronectes platessa* L.," *J. Exp. Mar. Biol. Ecol.,* 21, 179–189.

Pequegnat, J. E., Fowler, S. W., and Small, L. F. (1969). "Estimates of the Zinc Requirements of Marine Organisms," *J. Fish. Res. Board Can.,* 26(1), 145–150.

Pesch, G., Reynolds, B., and Rogerson, P. (1977). "Trace Metals in Scallops from within and around Two Ocean Disposal Sites," *Mar. Pollut. Bull.,* 8(10), 224–228.

Phelps, D. K. (1967). "Partitioning of the Stable Elements Fe, Zn, Sc, and Sm within the Benthic Community, Anasco Bay, Puerto Rico." In *Proceedings of the International Symposium on Radioecological Concentration Processes, Stockholm, 1966*, pp. 721-734.

Phelps, D. K., Santiago, R. J., Luciano, D., and Irizarry, N. (1969). "Trace Element Composition of Inshore and Offshore Benthic Populations." In *Proceedings of the 2nd National Symposium on Radioecology*. U.S. Atomic Energy Commission (Conf. 670503), pp. 509-526.

Phillips, A. H. (1917). "Analytical Search for Metals in Tortugas Marine Organisms." *Carnegie Inst. Wash. Paper Dept. Mar. Biol.* 251, pp. xi, 91-93.

Phillips, D. J. H. (1977a). "Effects of Salinity on the Net Uptake of Zinc by the Common Mussel *Mytilus edulis*," *Mar. Biol.*, 41, 79-88.

Phillips, D. J. H. (1977b). "The Common Mussel *Mytilus edulis* as an Indicator of Trace Metals in Scandinavian Waters. I: Zinc and Cadmium," *Mar. Biol.*, 43, 283-291.

Phillips, J. H. (1976a). "The Common Mussel *Mytilus edulis* as an Indicator of Pollution by Zinc, Cadmium, Lead and Copper. I: Effects of Environmental Variables on Uptake of Metals," *Mar. Biol.*, 38, 59-69.

Phillips, J. H. (1976b). "The Common Mussel *Mytilus edulis* as an Indicator of Pollution by Zinc, Cadmium, Lead and Copper. II: Relationship of Metals in the Mussel to Those Discharged by Industry," *Mar Biol.*, 38, 71-80.

Pillai, V. K. (1956). "Chemical Studies on Indian Seaweeds. I: Mineral Constituents," *Proc. Indian Acad. Sci.*, 44(1), Sec. B, 3-29.

Portmann, J. E. (1972). "The Levels of Certain Metals in Fish from Coastal Waters around England and Wales," *Aquaculture*, 1, 91-96.

Preston, A., Jeffries, D. F., Dutton, J. W. R., Harvey, B. R., and Steele, A. K. (1972). "British Isles Coastal Waters: the Concentrations of Selected Heavy Metals in Sea Water, Suspended Matter and Biological Indicators– a Pilot Survey," *Environ. Pollut.*, 3, 69-82.

Pringle, B. H., Hissong, D. E., Katz, E. L., and Mulawka, S. T. (1968). "Trace Metal Accumulation by Estuarine Mollusks," *J. Sanit. Eng. Div.*, 94, SA3, 455-475.

Pyle, T. E. and Tieh, T. T. (1970). "Strontium, Vanadium, and Zinc in the Shells of Pteropods," *Limnol. Oceanogr.*, 15(1), 153-154.

Ratkowsky, D. A., Thrower, S. J., Eustace, I. J., and Olley, J. (1974). "A Numerical Study of the Concentration of Some Heavy Metals in Tasmanian Oysters," *J. Fish. Res. Board Can.*, 31, 1165-1171.

Reish, D. J., Martin, J. M., Piltz, F. M., and Word, J. Q. (1976). "The Effect of Heavy Metals on Laboratory Populations of Two Polychaetes with Comparisons to the Water Quality Conditions and Standards in Southern California Marine Waters," *Water Res.*, 10, 299-302.

Renfro, W. C. (1973). "Transfer of ^{65}Zn from Sediments by Marine Polychaete Worms," *Mar. Biol.*, 21, 305-316.

Renfro, W. C., Fowler, S. W., Heyraud, M., and LaRosa, J. (1975). "Relative

Importance of Food and Water in Long-Term Zinc-65 Accumulation by Marine Biota," *J. Fish. Res. Board Can.,* **32,** 1339–1345.

Rice, T. R. (1963). "Review of Zinc in Ecology." In V. Schultuz and A. W. Klement, Jr., Eds., *Radioecology.* Reinhold, New York, pp. 619–631.

Riley, J. P. and Roth, I. (1971). "The Distribution of Trace Elements in Some Species of Phytoplankton Grown in Culture," *J. Mar. Biol. Assoc. U.K.,* **51,** 63–72.

Riley, J. P. and Segar, D. A. (1970). "The Distribution of the Major and Some Minor Elements in Marine Animals. I: Echinoderms and Coelenterates," *J. Mar. Biol. Assoc. U.K.,* **50,** 721–730.

Robertson, D. E. (1967). "Trace Elements in Marine Organisms," *Rapp. Am. BNWL,* **481-2,** 56–59.

Robertson, D. E., Rancitelli, L. A., Langford, J. C., and Perkins, R. W. (1972). *Battelle-Northwest Contribution to the IDOE Base-Line Study.* Battelle Pacific Northwest Laboratories, Richland, Wash., 46 pp.

Romeril, M. G. (1971). "The Uptake and Distribution of [65]Zn in Oysters," *Mar. Biol.,* **9,** 347–354.

Romeril, M. G. (1974). "Trace Metals in Sediments and Bivalve Mollusca in Southampton Water and the Solent," *Rev. Int. Oceanogr. Med.,* **33,** 31–47.

Romeril, M. G. (1977). "Heavy Metal Accumulation in the Vicinity of a Desalinization Plant," *Mar. Pollut. Bull.,* 8(4), 84–87.

Rosenberg, R. (1977). "Effects of Dredging Operations on Estuarine Benthic Macrofauna," *Mar. Pollut. Bull.,* 8(5), 102–104.

Saeki, M., Okano, S., and Mori, K. (1955). "Studies on the Radioactive Material in the Radiologically Contaminated Fishes Caught in the Pacific Ocean in 1954," *Bull. Jap. Soc. Fish.,* **20,** 902–906.

Saenko, G. N., Koryakova, M. D., Makienko, V. F., and Dobrosmyslova, I. G. (1976). "Concentration of Polyvalent Metals by Seaweeds in Vostok Bay, Sea of Japan," *Mar. Biol.,* **34,** 169–176.

Saltman, P. and Boroughs, H. (1960). "The Accumulation of Zinc by Fish Liver Slices," *Arch. Biochem. Biophys.,* **86,** 169–174.

Sastry, V. W. and Bhatt, Y. M. (1965). "Zinc Content of Some Marine Bivalves and Barnacles from Bombay Shores," *J. Indian Chem. Soc.,* **42,** 121–122.

Segar, D. A., Collins, J. D., and Riley, J. P. (1971). "The Distribution of the Major and Some Minor Elements in Marine Animals. II: Molluscs," *J. Mar. Biol. Assoc. U.K.,* **51,** 131–136.

Seymour, A. H. (1959). "The Distribution of Radioisotopes among Marine Organisms in the Western Central Pacific," *Pubbl. Staz. Zool. Napoli,* **31** (Suppl.), 25–33.

Seymour, A. H. and Nelson, V. A. (1973). "Decline of [65]Zn in Marine Mussels Following the Shutdown of Hanford Reactors." In *Radioactive Contamination of the Marine Environment.* International Atomic Energy Agency, Vienna, Austria, pp. 277–286.

Sherwood, M. J. and Mearns, A. J. (1977). "Environmental Significance of Fin Erosion in Southern California Demersal Fishes." In H. F. Kraybill, C. J. Dawe, J. C. Harshbarber, and R. G. Tradiff, (Eds.), Aquatic Pollutants and Biologic Effects with Emphasis on Neoplasia," *Ann. N.Y. Acad. Sci.*, **298**, 177–189.

Shulman, J., Brisbin, I. L. and Knox, W. (1961). "Effect of Temperature, Salinity, and Food Intake on the Excretion of Zn^{65} in Small Marine Fish," *Biol. Bull.*, **121**(2), 378.

Shuster, C. N., Jr., and Pringle, B. H. (1968). "Effects of Trace Metals on Estuarine Molluscs." In *Proceedings of the First Mid-Atlantic Industrial Waste Conference Nov. 13–15, 1967*, pp. 285–304. Available from Department of Civil Engineering, University of Delaware, Newark, Del.

Shuster, C. N., Jr. and Pringle, B. H. (1969). "Trace Metal Accumulation by the American Oyster, *Crassostrea virginica*," *1968 Proc. Nat. Shellfish. Assoc.*, **59**, 91–103.

Skipnes, O., Roald, T., and Haug, A. (1975). "Uptake of Zinc and Strontium by Brown Algae," *Physiol. Plant.*, **34**, 314–320.

Small, L. F. and Fowler, S. W. (1973). "Turnover and Vertical Transport of Zinc by the Euphausiid *Meganyctiphanes norvegica* in the Ligurian Sea," *Mar. Biol.*, **18**, 284–290.

Small, L. F., Fowler, S. W., and Keckes, S. (1973). "Flux of Zinc through a Macroplanktonic Crustacean." In *Radioactive Contamination of the Marine Environment*. International Atomic Energy Agency, Vienna, Austria, pp. 437–452.

Small, L. F., Keckes, S., and Fowler, S. W. (1974). "Excretion of Different Forms of Zinc by the Prawn, *Palaemon serratus* (Pennant)," *Limnol. Oceanogr.*, **19**(5), 789–793.

Stenner, R. D. and Nickless, G. (1974a). "Absorption of Cadmium, Copper and Zinc by Dog Whelks in the Bristol Channel," *Nature*, **247**, 198–199.

Stenner, R. D. and Nickless, G. (1974b). "Distribution of Some Heavy Metals in Organisms in Hardangerfjord and Skjerstadfjord, Norway," *Water, Air, Soil Pollut.*, **3**, 279–291.

Stenner, R. D. and Nickless, G. (1975). "Heavy Metals in Organisms of the Atlantic Coast of S. W. Spain and Portugal," *Mar. Pollut. Bull.*, **6**, 89–92.

Stickney, R. R., Windom, H. L., White, D. B., and Taylor, F. E. (1975). "Heavy-Metal Concentrations in Selected Georgia Estuarine Organisms with Comparative Food-Habit Data." In F. G. Howell, F. B. Gentry, and M. H. Smith, Eds., *Mineral Cycling in Southeastern Ecosystems*. U.S. Energy Resources Development Administration, pp. 257–267. Available as CONF-740513 from National Technical Information Service, Springfield, VA 22161.

Styron, C. E., Hagan, T. M., Campbell, D. R., Harvin, J., Whittenburg, N. K., Baughman, G. A., Bransford, M. E., Saunders, W. H., Williams, D. C., Woodle, C., Dixon, N. K., and McNeill, C. R. (1976). "Effects of Temperature and Salinity on Growth and Uptake of Zn-65 and Cs-137 for Six Marine Algae," *J. Mar. Biol. Assoc. U.K.,* **56,** 13–20.

Tabata, K. (1969a). "Studies on the Toxicity of Heavy Metals to Aquatic Animals and the Factors to Decrease the Toxicity. I: On the Formation and the Toxicity of Precipitate of Heavy Metals," *Bull. Tokai Fish. Res. Lab.,* **58,** 203–214.

Tabata, K. (1969b). "Studies on the Toxicity of Heavy Metals to Aquatic Animals and the Factors to Decrease the Toxicity. II: The Antagonistic Action of Hardness Components in Water on the Toxicity of Heavy Metal Ions," *Bull. Tokai Fish. Res. Lab.,* **58,** 215–232.

Taylor, D. D. and Bright, T. J. (1973). *The Distribution of Heavy Metals in Reef-Dwelling Groupers in the Gulf of Mexico and Bahama Islands.* Department of Oceanography, Texas A&M University, Rep. TAMU–SG–73–208, College Station, 249 pp.

Tennant, D. A., and Forster, W. D. (1969). "Seasonal Variation and Distribution of 65-Zn, 54-Mn, and 51-Cr in Tissues of the Crab *Cancer magister* Dana," *Health Phys.,* **18,** 649–659.

Thrower, S. J. and Eustace, I. J. (1973a). "Heavy Metal Accumulation in Oysters Grown in Tasmanian Waters," *Food Technol. Austr.,* **25,** 546–553.

Thrower, S. J. and Eustace, I. J. (1973b). "Heavy Metals in Tasmanian Oysters in 1972," *Austr. Fish.,* **32,** 7–10.

Ting, R. Y. (1971). "Distribution of Zn, Fe, Mn, and Sr in Marine Fishes of Different Feeding Habits." In *Proceedings of the Third National Symposium on Radioecology,* Vol. 2; *Radionuclides in Ecosystems,* pp. 709–720.

Topping, G. (1973). Heavy Metals in Shellfish from Scottish Waters," *Aquaculture,* **1,** 379–384.

Valiela, I., Banus, M. D., and Teal, J. M. (1974). "Response of Salt Marsh Bivalves to Enrichment with Metal-Containing Sewage Sludge and Retention of Lead, Zinc and Cadmium by Marsh Sediments," *Environ. Pollut.,* **7,** 149–157.

Vallee, B. L. (1959). "Biochemistry, Physiology and Pathology of Zinc," *Physiol. Rev.,* **39,** 443–490.

Van As, D., Fourie, H. O., and Vleggaar, C. M. (1973). "Accumulation of Certain Trace Elements in the Marine Organisms from the Sea around the Cape of Good Hope." In *Radioactive Contamination of the Marine Environment.* International Atomic Energy Agency, Vienna, Austria, pp. 615–624.

Van Weers, A. W. (1973). "Uptake and Loss of [65]Zn and ^{60}Co by the Mussel *Mytilus edulis* L." In *Radioactive Contamination of the Marine Environment.* International Atomic Energy Agency, Vienna, Austria, pp. 385–401.

Vattuone, G. M., Griggs, K. S., McIntyre, D. R., Littlepage, J. L., and Harrison, F. L. (1976). *Cadmium Concentrations in Rock Scallops in Comparison with Some Other Species.* U.S. Energy Resources Development Administration, UCRL 52022, pp. 1–11. Available from National Technical Information Service, Springfield, VA 22151.

350   Accumulation of Zinc by Marine Biota

Vinogradov, A. P. (1953). *The Elementary Chemical Composition of Marine Organisms.* Sears Foundation for Marine Research, Yale University, Mem. 2, New Haven, Conn., 647 pp.

Vinogradova, Z. A. and Koual'skiy, V. V. (1962). "Elemental Composition of the Black Sea Plankton," *Dokl. Acad. Sci. U.S.S.R., Earth Sci. Sec.,* **147**, 217–219.

Waldichuk, M. (1974). "Some Biological Concerns in Heavy Metals Pollution." In F. J. Vernberg and W. B. Vernberg, Eds., *Pollution and Physiology of Marine Organisms.* Academic Press, New York, pp. 1–57.

Walker, G. (1977). "Copper Granules in the Barnacle *Balanus balanoides,*" *Mar. Biol.,* **39**, 343–349.

Walker, G., Rainbow, P. S., Foster, P., and Crisp, D. J. (1975). "Barnacles: Possible Indicators of Zinc Pollution?" *Mar. Biol.,* **30**, 57–65.

Walker, G., Rainbow, P. S., Foster, P., and Holland, D. L. (1975). "Zinc Phosphate Granules in Tissue Surrounding the Midgut of the Barnacle *Balanus balanoides,*" *Mar. Biol.,* **33**, 161–166.

Watling, H. R. and Watling, R. J. (1976a). "Trace Metals in Oysters from Knysna Estuary," *Mar. Pollut. Bull.,* **7**(3), 45–48.

Watling, H. R. and Watling, R. J. (1976b). "Trace Metals in *Choromytilus meridionalis,*" *Mar. Pollut. Bull.,* **7**(5), 91–94.

Wharfe, J. R. and Van Den Broek, W. L. F. (1977). "Heavy Metals in Macroinvertebrates and Fish from the Lower Medway Estuary, Kent," *Mar. Pollut. Bull.,* **8**(2), 31–34.

Williams, R. B. and Murdock, M. B. (1969). "The Potential Importance of *Spartina alterniflora,* in Conveying Zinc, Manganese, and Iron into Estuarine Food Chains." In *Proceedings of the 2nd National Symposium on Radioecology.* U.S. Atomic Energy Commission (Conf. 670503), pp. 431–440.

Windom, H. L. and Smith, R. G. (1972). "Distribution of Iron, Magnesium, Copper, Zinc, and Silver in Oysters along the Georgia Coast," *J. Fish. Res. Board Can.,* **29**, 450–452.

Windom, H., Stickney, R., Smith, R., White, D., and Taylor, F. (1973). "Arsenic, Cadmium, Copper, Mercury, and Zinc in Some Species of North Atlantic Finfish," *J. Fish. Res. Board. Can.,* **30**(2), 275–279.

Wolfe, D. A. (1970). "Levels of Stable Zn and [65]Zn in *Crassostrea virginica* from North Carolina," *J. Fish. Res. Board Can.,* **27**(1), 47–57.

Wolfe, D. A. (1974). "The Cycling of Zinc in the Newport River Estuary, North Carolina." In F. J. Vernberg and W. B. Vernberg, Eds., *Pollution and Physiology of Marine Organisms,* Academic Press, New York, pp. 79–99.

Wort, D. J. (1955). "The Seasonal Variation in Chemical Composition of *Macrocystis integrifolia* and *Nereocystis leutkeana* in British Columbia Coastal Waters," *Can. J. Bot.,* **33**, 323–340.

Wright, D. A. (1976). "Heavy Metals in Animals from the North East Coast," *Mar. Pollut. Bull.,* **7**(2), 36–38.

Yoshii, G. (1956). "Studies on the Radioactive Samples (Especially *Katsuwonus vagans*) Collected by the *Shunkotsumaru* in the Pacific Ocean in 1954." In *Research on the Effects of the Influence of Nuclear Bomb Test Explosion*, Vol. II, 917–936

Young, D. R. and McDermott, D. J. (1975). *Trace Metals in Harbor Mussels.* Annual Report, Southern California Coastal Water Resources Project, El Segundo, June 30, pp. 139–142.

Young, E. G. and Langille, W. M. (1958). "The Occurrence of Inorganic Elements in Marine Algae of the Atlantic Provinces of Canada," *Can. J. Bot.,* **36**, 301–310.

Young, M. L. (1975). "The Transfer of ^{65}Zn and ^{59}Fe along a *Fucus serratus* (L.) *Littorina obtusata* (L.) Food Chain," *J. Mar. Biol. Assoc. U.K.,* **55**, 583–610.

Zafiropoulos, D. and Grimanis, A. P. (1977). "Trace Elements in *Acartia clausi* from Elefsis Bay of the Upper Saronikos Gulf, Greece," *Mar. Pollut. Bull.,* **8**(4), 79–81.

14

ZINC SPECIATION AND TOXICITY TO FISH

Gordon K. Pagenkopf

Department of Chemistry, Montana State University, Bozeman, Montana

1. Introduction   353
2. Data   354
3. Procedures and Results   354
4. Discussion   358
References   359

1. INTRODUCTION

The presence of complexing ligands such as hydroxide ion, carbonate ion, sulfate ion, and humic materials in natural waters plays a dominant role in regulating the speciation of a particular metal. When the alkalinity is high, the dominant species may be the soluble carbonate complex. Hydroxide ligation is very important for most metals if the solution pH is 7 or greater. As a consequence of these complexation reactions the concentration of the aquated metal ion is often less than 10% of the total metal in solution. Recognition of these species distributions has led to alternative methods of interpretating the toxicities of trace metals toward aquatic biota, particularly fish. Many studies have been designed to investigate the influence of water hardness on fish toxicity, but inadvertently the alkalinity was also varied and thus the degree of metal complexation varied.

Studies involving copper, which forms a very stable copper carbonate complex, indicate that the free metal ion and possibly the hydroxy complexes are much more toxic than the carbonate complex (Pagenkopf et al., 1974; Andrew et al., 1977). Comparable analysis of lead toxicity data for rainbow trout gives results similar to those for copper (Davies et al., 1976).

353

2. DATA

A sizable number of fish bioassay studies have utilized zinc as the toxicant. Not all of these studies have reported or completed detailed chemical analysis of the test waters, and thus the zinc species distributions cannot be calculated. The chemical reactions involving zinc and the other components commonly found in test waters are listed in Table 1. The interpretation of zinc toxicity to fishes involves the development of an equilibrium model to predict the zinc speciation in the test waters. If one or more of the zinc species is most toxic, a correlation should exist between the species concentration and survival time.

A majority of the zinc toxicity studies involve fathead minnows; however, data are available also for other fishes such as bluegills, zebrafish, goldfish, trout, salmon, and guppies. The water chemistry and the 96-hr LC_{50} values are listed in Table 2. The alkalinity is assumed equal to the bicarbonate ion concentration. These values, coupled with the pH and the dissociation constant for $CO_2 \cdot aq$, permit calculation of the total inorganic carbon concentration. In most cases the calcium and magnesium hardnesses are known. When the two are not reported separately, all of the hardness is assumed to be due to calcium. A majority of the tests are flow-through.

3. PROCEDURES AND RESULTS

The chemical species distributions were calculated using the chemical reactions listed in Table 1 and the analytical concentrations listed in Table 2. A majority

Table 1.   Equilibria Regulating Zinc Speciation

Reaction	Log K	Reference
$CO_2 \cdot aq = H^+ + HCO_3^-$	-6.34	Sillen and Martel (1964)
$HCO_3^- = H^+ + CO_3^{2-}$	-10.17	Sillen and Martel (1964)
$Ca^{2+} + CO_3^{2-} = CaCO_3 \cdot aq$	2.21	Pytkowicz and Hawley (1974)
$Ca^{2+} + SO_4^2 = CaSO_4 \cdot aq$	2.05	Garrels and Thompson (1962)
$Mg^{2+} + CO_3^{2-} = MgCO_3 \cdot aq$	2.05	Pytkowicz and Hawley (1974)
$Mg^{2+} + SO_4^2 = MgSO_4 \cdot aq$	2.10	Garrels and Thompson (1962)
$Zn^{2+} + OH^- = Zn(OH)^+$	6.31	Gubeli and Ste-Marie (1967)
$Zn^{2+} + 2OH^- = Zn(OH)_2 \cdot aq$	11.19	Gubeli and Ste-Marie (1967)
$Zn^{2+} + 3OH^- = Zn(OH)_3^-$	14.31	Gubeli and Ste-Marie (1967)
$Zn^{2+} + SO_4^{2-} = ZnSO_4 \cdot aq$	2.05	Owen and Gurry (1938)
$Zn^{2+} + CO_3^{2-} = ZnCO_3 \cdot aq$	5.0	Garrels and Christ (1965)
$ZnCO_3(s) = Zn^{2+} + CO_3^{2-}$	-10.0	Smith and Martell (1976)
$Zn(OH)_2(s) = Zn^{2+} + 2OH^-$	-16.8	Gubeli and Ste-Marie (1967)

Table 2. Zinc Bioassay Data[a]

Code Number[b]	$10^4 C_T$	$10^4 CA_T$	$10^4 Mg_T$	$10^4 SO_{4T}$	$10^5 Zn_T$	pH	Reference
F-1	34.4	13.5	6.20	3.64	15.3	7.70	Brungs (1969)
F-2	34.4	13.5	6.20	3.64	12.9	7.70	Brungs (1969)
F-3	34.6	13.8	6.90	3.50	7.68	7.70	Eaton (1973)
F-4	15.9	6.50	3.21	1.59	12.4	7.50	Eaton (1973)
F-5	16.1	6.90	3.34	1.66	15.1	7.65	Mount (1966)
F-6	30.1	13.0	6.32	3.14	12.5	7.75	Mount (1966)
F-7	34.1	14.6	7.10	3.51	23.5	7.70	Mount (1966)
F-8	8.1	3.60	1.80	0.89	7.20	8.0	Mount (1966)
F-9	7.35	3.20	1.60	0.79	7.82	7.6	Mount (1966)
F-10	9.95	4.24	2.06	3.63	19.1	6.3	Mount (1966)
F-11	8.53	3.63	1.76	3.54	21.1	6.1	Mount (1966)
F-12	15.3	6.53	3.17	8.20	28.3	5.85	Mount (1966)
F-13	16.2	6.90	3.37	7.23	38.2	6.00	Mount (1966)
F-14	33.4	14.3	6.95	14.1	44.3	6.10	Mount (1966)
F-15	32.8	14.0	6.81	13.9	54.3	6.10	Mount (1966)
F-16	8.53	5.75	1.76	1.52	21.0	7.15	Mount (1966)
F-17	9.95	5.70	2.06	1.57	9.5	7.25	Mount (1966)
F-18	15.8	10.6	3.27	1.77	19.1	7.00	Mount (1966)
F-19	15.7	10.6	3.29	1.78	19.1	6.95	Mount (1966)
F-20	29.4	19.8	6.09	3.47	29.1	6.75	Mount (1966)
F-21	30.8	20.7	6.39	3.38	30.8	7.20	Mount (1966)
F-22	15.8	—	—	—	26.8	5.00	Mount (1966)
F-23	15.8	—	—	—	9.50	8.60	Mount (1966)
F-24	3.21	1.20	0.60	0.03	1.33	7.5	Pickering and Henderson (1966)

355

Table 2. Continued

Code Number[b]	$10^4 C_T$	$10^4 CA_T$	$10^4 Mg_T$	$10^4 SO_{4T}$	$10^5 Zn_T$	pH	Reference
F-25	10.2	—	—	—	1.33	7.55	Pickering and Vigor (1965)
B-1	5.76	2.82	1.62	—	4.67	7.55	Cairns and Scheier (1957)
B-2	3.21	1.20	0.60	0.30	8.23	7.50	Pickering and Henderson (1966)
B-3	61.9	24.6	12.4	6.00	17.4	7.80	Pickering (1968)
R-1	49.6	30.0	2.0	3.64	6.12	7.80	Lloyd (1961)
R-2	7.74	4.68	0.31	0.57	2.75	7.09	Lloyd (1961)
R-3	50.0	50.4	—	—	7.19	7.8	Solbé (1974)
R-4	9.90	4.60	—	—	0.63	7.30	Nehring and Goettl (1974)
GF-1	3.21	1.20	0.60	0.30	9.83	7.50	Pickering and Henderson (1966)
G-1	3.21	1.20	0.60	0.30	1.94	7.50	Pickering and Henderson (1966)
Z-1	3.66	0.70	0.60	—	4.59	7.00	Skidmore (1965)
S-1	2.66	1.40	—	—	5.36	7.30	Sprague and Ramsey (1965)
S-2	2.66	1.50	—	—	0.84	7.30	Sprague (1964)
S-3	1.57	6.5	—	—	1.53	6.6	McLeay (1975)
J-1	8.88	2.4	—	—	2.29	7.58	Spehar (1976)
N-1	51.8	29.7	—	—	5.66	7.44	Solbé and Flook (1975)
K-1	9.46	4.80	—	—	1.03	7.30	Nehring and Goettl (1974)
L-1	7.92	3.50	—	—	0.98	7.30	Nehring and Goettl (1974)
M-1	9.02	4.00	—	—	1.47	7.30	Nehring and Goettl (1974)

[a] Concentrations are molar.

[b] F = fathead minnow; B = bluegill; R = rainbow trout; GF = goldfish; G = guppies; Z = zebrafish; S = salmonid; J = flagfish; N = stone loach; K = cutthroat trout; L = brown trout; M = brook trout.

Table 3. Major Zinc Species in Test Waters

Test	$10^5 Zn_T$	$10^5 Zn^{2+}$	$10^5 ZnOH^+$	$10^5 ZnCO_3$	pH
F–1	15.3	0.91	0.93	13.46	7.70
F–2	12.9	0.91	0.93	11.06	7.70
F–3	7.68	0.91	0.93	5.84	7.70
F–4	12.4	3.19	2.06	7.15	7.50
F–5	15.1	2.20	2.01	10.89	7.65
F–6	12.5	0.93	1.06	10.51	7.75
F–7	23.5	0.93	0.95	21.63	7.70
F–8	7.20	1.93	3.94	1.01	8.00
F–9	7.82	3.85	3.13	0.71	7.60
F–10	19.1	17.6	0.72	0.66	6.3
F–11	21.1	19.8	0.51	0.05	6.1
F–12	28.3	25.8	0.37	0.05	5.85
F–13	38.2	34.9	0.71	0.12	6.00
F–14	44.3	38.1	0.98	0.40	6.1
F–15	54.3	46.9	1.20	0.48	6.1
F–16	21.0	15.3	4.40	1.06	7.15
F–17	9.5	6.38	2.32	0.67	7.25
F–18	19.1	11.5	2.34	5.26	7.00
F–19	19.1	13.3	2.72	3.09	6.95
F–20	29.1	12.5	1.44	15.2	6.75
F–21	20.8	3.38	1.09	16.30	7.20
F–2	26.8	26.7	0.05	–	5.0
F–23	9.5	0.25	2.01	7.24	8.6
F–24	1.33	0.77	0.50	0.05	7.5
F–25	1.33	0.67	0.49	0.16	7.55
B–1	4.67	2.49	1.81	0.32	7.55
B–2	8.23	4.76	3.08	0.30	7.50
B–3	17.4	0.40	0.52	16.5	7.80
R–1	6.12	0.50	0.64	5.00	7.80
R–2	2.75	2.09	0.53	0.11	7.09
R–3	7.19	0.49	.64	6.06	7.80
R–4	0.63	0.41	0.17	0.05	7.3
GF–1	9.83	5.69	3.67	0.36	7.5
G–1	1.94	1.12	0.73	0.07	7.5
Z–1	4.59	3.74	0.76	0.08	7.0
S–1	5.36	3.71	1.51	0.12	7.3
S–2	0.84	0.58	0.24	0.02	7.3
S–3	1.53	1.41	0.12	–	6.6
J–1	2.29	1.14	0.88	0.24	7.58
N–1	5.66	1.13	0.66	3.87	7.44
K–1	1.03	0.67	0.27	0.08	7.3
L–1	0.98	0.65	0.26	0.06	7.3
M–1	1.47	0.97	0.39	0.11	7.3

of the calculations were completed with the program "COMICS" (Perrin and Sayce, 1967). This procedure does not consider heterogeneous-phase reactions. The species distributions for test waters that are supersaturated in $ZnCO_{3(s)}$ are obtained using the concentration of Zn^{2+} in equilibrium with $ZnCO_{3(s)}$. The major species are Zn^{2+}, $ZnOH^+$, and $ZnCO_3 \cdot aq$, with $Zn(OH)_2 \cdot aq$ and $ZnSO_4 \cdot aq$ representing a small fraction. The concentrations of the major species are summarized in Table 3. The calculation also include calcium and magnesium speciation, but these values are not summarized.

The pH values of the test waters have a standard deviation of approximately ±0.15 unit, which is equivalent to 41%. The LC_{50} values for duplicate tests vary in many cases by 30%. Temperature and ionic strength effects have been included where possible.

The technique often utilized to designate the presence of a precipitate is filtration through a 0.45-μ filter. The question that arises in conjunction with these studies and many others is, How much solid can be formed and still pass through a 0.45-μ filter? If a molecule has a cross-sectional diameter of 10 Å (this is large for $ZnCO_3$), an aggregation of 10^2 molecules could probably pass through the filter. How rapidly the aggregates form has not been well established at this point; however, the rate may be fairly fast, as indicated by the formation of $BaSO_{4(s)}$ in the turbidimetric determination of sulfate. Of course the rate is highly dependent on mechanical agitation and ionic medium. In these studies as much as 20 mg of $ZnCO_3$ is predicted to be suspended per liter of solution. With small particle size this amount of material would probably be nondetectable visually.

A portion of the $ZnCO_3$ present in solution is due to the intrinsic solubility. This amount is equal to the product of the solubility product constant $K_{sp(ZnCO_3)}$ and the formation constant for $ZnCO_3 \cdot aq$, K_1. There is a sizable variation in the value of K_1, from 10^3 to 10^5, and thus the intrinsic solubility lies within the range of 10^{-7} to 10^{-5} M. The higher value is equivalent to 1.25 mg $ZnCO_3/l$. Comparison of this value with the ones in Table 3 provides an indication of the degree of supersaturation.

4. DISCUSSION

Analysis of the species distributions in Table 3 indicates that a sizable fraction of the total zinc is complexed by carbonate in test waters of pH 7 or higher. The formation of a comparable species in copper toxicity studies renders the metal nontoxic (Pagenkopf et al., 1974). Zinc appears to behave in the same way. Two other major conclusions can also be drawn from the results in Table 3. The first of these is that fish are more tolerant to zinc at lower than at high pH. For example, comparison of the results for tests F–8, 9, 10, 11, 16, and 17, where

Zn^{2+} is the dominate species, indicates that fathead minnows are approximately three times more tolerant to zinc at pH 6.1 than at pH 8.0. The second conclusion involves the amount of calcium and magnesium present in the system. As the hardness is increased, the toxicity of zinc decreases; see tests F-10 through F-15.

Precipitation or coagulation of mucus on the gills of the fish appears to alter the gas exchange process and as a consequence creates hypoxia (Burton et al., 1972). The exact mechanism that accounts for the zinc toxicity and the observed variation with pH and hardness is not known; however, these studies identify a qualitative relationship between the Lewis acid character of H, Ca, Mg, and Zn ions and the variation in toxicity. If the gills possess Lewis base character, these four cations can conceivably coordinate or at least adsorb on the gill surfaces. Comparison of the stability constants for interaction of these ions with acetate anion provides an indication of the relative adduct strength. The formation constants for H^+, Ca^{2+}, Mg^{2+}, Zn^{2+}, and Cu^{2+} with acetate are $10^{4.8}$, $10^{0.6}$, $10^{0.8}$, $10^{1.6}$, and $10^{2.4}$, respectively (Sillen and Martell, 1964). At a constant pH an increase in the hardness increases the tolerance to zinc; in addition the concentration of calcium and magnesium is approximately a factor of 10 greater than the zinc concentration. This is in qualitative agreement with the acetate stability constants, the zinc complex being 10 times more stable than the calcium and magnesium complexes. As the hardness increases, these ions may compete with zinc for the sites on the gills.

Competition between hydrogen and zinc is also possible. The difference in stability constants is approximately 10^3, and thus hydrogen ion should influence the toxicity when its concentration is 10^{-3} times the zinc concentration. This is observed over the pH range of 8 to 6. It is also interesting to note that comparable copper toxicity occurs when the copper concentration is approximately one-tenth the zinc concentration. Again, this is in agreement with the relative acetate stability constants, the copper complex being 10 times more stable than zinc.

It should be emphasized that the preceding discussion is qualitative. Nevertheless, it may facilitate interpretation of the influence of speciation on the toxicity of trace metals to fish. The correlation between the observed toxicity and the concentrations of Zn^{2+} and $ZnOH^+$ is not as good as was observed in the copper studies.

REFERENCES

Andrew, R. W., Biesinger, K. E., and Glass, G. E. (1977). "Effect of Inorganic Complexing on the Toxicity of Copper to *Daphnia magna*," *Water Res.*, 11, 309-315.

Burton, D. T., Jones, A. H., and Cairns, J., Jr. (1972). "Acute Zinc Toxicity to Rainbow Trout (*Salmo gairdneri*), Confirmation of the Hypothesis That Death is Related to Tissue Hypoxia," *J. Fish. Res. Board Can.*, **29**, 1463–1466.

Brungs, W. A. (1969). "Chronic Toxicity of Zinc to the Fathead Minnow, *Pimephales promelas* Rafinesqul," *Trans. Am. Fish. Soc.*, **98**, 272–279.

Cairns, J., Jr. and Scheier, A. (1957). "The Effect of Temperature and Hardness of Water upon the Toxicity of Zinc to the Common Bluegill (*Lepomis macrochirus* RAF)," *Notul. Nat.*, **299**, 1–12.

Davies, P. H., Goettl, J. P., Jr., Sinley, J. R., and Smith, N. F. (1976). "Acute and Chronic Toxicity of Lead to Rainbow Trout *Salmo gairdneri*, in Hard and Soft Water," *Water Res.*, **10**, 199–206.

Eaton, J. G. (1973). "Chronic Toxicity of a Copper, Cadmium, and Zinc Mixture to the Fathead Minnow, *Pimephales promelas* Refinesque," *Water Res.*, **7**, 1723–1736.

Garrels, R. M. and Christ, C. L. (1965). *Solutions, Minerals, and Equilibria.* Harper and Row, New York, 450 pp.

Garrels, R. M. and Thompson, M. E. (1962). "A Chemical Model for Sea Water at 25°C and One Atmosphere Total Pressure," *Am. J. Sci.*, **260**, 57–66.

Gubeli, A. O. and Ste-Marie, J. (1967). "Stability of Hydroxide Complexes and Metal Hydroxide Solubility Products. I: Silver and Zinc," *Can. J. Chem.*, **45**, 827–832.

Lloyd, R. L. (1961). "The Toxicity of Mixtures of Zinc and Copper Sulfate to Rainbow Trout (*Salmo gairdneri* Richardson)," *Ann. Appl. Biol.*, **49**, 535–538.

McLeay, D. J. (1975). "Sensitivity of Blood Cell Counts in Juvenile Coho Salmon (*Oncorhynchus kisutch*) to Stresses Including Sublethal Concentrations of Pulp Mill Effluent and Zinc," *J. Fish. Res. Board Can.*, **32**, 2357–2364.

Mount, D. I. (1966). "The Effect of Total Hardness on Acute Toxicity of Zinc to Fish," *Air Water Pollut. Int. J.*, **10**, 49–56.

Nehring, R. B. and Goettl, J. P., Jr. (1974). "Acute Toxicity of a Zinc Polluted Stream to Four Species of Salmonids," *Bull. Environ. Contam. Toxicol.*, **12**, 464–469.

Owen, B. B. and Gurry, R. W. (1938). "The Electrolytic Conductivity of Zinc Sulfate and Copper Sulfate in Water at 25°," *J. Am. Chem. Soc.*, **60**, 3074–3078.

Pagenkopf, G. K., Russo, R. C., and Thurston, R. V. (1974). "Effect of Complexation on Toxicity of Copper to Fishes," *J. Fish. Res. Board Can.*, **31**, 462–465.

Perrin, D. D. and Sayce, I. G. (1967). "Computer Calculations of Equilibrium Concentrations in Mixtures of Metal Ions and Complexing Species," *Talanta*, **14**, 833–842.

Pickering, Q. H. (1968). "Some Effects of Dissolved Oxygen Concentrations upon the Toxicity of Zinc to Bluegill (*Lepomis macrochirus* RAF)," *Water Res.*, **2**, 187–194.

Pickering, Q. H. and Henderson, C. (1966). "The Acute Toxicity of Some Heavy Metals to Different Species of Warm Water Fishes," *Air Water Pollut. Int. J.*, **10**, 453–463.

Pickering, Q. H. and Vigor, W. N. (1965). "The Acute Toxicity of Zinc to Eggs and Fry of the Fathead Minnow," *Prog. Fish Cult.*, **27**, 153–157.

Pytkowicz, R. M. and Hawley, J. E. (1974). "Bicarbonate and Carbonate Ion-Pairs and a Model of Seawater at 25°C," *Limnol. Oceanogr.*, **19**, 223–234.

Sillen, L. G. and Martell, A. E. (1964). *Stability Constants.* Chemical Society, Spec Publ. 17, London.

Skidmore, J. F. (1965). "Resistance to Zinc Sulfate by the Zebrafish (*Braehydanio resio* Hamilton-Buchanan) at Different Phases of Its Life History," *Ann. Appl. Biol.*, **56**, 47–53.

Smith, R. M. and Martell, A. E. (1976). *Crictical Stability Constants*, Vol. 4: *Inorganic Complexes.* Plenum Press, New York, 257 pp.

Solbé, J. F. de L. G. (1974). "The Toxicity of Zinc Sulfate to Rainbow Trout in Very Hard Water," *Water Res.*, **8**, 389–391.

Solbé, J. F. de L. G. and Flook, V. A. (1975). "Studies on the Toxicity of Zinc Sulfate and Cadmium Sulfate to Stone Loach (*Noemacheilus barbatulus* L.) in Hard Water," *J. Fish. Biol.*, **7**, 631–637.

Spehar, R. L. (1976). "Cadmium and Zinc Toxicity to Flagfish (*Jordanella floridea*)," *J. Fish. Res. Board Can.*, **33**, 1939–1945.

Sprague, J. B. (1964). "Lethal Concentrations of Copper and Zinc for Young Alantic Salmon," *J. Fish. Res. Board Can.*, **21**, 17–26.

Sprague, J. B. and Ramsey, B. A. (1965). "Lethal Levels of Mixed Copper-Zinc Solutions for Juvenile Salmon," *J. Fish. Res. Board Can.*, **22**, 425–432.

15

ZINC AND PLANTS IN RIVERS AND STREAMS

Brian A. Whitton

Department of Botany, University of Durham, South Road, Durham, England

1.	Introduction	364
2.	Features of Lotic Environments with High Zinc Levels	364
	2.1. Occurrence of zinc	364
	2.2. Physical features	365
	2.3. Chemical features	365
3.	Floristic Composition of Vegetation	370
	3.1. Experimental study	370
	3.2. Northwest Miramichi	370
	3.3. Europe	372
	3.4. New Lead Belt, Southeast Missouri	374
	3.5. Other regions	375
	3.6. Author's observations	375
4.	Taxonomic Summary	377
5.	Adaptation to High Zinc Levels	377
6.	Morphological and Physiological Aspects	381
7.	Factors Influencing Toxicity	382
	7.1. Physical	382
	7.2. Chemical	383
8.	Accumulation	384
	8.1. Field populations	384
	8.2. Factors influencing accumulation	386
9.	Occurrence of *Cladophora* and *Stigeoclonium*	386
10.	Use of Plants to Monitor Zinc in Rivers	387

10.1. Bioassays   388
10.2. Measurement of tolerance of organisms from nature   389
10.3. Accumulation   389
11. Discussion   **390**
Acknowledgments   **392**
References   **392**

1. INTRODUCTION

The aim of this chapter is to establish what is known about the effects of zinc on plants in flowing waters, and to suggest objectives for future research. An attempt is made to generalize regarding the influence of highly elevated levels of zinc, much as has been done for the influence of temperature in thermal springs by authors such as Brock (1969) and Castenholz and Wickstrom (1975). Data collected by the present author during visits to mining sites in the period from 1974 to 1978, and hitherto unpublished, have been included.

There are many accounts in the literature of streams and rivers carrying elevated levels of zinc. In some cases these also include observations on the plants present, their species composition, abundance, or mineral composition. It is, however, impossible in most studies dealing with species composition or abundance to be sure that any observed effects are due to zinc and not to some other metal or potentially toxic factor. Only when zinc has been added deliberately to flowing water or used in laboratory studies can we be reasonably certain that it is the agent responsible. The apparently universal occurrence of elevated levels of cadmium in streams and rivers concurrently with those of zinc makes it particularly difficult to separate the effects of these two elements. This point is not stressed throughout the chapter, but it should be assumed that, in all unqualified comments on zinc in field situations, some component of the effects observed may well be due to cadmium.

2. FEATURES OF LOTIC ENVIRONMENTS
WITH HIGH ZINC LEVELS

2.1. Occurrence of Zinc

The concentrations and forms in which zinc occurs in fresh waters are reviewed in Part I of this volume, but a brief summary is given here of the aspects likely to be of particular importance for interpreting observations on plants.

Zinc may be in true solution, colloidal, or particulate. Most records quoted are either for the total amount of metal or that passing through some sort of filter.

The median concentration of zinc in 726 filtered samples of water taken from rivers and lakes in the United States was found by Hem (1972) to be 0.02 mg/l. Much higher values may result from past or present mining, milling, smelting, or industrial activities. High values reported include 22.8 mg/l at pH 6.0 (after passage through a 0.2-μm filter) (Say et al., 1977), 105 mg/l at pH 4.9 (Tyler and Buckney, 1973), and 193 mg/l in an acid mine drainage (pH 2.9: Hargreaves, 1977). A value of 3610 mg/l (at pH 5.0) in a seepage on a French smelter trip (P. J. Say and the present author, unpublished) appears to be a current record, but it seems probable that many more sites could be found with such high levels. In any case, saturated solutions (varying in concentration according to pH) are presumably of frequent occurrence whenever sediments or soil rich in zinc, and normally kept moist, become dried out.

2.2. Physical Features

Sites with elevated levels of zinc in water show considerable variety. To give a general impression of the types of habitat for plants, observations in the literature are combined here with ones made by the present author. Details of the most important sites studied are summarized in Table 1. It is hoped that these will be of use to others planning research on zinc and lotic ecology.

Flowing waters rich in zinc may range from tiny seepages on tips such as are widespread in the Pennines (north England) and parts of the French Pyrenees, to small rivers (e.g., Gueule/Guel, Belgium/Netherlands; Innerste, Germany; Nent, England), to very large rivers (Lot, southwest France). Where adits drain deep strata, flow at source, and also temperature, may be relatively constant throughout the year. In contrast, the flows of seepages from tips may vary according to rainfall (Kramer, 1976) and may be only intermittent. In regions with markedly seasonal climates, flow may be restricted to one part of the year, as with some sites in southern France and Sardinia.

Because many seepages and small streams rich in zinc are associated with mining activities, trees and shrubs may be sparse in the vicinity, and consequently illumination is high and the water may be subject to marked diurnal temperature fluctuation in summer. The surrounds of many of the adits in the Harz are deeply shaded, however, and the entrances to large adits elsewhere often provide good sites for studying the gradation from high light to deep shade.

2.3. Chemical Features

The suspended matter in zinc-rich streams may include particles rich in zinc as the result of mining for sphalerite or other ores. In other situations zinc-rich

Table 1. Main Areas Studied by Author for General Observations Included in Text

Country	Mine or Area	Whether Mining or Smelting Now	Rivers and Streams with Elevated Zinc	General Features of Area Relevant to Aquatic Plants	Key Background References
Belgium/Netherlands	La Calamine	−	Hohnbach, Gueule/Guel	Calcareous and often highly polluted by sewage effluents	Heimans (1936) Ernst (1974b)
England	Alston Moor Orefield	+	Numerous adits and small streams to Derwent, Nent, Rookhope Burn	Wide variety of fast-flowing waters, mostly unshaded and in pH range 6.0 to 8.0	Dunham (1948) Raistrick and Jennings (1965) Say et al. (1977)
France	Huelgoat	−	D'Argent	Seepages in area with old mine tips	
	Largentière	+	La Ligne, La Lande	Calcareous region with adits and seepages reaching several small rivers; sites + and − sewage pollution	
	Pic d'Araillé	−	Ruisseau de Larrode	Adits with high metal levels, entering streams; unshaded, but deep winter snow	Ernst (1974b)
	Planioles (Figeac)	−	Drainages from tip enter fen woodland	Drainage highly calcareous and dependent only on tip; good site for ecosystem/accumulation studies at high pH	

Rennes	Ruisseau de la Douettée	−	Seepages in area with old mine tips	
Soussa	Tributary of Gave de Cauterets	−	Seepages from old smelter tip produce low-pH values and very high zinc; area now a camp-site ("La Galène")	
St. Salvy	Stream at Noailhac	+	New mine, so presumably high zinc in stream new	
Villemagne	Trevezel	−	Adit and seepages; main stream with heavy deposits from tip subject to extreme erosion	
Viviez-Decazeville	Seepages, tailings, effluents to Rieu Mort and Rieu Lot	+	Terrestrial vegetation showing some recovery from earlier influence of smelting, but area still subject to extreme pollution; Rieu Lot probably one of largest rivers in world where zinc is the main pollutant	Agence Financière de Bassin Adour-Garonne (1975)
Germany	Harz	Adits and seepages; feed streams, eventually to Innerste	+	Some adits deeply shaded, but main areas of seepages in full light; latter subject to disturbance in 1977 because tips being removed	Alicke (1974) Gundlach and Steinkamp (1973)
	(near) Altenbrück	Sülz	+	River with high zinc combined with varying levels of Sewage pollution	Morrison (1976) Schubert (1954)

Table 1. Continued

Country	Mine or Area	Whether Mining or Smelting Now	Rivers and Streams with Elevated Zinc	General Features of Area Relevant to Aquatic Plants	Key Background References
Italy	Iglésias	+	Streams near S. Giovanni, S. Benedotto	Extensive area heavily influenced by mining; high zinc streams both + and − suspended matter, mostly moderately calcareous; probably one of longest histories of mining in world	
United States	(near) Knoxville	+	Seepages, streams, and tailing ponds. eventually to Mossy Creek		
	New Lead Belt, Missouri	+	Effluents from tailings ponds and seepages around smelters	Pollution recent; main streams calcareous, mostly with some shade	See text
	Old Lead Belt, Missouri	−	Seepages from Elvins tailings, eventually to Flat River	Open area of great potential scientific interest for research on an extreme environment	Kramer (1976)
Wales	Cardiganshire	−	Adits and seepages, eventually to Ystwyth	Long history of mining, with many aquatic and terrestrial observations	See text
Ireland	Avoca	+	Acid mine drainages to Avoca	High zinc and copper; main river covered in iron oxide sediment	Platt (1975)

suspended matter arises *in situ* as the result of precipitation associated with chemical changes in the water. Zinc sedimentation from a water column may take place by binding onto clay particles (Pita and Hyne, 1975) or hydrous oxides of iron (Jenne, 1968). The precipitation of iron may lead to coprecipitation of other elements (Jackson and Nichol, 1975). At present it is far from easy to generalize regarding the extent to which such processes may lead to the deposition of particulate zinc on rock and plant surfaces, or the influence that plants themselves may have on such deposition. There are, in any case, no data comparing the influence of particulate versus soluble zinc on aquatic plants. In the River Avoca, Ireland, whose vegetation is discussed in Section 3.6, the entry of acid drainages leads to surfaces becoming smothered with iron-rich (29% Fe) sediment, yet almost all the zinc (1.34 mg/l at pH 4.2) remains in solution (present author, unpublished). In contrast, Harding and Whitton (1978) found that 70.3% of the zinc passing through the Derwent Reservoir, England, becomes deposited on sediments, although only 19% of the zinc in the inflow river would not go through a 0.2-μm filter. No studies have been made of the influence of submerged plants on the solubility of zinc, but it would be reasonable to suspect that a diurnal cycle in the levels of soluble zinc may occur in some rivers with pH values above about 7.0, as an indirect result of the influence of plant photosynthesis on carbon dioxide levels and hence pH.

The levels of zinc present in solution are markedly influenced by pH (Jurinak and Inouye, 1962) and alkalinity (Ernst et al., 1975), and, as might be expected, zinc enrichment of springs associated with the oxidation of sulfide ores to sulfate tends to be much higher than that of springs associated with other minerals. The data on long-term changes in the zinc levels of the various types of spring are negligible, so it is not possible to speculate on the extent to which a specialized vegetation may indicate the present water chemistry or is a reflection of more extreme environments in the past. It would appear that streams and rivers with high levels of zinc always show the presence of other chemical factors that may have marked effects on plants. The association of cadmium with zinc has already been mentioned, but elevated levels of other heavy metals such as lead or nickel, or of fluoride, may also occur. Although the association of zinc and cadmium is general, the ratios of the two may differ widely, as may the ratios of each element in the water column as compared with the stream sediment.

It is possible to do little more than guess at what other factors may be of widespread significance for plants in zinc-rich sites. It seems probable, however, that oxygen levels will often be low near the source of seepages and springs, especially those associated with sulfide ores. Kramer (1976) reported a range of 2.62 to 3.67 mg O_2/l (at about 15°) in a seepage at Flat River Creek in the Old Lead Belt, Missouri.

3. FLORISTIC COMPOSITION OF VEGETATION

3.1. Experimental Study

Only one experimental study of the influence of zinc on lotic vegetation has been reported, that of Williams and Mount (1965) on periphyton communities developing in outdoor channels. As there were only four channels, the results are somewhat difficult to interpret. Some observations, however, seem directly attributable to the effect of increased zinc levels, and also resemble observations made at some sites with high zinc levels.

The photosynthetic organisms highly tolerant of zinc included:

Myxophyta	Various, especially narrow sheathed forms
Euglenophyta	*Euglena acus, Trachelomonas volvocina*
Chrysophyta	*Chrysococcus major*
Bacillariophyta	*Cymbella tumida, Nitzschia linearis, Synedra ulna*
Chlorophyta	*Chlamydomonas snowii, Oocystis lacustris, Spirogyra* sp.

The control channel supported the largest number of species, the main dominant, *Cladophora glomerata,* being absent from all other channels. Probably the absence of some species from the other channels was simply an indirect result of the absence of *Cladophora.* The composition of the two channels with the highest zinc levels was, however, quite different, these being dominated by a mat with blue-green algae, sheathed bacteria, and fungi abundant. The organisms producing the dominant populations in these channels seemed to have relatively heavy secretions of slime. These included the flagellate *Chlamydomonas snowii,* which, although gelatinous, moved up and down in the interstices of the mat as the light penetration changed, going to deeper layers during periods of intense solar radiation. Two fungal genera, *Alternaria* and *Leptomitus,* dominated the community with the highest zinc, together with a number of imperfect fungi.

3.2. Northwest Miramichi

Among field studies of mining pollution and lotic plants, one of the most detailed is that on the Northwest Miramichi River system in New Brunswick, Canada (Besch and Roberts-Pichette, 1970; Besch et al., 1972). Since both zinc and copper were present at elevated levels, it is not possible to quantify the role of zinc alone. The sites provided a variety of conditions, including one where zinc ranged from 6.4 to 65.5 mg/l and copper from 0.33 to 12.1 mg/l (at pH

4.0 to 5.5). The composition and the abundance of riparian vascular plants were influenced markedly by the metal pollution. None was actually favored by the pollution; all showed some damage or decrease in abundance. Besch and Roberts-Pichette distinguished three different degrees of severity:

1. Extremely high pollution: bank gravels remain barren.
2. Medium to high pollution: plant cover is reduced on the bank gravels, with *Equisetum arvense* the most resistant species and various monocotyledons (mainly Cyperaceae and Gramineae) also present;
3. Low pollution: plant cover on bank gravels slightly or not at all reduced; dicotyledons present; trees and shrubs show no signs of damage.

The authors concluded that submerged species are the most sensitive of all vascular plants since they were absent from one site that received only slight metal pollution. This site did receive strong acid pollution, but the authors pointed out that submerged species can occur in water at even lower pH values (Vallin, 1953).

A study (Besch et al., 1972) of the same river system showed that species diversity was less at polluted sites; at the two sites with the lowest diversity only *Mougeotia* sp. and diatoms were frequent. The tolerances of some of the more common diatoms in this river system are summarized in Table 2. The authors

Table 2. Tolerance of Common Diatoms to Combined Zinc and Copper in the Northwest Miramichi Area

Zinc Concentration in Water (mg/l)	Species
> 10	*Achnanthes microcephala, Eunotia exigua, Fragilaria virescens, Pinnularia interrupta* f. *biceps*
1–2	*Synedra ulna*
~ 1	*Fragilaria crotonensis, F. intermedia, Synedra rumpens*
0.1–0.2	*Achnanthes linearis, A. minutissima, Anomoeneis serians, Tabellaria fenestrata, T. flocculosa*
< 0.1	*Achnanthes deflexa, Ceratoneis arcus, Cymbella affinis, C. microcephala, Diatoma hiemale, Eunotia pectinalis, E. veneris, Gomphonema angustatum, G. intricatum* f. *pumila*

[a] Data of Besch et al. (1972).

emphasized that this indicator system applied only to streams of the "soft water zone" of the Atlantic Provinces of Canada, which do not receive organic pollution.

3.3. Europe

There are numerous areas in Europe with streams and rivers enriched by zinc as the result of past or present mining activities. The guide to *Carte Métallogenique de l'Europe*, for instance, lists 140 sites for zinc and/or lead mining on sheet 4 (Belgium, England, France, Ireland, northern Spain, Wales), and these do not include major smelters away from mining (e.g., Viviez, France). At some areas with deep ores where mining has commenced recently, all zinc enrichment is apparently concentrated on a single stream (e.g., St. Salvy, France). In other areas where deposits are more accessible and have been mined over many centuries, there may be dozens of streams with elevated zinc levels (e.g., Harz).

At no European site has any general description been made of the lotic vegetation, but many accounts make some mention of the flora of streams and rivers with elevated zinc levels. Carpenter (1924) reported one of the first such studies in a survey of the River Rheidol (Cardiganshire, Wales) below old lead mines, a river shown by later authors to be polluted by high levels of zinc. In 1919 the vegetation of the river below the mine was limited to slight coatings of mosses and liverworts, together with some growths of the red algae *Batrachospermum* and *Lemanea*. Most of the mines influencing the river closed in 1921, and by the next year there was a partial restoration of the flora. In a later study of the influence of this mine and others in the area, Reese (1937) was more critical as to what factors might actually be responsible for the observed differences. For instance, in one case she suggested that it was the silt deposited on the stream bed that was the chief factor involved in determining these differences. Other aspects of these early studies have been reviewed by Whitton (1970b).

More recent studies of the Cardiganshire rivers have been made by Jones (1958) and McLean and Jones (1975). The former reported that the Ystwyth was still seriously polluted by zinc 35 years after mining had ceased, with 40 to 90 mg Zn/l present at one site. Mosses and the typical rich growths of green algae and diatoms had disappeared. The account of McLean and Jones indicates that the levels of zinc had probably dropped by 1972-1973, with 4.1 mg total Zn/l and 2.1 mg soluble Zn/l at the site with the highest concentrations. Nevertheless, floristic differences were still obvious when the more polluted sites were compared with less polluted ones. Abundant growths of *Hormidium rivulare* growing alone in this region could serve as an indication of high zinc levels in the water. The only bryophyte at the most polluted site was *Scapania undulata*, whereas the flora at other sites included *Fontinalis squamosa, Rhacomitrium aciculare, Eurhynchium* (= *Rhynchostegium*) *riparioides,* and *Scorpidium scorpioides.* Filamentous greens were the only macroscopic algae in the more polluted streams, with *H. rivulare* the most abundant, followed by *Ulothrix* sp. Species of *Microspora, Spirogyra,* and *Zygnema* were absent from these areas.

A brief note on the algal flora of a stream in the Northern Pennine Orefield showing a gradient of zinc concentrations (30 to 1.5 mg/1 passing through a filter) was given by Say and Whitton (1978). On passing downstream the total number of species recorded over 2 years increased as the level of zinc decreased. Some species (e.g., *Neidium alpinum, Plectonema gracillimum*) were found only at the top end of the gradient, whereas others (e.g., *Cymbella ventricosa, Nitzschia amphibia*) grew only toward the lower end. Certain filamentous algae (*Hormidium rivulare, Ulothrix moniliformis, Mougeotia* sp. about 6 μm wide) were present at all levels of zinc. In general, *H. rivulare* is the most widespread dominant in streams with the highest zinc levels (Whitton, 1977), though *Mougeotia* sp. is usually dominant in calcareous waters (at high pH values). Although absent at the highest zinc levels, gelatinous chrysophytes are more abundant in streams of the Northern Pennine Orefield with intermediate zinc levels than at low levels. In several zinc-polluted streams which carry a reduced algal flora, *Chrysonebula* sp. may form conspicuous growths throughout the summer, whereas in other steams macroscopic growths of this alga are restricted to late winter (Whitton, 1975). Reduced competition with other algae growing in and over its mucilage may be the key factor influencing its development here.

Although the levels of zinc in acid mine drainages are often very high, the literature is not reviewed here, as it is uncertain just what contribution this zinc makes to the ecology of drainages. At the site with the highest zinc recorded (193 mg/1, pH 2.9), Hargreaves et al. (1975) found four algae: *Euglena mutabilis, Eunotia exigua, Chlamydomonas applanata* var. *acidophila, Zygogonium ericetorum.* These are among the most widespread species in all acid drainages, not just those with high zinc levels. The possibility should also be borne in mind that elevated zinc levels may play a role in some streams subject to pollution by acid rain.

Although there is an extensive literature on the phytosociology of terrestrial plants on zinc-rich soils, little mention is made of moister habitats. Heimans has, however, described in a series of papers (e.g., 1961) the vegetation on zinc-containing soils along the River Geul in southern Netherlands. The part of the valley occupied by the zinc plants, such as *Viola calaminaria,* was sharply delimited by the line marking the highest level to which the Geul rises in flood. The soils here were rich in zinc because of earlier mining and smelting activities enriching the river. It is uncertain to what extent these communities still exist in view of the present highly eutrophic condition of the river and fertilization of the fields, but Ernst (1964) listed *V. calaminaria* among the most frequent species on the banks of this river at Epen.

3.4. New Lead Belt, Southeast Missouri

Observations on waters with elevated levels of zinc (and lead) in the New Lead Belt, Missouri, have been included in various accounts by researchers at Rolla. The present author was shown many of the sites in June 1977 by two of these, N. L. Gale and B. G. Wixson, and further observations made then are included here. The normal streams and small rivers typical of this part of southeast Missouri are in general moderately calcareous, not seriously contaminated by sewage effluents, and probably very rich in algal species. They were apparently not subject anywhere to zinc enrichment before 1955, when initial lead discoveries were verified, and most contamination started only in the late 1960s.

Gale et al. (1973a) reported that a stalked diatom, *Cymbella,* formed "blooms" in spring in virtually all streams in the area, both natural and polluted. In many polluted streams these diatom growths formed the first stages of characteristic slime mats. Wixson and Bolter (1972) reported that three diatom genera are useful indicators of mine water discharge, *Synedra* and *Navicula* being tolerant and *Cymbella* intolerant. This different interpretation of the role of *Cymbella* may perhaps have been due to the fact that there is a range of species with differing behavior or to the fact that observations were made in summer rather than spring. Prolific growths of grey slime mats, composed of *Sphaerotilus,* algae, protozoa, and other organisms, developed on the bottom of streams receiving effluents. Once dolomite and other materials became trapped in the slime, photosynthesis was prevented, and the mats became anaerobic, eventually detaching from the substratum.

In some polluted streams algal growths developed to such an extent as to become a problem (Gale et al., 1973a); in addition to a range of diatoms, these authors noted *Cladophora, Oscillatoria, Mougeotia, Zygnema,* and *Spirogyra* as important. The algae present in June 1977 among flocs in a heavily contaminated stream below the smelter at Glover included the following, in decreasing order of abundance: *Ulothrix* sp., 9 μm wide; *Phormidium* and sheathed Oscillatoriaceae forms (all very narrow); *Cosmarium* sp.; *Mougeotia* sp., 7 μm; *Cymbella* spp.; *Scenedesmus acuminatus; Euglena* sp.; *Ankistrodesmus* spp.; and small cryptomonads. In comparison with an upstream, less polluted site, *Synedra ulna* and *Mougeotia* sp., 20 μm, were absent. A downstream site where a sewage effluent had become mixed with the metal-rich water showed some differences from the water contaminated only by metals: unicellular forms of *Scenedesmus* were dominant and a palmelloid *Chlamydomonas* frequent; *Mougeotia* sp., 6 μm, and *Cosmarium* sp. were absent.

Other studies have focused on the vegetation of a meander system built in 1972 below a large tailings pond at the Buick Mine (Jennett and Wixson, 1975). Here macroscopic species form the bulk of the vegetation during the growing season, with *Cladophora, Rhizoclonium, Hydrodictyon, Spirogyra,* and *Potamo-*

geton common (Gale and Wixson, 1977). The occurrence of the first two genera in water with elevated zinc levels is discussed below.

In contrast to the vegetation of polluted streams, Gale et al. (1973b) found that the algal growth in tailings ponds was consistently restricted to single-cell species, with *Chlorella* and *Chlamydomonas* dominant. In June 1977 the mill effluent pond at Fletcher had a sparse plankton dominated by unicellular *Scenedesmus.* The submerged surfaces of twigs and other debris sometimes had a cover of *Phormidium, Euglena mutabilis,* and *Nitzschia* spp.

It is difficult to judge to what extent the observations made on the polluted streams in the New Lead Belt reflect the effects of zinc and other heavy metals and to what extent they are due to enrichment with inorganic nutrients, especially phosphate. It is clear that in some cases there has been a marked change in the algal flora, and probably also a decrease in species diversity. The development of high-standing crops has no doubt been favored by high phosphate levels. The possibility should also be considered that any decreases among grazers such as molluscs and larger crustacea as a result of pollution have not been compensated by increases in more resistant grazers.

3.5. Other Regions

There are few data for streams and rivers outside Europe and North America. Apart from studies on acid mine drainages and on rivers grossly polluted by various effluents, where zinc is just one of many potential toxic substances, the only reports are from India (Gopal et al., 1975; see below) and Australia. Weatherley et al. (1967) noted several emergents—*Typha angustifolia,Phragmites communis* (= *australis*), and species of *Juncus*—in a stretch of the Molonglo River (near Canberra, Australia) that was highly polluted by zinc. In creeks entering the South Esk River, Tasmania, and polluted by high levels of Zn, Cu, and Pb, benthic algae were uncommon, except where diluted by unpolluted water (Tyler and Buckney, 1973). At one such site, *Ulothrix* sp. grew healthily in a pool containing 4 mg Zn/l.

3.6. Author's Observations

A few further observations by the present author are added here. The documentation for them is included in a series of papers in preparation, and the raw data are held on computer files at Durham University. They are based on visits to the areas in Table 1, together with other less important sites in Europe, and include 15% of the sites on sheet 4 of the *Carte Métallogenique de l'Europe* (see Section 3.3).

Photosynthetic organisms are visually obvious in summer at almost every site, whatever the level of zinc. In some seepages and springs with very high zinc levels, such obvious growths consist of only three to six species. Filamentous green algae are usually the dominants near the source, but other communities also occur. One especially characteristic community, occurring in seepages below old tips with pH values within or near the 6.5 to 7.0 range, has the same dominants and almost the same associated species at sites in both Europe and the United States. The dominants are *Plectonema gracillimum* and *Dicranella varia*, the latter existing almost entirely as protonema. Associated species include *Achnanthes minutissima*, *Mougeotia* sp., and *Stichococcus* sp. This community exists, for instance, adjacent to the well-known sites for terrestrial phytosociological studies at Einersberger Zentrale in the Harz (see Ernst, 1974b, p. 5) and at the site in the Old Lead Belt, Missouri, whose chemistry is described by Kramer (1976).

Rivers and streams combining high levels of zinc and extensive deposition of iron oxide vary markedly in the extent to which they develop plant growths. In some situations where zinc and cadmium are probably the only soluble substances at levels potentially toxic to plants, the algal standing crop may be so high that the whole streambed is covered during daylight with oxygen bubbles. On the other hand, in a few relatively fast-flowing streams and rivers combining high levels of Zn, high levels of other heavy metals like Cu, Cr, or Ni, and the deposition of iron oxide, visually obvious growths of photosynthetic organisms may be nearly or completely absent. Such sites include parts of the Rieu Mort (France), the River Avoca (Ireland), and the "Iglésias Mining River" (Sardinia); see Table 1. Ponds and stretches of river that are slow flowing, but otherwise combine the same environmental parameters, may lack visually obvious growths of algae or mosses, but apparently always have a cover of one or a very few angiosperms. In Europe, at least, such sites are usually dominated by *Phragmites australis* and/or *Typha latifolia* (e.g., lagoons near Iglésias).

There are apparently no accounts of the influence of zinc (as opposed to mixtures of toxic substances) on the development of "sewage-fungus" communities in rivers. Inspection by the present author of the Zellbach, a tributary of the Innerste, in September 1977 showed that the vegetation of a fast-flowing stretch of this river combining high zinc (1.6 mg/1 passing through a 0.2-μm filter, pH 7.3) and heavy sewage pollution consisted amost entirely of heterotrophs. Observations on other sites with high organic pollution, but without very high zinc levels, suggest that the near absence of algae was exceptional. In the absence of analyses of the organic content of the water, such an observation is of course highly subjective. The possibility that high zinc levels may extend considerably the heterotroph-dominated zone below a sewage effluent is nevertheless of sufficient interest to require further study.

4. TAXONOMIC SUMMARY

All the larger taxa of photosynthetic organisms of any quantitative importance in fresh waters have a few species tolerant of elevated zinc levels (Table 3). This is in marked contrast to thermal alkaline springs, which are dominated by one phylum, the blue-green algae (Castenholz and Wickstrom, 1975), or thermal acid springs, where only a single species, *Cyanidium caldarium,* is capable of growing at the higher temperatures. The situation in highly acidic streams at lower temperatures (Hargreaves et al., 1975) is more like that in high-zinc streams, although blue-green and perhaps also red algae are absent in the former.

5. ADAPTATION TO HIGH ZINC LEVELS

Are the species present in flowing waters with high zinc levels represented by population genetically adapted for tolerance, or are they species that possess this tolerance whether or not they actually grow in an environment with a high zinc concentration? In the case of three species of *Hormidium, H. rivulare, H. flaccidum* and *H. fluitans,* the algae growing at high zinc levels have acquired genetic tolerance of these levels (Say et al., 1977). All three species are widespread both in waters that are free of zinc pollution and in waters that have very high zinc levels. Although populations at high zinc levels had aquired tolerance, laboratory assays did not show any detectable increase in the zinc requirement for optimum growth. It was not possible to give an exact threshold of zinc above which populations were adapted, since other environmental factors influence the toxicity of the zinc. It seems reasonable, however, to conclude for *H. rivulare* that with increasing levels of zinc above 0.2 mg/1 there is an increasing likelihood that a population will show genetic adaptation for tolerance, and that nearly all populations with pH $\geqslant$ 5.0 and zinc $\geqslant$ 0.8 mg/1 are so adapted. The authors pointed out that their assay techniques would not be sensitive enough to show adaptation to levels lower than 0.2 mg 1 Zn/l, should this occur. Levels of zinc in rainfall often approach the threshold that might be expected to lead to adaptation in *Hormidium* populations growing on rocks and largely dependent on rain for their water supply.

The situation with *Stigeoclonium tenue* is similar (Harding and Whitton, 1976). The alga is widespread and often abundant in waters with and without zinc pollution. At sites with mean zinc levels in water (passing through a filter) of about 0.2 mg/l and above, populations showed increased tolerance to the metal in comparison with populations from sites with lower zinc levels. Experiments with *Microthamnion strictissimum* and *Mougeotia* sp. have both indi-

Table 3. Examples of Photosynthetic Plants with Populations at Lotic Sites Where Levels of Zinc Are Sufficiently High to Prevent Growth of Many Species That Would Otherwise Be Expected to Grow There

Organism	Location
Myxophyta	
Phormidium autumnale Gom.	Riou Mort, France
Plectonema gracillimum	Elvins tailings, Old Lead Belt
Schizothrix—various forms	Adit at Villemagne, France
Rhodophyta	
Lemanea fluviatilis Ag.	Harding and Whitton (1978)
Euglenophyta	
Euglena mutabilis Schmitz	Near outflow of Fletcher pond, New Lead Belt
Chrysophyta	
Chrysonebula (near) *holmesii* (Lund)	Whitton (1965)
Hydrurus foetidus (Vill.) (Kirchn.)	Gillgill Burn, Cumbria
Xanthophyta	
Tribonema sp.	Tailings pond stream, Viviez, France
Cryptophyta	
Several small-celled flagellates	Stream by Asarco smelter, New Lead Belt
Bacillariophyta	
Achnanthes microcephala (Kütz.) Grunow	Besch et al. (1972)
Achnanthes minutissima	Elvins tailings, Old Lead Belt
Caloneis bacillum (Lyngb.) Heib.	Huelgoat
Diatoma hiemale (Lyngb.) Heib. var. *mesodon* (Ehr.) Grunow	Flush at Carnoet, Huelgoat
Eunotia exigua (Bréb.) Grunow	Tigroney, Avoca
Eunotia tenella (Grunow) Hust.	Gilgill Burn, Cumbria
(*Gomphonema parvulum* (Kütz.) Grunow	River Sülz
Neidium alpinum	Say and Whitton (1978)
Pinnularia microstauron (Ehr.) Cleve	Tigroney Grass Level, Avoca
Pinnularia subcapitata Greg.	River Laute, Harz
Conjugatophyta	
Species of *Cosmarium, Cylindrocystis, Mesotaenium, Mougeotia, Spirogyra*	
Chlorophyta	
Chlamydomonas—various species, usually in palmelloid condition, or at least with a	

378

Table 3 Continued

Organism	Location
gelatinous surrounding layer	
Draparnaldia glomerata (Vauch.) Ag.	Gillgill Burn, Cumbria
Haematococ cus pluviatilis Flot. em Wille	La Ligne Rieu, Largentière
Hormidium flaccidum A. Braun	Say et al. (1977)
Hormidium fluitans (Gay) Heering)	Say et al. (1977)
Hormidium rivulare Kütz.	Say et al. (1977)
Microspora sp.	
Microthamnion kuetzingianum Näg.	Gillgill Burn, Cumbria
Stichococcus sp.	Einersberger Zentrale, Harz
Stigeoclonium tenue Kütz.	Harding and Whitton (1976)
Lichens	
Verrucaria sp.	
Mosses	
Dicranella varia (Hedw.) Schimp.	Seepage on tip near Knoxville
Ditrichum cylindricum (Hedw.) Grout Grout	Seepage at La Galène, Soussu, France
Philonotis fontana (Hedw.) Brid.	Shimwell and Laurie (1972)
Pohlia carnea (Schimp.) Lindb.	Huelgoat
Rhynchostegium riparioides (Hedw.) C. Jens.	River Sülz, Germany
Liverworts	
Scapania undulata (L.) Dum	McLean and Jones (1975)
Solenostoma sphaerocarpum (Hook.) Steph.	Stream on Pic d'Arrailé, France
Vascular cryptogams	
Equisetum arvense L.	Adit at Largentière
Dicotyledons	
Rumex acetosa L.	Einersberger Zentrale, Harz
Monocotyledons	
Agrostis stolonifera L.	Einersberger Zentrale, Harz
Juncus spp.	
Phragmites australis (Cav.) Trin.	Iglésias
Scirpus americanus Pers.	Elvins tailings, Old Lead Belt
Typha anqustifolia L.	Weatherley et al. (1967)
Typha latifolia L.	Iglésias

cated that populations of these species taken from high-zinc sites have acquired genetic tolerance (present author, unpublished). It thus seems probable that most filamentous green algae growing at sites with high zinc levels will prove to be genetically tolerant populations of species which normally do not show such tolerance. There have been no parallel studies on freshwater diatoms, the phylum with the largest total of lotic species at high-zinc sites, but Jensen et al. (1974) showed significant intraspecific differences in the tolerance of the marine diatom *Skeletonema costatum* to zinc pollution.

Bradshaw (1975) commented that no species has yet been found which is preadapted to heavy metals and possesses tolerance throughout all its populations. He stressed that species that occupy toxic-metal-contaminated soils apparently do so only because they evolve tolerance. That this will prove to be universal for angiosperms is, however, doubtful, for McNaughton et al. (1974) provided what appears to be convincing evidence for tolerance in *Typha latifolia* to combined Zn, Cd, and Pb without the evolution of tolerant populations. In view of the biochemical and cytological diversity of algae and other plants of waters with high zinc levels, it would be rash to assume that evolution of tolerant populations is always a requirement for success. The red alga *Lemanea fluviatilis* is one species that requires critical study, since field transplants (Harding, 1978; Harding and Whitton, 1978) and laboratory growth studies (present author, unpublished) both indicate that plants from a low-zinc site ($\bar{x}$ = 0.02 mg Zn/1) can tolerate much higher levels of zinc.

There are no data for flowing waters of the sort reviewed by Bradshaw (1975) for terrestrial plants to indicate how long it may take to evolve tolerant populations. It would certainly be interesting to compare the flora of regions where waters have probably been in contact with high levels of zinc for thousands of years, such as parts of southwest Sardinia, with that of ones which have only recently become polluted as the result of deep ores being brought to the surface in new mine operations.

Laboratory populations of algae sometimes show morphological variation according to the level of zinc present, and in at least one case the response to high zinc might be considered a form of environmental adaptation (see Section 6). Another instance may be that reported by De Filippis and Pallaghy (1976c). Cultures of *Chlorella* sp. (Emerson strain) incubated with ^{65}Zn/1 mM ZnCl$_2$ developed tolerance after approximately 40 cell divisions, and almost regained the rates of cell division shown by control cultures. Tolerance was accompanied by an inhibition of a temperature-sensitive component of zinc uptake and by a reduction in the number of exchange sites available for zinc in the cell walls. The authors apparently assume that tolerance was simply an environmental response, although, with the extremely high level of zinc used, the possibility of genetic adaptation should not be ruled out.

6. MORPHOLOGICAL AND PHYSIOLOGICAL ASPECTS

Most species producing large populations in high-zinc streams appear to be morphologically similar to populations from low-zinc streams, but there are some differences. An increased frequency of geniculations has been observed in some populations of *Hormidium rivulare* (Say et al., 1977), as both an environmental and a genetic response to high zinc. These structures consist of two adjacent bends in a filament surrounded by mucilage. Several of the floristic accounts reviewed above mention mucilaginous species in high-zinc streams, where they seem rather more frequent than in otherwise similar streams lacking zinc enrichment. Most forms of blue-green algae are sheathed, belonging to either *Plectonema* or *Schizothrix,* although unsheathed forms can occur, for example, *Phormidium autumnale* from Riou Mort (15.7 mg Zn/l passing through a filter, pH 6.7) (present author, unpublished).

In laboratory culture of *Stigeoclonium tenue,* basal growth increasingly predominates over "upright" growth as zinc concentrations are increased (Harding and Whitton, 1977). At the field site with the highest zinc level (20 mg/l) at which this alga was found by Harding and Whitton (1976), it existed in a predominantly basal form. In culture at lower zinc levels it developed typical upright growth. Among mosses the proportion of protonema to adult gametophyte is often much greater in high- than in low-zinc streams (present author, unpublished).

A general review of the influence of zinc on the physiology of aquatic plants is outside the scope of this chapter, but key observations are mentioned briefly in order to consider possible implications for the ecology of streams and rivers. Zinc is apparently essential for all plants (Eyster, 1964), although in most cases this requirement is fulfilled by low environmental levels. For instance, stocks of *Stigeoclonium tenue* could be maintained in a medium with less than 0.002 mg Zn/l (Harding and Whitton, 1976). No zinc-tolerant population of either *S. tenue* or *Hormidium rivulare* has been found with an increased requirement for zinc. However, a population of *Euglena mutabilis* taken from a stream with 1.1 mg Zn/l (at pH 2.6) showed increased total growth in culture at all pH values between 1.75 and 6.0 when the level of zinc was raised from 0.08 to 1-10 mg/l (J. W. Hargreaves and the present author, unpublished). There are no records of zinc deficiency for aquatic plants, but no studies have been carried out on rivers where zinc levels are known to be especially low.

In a study of phytoplankton photosynthesis in two Swiss lakes, Gächter (1976) showed that, if the background zinc level was not increased by more than 5×10^{-8} mol Zn/l (an approximate doubling), photosynthesis was not affected. However, when this concentration of zinc was included with very low concentrations of other heavy metals, photosynthesis was significantly reduced because of a synergistic effect of the combined metals. De Filippis and Pallaghy (1976b) found that 1 mM ZnCl$_2$ led to a reduction in chlorophyll during the early stages of growth in culture of *Chlorella* sp. (Emerson strain). Nash (1975) reported that terrestrial lichen photosynthesis was depressed by much lower concentrations of zinc than those affecting respiration.

Although photosynthesis may be relatively sensitive to zinc, it is clear that some aquatic plants carry out photosynthesis under conditions of high external zinc. It is not so clear whether the same is true for nitrogen fixation. The only published record for a heterocystous blue-green alga from such an environment involves *Nodularia spumigena* (Gopal et al., 1975). This alga formed a covering on moist soil a short distance away from the effluent of a zinc smelter at Debari, India. In culture the alga tolerated up to 2 mg Zn/l, although any addition above the basal level to the medium led to a decrease in total yield. Gopal et al. nevertheless suggested that this strain was more tolerant than other blue-green algae (from culture collections). Although there are no other accounts of heterocystous algae from high-zinc environments, *Anabaena, Nostoc,* and *Scytonema* are abundant on high-copper soils in Rhodesia (Wild, 1968).

Non-heterocystous blue-green algae are widespread in streams with high zinc levels, especially where the pH is above 6.0 (present author, unpublished). Most forms are sheathed (see above), and so might be suspected of being able to carry out nonheterocystous nitrogen fixation. Assays for nitrogen fixation at Einersberger Zentrale by thick growths of *Plectonema gracillimum,* perhaps the most widespread blue-green alga in high-zinc streams, proved, however, to be negative (D. W. Lorsch and the present author, unpublished). It therefore seems probable that nitrogen fixation by blue-green algae is unlikely to be important in high-zinc streams. This may perhaps indicate that in such waters the levels of combined nitrogen are relatively high, rather than implying any selective effect of zinc on nitrogen fixation, but the data are too few to comment further.

7. FACTORS INFLUENCING TOXICITY

7.1. Physical

Patrick (1971) reported that the toxicity of zinc to *Nitzschia linearis* in culture increased as the temperature rose from 22 to 30°. Toxicity also increased with rising temperature in *Cyclotella meneghiniana,* but it decreased in *Scenedesmus quadricauda* and *Chlamydomonas* sp. (Cairns et al., 1978). Fisher and Wurster (1973) found three species of phytoplankton to be most sensitive to chlorinated biphenyls when existing in suboptimal environments, but the limited data for zinc and temperature do not indicate any such obvious pattern. Although there have been no accounts of the influence of light on the toxicity of zinc, Whitton (1968) reported for three other heavy metals and *Anacystis nidulans* that each metal was more toxic in the dark, perhaps as a result of increased extracellular production in the light. If so, this would contrast with observations of De Filippis and Pallaghy (1976a) on *Chlorella* sp. (Emerson strain) that zinc appeared to inhibit the release of extracellular glycollate.

7.2. Chemical

The toxicity of zinc to a population of *Hormidium rivulare* isolated from an acid mine drainage (pH 3.1: Hargreaves and Whitton, 1976a) was least at the optimun pH for growth, 3.5 to 4.0 (Hargreaves and Whitton, 1976b); toxicity increased markedly at higher pH values. Nevertheless, the influence of pH on zinc toxicity was apparently less than that on copper toxicity. Although copper was much more toxic than zinc at pH 6.0, zinc was more toxic than copper at pH 3.5. A zinc-sensitive and a zinc-tolerant population of the same species isolated from streams with pH values of 4.4 and 6.8, respectively, both showed markedly increased toxicity of zinc with increased pH (Say and Whitton, 1977). Gächter (1976) found that zinc toxicity to phytoplankton photosynthesis in the Alpnachersee was highest when the pH was highest (pH 8.7). The influence of pH on zinc toxicity to *Stigeoclonium tenue* differed, however, at least over the pH range 6.1 to 7.6 (Harding and Whitton, 1977). The only detectable effect on a sensitive population (from pH 7.4) was a slight decrease in toxicity between pH 7.1 and 7.6. For a tolerant population (from pH 7.7) there was an obvious decrease in toxicity over the range 6.6 to 7.6. Probably this was not simply due to precipitation at higher pH values, as it was evident when zinc passing through a 0.2-μm filter was used as a criterion, provided that the level of calcium in the medium was kept down to 1 mg/1.

By comparing the laboratory tolerance to zinc of isolates of three species of *Hormidium* from a wide range of stream sites with the water chemistries at those sites, Say et al. (1977) were able to suggest two groups of chemical factors that influence the toxicity of zinc in the field. Factors apparently reducing zinc toxicity include magnesium, calcium, various hardness factors, and, for *H. rivulare* at least, an increase in phosphate and a decrease in pH. Cadmium and lead appear to increase toxicity. The influence of calcium, phosphate, pH, and cadmium on *H. rivulare* was later studied experimentally (Say and Whitton, 1977), and in each case the result corresponded to that suggested by analysis of field data, as already described for pH. Magnesium and calcium both reduced the toxicity of zinc, but the behavior of the two elements suggested that the mechanisms by which they did this might not be the same. Whereas calcium reduced the toxicity of zinc to sensitive and tolerant populations, magnesium was relatively much more effective for the tolerant populations, a result similar to that found by Harding and Whitton (1977) for *Stigeoclonium tenue*. Inspection of the data for the streams from which the algae were taken does not indicate that this difference can be interpreted as an artifact associated with the varying magnesium levels in these streams. Accounts in the literature suggest that calcium reduces the toxicity of zinc to most aquatic organisms (see Whitton and Say, 1975).

Increased phosphate levels led to a marked reduction in zinc toxicity to zinc-tolerant populations of *Hormidium rivulare*, moderate reduction in toxicity to zinc-sensitive *H. rivulare* (Say and Whitton, 1977) and zinc-tolerant *Stigeoclonium tenue* (Harding and Whitton, 1977), but no detectable effect on zinc-sensitive *S. tenue*. Phosphate at high concentrations, but not nitrate, reduced the toxicity of zinc (at pH 6.0) to laboratory cultures of *Plectonema boryanum* and *Chlorella vulgaris* (Rana and Kumar, 1974). Various authors dealing with angiosperm metabolism have noted interactions between zinc and phosphate (Burleson and Page, 1967; Ernst, 1968; Motsara, 1973; Wallace et al., 1973; De Filippis and Pallaghy, 1975). It seems therefore that the influence of phosphate in reducing zinc toxicity is widespread, though not universal; at least in the case of zinc-tolerant populations of *H. rivulare* the effect takes place at sufficiently low phosphate levels to indicate that it may be a key factor influencing their survival in the field.

The toxic effects of zinc and cadmium to *H. rivulare* were found to be synergistic (Say and Whitton, 1977), a response similar to that observed by Hutchinson and Czyrska (1972) for the pond species *Lemna valdiviana*. With *H. rivulare*, cadmium was 34 times more toxic than zinc to a zinc-plus-cadmium-sensitive population, and 15.5 times more toxic than zinc to a zinc-plus-cadmium-resistant population. The results suggest that any level of cadmium above 0.01 mg/1 should be suspected of producing a significant increase in the toxicity of any zinc present.

8. ACCUMULATION

8.1. Field Populations

There are many data on the levels of zinc in aquatic plants from both low- and high-zinc environments. Sometimes the data are combined with levels in the water to give enrichment ratios (defined by Brooks and Rumsby, 1965). It is difficult, however, to compare many of the results because the approaches vary widely. Some authors take the whole organism together with any associated silt or epiphytes. This was presumably the case for the *Cladophora* with 250,000 µg Zn/g* in an abandoned tailings area reported by Wixson and Gale (1975). Such data indicate the ability of a whole community to accumulate a heavy metal and are probably significant when considering sources of metals for grazers. Other authors have attempted to wash the plants. Here the observations are valuable for comparison if a plant can be freed of surface material without too much disturbance; if a species has a mucilaginous covering or a tendency to acquire a firm inorganic crust, a full account of the methods used for cleaning is essential. When species are to be used as environmental monitors, it is essential to give exact details not only of cleaning, but also of the fraction used.

Adams et al. (1973) included zinc in an extensive series of analyses of vascular plants from the Delaware, Susquehanna, and Allegheny river watersheds. Of nine species studied in more detail, six (*Potamogeton illinoensis, Elodea canadensis, E. nuttallii, Vallisneria americana, Eleocharis acicularis, Myriophyllum exalbescens*) showed statistically high variation between sites for zinc, suggesting that these species may prove useful as indicators of environmental zinc levels. The environmental zinc levels are not given, but the content of the species with the lower levels lies in the range 50 to 100 µg/g. The zinc contents of algal samples from the Danube River and Danube Canal were considerably lower (Rehwoldt et al., 1975), ranging from 3.81 to 9.11 µg/g.

Dietz (1973) reported the enrichment ratios for zinc and other metals of three flowering plants in the Ruhr and its tributaries. He pointed out that the enrichment ratios for the mosses were higher than those for flowering plants. The zinc contents of various bryophytes were measured during a study of the Clarach River system, Cardiganshire (McLean and Jones, 1975). The composition of the water at sites from which bryophytes were taken ranged from negligible zinc enrichment up to 1.2 mg total Zn/l, while the contents of the plants ranged from 125 to 1950 µg/g. Shimwell and Laurie (1972) gave analyses of plants on old lead-zinc spoil heaps in the Southern Pennines, England. Although the sites were regarded as terrestrial, several of the mosses can also occur in small, zinc-rich streams (see above). Species like *Dicranella varia,* which lack cuticles and absorb water all over the gametophyte body, showed a higher content of both zinc and lead than species like *Philonotis fontana,* with a more or less continuous cuticle. Here absorption takes place toward the base of the gametophyte. With *P. fontana* there was 25 times more zinc in old than new shoots (2400 vs. 297 µg/g). The zinc contents of algae and mosses are discussed further in Section 10.3.

---

*All data on composition given here refer to dry weight. The various temperatures used for drying by different authors could lead to differences of up to about 10% in comparable tissues.

8.2. Factors Influencing Accumulation

If accurate predictions are to be made of the influence of zinc on the ecology of a river, it is important to establish the extent to which zinc uptake by plants is passive or active. Pickering and Puia (1969) found that uptake of zinc by *Fontinalis antipyretica* was effected initially by an exchange adsorption process, but that subsequent uptake into the cell was an active metabolic process. Most other studies have dealt with phytoplankton or marine species, so only brief mention is made here. Davies (1973) interpreted the pattern of ^{65}Zn uptake by the marine diatom *Phaeodactylum tricornutum* as indicating that uptake is passive and diffusion controlled. Light-dependent uptake of ^{65}Zn by the green flagellate *Dunaliella tertiolecta* took place only in starved cells (Parry and Hayward, 1973). Bryan (1976) suggested that differing relationships between uptake and light in marine seaweeds can be explained in terms of an indirect effect of photosynthesis on the internal pH of the plant and on the synthesis of more binding sites. De Filippis and Pallaghy (1976c) showed that a large initial uptake of ^{65}Zn by *Chlorella* sp. (Emerson strain) from a medium containing 65 mg Zn/l was similar at 25° and 2°, and suggested that it probably represented entry into cell walls. After 1 day the zinc content reached 10,000 μg/g, but uptake subsequently did not keep pace with cell division, resulting in a considerable reduction of zinc content. In contrast to most of these studies, Rana and Kumar (1974) interpreted a correlation between growth and zinc uptake by *Plectonema boryanum* and *Chlorella vulgaris* as an indication of active uptake, but their results seem equally open to interpretation as passive uptake.

The presence of mucilage, the cation exchange capacity of any mucilage, and the physiognomic form of the whole community, may all be expected to influence the rate of zinc uptake by plants. Rose and Cushing (1970) showed that the major site of sorption of ^{65}Zn by matlike periphyton communities from the Columbia River was the upper surface, with a diffusion gradient existing within the community.

9. OCCURRENCE OF *CLADOPHORA* AND *STIGEOCLONIUM*

The influence of zinc on the distribution of *Cladophora glomerata* and *Stigeoclonium tenue* is of special interest, since these are among the plants best known to people concerned with river management. Most accounts indicate that *Cladophora* is very sensitive to heavy metals, including zinc. Thomas (1944), 1962) and Schanz and Thomas (1978) reported that for both *Cladophora* and the closely related *Rhizoclonium* (taken from lake water) the toxic level of zinc is somewhat less than 0.1 mg/1, while in one medium tested by Whitton (1967) zinc proved even more toxic than was found by Thomas. As mentioned above, Williams and Mount (1965) found that *Cladophora* was especially sensitive to zinc in experimental channels. Among 37 populations of green algae taken from flowing waters, *C. glomerata* ranked 36 = in order of resistance to zinc (i.e., extremely sensitive) (Whitton, 1970a). Bellis (1968), however, found good growth of the same species in a medium containing 2 mg Zn/l: he attributed it to the high level of chelating agent (ethylenediaminetetra-acetic acid, EDTA) in the medium.

Heavy growths of *C. glomerata* occur in just a few river sites where zinc is present at elevated levels. The presence of *Cladophora* in a zinc-enriched meander system in the New Lead Belt, Missouri (Gale and Wixson, 1977), was mentioned above. *Cladophora glomerata* is one of the dominants in fast-flowing reaches of the River Gueule in Belgium (present author, unpublished), where the water is subject to zinc pollution as a result of former mining and smelting. Here the water is both highly calcareous and rich in key nutrients as the result of gross pollution by sewage effluents.

In contrast to *Cladophora,* almost all reports for *Stigeoclonium* indicate that the latter can be highly tolerant of zinc, and probably also other heavy metals. An exception is McLean (1974), who reported that *S. tenue* showed no greater tolerance than *Cladophora glomerata* at sites in south Wales, and suggested that zinc and lead would not be tolerated in the absence of organic pollution. *Stigeoclonium tenue* has been found, however, from sites that lack organic pollution but have levels as high as 20.0 mg Zn/l passing through a filter (Harding and Whitton, 1976), although, as described above, the populations present at elevated levels are tolerant ones. Apparently *S. tenue* never forms such mass growths as *Hormidium* or *Mougeotia* at the highest zinc levels (present author, unpublished). At intermediate levels (about 1 mg soluble zinc/l) and at sites where there is at least slight key plant nutrient enrichment from sewage effluents or fertilizer runoff, *S. tenue* very frequently covers the whole substratum of rocky, fast-flowing streams and rivers. Harding and Whitton (1976) reported that in one river with relatively high zinc levels this alga was abundant throughout both spring and summer even though the nutrient input from sewage effluents was low. This is an apparent contrast to the behavior of *S. tenue* in low-zinc rivers reported by McLean and Benson-Evans (1974), where the alga remained conspicuous in summer only at sites with a high organic (sewage effluent) content.

10. USE OF PLANTS TO MONITOR ZINC IN RIVERS

So far no aquatic parallels to *Viola calaminaria* have been found that are confined to sites with high levels of zinc. If any such species should occur, it would almost certainly prove to be sufficiently rare, even among high-zinc sites, to be of little value as an indicator. On the other hand, it is becoming possible to recognize lotic zinc communities, just as phytosociologists have done for terrestrial environments. Although such communities are likely to be important in pure and applied research, it is doubtful whether their recognition will prove of much use as a direct aid in prospecting or monitoring pollution. It is suggested that plants may be of value as monitors in four main ways:

1. Using the dominance of *Cladophora* or *Stigeoclonium* as an indicator of the biological effects of any known zinc pollution (see Section 9).
2. Use of laboratory bioassays to monitor the toxicity of any zinc present in river water or to predict the effects of future changes in zinc level.
3. Measurement of the tolerance of organisms taken from nature.
4. Use of species that can be sampled from a river and whose metal content can be measured in an easily reproducible manner.

10.1. Bioassays

Most of the now widespread bioassay techniques developed for algae have at some time been used to monitor zinc. For instance, Rachlin and Farran (1974) suggested that measurement of the level of zinc required to bring about a 50% reduction in growth rate is a useful criterion, and showed that with a strain of *Chlorella vulgaris* the level required was 2.4 mg $Zn/1$. A modification of the "Algal assays procedure bottle test" was used by Bartlett et al. (1974), with *Selenastrum capricornutum* as the assay organism, to test the effects of Cu, Zn, and Cd in artificial media and to assay the waters from various parts of the Couer d'Alene River, Idaho. As environmental factors influence the toxicity of zinc in different ways, any future bioassay "package" involving algae will need at least three different species.

Bioassays for zinc (or other heavy metals) should make use of strains whose original environmental history is known. Most of the experimental accounts in the literature on the influence of zinc make use of organisms for which little if anything is known about their original habitats. (This of course applies not only to algae but also to the extensive literature on invertebrates and fish.) In some physiological studies (e.g., those of *Chlorella* sp. by De Filippis and Pallaghy, 1967) the level of zinc is so high that it seems probable that the strain either came originally from a high-zinc environment or has subsequently become adapted genetically for zinc tolerance. Strains used for bioassay should be stable genetically, or the design of the assay should give no opportunity for any resistant mutant to make itself apparent in the results.

10.2. Measurement of Tolerance of Organisms from Nature

It is evident from the account of adaptation given above that some filamentous green algae produce populations whose tolerance is closely related to the particular environment in which they are growing. Laboratory assays of such organisms may serve to indicate the influence of zinc in streams and rivers where the zinc concentration varies with time, as below old spoil tips (Say et al., 1977). If a stream site receives water from only one source, the tolerance of a population reflects the maximum toxicity of zinc encountered over a long period at the site in question. If there are several sources, care is needed in the interpretation of results; there may be populations with differing zinc resistance at one site, for example, with *Hormidium rivulare* (P. J. Say and the present author, unpublished).

10.3. Accumulation

Various authors have suggested that the accumulation of heavy metals by aquatic plants may be used to monitor changing environmental levels or to prospect for high ones, and these approaches are now well developed and in some cases applied for practical purposes. Certain species of bryophyte and macroscopic algae such as *Cladophora* and *Lemanea* seem of most potential use as monitors. The most detailed studies are those of Empain (1976a, 1976b), who analyzed the heavy metal contents (including zinc: 1976b) of several species of bryophyte from the Somme, Meuse and Sambre rivers, Belgium, and concluded that:

1. Bryophytes may "integrate" variations in the heavy metal content of the water.
2. Bryophytes accumulate metals present in trace concentrations, thus aiding in their detection and measurement.
3. Analysis provides an estimate of the total metal content of the water that is available for uptake.

Empain reported (1976b) a good correlation between the zinc contents of *Platyhypnidium* (= *Rhynchostegium*) *riparioides, Cinclidotus nigricans,* and *Fontinalis antipyretica* and the level of the water. The contents ranged from 200 to 7500 μg Zn/g (dried at 40°). He concluded (p. 11) that the mosses were not saturated by the levels of heavy metals present at any site. Of these species, *P. riparioides* seems of most potential use as a monitor; it is also tolerant of a variety of other toxic factors (Empain, 1978).

Keeney et al. (1976) have suggested that *Cladophora glomerata* is a potentially useful monitor of concentrations of heavy metals. This alga has been shown in various studies to concentrate heavy metals from the surrounding water with a relatively high enrichment ratio, and Keeney et al. reported that *Cladophora* concentrated metals from the surrounding water with a reasonably constant enrichment ratio for each metal. The present author has also found a linear relationship between zinc and lead in *Cladophora* growing in the River Wear and the levels in the surrounding water at the time of collection. The enrichment ratio over a range of zinc concentrations in the water from 0.01 to 0.35 mg Zn/l was approximately 1300. It seems likely that zinc uptake is influenced by other chemical factors (Keulder, 1975) as much as is toxicity, so the enrichment ratio would also be expected to depend on these factors.

A plot of (the logarithm of) zinc in *Lemanea fluviatilis* against (the logarithm of) zinc in water showed a linear relationship over the range 0.01 to 1 mg Zn/l (Harding and Whitton, 1978). The laboratory data of McLean and Jones (1975) on uptake by *Fontinalis antipyretica* and *Scapania undulata* show a linear relationship between plant and medium over a certain zinc range, followed by a plateau, and finally another sharp rise with increasing zinc levels in the medium. The authors suggested that the latter rise might be due to the plant being swamped by an uncontrollable passive uptake mechanism. It is clear that much more background information is needed if the full potential of plant composition studies is to be developed for assaying zinc and other heavy metals in rivers.

11. DISCUSSION

It may be helpful to give a general impression of current knowledge about zinc and plants in streams and rivers, and also to speculate briefly on a few aspects. Most of the comments should be treated as provisional, but they should help readers to understand the significance of their own local sites for general research.

It seems doubtful whether zinc ever reaches sufficiently high levels to be toxic to every photosynthetic plant. Unless some other unfavorable factor occurs simultaneously, a few species are always present. Nevertheless, at sites with very high zinc levels it is usually easy to suggest species that might be expected if there were no zinc enrichment. At somewhat lower zinc levels the situation is rather different. It may be difficult to recognize much floristic difference from streams lacking any zinc enrichment. However, assays with filamentous green algae indicate that, with this group at least, populations not adapted for tolerance may be sensitive at levels as low as 0.2 mg Zn/l (and possibly even lower).

Although the number of aquatic photosynthetic species capable of growing at very high zinc sites is limited, their taxonomic diversity is marked. Diatoms usually provide the largest number of species, but this is also true for perhaps the majority of sites without zinc enrichment. It may be stated that, in general, each major taxon has produced about the same proportion of species capable of growing at high zinc levels as is present in a general lotic flora. The under-representation of dicotyledons in Table 3 is probably a reflection of the fact that suitable high-zinc environments, such as slow-flowing and not too silted rivers, have yet to be found. There is certainly a wide range of dicotyledons in terrestrial high-zinc habitats (Antonovics et al., 1971; Ernst, 1974b). At least as far as zinc is concerned, there seems no evidence to support the conclusion of Griffiths et al. (1975) that "it is quite clear that microscopic algae are less sensitive to toxic effects of metal pollution than higher plants and animals."

The taxonomic diversity of plants in lotic, high-zinc environments and their morphological similarity to populations in low-zinc environments suggest that it may be a relatively easy evolutionary step to acquire tolerance of zinc. Although many more species may be potentially able to produce tolerant populations, the acquisition of mechanisms for tolerance may make the species sufficiently uncompetitive that there is no suitable niche in high-zinc environments. The fact that representatives of filamentous blue-green algae, *Mougeotia* and *Ulothrix* in high-zinc streams are all narrow forms, whereas broader species are widespread in other streams, could be regarded as circumstantial evidence in favor of the hypothesis. Further speculation clearly requires information on mechanisms of tolerance in aquatic plants and on *in situ* growth rates.

There are certain features which, though far from universal in high-zinc streams, have been reported from several different sites, and for which the explanation is not immediately obvious. Several accounts mention the development of mixed communities embedded in slime and having abundant heterotrophs under very high zinc conditions. It seems possible that in all these situations there is a high input of planktonic or non-adapted species, which then become killed in the high-zinc stream. It is also possible to speculate that this is a response of the whole ecosystem to a greater toxicity of zinc to photosynthetic than heterotrophic organisms.

The high standing crop of plants present in some streams and rivers with high zinc levels was mentioned in the account of the Missouri New Lead Belt. Two hypotheses were raised for the Missouri streams: that the high crops are brought about by high phosphate levels, or that certain key grazers have been eliminated selectively by zinc. It seems possible that both factors may at times be important. The fact that some springs have very high standing crops at source, but low crops a short way downstream, may perhaps be due to depletion of a key nutrient. The persistence of large *Stigeoclonium tenue* crops through the summer in rivers having high zinc but lacking a high input of sewage effluent may perhaps reflect a reduction in the influence of grazers.

One further possible feature is based solely on the author's observations, and is mentioned only to help others make critical studies. Some mosses and flowering plants seem to occur in more truly lotic situations where the levels of zinc are high than they would otherwise do. This is apparently the situation with, for instance, *Dicranella varia, Ditrichum cylindricum, Agrostis stolonifera,* and *Rumex acetosa.*

It seems evident that any comment on the levels of zinc present in rivers is of relatively little value for studies of plants without some indication of other environmental parameters. At present, Pb, Ca, Mg, and P would appear to be the most important. The influence of phosphate in lowering zinc toxicity may be particularly significant in many high-zinc streams, where phosphate is perhaps often present at relatively high levels in comparison with natural streams lacking any influence of sewage or fertilizer.

The accounts of streams and small rivers that have occupied much of this chapter may seem of only academic interest to engineers and administrators. The diversity of these sites, however, provides the researcher with a great variety of combinations of physical and chemical conditions. This circumstance should eventually make it possible to quantify the effects of zinc in detail, and thus aid in the study of large rivers receiving not only zinc but a variety of other toxic substances.

ACKNOWLEDGMENTS

Many of the previously unpublished data were obtained during work carried out under Contract 074-74-1 ENV. UK of the European Communities Environmental Research Programme. Drs. J. P. C. Harding, B. P. Jupp and P. J. Say aided various stages of this research. I am also most grateful to the many people who helped us to locate field sites or provided other useful information. These include J. H. Davis (Amax Lead Company of Missouri), G. Friedrich (Landesanstalt für Wasser und Abfall Nordrhein-Westfalen), N. L. Gale (University of Missouri-Rolla), H. Gies (Preussag, Goslar), R. W. Holton (University of Knoxville), D. W. Lorsch (University of Hamburg), D. Mollenhauer (Forschungsinstitut Senckenberg), J. W. Platt (Avoca Mines, Ltd), F. W. Smith (formerly Weardale Lead Mining Co.), A. Weber (University of Hamburg), H. Wedow (U.S. Geological Survey), and B. G. Wixson (University of Missouri-Rolla). J. R. Carter aided with the identification of diatom records reported here for the first time. Dr. K. W. Besch gave permission to reproduce the data in Table 2.

REFERENCES

Adams, F. S., Cole, H., Jr., and Lowry, L. B. (1973). "Element Constitution of Selected Aquatic Vascular Plants from Pennsylvania: Submersed and Floating Leaved Species and Rooted Emergent Species," *Environ. Pollut.,* 5, 117–147.

Agence Financière de Bassin Adour-Garonne (1975) *Résultats des Analyses sur le Lot de 1968 à 1974.* Agence de Bassin Adour-Garonne, 84 rue de Férétra, 31078 Toulouse.

Alicke, R. (1974). "Die hydrochemischen Varhältnisse im Westharz in ihrer Beziehung zur Geologie und Petrographie," *Clausthaler Geol. Abh.* **20**, 223 pp.

Antonovics, J., Bradshaw, A. D., and Turner, R. G. (1971). "Heavy Metal Tolerance in Plants," *Adv. Ecol. Res.,* **7**, 1–85.

Bachmann, R. W. (1961). "Zinc^{65} in Studies of the Freshwater Zinc Cycle." In V. Schultz and W. Klement, Eds., *Radioecology,* Vol. 1. Proceedings of a Symposium on Radioecology, Colorado. Reinhold, New York, 746 pp., pp. 485–496.

Bartlett, L., Rabe, F. W., and Funk, W. H. (1974). "Effect of Copper, Zinc and Cadmium on *Selenastrum capricornutum,*" *Water Res.,* **8**, 179–185.

Bellis, V. J. (1968). "Unialgal Cultures of *Cladophora glomerata* (L.) Kutz. II: Response to Calcium-Magnesium Ratio and pH of the Medium." In *Proceedings of the 11th Conference on Great Lakes Research,* 11–15.

Besch, K. W. and Roberts-Pichette, P. (1970). "Effects of Mining Pollution on Vascular Plants in the Northwest Miramichi River System," *Can. J. Bot.,* **48**, 1647–1656.

Besch, K. W., Ricard, M., and Cantin, R. (1972). "Benthic Diatoms as Indicators of Mining Pollution in the Northwest Miramichi River System, New Brunswick, Canada," *Int. Rev. Ges. Hydrobiol.,* **57**, 39–74.

Boelen, Ch. and de Boeck, R. (1976). "La Pollution Métallique de la Vesdre," *Tribune CEBEDEAU (Belgium),* Nos. 391 and 392, pp. 230–239.

Bradshaw, A. D. (1975). "The Evolution of Metal Tolerance and Its Significance for Vegetation Establishment on Metal Contaminated Sites." In T. C. Hutchinson, Ed., *Symposium Proceedings, International Conference on Heavy Metals in the Environment, Toronto, Canada,* Vol II, Part 2, pp. 599–622.

Brock, T. D. (1969). "Microbial Growth under Extreme Conditions," *Symp. Soc. Gen. Microbiol.,* **19**, 15–41.

Brooks, R. R. and Rumsby, M. D. (1965). "The Biogeochemistry of Element Uptake by Some New Zealand Bivalves," *Limnol. Oceanogr.,* **10**, 521–527.

Bryan, G. W. (1976). "Some Aspects of Heavy Metal Tolerance in Aquatic Organisms." In A. P. M. Lockwood, Ed., *Effects of Pollutants on Aquatic Organisms.* Cambridge University Press, Cambridge, pp. 7–34.

Burleson, C. A. and Page, N. R. (1967). "Phosphorus and Zinc Interactions in Flax," *Soil Sci.,* **31**, 510–513.

Cairns, J., Jr, Buikema, A. L., Jr, Heath, A. G., and Parker, B. C. (1978). *Effects of Temperature on Aquatic Organism Sensitivity to Selected Chemicals.* Virginia Water Resources Research Center, Virginia Polytechnic Institute and State University, Bull. 106, Blacksburg, 88 pp.

Carpenter, K. E. (1924). "A Study of the Forms of Rivers Polluted by Lead Mining in the Aberystwyth District," *Ann. Appl. Biol.,* 11, 1-23.

Castenholz, R. W. and Wickstrom, C. E. (1975). "Thermal Streams." In B. A. Whitton, Ed., *River Ecology,* Blackwell, Oxford, 725 pp., pp. 264-285.

Cushing, C. E. and Rose, F. L. (1970). "Cycling of [65]Zn by Columbia River Periphyton in a Closed Lotic Microcosm," *Limnol. Oceanogr.,* 15, 762-767.

Cushing, C. E. and Watson, D. G (1968). "Accumulations of ^{32}P and [65]Zn by Living and Killed Plankton," *Oikos,* 19, 143-145.

Davies, A. G. (1973). "The Kinetics of and a Preliminary Model for the Uptake of Radio-zinc by *Phaeodactylium tricornutum* in Culture." In *Radioactive Contamination of the Marine Environment,* International Atomic Energy Agency, Vienna, pp. 403-420.

De Filippis, L. F. and Pallaghy, C. K. (1975). "Localization of Zinc and Mercury in Plant Cells," *Micron,* 6, 111-120.

De Filippis, L. F. and Pallaghy, C. K. (1976a). "The Effect of Sublethal Concentrations of Mercury and Zinc on *Chlorella.* I: Growth Characteristics and Uptake of Metals," *Z. Pflanzenphysiol.,* 78, 197-207.

De Filippis, L. F. and Pallaghy, C. K. (1976b). "The Effect of Sublethal Concentrations of Mercury and Zinc on *Chlorella.* II: Photosynthesis and Pigment Composition," *Z. Pflanzenphysiol.,* 78, 314-322.

De Filippis, L. F. and Pallaghy, C. K. (1976c). "The Effect of Sublethal Concentrations of Mercury and Zinc on *Chlorella.* III. Development and Possible Mechanisms of Resistance to Metals," *Z. Pflanzenphysiol.,* 79, 323-335.

Dietz. F. (1973). "The Enrichment of Heavy Metals in Submerged Plants." In S. H. Jenkins, Ed., *Advances in Water Pollution Research: Proceedings of the 6th International Conference.* Pergamon Press, Oxford, 946 pp., pp. 53-62.

Dunham, K. C. (1948). *Geology of the Northern Pennine Orefield.* I. *Tyne to Stainmore.* Memoirs of the Geological Survey., H. M. Stationery Office, London, 357 pp.

Empain, A. (1976a). "Estimation de la Pollution par Métaux Lourds dans la Somme par l'Analyse des Bryophytes Aquatiques," *Bull. Fr. Piscicult.,* 48, 138-142.

Empain, A. (1976b). "Les Bryophytes Aquatiques Utilises comme Traceurs de la Contamination en Métaux Lourds des Eaux Douces," *Mém. Soc. R. Bot. Belg.,* 7, 11, 141-156.

Empain, A. (1978). "Relations Quantitatives entre les Populations de Bryophytes Aquatiques et la Pollution des Eaux Courantes. Définition d'un Indice de Qualité des Eaux," *Hydrobiologia,* 60, 49-74.

Ernst, R., Allen, H. E., and Mancy, K. H. (1975). "Characterization of Trace Metal Species and Measurement of Trace Metal Stability Constants by Electrochemical Techniques," *Water Res.,* 9, 969-979.

Ernst, W. (1964). "Ökologisch-soziologische Untersuchungen in den Schwermetallpflanzengsellschaften Mitteleuropas unter Einschluss der Alpen," *Abh. Landesmus. Naturk. Münster*, **27**(1), 1–54.

Ernst, W. (1966). "Ökologisch-soziologische Untersuchungen an Schwermetallpflanzengesellshaften Sudfrankreichs und des ostlichen Harzvorlandes," *Flora (Jena)*, **156**, 301–318.

Ernst, W. (1968). "Der Einfluss der Phosphatversorgung sowie die Wirkung von lonogenem und chelatisiertem Zink auf die Zink- und Phosphataufnahme einiger Schwermetallpflanzen," *Physiol. Plant.*, **21**, 323–333.

Ernst, W. (1974a). "Mechanismen der Schwermetallresistenz." In *Verhandlungen Gesellshaft für Ökologie*, Erlangen, pp. 189–197.

Ernst, W. (1974b). *Schwermetallvegetation der Erde* (= *Geobotanica Selecta* V). Gustav Fischer, Stuttgart, 194 pp.

Eyster, C. H. (1964). "Micronutrient Requirements of Algae." In D. F. Jackson, Ed., *Algae and Man*. Plenum Press, New York, pp. 86–118.

Failla, M. L., Benedict, C. D., and Weinberg, E. D. (1976). "Accumulation and Storage of Zn^{2+} by *Candida utilis*," *J. Gen. Microbiol.*, **94**, 23–26.

Fisher, N. S. and Wurster, C. F. (1973). "Individual and Combined Effects of Temperature and Polychlorinated Biphenyls on the Growth of Three Species of Phytoplankton," *Environ. Pollut.*, **5**, 205–212.

Funk, W. H., Rabe, F. W., Filby, R., Parker, J. I., and Winner, J. E. (1975). *Biological Impact of Combined Metallic and Organic Pollution in The Coeur d'Alene-Spokane River Drainage System*. National Technical Information Service, PB-222 946, 187 pp.

Gächter, R. (1976). "Untersuchungen über die Beeinflussung der planktischen Photosynthese durch anorganische Metallsalze im eutrophen Alpenachersee und der mesotrophen Horwer Bucht," *Schweiz. Z. Hydrol.*, **39**, 97–119.

Gale, N. L. and Wixson, B. G. (1977). "Water Quality–Biological Aspects." In B. G. Wixson, Ed., *The Missouri Lead Study*, Vol. I. Final Report to the National Science Foundation Research Aplied to National Needs, University of Missouri, Rolla and Columbia, pp. 400–543.

Gale, N. L., Hardie, M. G., Jennett, J. C., and Aleti, A. (1973a). "Transport of Trace Pollutants in Lead Mining Wastewaters." In D. D. Hemphill, Ed., *Trace Substances in Environmental Health*–VI, pp. 95–106.

Gale, N. L., Wixson, B. G., Hardie, M. G., and Jennett, J. C. (1973b). "Aquatic Organisms and Heavy Metals in Missouri's 'New Lead Belt,'" *Wat. Resour. Bull.*, **9**, 673–688.

Gopal, T., Rana, B. C., and Kumar, H. D. (1975). "Autecology of the Blue-green alga *Nodularia spumigena* Mertens," *Nova Hedwigia*, **26**, 225–232.

Griffiths, A. J., Hughes, D. E., and Thomas, D. (1975). "Some Aspects of Microbial Resistance to Metal Pollution." In M. J. Jones, Ed., *Minerals and the Environment*. Institution of Mining and Metallurgy, London, 803 pp., pp. 387–394.

Gundlach, H. and Steinkamp, K. (1973). "Geochemische Prospektion im Ober-harz, einem alten Bergbaugebiet," *Z. Dsch. Geol. Ges.,* **124**, 37–49.

Harding, J. P. C. (1978). "Studies on Heavy Metal Toxicity and Accumulation in the Catchment Area of the Derwent Reservoir." Ph. D. thesis, University of Durham, England, 482 pp.

Harding, J. P. C. and Whitton, B. A. (1976). "Resistance to Zinc of *Stigeoclonium tenue* in the field and the Laboratory," *Br. Phycol. J.,* **11**, 417–426.

Harding, J. P. C. and Whitton, B. A. (1977). "Environmental Factors Reducing the Toxicity of Zinc to *Stigeoclonium tenue,*" *Br. Phycol. J.,* **12**, 17–21.

Harding, J. P. C. and Whitton, B. A. (1978). "Accumulation of Heavy Metals by *Lemenea* in European Rivers Affected by Mining," *Br. Phycol. J.,* **13**, 200–201.

Hargreaves, J. W. (1977). "Ecology and Physiology of Photosynthetic Organisms in Highly Acid Streams." Ph. D. thesis, University of Durham, England, 336 pp.

Hargreaves, J. W. and Whitton, B. A. (1976a). "Effect of pH on Growth of Acid Stream Algae," *Br. Phycol. J.,* **11**, 215–223.

Hargreaves, J. W. and Whitton, B. A. (1976b). "Effect of pH on Tolerance of *Hormidium rivulare* to Zinc and Copper," *Oecologia,* **26**, 235–243.

Hargreaves, J. W., Lloyd, E. J. H., and Whitton, B. A. (1975). "Chemistry and Vegetation of Highly Acidic Streams," *Freshwater Biol.,* **5**, 563–576.

Harvey, R. S., and Patrick, R. (1967). "Concentration of ^{137}Cs, ^{65}Zn and ^{85}Sr by Freshwater Algae," *Biotech. Bioeng.,* **9**, 449–456.

Heimans, J. (1936). "De Herkomst van de Zinkflora aan de Geul," *Ned. Kruidk. Arch.,* **46**, 878–897.

Hem, J. D. (1972). "Chemistry and Occurrence of Cadmium and Zinc in Surface water and Groundwater," *Water Resour. Res.,* **8**, 661–679.

Hutchinson, T. C. and Czyrska, H. (1972). "Cadmium and Zinc Toxicity and Synergism to Floating Aquatic Plants." In *Water Pollution Research in Canada 1972: Proceedings of the 7th Canadian Symposium on Water Pollution Research,* pp. 59–65.

Jackson, R. G. and Nichol, I. (1975). "Factors Affecting Trace Element Dispersion in the Yellowknife Area, N.W.T., Canada," *Verh. Int. Verein. Theor. Angew. Limnol.,* **19**, 308–316.

Jenne, E. A. (1968). "Controls on Mn, Fe, Co, Ni, Cu and Zn Concentrations in Soil and Water: the Significant Role of Hydrous Mn and Fe Oxides." In R. A. Baker, Ed., *Trace Inorganics in Water.* American Chemical Society, Advances in Chemistry Series 73, Washington, D.C.

Jennett, J. C. and Wixson, B. G. (1975). "The New Lead Belt: Aquatic Metal Pathways-Control." In T. C. Hutchinson, Ed., *Symposium Proceedings, International Conference on Heavy Metals in the Environment, Toronto, Canada,* Vol. II., Part 1, pp. 247–256.

Jensen, A., Rystad, B., and Melsom, S. (1974a). "Heavy Metal Tolerance of Marine Phytoplankton. 1: The Tolerance of Three Algal Species to Zinc in Coastal Sea Water," *J. Exp. Mar. Biol. Ecol.*, **15**, 145–157.

Jensen, A., Rystad, B., and Melson, S. (1974b). "Heavy Metal Tolerance of Marine Phytoplankton. II: The Tolerance of Three Algal Species to Zinc in Coastal Seawater," *J. Exp. Mar. Biol. Ecol.*, **22**, 249–256.

Jones, J. R. E. (1958). "A Further Study of the Zinc-Polluted Ystwyth," *J. Anim. Ecol.*, **27**, 1–14.

Jurinak, J. J. and Inouye, T. S. (1962). "Some Aspects of Zinc and Copper Phosphate Formation in Aqueous Systems," *Soil Sci. Soc. Am. Proc.*, **26**, 144–147.

Keeney, W. L., Breck, W. G., Van Loon, G. W., and Page, J. A. (1976). "The Determination of Trace Metals in *Cladophora glomerata*—*C. glomerata* as a Potential Biological Monitor," *Water Res.*, **10**, 981–984.

Keulder, P. C. (1975). "Influence of the Clay Types Illite and Montmorillonite on the Uptake of ^{65}Zn by *Scenedesmus obliquus.*" *J. Limnol. Soc. S. Afr.*, **1**, 33–35.

Kramer, R. L. (1976). M. Sc. thesis, University of Missouri, Rolla.

McLean, R. O. (1974). "The Tolerance of *Stigeoclonium tenue* (Kütz.) to Heavy Metals in South Wales," *Br. Phycol. J.*, **9**, 91–95.

McLean, R. O. and Benson-Evans, K. (1974). "The Distribution of *Stigeoclonium tenue* Kütz. in South Wales in Relation to Its Use as Indicator of Organic Pollution," *Br. Phycol. J.*, **9**, 83–89.

McLean, R. O. and Jones, A. K. (1975). "Studies of Tolerance to Heavy Metals in the Flora of the Rivers Ystwyth and Clarach, Wales," *Freshwater Biol.*, **5**, 431–444.

McNaughton, S. J., Folsom, T. C., Lee, T., Park, F., Price, C., Roeder, D., Schmitz, J., and Stockwell, C. (1974). *Ecology*, **55**, 1163–1165.

Morrison, T. A. (1976). "Some Historical Notes on Mining in the Harz Mountains, Germany," *Trans. Inst. Min. Metall.*, **85**, A51–A56.

Motsara, M. R. (1973). "On the Effects of Phosphorus on Zinc Uptake by Barley," *Plant Soil*, **38**, 381–392.

Nash, T. H. III (1975). "Influence of Effluents from a Zinc Factor on Lichens," *Ecol. Monogr.*, **45**, 183–198.

Overnell, J. (1975). "The Effect of Some Heavy Metal Ions on Photosynthesis in a Freshwater Alga," *Pesticide Biochem. Physiol.*, **5**, 19–26.

Parry, G. D. R. and Hayward, J. (1973). "The Uptake of ^{65}Zn by *Dunaliella tertiolecta* Butcher," *J. Mar. Biol. Assoc., U.K.*, **53**, 915–922.

Pasternak, K. (1973). "The Spreading of Heavy Metals in Flowing Waters in the Region of Occurrence of Nature Deposits and of the Zinc and Lead Industry," *Arch. Hydrobiol.*, **15**, 145–166.

Pasternak, K. (1974). "The Influence of the Pollution of a Zinc Plant at Miasteczko Slaskie on the Content of Microelements in the Environment of Surface Waters," *Acta Hydrobiol.,* **16**, 273–297.

Patrick, R. (1971). Report in *Water Quality Criteria, 1972.* National Academy of Sciences–National Academy of Engineering, Washington, D.C. (Quoted by Cairns et al., 1978).

Pickering, D. C. and Puia, I. L. (1969). "Mechanism for Uptake of Zinc by *Fontinalis antipyretica,*" *Physiol. Plant.,* **22**, 653–661.

Pita, F. W. and Hyne, N. J. (1975). "The Depositional Environment of Zinc, Lead and Cadmium in Reservoir Sediments," *Water Res.,* **9**, 701–706.

Platt, J. W. (1975). "Environmental Control at Avoca Mines Ltd., Co. Wicklow, Ireland." In M. J. Jones, Ed., *Minerals and the Environment.* Institution of Mining and Metallurgy, England, 803 pp., pp. 731–758.

Rachlin, J. W. and Farran, M. (1974). "Growth Responses of the Green Alga *Chlorella vulgaris* to Selective Concentrations of Zinc," *Water Res.,* **8**, 575–577.

Raistrick, A. and Jennings, B. (1965). *A History of Lead Mining in the Pennines.* Longmans, London, 347 pp.

Rana, B. C. and Kumar, H. D. (1974). "The Toxicity of Zinc to *Chlorella vulgaris* and *Plectonema boryanum* and Its Protection by Phosphate," *Phykos,* **13**, 60–66.

Reese, M. J. (1937). "The Microflora of the Non-calcareous Streams Rheidol and Melindwr, with Special Reference to Water Pollution from Lead Mines in Cardiganshire," *J. Ecol.,* **25**, 385–407.

Rehwoldt, R., Karimian-Teherani, D., and Altmann, H. (1975). "Measurement and Distribution of Various Heavy Metals in the Danube River and Danube Canal Aquatic Communities in the Vicinity of Vienna, Austria," *Sci. Total Environ.,* **3**, 341–348.

Rose, F. L. and Cushings, C. E. (1970). "Periphyton: Autoradiography of Zinc-65 Adsorption," *Science,* **168**, 576–577.

Say, P. J. and Whitton, B. A. (1977). "Influence of Zinc on Lotic Plants. II: Environmental Effects on Toxicity of Zinc to *Hormidium rivulare,*" *Freshwater Biol.,* **7**, 377–384.

Say, P. J. and Whitton, B. A. (1978). "Chemistry and Benthic Algae of a Zinc-Polluted Stream in the Northern Pennines," *Br. Phycol. J.,* **13**, 206.

Say, P. J., Diaz, B. M., and Whitton, B. A. (1977). "Influence of Zinc on Lotic Plants. I: Tolerance of *Hormidium* Species to Zinc," *Freshwater Biol.,* **7**, 357–376.

Schanz, F. and Thomas, E. A. (1978). "Cultures of Cladophoraceae in Water Pollution Problems," *Mitt. Int. Verein. Theor. Angew. Limnol.,* **21**, 57–64.

Schubert, R. (1954). "Zur Systematik und Pflanzengeographie der Charakterpflanzen der Mitteldeutschen Schwermetallpflanzengesellschaften," *Wiss. Z. Martin-Luther-Univ. Halle-Wittenberg,* **3**(4), 863–882.

Shimwell, D. W. and Laurie, A. E. (1972). "Lead and Zinc Contamination of Vegetation in the Southern Pennines," *Environ. Pollut.*, **3**, 291-301.

Thomas, E. A. (1944). "Uber eine blasenbildende Krankheit von kultivierten grunen Fadenalgen (*Cladophora* und *Rhizoclonium*)," *Vierteljahrssch. Naturforsch. Ges. Zürich*, **106**, 277-288.

Thomas, E. A. (1962). "Zink im Trinkwasser als Algengift." *Arch, Midrobiol.*, **42**, 246-253.

Tyler, P. A. and Buckney, R. T. (1973). "Pollution of a Tasmanian River by Mine Effluents. I: Chemical Evidence," *Int. Rev. Ges. Hydrobiol.*, **58**, 873-883.

Vallun, S. (1952). "Zwei azidotrophe Seen im Küstengebiet von Nordschweden," *Rep. Inst. Freshwater Res. Drottningholm*, **34**, 167-189.

Wallace, A., El Gazzar, A., and Alexander, G. V. (1973). "High Phosphorus Levels on Zinc and Other Heavy Metal Concentrations in Hawkeye and P154619-5-1 Soyabeans," *Comm. Soil Sci. Plant Anal.*, **4**, 343-345.

Weatherley, A. H., Beevers, J. R., and Lake, P. S. (1967). "The Ecology of a Zinc-Polluted River." In A. H. Weatherley, Ed., *Australian Inland Waters and Their Fauna: Eleven Studies.* Australian National University Press, Canberra.

Weber, C. I. (1973). "Recent Developments in the Measurement of the Response of Plankton and Periphyton to Changes in Their Environment." In G. E. Glass, Ed. *Bioassay Techniques and Environmental Chemistry.* Ann Arbor, Mich, 499 pp.

Whitton, B. A. (1967). "Studies on the Growth of Riverain *Cladophora* in Culture," *Arch. Mikrobiol.*, **58**, 21-29.

Whitton, B. A. (1968). "Effect of Light on Toxicity of Various Substances to *Anacystis nidulans,*" *Plant Cell. Physiol.*, **9**, 23-26.

Whitton, B. A. (1970a). "Toxicity of Zinc, Copper and Lead to Chlorophyta from Flowing Waters," *Arch. Mikrobiol.*, **72**, 353-360.

Whitton, B. A. (1970b). "Toxicity of Heavy Metals to Freshwater Algae: a Review," *Phykos*, **9**, 116-125.

Whitton, B. A. (1970c). "Biology of *Cladophora* in Freshwaters," *Water Res.*, **4**, 457-476.

Whitton, B. A. (1975). "Algae." In B. A. Whitton, Ed., *River Ecology*, Blackwell, Oxford, 725 pp., pp. 3-105.

Whitton, B. A. (1977). "Influence of Zinc on Stream Algae," *Br. Phycol. J.*, **12**, 123.

Whitton, B. A. and Say, P. J. (1975). "Heavy Metals." In B. A. Whitton, Ed., *River Ecology*. Blackwell, Oxford, 724 pp., pp. 286-311.

Wild, H. (1968). "Geobotanical Anomalies in Rhodesia. I: The Vegetation of Copper Bearing Soils," *Kirkia*, **7**, 1-71.

Williams, L. G. and Mount, D. I. (1965). "Influence of Zinc on Periphytic Communities," *Am. J. Bot., 52*, 26–34.

Wixson, B. G. and Bolter, E. (1972). "Evaluations of Stream Pollution and Trace Substances in the New Lead Belt of Missouri." In D. D. Hemphill, Ed., *Trace Substances in Environmental Health*–V, pp. 143–152.

Wixson, B. G. and Gale, N. L. (1975). "Some Limnological Effects of Lead and Associated Heavy Metals from Mineral Production in Southeast Missouri, U.S.A." In International Symposium, *Interactions between Water and Living Matter,* Odessa, USSR, 10 pp.

16

ZINC UPTAKE
AND ACCUMULATION
BY AGRICULTURAL CROPS

P. M. Giordano
J. J. Mortvedt

Soils and Fertilizer Research Branch, National Fertilizer Development Center, Tennessee Valley Authority, Muscle Shoals, Alabama

1. Introduction	402
2. Forms of Zinc in Soils	402
3. Mobility and Movement of Zinc to Plant Roots	403
4. Mechanism of Zinc Uptake	403
5. Soil and Climatic Factors Affecting Zinc Uptake	404
6. Responses of Plant Genotypes to Zinc	405
7. Interactions with Other Nutrients	406
7.1. Zinc-phosphorus	406
7.2. Zinc-nitrogen	407
7.3. Zinc-iron	408
8. Diagnosis of Deficiences: Soil and Tissue Tests	408
9. Sources, Rates, and Application Methods	409
10. Summary	411
References	412

1. INTRODUCTION

The requirement of zinc for normal plant growth was first recognized in the late nineteenth century, but acceptance of this element as an essential plant nutrient did not occur until the early 1930s. Since then zinc deficiency has been identified throughout the world on numerous crops grown on widely varying soils under a variety of management practices. This chapter summarizes and condenses information in earlier reviews and provides an updating on factors affecting zinc availability and uptake by crops, differential plant response, nutrient interactions, and diagnosis and correction of zinc deficiency. For more complete treatment of these subjects, several excellent reviews are available (Thorne, 1957; Hodgson, 1963; Chapman, 1966; Mortvedt et al., 1972; Lindsay, 1972).

2. FORMS OF ZINC IN SOILS

Most agricultural soils contain zinc in the range of less than 10 to greater than 300 ppm, depending on parent material. Sphalerite (ZnS) is the most prevalent zinc-containing mineral in the lithosphere, although zinc is often found in the mineral fraction of soils in crystal clay lattices by isomorphous substitution or occlusion. According to Lindsay (1972), more soluble compounds such as zincite (ZnO) and smithsonite ($ZnCO_3$) do not persist as such in soils and hence are suitable as zinc fertilizers. Results of phase equilibria studies indicate that the predominant zinc species in soil below pH 7.7 is Zn^{2+}, and above this pH level is the neutral species, $Zn(OH)_2$. On the basis of Lindsay's studies, the zincate ion [$Zn(OH)_4{}^{2-}$] is not present in significant quantities in soils and has little bearing on zinc solubility in soils. Contrasting results were reported in reviews by Thorne (1957) and Mikkelsen and Kuo (1977).

Aside from the zinc in soil minerals, the element is surface-adsorbed by calcite, dolomite, organic matter, metal oxides, and clays. The mechanisms are discussed by Mikkelsen and Kuo (1977) and Hodgson (1963). Surface precipitation, though a separate phenomenon, is not easily distinguished from surface adsorption. Mikkelsen and Kuo contend that surface-adsorbed zinc is the precursor of precipitated zinc when excessive amounts of the element are applied to soils.

As a consequence of the above immobilizing reactions, concentrations of zinc in soil solution are usually low, especially in alkaline soils (Hodgson et al., 1966). Values of about 75 ppb Zn were found in solutions of several acid soils from New York, while values of less than 2 ppb Zn were detected in calcareous soils from Colorado. Approximately 60% of the solution zinc was complexed in the soluble organic matter fraction. Concentrations of zinc in soil solution after flooding are normally higher, usually ranging from 40 to several hundred parts per billion. Probably this contributes to the lower zinc uptake and more severe zinc deficiency of rice (*Oryza sativa* L.) in nonflooded than in flooded soils, reported by Giordano and Mortvedt (1972, 1974).

3. MOBILITY AND MOVEMENT OF ZINC TO PLANT ROOTS

Although most soils have adequate indigenous supplies of zinc, plant uptake is dependent on solubility and movement to roots. Of the two contributing processes, convection was shown to account for 5% of the zinc transported to corn roots, with diffusion supplying the remainder (Elgawhary et al., 1970). Thus diffusion is a critical phenomenon in regulating zinc uptake by plants as well as in determining the effectiveness of various zinc fertilizers. The intensity-capacity factor is also involved since both the labile solid phase and soil solution contribute to the zinc supply (Lindsay, 1972).

Chelating or complexing agents of natural or synthetic origin influence zinc diffusion by forming a larger chelated ion, by decreasing the reaction of zinc with soil, and by increasing zinc concentrations in solution (Wilkinson, 1972). Chelated zinc is often more effective than inorganic zinc for many upland crops, but Mikkelsen and Kuo (1977) and Giordano (1977) claim the opposite for flooded rice. A possible explanation is greater diffusibility of chelates, which transport zinc downward and beyond the effective root zone of small rice plants. In addition, exchange reactions of zinc in the chelate molecule with iron and calcium could lessen chelate effectiveness under reduced soil conditions.

The significance of root interception of zinc is obscure, and the degree to which this mechanism contributes to plant uptake has not been satisfactorily resolved (Wilkinson, 1972). Roots growing in soil contact mineral surfaces approximately equal in area to the roots. Accordingly, the processes of diffusion and convection do not function in uptake by root interception. Hence difficulty in accurate measurement by various workers has resulted in either over- or underestimating the role of root interception in ion uptake.

4. MECHANISM OF ZINC UPTAKE

The controversy in the literature as to whether zinc uptake by plants is a physical (passive) or a metabolically mediated (active) process was reviewed by Moore (1972) and Wallace (1975). Much of the problem arises from variations in the experimental procedures employed by the investigators. Such parameters as substrate zinc concentration, duration of absorption, and competing ions dictate the results obtained in short-term uptake studies in culture solution. Giordano et al. (1974) reported that uptake of zinc by rice was retarded by metabolic inhibitors when substrate zinc concentrations approximated those found in soil solution. The presence of 10^{-4} M $FeCl_2$ or $MnCl_2$ in absorption solutions depressed zinc uptake significantly, while $MgCl_2$ reduced the concentrations of zinc in roots but enhanced translocation to shoots. Most evidence suggests that uptake of zinc is predominantly an active process, but many factors may interact to influence interpretation.

5. SOIL AND CLIMATIC FACTORS AFFECTING ZINC UPTAKE

Aside from naturally occurring reactions in soils which limit plant uptake of zinc, other parameters interact and influence availability. Perhaps the most influential single factor is soil pH. Although zinc deficiency in crops is not a major concern on most acid soils, the problem can exist on sandy soils of low total zinc content or with severe leaching. Since the solubility of zinc decreases as pH increases, deficiency is most prevalent on calcareous soils of pH 7.4 or higher. Accordingly, liming acid soils, especially those that are coarse textured, may induce zinc deficiency, most often in the pH range of 6 to 8, where zinc solubility is at a minimum. Flooding of acid soils for rice production also results in a rise in pH toward neutrality, which may induce zinc deficiency in soils with low native supplies of the element. Although excessive zinc is not a widespread problem, Chapman (1966) indicates that certain acid peats and acidification of some soils through continued use of acid-forming fertilizers can lead to zinc toxicity of sensitive crops.

Organic matter constitutes an important reserve of available zinc for crops. Zinc deficiency has been associated with irrigated soils from which topsoil has been removed for land leveling. The exposed subsoil is generally of lower zinc content, is higher in pH, and may contain free $CaCO_3$. Supplying organic matter by growing corn or green manure crops is often beneficial to supplement the available zinc supply. Conversely, adding organic matter to flooded neutral or alkaline rice soils is detrimental in that resulting bicarbonate aggravates zinc deficiency (Forno et al., 1975).

Light and temperature effects on yield response to zinc have been well documented. Ozanne (1955) noted a greater response by subterranean clover (*Trifolium subterranean* L.) under conditions of longer day length and greater light intensity. Lucas and Knezek (1972) reported that zinc deficiency in Michigan often is more severe when light intensity is moderate and soils are cool and wet.

In soils with low levels of available zinc, severe deficiency symptoms are frequently reported in early spring, when temperatures are cool and soils wet, but by mid-July symptoms usually disappear. This pattern of events is probably related to one factor or a combination of factors. Since organic matter is a source of zinc for plants, cool soils may reduce the rate of microbial release of available zinc. A second factor is delayed root growth at low temperature, which would limit zinc uptake. Bauer and Lindsay (1965) and Giordano and Mortvedt (1978) showed no differences in extractable zinc at different soil temperatures.

Interactions of soil temperature and rate and source of phosphorus have been investigated relative to the development of zinc deficiency. Martin et al. (1965) found that phosphorus-induced zinc deficiency in tomatoes (*Lycopersicon esculentum*) occurred only at low soil temperatures in soils containing moderate levels zinc; no temperature effects were observed with extremely low available zinc. Sharma et al. (1968) and Mikkelsen and Kuo (1977) found low temperature (16°C) to inhibit zinc uptake by rice, and neither phosphorus nor zinc additions appreciably increased zinc concentration or uptake. Giordano and Mortvedt (1978) reported the uptake of zinc to be greater with triammonium pyrophosphate than monoammonium phosphate as a source of phosphorus for corn, but soil temperature had no differential effect. The fact that uptake of zinc was greater at 32 than at 16°C appeared to reflect increased growth. The P/Zn ratios were higher in 18- than in 36-day-old plants, suggesting that uptake of phosphorus was more rapid than that of zinc during early growth. Furthermore, growth depression associated with cool, wet conditions may be attributed to insufficient early absorption of both phosphorus and zinc during the seedling stage, as shown by a higher rate of phosphorus uptake from warm soil.

6. RESPONSES OF PLANT GENOTYPES TO ZINC

It has long been recognized that plants differ in their capacities to absorb and utilize zinc, as well as to tolerate excess amounts. This subject was reviewed in depth by Brown et al. (1972). A classic example of differential response among genotypes is illustrated by Saginaw and Sanilac navy beans (*Phaseolus vulgaris* L.). In Michigan the Saginaw cultivar was more efficient and tolerant of excess zinc than Sanilac. Accordingly, Sanilac was more susceptible to zinc deficiency than Saginaw, although with adequate zinc Sanilac had a higher yield potential. Later studies attributed the difference in response to greater accumulation of iron and phosphorus in Saginaw tops (Brown et al., 1972).

More recently, Rao et al. (1977) studied three soybean (*Glycine max* L.) cultivars that differed in the capacity to absorb zinc. In addition to total uptake differences, the distribution of zinc in plant parts varied among the cultivars. Furthermore, variation in the recovery of native (*P* value) and labile (*L* value) zinc among genotypes corresponded well with reported tolerances to zinc deficiency.

Giordano and Mortvedt (1969) obtained a differential response among several corn hybrids to zinc and phosphorus applied to a zinc-deficient soil in factorial combination. No relationship was found between P/Zn ratios in corn forage and degree of responsiveness to zinc. They concluded that P/Zn ratios are not definitive in assessing zinc nutrition when phosphorus or possibly other nutrients are limiting. Also, Terman et al. (1975) contend that differences in the absorption of zinc or other nutrients by corn hybrids are influenced through genetic effects on growth rates and yield potentials.

Differential response of rice cultivars to zinc was reported by Giordano and Mortvedt (1974). They observed that differences in phosphorus and zinc uptake occurred among cultivars grown in flooded and nonflooded soil, but these were not related to the relative response to zinc. Very early maturing genotypes appeared more susceptible to zinc stress than late maturing types.

7. INTERACTIONS WITH OTHER NUTRIENTS

Adequate zinc nutrition in plants depends on several factors in addition to ability of the soil to supply zinc. Other growth factors include the rate of zinc absorption, distribution of zinc to functional sites, and mobility within the plant. Moreover, interactions of zinc with other elements modify plant nutrition by affecting some growth factors. An understanding of these interactions is important in providing adequate zinc nutrition.

Optimum growth usually occurs within a limited range of nutrient concentrations. Therefore it is important in soil management and fertilization practices that concentrations of plant nutrients be maintained in these ranges. An interaction in plant growth can be defined as a mutual or reciprocal action of one element upon another, that is, the effect of one nutrient is dependent on or affected by the level or concentration of another. Several main interactions of zinc with other elements are briefly discussed in this section. The literature on interactions was fully reviewed and discussed by Olsen (1972).

7.1. Zinc-Phosphorus

Interaction of phosphorus and zinc has been widely studied, with the earliest work reported by Barnette et al. (1936). This interaction is usually associated with high levels of available phosphorus or with applications to soils that are low in available zinc. High uptake of phosphorus antagonizes or reduces uptake of zinc frequently to the point of zinc deficiency. This disorder, termed a phosphorus-induced zinc deficiency, can usually be prevented or corrected by applying zinc to the soil.

Actual causal relationships and mechanisms of this interaction are not known. Four possible causes have been studied: (1) phosphorus-zinc interaction in the soil, (2) decreased rate of zinc translocation in plants due to high levels of phosphorus, (3) dilution in zinc concentration in plant top growth due to growth response to phosphorus, and (4) high phosphorus concentrations in plants which interfere with metabolic functions of zinc in the cells.

Experimental results have shown that phosphorus-zinc interactions probably do not occur in the soil. It has been hypothesized that insoluble $Zn_3(PO_4)_2$ formed after phosphorus applications to soil, resulting in a decreased zinc concentration in soil solution. Boawn et al. (1957) and others have shown, however, that the zinc in $Zn_3(PO_4)_2$ is available to plants if finely divided and mixed with the soil. Interpretations of results concerning the effect of phosphorus on zinc uptake or translocation are conflicting. Most data show that phosphorus and zinc are mutually antagonistic, so application of one nutrient results in decreased uptake of the other. Decreased growth results only when either element is near the deficiency range, and in such cases a deficiency can be prevented by applying both elements.

When rate of plant growth exceeds rate of uptake of a nutrient, concentrations of that nutrient decrease in the plant tissue. With marginal zinc supply, increased plant growth with adequate phosphorus can result in decreased plant zinc concentrations to the point of zinc deficiency. With adequate zinc supply in the soil, there should be sufficient zinc for optimum growth even when plant zinc concentrations are diluted because of increased growth from added phosphorus.

Results of several studies concerning the physiological effects of phosphorus on zinc metabolism have been reported, but they have not elucidated this relationship. Sharma et al. (1968) found that the largest reduction in zinc in plants occurred with the lowest increment of applied phosphorus. Although a growth response to phosphorus may have occurred, they suggested that the main effect of applied phosphorus was a physiological inhibition in the translocation of zinc from roots to tops. Thorne (1957) reviewed studies on the physiological role of zinc in plant nutrition.

7.2. Zinc-Nitrogen

Increasing the nitrogen supply usually results in increased plant growth, which can dilute the zinc concentration. Viets et al. (1957) and Giordano et al. (1966) reported that the effects of nitrogen on the zinc concentrations in plants are also related to nitrogen source effects in decreasing soil pH; this would increase zinc uptake. Boawn et al. (1960) found the same pH effect (increasing zinc uptake) in grain sorghum [*Sorghum bicolor* (L.) *Moench*] and potatoes (*Solanum tuberosum* L.). They also reported that zinc concentrations and uptake by sugar beets (*Beta vulgaris* L.) increased with nitrogen application rate. Others have suggested that fertilizer nitrogen may enhance zinc uptake by increasing protein nitrogen, which retains zinc, or by increasing the exchange capacity of roots, thus stimulating the absorptive capacity for zinc (Lindsay, 1972).

7.3. Zinc-Iron

These two micronutrients apparently are mutually antagonistic in plant nutrition. Lingle et al. (1963) found that zinc interfered with the absorption and translocation of iron in soybeans. Brown and Tiffin (1962) also showed that added zinc resulted in iron chlorosis in corn and millet (*Setaria italica* L.). Mortvedt and Giordano (1971) demonstrated mutual iron-zinc antagonism in grain sorghum grown on calcareous soil. Zinc concentration and uptake decreased with increased rates of FeEDDHA [ferric chelate of ethylenediamine (di-*o*-hydroxyphenylacetic acid)], and forage yields decreased at the highest iron rate, with tissue zinc concentrations decreasing to 12 ppm. In contrast, soil applications of zinc in the absence of applied iron also resulted in decreased forage yields. Therefore iron fertilization of soils that are marginal in zinc can result in zinc deficiencies, and vice versa.

8. DIAGNOSIS OF DEFICIENCIES: SOIL AND TISSUE TESTS

Both soil tests and plant analyses have been used to diagnose zinc deficiencies in crops. Zinc concentrations in plant tissue generally reflect the amounts of available zinc in soils. Interpretation of these analytical results requires a knowledge of crop species and cultivar, physiological age, size and plant part, and concentrations of other elements, such as phosphorus and iron, which interact with zinc.

Concentrations of zinc in plants vary widely; the sufficiency range usually is considered to be 20 to 150 ppm. Concentrations below 20 ppm Zn generally indicate a marginal or deficient supply of zinc for the crop, while those in excess of 400 ppm are in the toxic range (Chapman, 1966). Because crop species and sometimes cultivars within a given species vary in zinc tolerance, critical concentration values cannot be given for all crops. Results published by Cox and Wear (1977) of a regional project in the southeastern United States showed that critical zinc concentrations both in ear leaves of corn and in rice leaves when panicles were 2 mm long were estimated to be 15 ppm. Jones (1972) discussed all phases of plant analyses for zinc and other micronutrients, including sampling, preparation, procedures, and interpretation of results.

Several soil tests for available zinc have been developed. Amounts of soil zinc extracted by 0.1 N HCl have been used extensively to estimate levels of plant-available zinc, especially in soils of similar pH and texture (Wear and Sommer, 1948). The critical level for 0.1 N HCl extractable zinc was 1.0 ppm. The double-acid procedure (0.05 N HCl + 0.025 N H_2SO_4) was developed for estimation of plant-available phosphorus in North Carolina, but it is now also calibrated for other plant nutrients: K, Ca, Mg, Mn, and Zn (Cox and Kamprath, 1972). Thus one extractant can be used for a number of soil tests, resulting in increased efficiency.

Lindsay and Norvell (1968) developed the DTPA soil test for zinc and other metallic micronutrients, primarily for use on calcareous soils. Because zinc deficiencies are found mainly on nearly neutral to alkaline soils, this test is widely used. The zinc levels used in interpreting this soil test are as follows: low, 0 to 0.5 ppm; marginal, 0.5 to 1.0 ppm; and adequate, greater than 1 ppm. Another soil test for zinc is the EDTA-ammonium carbonate test (Trierweiler and Lindsay, 1969). This chelate extractant has been used on acid to neutral soils for available zinc and other micronutrients. Cox and Wear (1977) reported results obtained when the two acid extractants and DTPA were used to estimate available zinc for corn in southeastern U.S. soils. Critical zinc levels were 1.4 ppm for 0.1 N HCl, 0.8 ppm for the double acid, and 0.5 ppm for DTPA.

Soil tests also can be used to detect buildup of zinc and thus estimate residual effects of applied zinc, Boawn (1974) showed that a $ZnSO_4$ application of 5.6 kg Zn/ha was adequate for 2 years, while a rate of 11.2 kg Zn/ha was adequate for 4 years. Buildup of zinc due to overfertilization with the element or to application of pesticides or fertilizers containing zinc can be detected, possibly before toxicity levels are reached. Zinc recommendations should be reevaluated after several annual applications if soil test results show a buildup of available zinc. Corrective actions such as liming soils and changing management practices can then be taken to prevent decreases in crop yields.

Reviews of the literature on soil tests for zinc were published by Cox and Kamprath (1972), Viets and Lindsay (1973), and Mortvedt (1977). Concepts of soil testing methods, procedures for calibrating and interpreting results, and descriptions of soil tests are included in these reviews.

9. SOURCES, RATES, AND APPLICATION METHODS

The most commonly used inorganic zinc sources are $ZnSO_4$ and ZnO, although a number of other sources are also sold as zinc fertilizers (Table 1). These fertilizers are usually the least expensive per unit of zinc. Zinc frits are prepared by fusing iorganic zinc sources into a silicate matrix. The molten mix is quenched, dried, and milled, and the solubility of the zinc is controlled by particle size. Fritted zinc products are generally used on acid, sandy soils in high-rainfall areas.

Table 1. Some Sources of Fertilizer Zinc

Source	Formula	Zinc (approx. %)
Zinc sulfate monohydrate	$ZnSO_4 \cdot H_2O$	35
Zinc sulfate heptahydrate	$ZnSO_4 \cdot H_2O$	23
Basic zinc sulfate	$ZnSO_4 \cdot 4Zn(OH)_2$	55
Zinc oxide	ZnO	78
Zinc carbonate	$ZnCO_3$	52
Zinc sulfide	ZnS	67
Zinc frits	(Silicates)	1–15
Zinc phosphate	$Zn_3(PO_4)_2$	51
Zinc chelates	$Na_2 Zn\text{-}EDTA$	14
	$NaZn\text{-}NTA$	13
	$NaZn\text{-}HEDTA$	9
Zinc polyflavonoid	–	10
Zinc ligninsulfonate	–	5

Synthetic chelates are formed by combination of a chelating agent with divalent zinc. Although several types of chelates are available, Zn-EDTA is the most popular product. Chealtes range from 5 to 20 times higher in cost per unit of zinc than inorganic sources. Natural organic complexes are made by reacting zinc salts with organic by-products of the wood pulp industry. The two general classes of these complexes, which usually are less effective than synthetic chelates but also less costly, are lignin sulfonates and polyflavonoids.

Application rates vary widely with zinc source, method of application, and crop. Some typical recommended rates for various crops are listed in Table 2. Generally, rates are lower with band than with broadcast application, and chelates usually are more effective than inorganic sources per unit of applied zinc. Effectiveness ratios of zinc chelates and inorganic zinc sources may range from 3:1 to 10:1, depending on the crop species, zinc rate, and method of application.

Corn is the most widely zinc-fertilized crop in the United States. Zinc fertilizers are also recommended for wheat (*Triticum* spp.), soybeans, grain sorghum, snap beans (*Phaseolus vulgaris* L.), potatoes, onions (*Allium cepa* L.), and a number of other vegetables and tree fruits. Zinc is also recommended for rice, and this crop has the largest percentage of its total acreage fertilized with zinc in the United States.

Foliar applications usually are effective in correcting zinc deficiencies, but more than one application rate may be required, thus increasing costs. Salt concentrations of foliar sprays must not be too high; otherwise leafburn will

Table 2. Some Recommended Rates and Methods of Zinc Application for
Various Crops

Crop	Zinc Source	Rate of Zinc	Method of Application
Citrus	$ZnSO_4$, ZnO	6 g/1	Foliar
Corn, snap beans, sorghum	$ZnSO_4$, ZnO	2–10 kg/ha	Broadcast or banded
Corn, snap beans, sorghum	Zn chelates	0.2–4 kg/ha	Banded
Onions	$ZnSO_4$, ZnO	3 kg/ha	Banded
Onions, potatoes	Zn chelates	0.2 kg/ha	Foliar
Peaches	$ZnSO_4$	0.3 g/1	Foliar
Pecans	$ZnSO_4$	0.7–1.8 kg/tree	Broadcast
Rice	$ZnSO_4$, $ZnCl_2$ ZnO	5–10 kg/ha	Broadcast
Rice	Zn chelates	1 kg/ha	Broadcast
Wheat	$ZnSO_4$, ZnO	1–5 kg/ha	Broadcast

result. Addition of wetting agents to foliar sprays increases the amount of spray
that adheres to the leaves. Early morning applications usually result in increased
absorption because relative humidity is high, stomata are open, and photosynthesis is taking place.

Foliar spray applications of zinc generally are recommended for tree crops,
although soil applications have been effective in pecan (*Carya illinoensis*)
orchards. Dormant sprays, applied in late winter or early spring, also are widely
used. Foliar zinc sprays are used mainly as emergency treatments after deficiency symptoms have appeared in field and vegetable crops.

10. SUMMARY

Zinc uptake and accumulation vary widely in agricultural plants and are related
to soil pH and texture, soil organic matter content, temperature and light intensity, crop species, and cultivars within a species. Interactions with other plant
nutrients, especially P, N, and Fe, affect zinc uptake by plants.

Both plant tissue analyses and soil tests can be used to diagnose zinc deficiencies in crops. A number of both inorganic and organic sources of zinc are
effective in correcting zinc deficiencies in agricultural crops. Recommendations
regarding sources, rates, and methods of application should be followed for
specific crops.

REFERENCES

Barnette, R. M., Camp, J. P., Warner, J. D., and Gall, O. E. (1936). "Use of Zinc Sulfate under Corn and Other Field Crops," *Fla. Agr. Exp. Sta. Bull.* 293, p. 3.

Bauer, A. and Lindsay, W. L. (1965). "The Effect of Soil Temperature on the Availability of Indigenous Soil Zinc," *Soil Sci. Soc. Am. Proc.,* 29, 413–420.

Boawn, L. C. (1974). "Residual Availability of Fertilizer Zinc," *Soil Sci. Soc. Am. Proc.,* 38, 800–803.

Boawn, L. C., Viets, F. G., Jr., and Crawford, C. L. (1957). "Plant Utilization of Zinc from Various Types of Compounds and Fertilizer Materials," *Soil Sci.,* 83, 219–227.

Boawn, L. C., Viets, F. G., Jr., Crawford, C. L., and Nelson, J. L. (1960). "Effect of Nitrogen Carrier, Nitrogen Rate, Zinc Rate, and Soil pH on Zinc Uptake by Sorghum Potatoes and Sugar Beets," *Soil Sci.,* 90, 329–337.

Brown, J. C. and Tiffin, L. O. (1962). "Zinc Deficiency and Iron Chlorosis Dependent on the Plant Species and Nutrient-Element Balance in Tulare Clay," *Agron. J.,* 54, 356–363.

Brown, J. C., Ambler, J. E., Chaney, R. L., and Foy, C. D. (1972). "Differential Response of Plant Genotypes to Micronutrients." In J. J. Mortvedt, P. M. Giordano, and W. L. Lindsay, Eds., *Micronutrients in Agriculture.* Soil Science Society of America, Madison, Wis., pp. 389–418.

Chapman, H. D. (1966). *Diagnostic Criteria for Plants and Soils.* University of California Press, Riverside, pp. 484–499.

Cox, F. R. and Kamprath, E. J. (1972). "Micronutrient Soil Tests." In J. J. Mortvedt, P. M. Giordano, and W. L. Lindsay, Eds., *Micronutrients in Agriculture.* Soil Science Society of America, Madison, Wis., pp. 289–317.

Cox, F. R. and Wear, J. I. (1977). *Diagnosis and Correction of Zinc Problems in Corn and Rice Production.* North Carolina State University, South. Coop. Ser. Bull. 222, Raleigh, 73 p.

Elgawhary, S. M., Lindsay, W. L., and Kemper, W. D. (1970). "Effect of EDTA on the Self-Diffusion of Zinc in Aqueous Solution and in Soil," *Soil Sci. Soc. Am. Proc.,* 34, 66–70.

Forno, D. A., Yoshida, S., and Asher, J. (1975). "Zinc Deficiency in Rice: I: Soil Factors Associated with the Deficiency," *Plant Soil,* 42, 537–550.

Giordano, P. M. (1977). "Efficiency of Zinc Fertilization for Flooded Rice," *Plant Soil,* 48, 673–684.

Giordano, P. M. and Mortvedt, J. J. (1969). "Response of Several Corn Hybrids to Level of Water-Soluble Zinc in Fertilizers," *Soil Sci. Soc. Am. Proc.,* 33, 145–148.

Giordano, P. M. and Mortvedt, J. J. (1972). "Rice Response to Zn in Flooded and Nonflooded Soil," *Agron. J.,* 64, 521–524.

Giordano, P. M. and Mortvedt, J. J. (1974). "Response of Several Rice Cultivars to Zinc," *Agron. J.,* **66,** 220–223.

Giordano, P. M. and Mortvedt, J. J. (1978). "Response of Corn to Zn in Ortho- and Pyrophosphate Fertilizers, as Affected by Soil Temperature and Moisture," *Agron. J.,* 70, 531–534.

Giordano, P. M., Mortvedt, J. J., and Papendick, R. I. (1966). "Response of Corn (*Zea mays* L.) to Zinc as Affected by Placement and Nitrogen Source," *Soil Sci. Soc. Am. Proc.,* **30,** 767–770.

Giordano, P. M., Noggle, J. C., and Mortvedt, J. J. (1974). "Zinc Uptake by Rice as Affected by Metabolic Inhibitors and Competing Cations," *Plant Soil,* 41, 637–646.

Hodgson, J. F. (1963). "Chemistry of the Micronutrient Elements in Soils." In A. G. Norman, Ed., *Advances in Agronomy,* Vol. 15. Academic Press, New York, pp. 119–159.

Hodgson, J. F., Lindsay, W. L., and Trierweiler, J. F. (1966). "Micronutrient Cation Complexing in Soil Solution. II: Complexing of Zn and Cu in Displaced Solution from Calcareous Soils," *Soil Sci. Soc. Am. Proc.,* 30, 723–726.

Jones, J. B., Jr. (1972). "Plant Tissue Analysis for Micronutrients." In J. J. Mortvedt, P. M. Giordano, and W. L. Lindsay, Eds., *Micronutrients in Agriculture.* Soil Science Society of America, Madison, Wis., pp. 319–346.

Lindsay, W. L. (1972a). "Inorganic Phase Equilibria of Micronutrients in Soils." In J. J. Mortvedt, P. M. Giordano, and W. L. Lindsay, Eds., *Micronutrients in Agriculture.* Soil Science Society of America, Madison, Wis., pp. 41–57.

Lindsay, W. L. (1972b). "Zinc in Soils and Plant Nutrition." In N. C. Brady, Ed., *Advances in Agronomy,* Vol. 24. Academic Press, New York, pp. 147–186.

Lindsay, W. L. and Norvell, W. A. (1968). *Development of a DTPA Micronutrient Soil Test.* Agronomy Abstracts, American Society of Agronomy, Madison, Wis., p. 84.

Lingle, J. C., Tiffin, L. O., and Brown, J. C. (1963). "Iron Uptake-Transport of Soybeans as Influenced by Other Cations," *Plant Physiol.,* 38, 71–76.

Lucas, R. E. and Knezek, B. D. (1972). "Climatic and Soil Conditions Promoting Micronutrient Deficiencies in Plants." In J. J. Mortvedt, P. M. Giordano, and W. L. Lindsay, Eds., *Micronutrients in Agriculture.* Soil Science Society of America, Madison, Wis., pp. 265–288.

Martin, W. E., McLean, J. G., and Quick, J. (1965). "Effect of Temperature on Phosphorus Induced Zinc Deficiency," *Soil Sci. Soc. Am. Proc.,* 29, 411–413.

Mikkelsen, D. S. and Kuo, S. (1977). *Zinc Fertilization and Behavior in Flooded Soils.* Commonwealth Bureau of Soils, Spec. Publ. 5, 59 p.

Moore, D. P. (1972). "Mechanisms of Micronutrient Uptake by Plants." In J. J. Mortvedt, P. M. Giordano, and W. L. Lindsay, Eds., *Micronutrients in Agriculture.* Soil Science Society of America, Madison, Wis., pp. 171–198.

17

ZINC TOLERANCE BY PLANTS

Werner Mathys

Hygiene-Institut, Universität Münster, Münster, West Germany

1. Introduction	415
2. The Habitat	417
3. Uptake and Localization of Zinc	418
4. Physiology of Zinc Tolerance	420
4.1. Determination of the degree of tolerance	420
4.2. Enzymes and heavy metal tolerance	422
4.3. Zinc tolerance and complexing agents	427
5. The Mechanism of Zinc Tolerance	430
References	432

1. INTRODUCTION

Zinc, copper and other heavy metals are necessary for many enzymatic and physiological processes occurring within plants (Hewitt, 1958; Vallee and Wacker, 1970). The amounts required are very low, and normally the small quantities of heavy metals present in undisturbed soils are sufficient for plant life.

At many places all over the world, however, poisoning concentrations of heavy metals prevent normal plant growth. Zinc, in particular, is a very common element in contaminated areas. Partly it is of natural origin (Ernst, 1974); partly the affected sites document human action in the past and at the present time. Increasing industrialization is associated with a greater demand for heavy metals and greater pollution (e.g., by smelters or mine wastes).

415

In the course of evolution a special plant population has been established on these polluted sites. Even small areas are sharply separated from the neighboring vegetation. Some species, such as *Viola calaminaria, Minuartia verna, Thlaspi alpestre,* and *Armeria maritima,* are strongly restricted to these habitats and are never found on nearby unpolluted pastures. Other species (e.g., *Agrostis tenuis, Festuca ovina,* and *Anthoxanthum odoratum*) are usually common members of the normal plant cover of the region. A detailed study on the ecological problems of heavy metal populations has been made by Ernst (1974).

A common characteristic of all species growing on zinc-contaminated habitats is the evolution of zinc-tolerant ecotypes (Baumeister, 1967; Antonovics et al., 1971). Zinc tolerance is one of the many possible tolerances against metals: resistance to aluminum (Clarkson, 1966), lead (Wilkins, 1960), copper (McNeilly, 1968), cadmium (Coughtrey and Martin, 1977), nickel (Sasse, 1976), chromium (Lyon et al., 1969), and other metals is also well known. Zinc tolerance is the most intensively studied tolerance (for reviews see Antonovics et al., 1971; Antonovics, 1975).

The evolution for zinc tolerance can be very rapid (Wu and Bradshaw, 1972; Gartside and McNeilly, 1974a, 1974b; Bradshaw, 1975). The evolution of zinc-tolerant races of *Festuca ovina* and *Agrostis canina* growing under galvanized fences in Britain took only 25 years (Bradshaw et al., 1965). Only a few years were required for the evolution of tolerant *Agrostis tenuis* and *A. canina* near a zinc smelter (Ernst et al., 1975) and of lead-tolerant *Marchantia polymorpha* (Briggs, 1972). The evolutionary processes in mine populations of grasses have been intensively investigated by Bradshaw (1952, 1975), Antonovics and Bradshaw (1970), and McNeilly (1968).

Tolerant ecotypes are distinguished from the sensitive ones not only by their degree of tolerance, but also, for example, by their biomass production (Cook et al., 1972; Mathys, 1975a, 1977), their requirement for nutrients (Jowett, 1959), and the development of self-fertility (Antonovics, 1968; Lefèbre, 1970, 1975). Gene flow between tolerant and nontolerant populations is very low (Bradshaw et al., 1969; Khan, 1969; Antonovics, 1975), and therefore tolerant species are often isolated from their sensitive ecotypes. Such isolation is the rule for species growing only on mine wastes and thus separated by long distances from other races (e.g., *Armeria maritima;* Lefèbre, 1975). This species (and *Thlaspi alpestre* and *Viola calaminaria*) could prevent the competition of other plants only by establishing a zinc tolerance and by restricting itself to contaminated areas. Such species have to be considered as glacial relicts (Ernst, 1974).

Investigations of the genetics of tolerant plants showed that zinc tolerance in *Anthoxanthum odoratum* and *Agrostis tenuis* is dominant and directional and is controlled by a polygenic system (Gartside and McNeilly, 1974a, 1974b). Urquhart (1971) demonstrated a high heritability for lead tolerance in *Festuca ovina;* Bröker (1963), for zinc resistance in *Silene inflata.* Unfortunately there is a lack of studies crossing zinc-tolerant races with species tolerant of other heavy metals.

2. THE HABITAT

Analysis of heavy-metal-contaminated soils shows the extreme conditions under which tolerant species have to survive (Table 1). Zinc is often associated with

Table 1. Heavy Metal Contents of Some Contaminated Soils in Western Germany (μmol/g dry soil)

Habitat	Metal			
	Zn	Cu	Pb	Cd
Zinc mine (12th century)	682	0.7	58.0	0.09
Zinc-copper waste	370	38.0	4.2	0.01
Zinc mine (20th century)	275	1.7	10.3	2.40
Zinc smelter	75	3.8	0.9	1.40
Copper waste	9.2	75.5	9.2	0.00
Pasture soil	2.8	0.9	0.0	0.00

Table 2. Heavy Metal Contents (μmol/g dry weight) in the Shoots of Some Species Growing on Zinc-Contaminated Soils

Species	Metal					
	Zn	Pb	Cu	Cd	Fe	Mn
Old mine waste						
Agrostis tenuis	5.7	0.35	0.25	0.01	2.9	1.2
Festuca ovina	17.1	0.45	0.11	0.02	1.3	0.2
Silene cucubalus	27.3	0.21	0.27	0.41	N.D.	N.D.
Minuartia verna	65.7	1.40	0.38	0.50	N.D.	N.D.
Thlaspi alpestre	185.6	3.50	0.34	2.10	N.D.	N.D.
Cardaminopsis halleri	382.4	3.20	0.12	0.50	2.2	0.5
Philonotis fontana	173.4	5.32	0.40	0.41	10.7	1.8
Bryum bimum	297.6	4.86	0.31	0.54	18.9	2.1
Zinc smelter						
Agrostis tenuis	6.8	0.01	1.1	1.32	0.9	0.4
Agrostis canina	5.4	0.01	1.4	1.00	1.6	0.3

lead, and combinations of zinc with copper are also documented. A great part of the heavy metals is easily soluble and can be taken up by the plants (Ernst et al., 1974). New human-made habitats such as smelters and wastes contain various combinations of heavy metals; in most of them, however, zinc is the main constituent.

The good solubility of zinc is well demonstrated by the analysis of plants from polluted sites (Table 2). On old mine wastes with their typical community of tolerant plants great differences in uptake pattern are noticeable. The grasses *Agrostis tenuis* and *Festuca ovina* accumulate only a small portion of the heavy metals taken up by the dicotyls. *Cardaminopsis halleri, Thlaspi alpestre,* and the mosses accumulate especially high amounts of zinc. All species tested contain concentrations of zinc in their leaves that are toxic to nontolerant ones. This may be an indication that restriction of uptake is not the mechanism for tolerance.

3. UPTAKE AND LOCALIZATION OF ZINC

The above-mentioned uptake of heavy metals from the habitat can be confirmed in culture experiments. All metal-tolerant species take up large amounts of zinc; an exclusion mechanism, like that for copper-tolerant *Chlorella vulgaris,* cannot be detected (Foster, 1977). A comparison of the uptakes of tolerant and non-tolerant populations of *Agrostis tenuis* (Mathys, 1973) revealed no restriction of

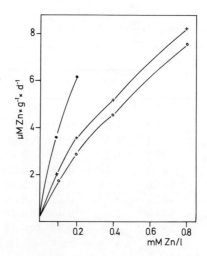

Figure 1. Zinc uptake (μmol Zn g day) of zinc-tolerant (o), copper-tolerant (x), and sensitive (•) populations of *Agrostis tenuis.*

zinc from the shoots of tolerant ecotypes. The tolerant race accumulates appreciable amounts of zinc, depending on the zinc content of the nutrient medium (Figure 1). Similar uptake patterns have been reported by Ernst (1974) for *Silene cucubalus* and *Thlaspi alpestre,* by Reilly et al. (1970) for copper resistant *Becium homblei,* by Wu and Antonovics (1975b) for zinc tolerant *Agrostis stolonifera,* by Baker (1978) for zinc resistant *Silene maritima.* In *S. maritima,* however, the tolerant ecotype took up lower amounts of zinc than the sensitive one; translocation of zinc into the shoots seemed in some way restricted. Zinc and copper uptake by *A. stolonifera* tolerant to both zinc and copper showed no effect of zinc on copper uptake (Wu and Antonovics, 1975b). Similar results have been reported by Veltrup (1975) for heavy metal uptake by *Hordeum vulgare.* The mechanisms for zinc and copper uptake seem to be independent (see also Wainwright and Woolhouse, 1977), perhaps explaining the different mechanisms of zinc and copper tolerance.

The main site of zinc accumulation in *Agrostis tenuis* is the root (Mathys, 1973; Turner, 1970). This fact has led to the hypothesis that the root may act as a protective organ against zinc (Turner, 1970; Turner and Gregory, 1967; Turner and Marshall, 1971, 1972). Experiments of these authors and of Peterson (1969) dealing with the intracellular distribution of zinc within the roots of *A. tenuis* show nearly 90% of total zinc to be located in the cell wall of the tolerant ecotype. For German populations of *A. tenuis* this distribution cannot be confirmed (Table 3). The zinc-tolerant ecotype of German *A. tenuis* accumulates only 51% of zinc in the cell wall of the root; the amount of soluble zinc in the leaves reaches nearly 70% (Mathys, 1973). Yet a greater part of the zinc is bound to the cell wall of the tolerant race than to the cell wall of the sensitive ecotype. A single, primary role of the root in the evolution of zinc resistance, however, has to be denied.

Table 3. Percentage Distribution of Zinc with the Cells of Roots of *Agrostis tenuis* from Zinc Cultures

Ecotype	Vacuole + Plasma	Cell Wall
Zinc-tolerant (Germany)	48.4	51.6
Zinc-tolerant[a] (U.K)	9.4	90.6
Copper-tolerant (Germany)	66.0	34.0
Copper-tolerant[a] (U.K.)	43.8	56.2
Sensitive (Germany)	67.0	33.0

[a] Data from Turner (1970).

4. PHYSIOLOGY OF ZINC TOLERANCE

4.1. Determination of the Degree of Tolerance

When comparing the physiologies of tolerant and sensitive races, it is of great importance to know the exact degree of tolerance. Different methods of determination have been developed: the test for protoplasmatic resistance (Repp, 1963; Rüther, 1967; Gries, 1966) and the rooting test (Wilkins, 1957; Jowett, 1964).

The rooting test compares the length of roots in zinc solutions with the length of roots in solutions without zinc. The protoplasmatic resistance method allows one to test every single cell of an organism for its tolerance by putting slices of plant organs into heavy-metal-containing solutions and estimating the number of surviving cells. Both methods permit quick determination of the tolerance index. In particular, the rooting test has been widely used in investigations of the evolution and genetics of grasses, and has recently been applied to leaves of dicotyls (Wu and Antonovics, 1975a).

From tests of German populations of *Agrostis tenuis* (Figure 2) the evolution of zinc tolerance in all races growing on contaminated soils is well documented. Plants sampled near a zinc smelter have achieved the highest degree of tolerance, perhaps because of the high amounts of water-soluble zinc and because of zinc

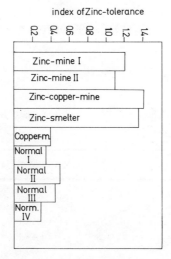

Figure 2. Index of zinc tolerance (mean root length in zinc culture : mean root length in the control) of *Agrostis tenuis* from different habitats.

air particles. Tolerance is highly specific (Bradshaw et al., 1965; Mathys, 1973). Cotolerances to heavy metals *not* occurring in the original habitats are not very common; a possible cotolerance has been described only for zinc and nickel (Bradshaw et al., 1965). As a rule, there is never a cotolerance between zinc and copper, indicating completely different mechanisms.

In addition to the two tests mentioned above, the measurement of total biomass production (Mathys 1975, 1977) is a very useful way to test for tolerance. Its advantage is to elucidate long-term effects. Sensitive plants usually diminish their production (Table 4); tolerant ones not only endure the heavy metal load but also are highly stimulated. Tolerant races of *Silene cucubalus* augment their biomass up to 300%, *Philonotis fontana* to 500%, and *Thlaspi alpestre* to more than 1000%. Appreciable amounts of zinc seem to be necessary for optimal growth. *Thlaspi alpestre* is barely able to live without high zinc nutrition. This species seems, furthermore, to have a natural resistance to zinc, which is augmented to a very high degree in plants growing in zinc-contaminated areas. Real cotolerances can be shown in this species (Table 5) for zinc and nickel and zinc and cadmium. Also there seems to be a slight tolerance to cobalt. Only copper

Table 4. Biomass Production (Percentage of Control Biomass) of Several Species after a Growth Period of 4 Weeks in a Zinc Culture[a]

Habitat	Species				
	Silene cucubalus	*Rumex acetosa*	*Agrostis tenuis*	*Thlaspi alpestre*	*Philonotis fontana*
Zinc-contaminated					
Germany I	377	131	108	—	—
Germany II	229	—	—	1000	—
Germany III	—	160	160	700	500
France	320	—	—	—	—
Copper-contaminated					
Germany	61	—	87	—	—
Nickel-contaminated					
France	70	—	—	—	—
Not contaminated					
Germany I	—	38	90	—	50
Germany II	52	—	—	—	—
Austria	41	—	—	—	—
France I	47	—	—	—	—
France II	—	—	—	74	—
France III	—	—	—	87	—

[a] 2.0 mM Zn for *Thlaspi* and *Philonotis*, 0.4 mM for the other species.

Table 5. Biomass Production (Grams per Plant) of *Thlaspi alpestre* after a 4-Week Application of Some Heavy Metals to the Nutrient Solution

Metal (m*M*)	Zinc-Tolerant	Sensitive
Zinc, 2.0	10.8	4.8
Nickel, 0.17	2.0	2.5
Nickel, 0.7	1.5	+[a]
Nickel, 1.7	0.6	+
Cadmium, 0.18	1.4	+
Cobalt, 0.68	0.7	1.0
Copper, 0.16	+	+
None	1.0	5.9

[a] + = death.

proves to be very toxic; there is no cotolerance. Tolerance against nickel is as high as in serpentine plants (Sasse, 1976), which have a resistance against zinc, too.

4.2. Enzymes and Heavy Metal Tolerance

There are some data on the physiology of heavy metal resistance in bacteria and yeasts (Horii et al., 1956; Murayama, 1961; Ashida and Nakamura, 1959; Ashida, 1965). It was shown that these organisms can evolve heavy-metal-resistant enzymes. Some yeasts achieve tolerance by precipitation of the metal in the cell wall (Ashida et al., 1963).

In higher plants comparable evolutionary steps are not detectable (Mathys, 1975b). Although the plants have different degrees of tolerance, there are no differences in their leaf enzymes in regard to tolerance (Table 6). The tolerant populations have enzymes with the same sensitivity to metals as do nontolerant ecotypes. Similar results are reported by Wu et al. (1975a) for malate dehydrogenase of *Agrostis stolonifera*.

However, it is well documented that there are great differences in the sensitivity of the enzymes to the various metals (Figures 3 to 5). The most sensitive enzyme is nitrate reductase; the most toxic element is copper. The threshold values for the other enzymes are much higher (500-fold and more) than the value for nitrate reductase.

The ATPases and phosphatases also react in the same manner (Figure 6). Application of zinc causes a reduction of activity in both tolerant and sensitive clones. The threshold values are very high, reaching 25 m*M* Zn for 50% inhibition of ATPase. At 10 m*M*, zinc stimulates the ATPases of all ecotypes. Veltrup

Table 6. Concentrations (m*M*) of Zinc Inhibiting 50% of Activity of Different Enzymes in Leaves of Zinc-Tolerant and Non-Zinc-Tolerant Ecotypes of *Silene cucubalus*

	Enzyme[a]						
Ecotype	NR	MDH	ICDH	G6P-DH	ENO	AcP	ATPase
Zinc-tolerant	0.006	1.40	0.17	0.36	0.06	10.0	25.0
Copper-tolerant	0.006	1.46	0.16	0.35	0.05	9.8	21.0
Sensitive	0.006	1.46	0.17	0.35	0.06	9.7	25.0

[a] NR = nitrate reductase; MDH = malate dehydrogenase; ICDH = isocitrate dehydrogenase; G6P-DH = glucose-6-phosphate dehydrogenase; ENO = enolase; AcP = acid phosphatase; ATPase = adenosine triphosphatase.

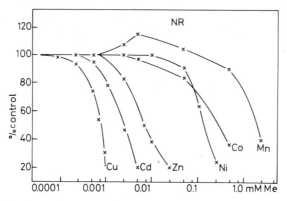

Figure 3. Effect of some heavy metals on activity of nitrate reductase (NR) in leaves of a zinc-tolerant population of *Silene cucubalus*. Control value = 100%.

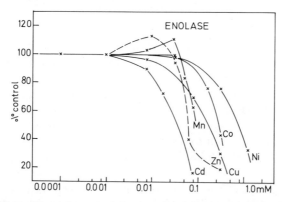

Figure 4. Effect of some heavy metals on activity of enolase in leaves of a zinc-tolerant population of *Silene cucubalus*. Control value = 100%.

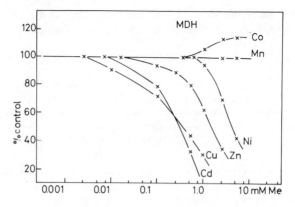

Figure 5. Effect of some heavy metals on activity of malate dehydrogenase (MDH) in leaves of a zinc-tolerant population of *Silene cucubalus*. Control value = 100%.

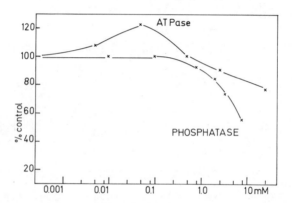

Figure 6. *In vitro* effect of zinc (mM Zn) on activity of ATPase and phosphatase in leaves of a zinc-tolerant population of *Silene cucubalus*. Control values = 100%.

(1975) found stimulation by zinc for ATPases of *Hordeum vulgare* and discussed a participation of ATPases in ion uptake. Cox and Thurman (1978) failed to detect tolerant acid phospatases in *Anthoxanthum odoratum*, yet the phosphatases are distinguished by their K_i values. It is assumed that nontolerant ecotypes form more stable complexes with zinc than do tolerant ones.

When testing intact root tips of *Agrostis tenuis*, Woolhouse (1969) detected different reactions to lead in tolerant and sensitive forms. Phosphatases of lead-tolerant *A. tenuis* are less inhibited than those of sensitive plants. This somewhat puzzling behavior can be confirmed for intact leave slices of *Silene cucubalus* (Figure 7). Only the ATPase of the tolerant race is stimulated by zinc *in vivo;* inhibition of the ATPase of the tolerant plants occurs only at very high concen-

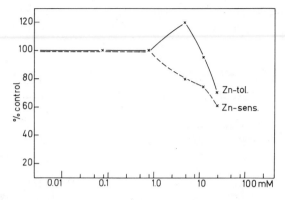

Figure 7. *In vivo* effect of zinc (m*M* Zn) on activity of ATPases in leaf slices of zinc-tolerant and sensitive populations of *Silene cucubalus*. Control values = 100%.

trations of zinc. Veltrup (1978) reported that the ATPases in *Hordeum vulgare* are stimulated by copper only *in vivo*, not *in vitro*. The role of an intact cell system is thus emphasized; there are great differences between *in vivo* and *in vitro* testing.

In vivo enzyme activities document as well as does, for example, biomass production the tolerance of the zinc forms. For nitrate reductase (Table 7) concentrations as low as 0.2 m*M* Zn cause a repression of activity in the non-zinc-tolerant populations. The nitrate reductase of the tolerant forms is enhanced to a high degree and is scarcely inhibited even at large zinc excess. Remarkable effects are also observed for the other enzymes. Isocitrate dehydrogenase of resistant populations is stimulated; that of sensitive ecotypes, depressed (Table 8). A reduction of enzymatic activity in the tolerant plants occurs for glucose-6-

Table 7. *In vivo* Activity of Nitrate Reductase (μmol NO_2^-/h x g fresh weight) in Leaves from *Silene cucubalus* at Various Levels of Zinc in the Culture Medium

Zinc (mmol/1)	Ecotype				
	Zinc-Tolerant	Zinc-Tolerant	Nickel-Tolerant	Copper-Tolerant	Sensitive
0.0	3.2	2.0	1.8	5.2	6.3
0.2	3.9	3.0	1.5	2.0	2.1
0.4	5.5	2.9	0.8	1.2	0.9
0.8	4.5	N.D.	+[a]	+	+
1.6	2.6	N.D.	+	+	+

[a] + = death.

Table 8.   Effects of Zinc (0.4 mmol/1 nutrient solution) on the Relative Activity of Some Enzymes in the Leaves of Zinc-Tolerant and Non-Zinc-Tolerant Ecotypes of *Silene cucubalus*
Control values = 100%.

Ecotype	Enzyme[a]						
	MDH	ICDH	G6P-DH	GOT	CA	POD	Gl-DH
Zinc-tolerant	95.3	126.8	54.9	90.3	217.9	60.6	94.4
Zinc-tolerant	100.4	137.4	70.6	82.5	362.5	61.7	N.D.
Copper-tolerant	102.1	93.4	133.7	178.5	97.4	97.8	100.0
Nickel-tolerant	94.3	N.D.	228.9	144.5	100.0	129.6	N.D.
Sensitive	88.3	83.0	137.9	107.7	101.9	168.9	126.3

[a]MDH = malate dehydrogenase; ICDH = isocitrate dehydrogenase; G6P-DH = glucose-6-phosphate dehydrogenase; GOT = glutamate oxaloacetate transaminase; CA = carbonic anhydrase; POD = peroxidase; Gl-DH = glutamate dehydrogenase.

phosphate dehydrogenase and peroxidase, whereas the non-zinc-tolerant forms augment their enzyme activities. Peroxidase has been used as an unspecific stress indicator (Wakiuchi et al., 1971; Keller, 1974) and has demonstrated in our experiments a poisoning of the sensitive plants, as has glucose-6-phosphate dehydrogenase. On the other hand, these enzymes indicate a disturbed metabolism in the tolerant forms in the control.

Carbonic anhydrase, an indicator enzyme for the zinc level in the cytoplasm (Randall and Bouma, 1973; Ohki, 1976; Wood and Silby, 1952), offers strong proof of a real need for zinc in the tolerant races: application of zinc causes a significant enhancement of activity in the tolerant plants, whereas the activity of the sensitive populations remains nearly unaffected. The zinc supply in control cultures (with sufficient zinc for normal ecotypes) seems to be too low.

Only a slight effect of zinc is noticed for malate dehydrogenase. This enzyme has a great compatability with zinc *in vitro* and is scarcely affected *in vivo*. As isozymes of malate dehydrogenase do not change very much during the development of plants, they are suitable for comparative investigations of different ecotypes (Longo and Scandalios, 1969; Rocha and Ting, 1970; O'Sullivan and Wedding, 1972). In *Silene cucubalus* (Figure 8) tolerant races seem to have more isozymes in seed and leaves than do non-zinc-tolerant ones. Yet the possibility cannot be excluded that these are only ecological variations without correlation to zinc tolerance. Such variations of esterases are reported by Wu et al. (1975b) for *Agrostis tenuis* and of phosphatases by Cox and Thurman (1978). No correlation between distribution of isozymes and tolerance could be detected in these cases.

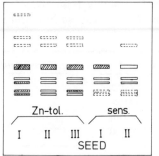 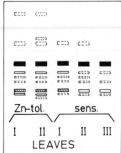

Figure 8. Isozymes of malate dehydrogenase in seed and leaves of zinc-tolerant and sensitive ecotypes of *Silene cucubalus.*

4.3. Zinc Tolerance and Complexing Agents

As the existence of tolerant enzymes in higher plants has to be denied and as *in vivo* experiments have shown the importance of an intact cell system, zinc tolerance can be interpreted only in terms of mechanisms preventing a buildup of heavy metals in the plasma, either by complexation or by special transport.

The development of tolerance by the production of complexing agents is known for selenium (Rosenfeld and Beath, 1964), fluoride (Preuss et al., 1970), and copper (Ashida et al., 1963). Reilly (1972) assumed amino acids to be involved in copper tolerance. Oxalate has been discussed as a complexing agent for chromium (Lyon et al., 1969).

In *Silene cucubalus* the tall, fast-growing sensitive populations produce the highest concentrations of oxalate (Table 9). The heavy-metal-resistant plants contain somewhat lower amounts. The water-soluble part, however, is higher within the zinc-tolerant plants; thus relatively more oxalate is available for possible complexation of zinc. In *Rumex acetosa* nearly all oxalate is water soluble. An application of zinc causes drastic changes in the production of oxalate. Synthesis of this organic acid is severely depressed in the non-zinc-tolerant forms of *S. cucubalus,* but not of *R. acetosa.* Within the zinc-tolerant ecotypes the addition of zinc stimulates the production of oxalate in both *S. cucubalus* and *R. acetosa.* A correlation between oxalate content and zinc tolerance, however, cannot be found; there are only ecotype-specific variations.

Analysis of malate, on the other hand, reveals great quantitative differences between zinc-tolerant and non-zinc-tolerant populations (Mathys, 1975a, 1977). Malate content is dependent on state of development (Table 10). In all green organs of *Silene cucubalus* substantially more malate is found in the zinc-tolerant ecotype than in the sensitive one. The leaves are the main site of accumula-

Table 9. Oxalate Contents (μmol total oxalate/g fresh weight) in Leaves of Some Ecotypes of *Silene cucubalus* and *Rumex acetosa* in Cultures with 0.001 m*M* Zn (control) and 0.4 m*M* Zn

Ecotype	Control		0.4m*M* Zn	
	Total Oxalate	Percentage Water-Soluble Oxalate	Total Oxalate	Percentage Water-Soluble Oxalate
Silene cucubalus				
Zinc-tolerant I	58.9	89.9	95.7	84.0
Zinc-tolerant II	68.6	93.9	75.8	95.6
Zinc-tolerant III	31.0	90.3	34.3	87.2
Copper-tolerant	65.2	71.6	46.8	75.4
Nickel-tolerant	52.2	61.3	47.8	78.0
Sensitive I	79.6	68.3	40.1	75.8
Sensitive II	96.5	71.0	57.3	83.1
Rumex acetosa				
Zinc-tolerant I	160.3	93.0	190.2	93.0
Zinc-tolerant II	159.6	91.4	178.9	96.8
Sensitive	148.8	92.5	142.0	95.6

Table 10. Malate Contents (μmol/g fresh weight) in Organs of a Zinc-Tolerant and a Sensitive Population of *Silene cucubalus* ($\pm$ SE)

Organ	Ecotype	
	Zinc-Tolerant	Sensitive
Leaves		
Not unfolded	2.8 ± 0.9	0.6 ± 0.4
2nd–5th pair	5.9 ± 1.0	0.9 ± 0.5
6th–10th pair	8.9 ± 0.2	1.7 ± 0.3
11th–15th pair	13.9 ± 0.4	2.3 ± 0.4
Stem	4.5 ± 0.3	0.9 ± 0.1
Root	1.8 ± 0.1	1.0 ± 0.1
Seed	1.5 ± 0.2	0.3 ± 0.0

tion; here more than five times as much malate is produced in the tolerant form as in the sensitive. It is assumed that there may be a connection between number of isozymes of malate dehydrogenase, malate synthesis, and zinc tolerance.

A general statement can be formulated only by comparing many ecotypes of different species for a common basis. It can clearly be demonstrated (Table 11) that *all* zinc-tolerant ecotypes of all species tested contain much more malate in their leaves or stems than do the non-zinc-tolerant. Malate is hence a common property of *zinc* tolerance in herbage plants and perhaps in some mosses (Mathys, 1975a). Plants resistant to other metals (copper or nickel) behave like non-tolerant forms; they produce little malate.

The difference in malate content between zinc-tolerant and nontolerant populations is not the same for all species. There seems to be a direct correlation between degree of tolerance (Table 11) and surplus production of malate. *Thlaspi alpestre,* the species with the highest degree of tolerance, produces the most malate. In this species a second mechanism may have evolved. Like all crucifers (Kjaer, 1960; Jiracek and Krulich, 1971; Kutacek and Kralova, 1972),

Table 11. Malate Contents (μmol/g fresh weight) in Leaves of Some Ecotypes of Different Species and Indices of Tolerance (Biomass Production in Zinc Culture: Biomass Production in Control)

Species	Index of Tolerance	Malate Content
Silene cucubalus		
Zinc-tolerant I	3.8	8.8 ± 0.2
Zinc-tolerant II	2.3	8.9 ± 0.4
Zinc-tolerant III	1.3	5.8 ± 0.1
Zinc-tolerant IV	3.2	7.6 ± 0.6
Copper-tolerant	0.6	1.6 ± 0.3
Nickel-tolerant	0.7	1.6 ± 0.0
Sensitive I	0.5	2.2 ± 0.3
Sensitive II	0.4	0.8 ± 0.0
Sensitive III	0.5	0.8 ± 0.0
Rumex acetosa		
Zinc-tolerant I	1.3	8.2 ± 0.2
Zinc-tolerant II	1.2	6.7 ± 0.1
Sensitive	0.4	1.7 ± 0.8
Agrostis tenuis		
Zinc-tolerant I	1.1	7.8 ± 0.1
Zinc-tolerant II	1.4	8.4 ± 0.4
Sensitive	0.9	6.0 ± 0.2
Thlaspi alpestre		
Zinc-tolerant	10	73.6 ± 3.7
Sensitive I	0.7	40.3 ± 4.0
Sensitive II	0.9	29.7 ± 3.8

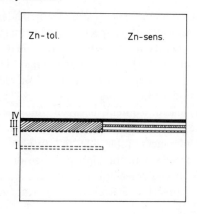

Figure 9. Chromatography of mustard oil glucosides in a zinc-tolerant and a sensitive population of *Thlaspi alpestre.*

T. alpestre contains large amounts of mustard oil glucosides (Figure 9). The tolerant ecotypes contain twice as much as the sensitive populations (Mathys, 1975a, 1977). Precursors of these mustard oils (e.g., amino acids) may be involved in the general resistance of this species, but further investigations of this aspect are needed. There are also indications (Sasse, 1976) that mustard oil precursors may help to explain nickel resistance.

5. THE MECHANISM OF ZINC TOLERANCE

Malate seems to be the key to understanding the physiological mechanism of zinc tolerance. There are at least two possible ways in which malate can effect this tolerance (Mathys, 1977). The first requires complexation of zinc immediately after its uptake into the plasma (Figure 10); the second, a special transport system. The stability constant (Bjerrum et al., 1957) for the zinc-malate complex is high enough to render possible a diminution of zinc to a low level in the plasma, provided that there is sufficient malate to make possible a 1:1 complex. The more malate is available, the more the ionic (= toxic) zinc will decrease. Only the zinc-tolerant ecotypes contain such high values of malate that the concentration of the organic acid exceeds the content of zinc.

Thus formation of a zinc-malate complex is sufficient to explain zinc tolerance and the different degrees of tolerance that have been observed. It is assumed that the surplus production can be easily achieved by the plants by only slight modification of enzymes, for instance, malate dehydrogenase isozymes. Evolution of tolerance can occur rapidly as no fundamental variations have to be made.

Furthermore, tolerance can also be rendered possible by special transport, as malate is involved in many transport processes (Lüttge, 1973; Cram, 1974). A combination of these two possibilities is thought to provide a good model of the mechanism of zinc tolerance (Figure 10).

If one assumes that the high concentrations of malate are mainly located within the cytoplasm (Torii and Laties, 1966) and can permeate easily between compartments (Lips and Beevers, 1966; Cram and Laties 1974), malate may act as follows. The first step in ion (zinc) uptake is transport through the plasmalemma. Here ATPases may be involved in some way (Veltrup, 1975). This transport leads to an accumulation of zinc first in the plasma, where it can react with zinc-dependent enzymes. The efficiency of the second step in ion uptake, uptake through the tonoplast, determines whether zinc will be eliminated from the plasma into the vacuole. If (in the tolerant ecotypes) the turnover for zinc is higher at the tonoplast than at the plasmalemma, the consequence is a constant diminution of zinc in the plasma. Additionally, in the zinc-tolerant plants with their higher amounts of malate, transport through the cytoplasm may be more effective and, because of complexation, more innocuous. Thus the concentration of ionic zinc may be kept small within the plasma without altering the total uptake. As ion uptake is different for zinc and copper (Veltrup, 1978; Wu and Antonovics, 1975b), specifity of resistance is easily explained.

The elimination of ionic zinc from the plasma seems to be so effective that in control cultures zinc-tolerant plants suffer from an actual deficit of zinc (see the discussions of nitrate reductase, carbonic anhydrase, and biomass production in Section 4.2). The malate that is transported bound to zinc into the vacuole does not stay there. As the zinc-malate complex is weaker than many other complexes (e.g., zinc-oxalate), zinc is transferred from malate to oxalate or other

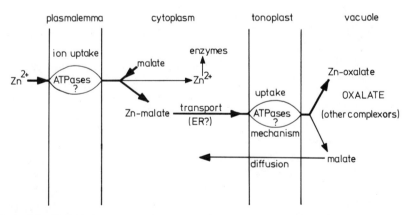

Figure 10. Model of a mechanism of zinc tolerance in herbaceous plants (explanation in text).

complexors. Malate can then diffuse back into the plasma for a new complexation of zinc. Thus malate functions as a "transport vehicle" for zinc. The only requirement the complexing agent (e.g., oxalate or precursors of mustard oils) has to fulfill is to have a greater affinity for zinc than does malate. Oxalate functions as a "terminal acceptor," as a sink for large amounts of zinc. The complexors in the vacuole effect a constant concentration gradient for unidirectional transport of zinc from the plasma into the vacuole.

This model of a mechanism for zinc tolerance may be applicable to many species. In some species, such as *Agrostis tenuis* (Turner 1970) and *Silene maritima* (Baker 1978), a restricted uptake or translocation of zinc may be an additional explanation for zinc tolerance.

REFERENCES

Antonovics, J. (1968). "Evolution in Closely Adjacent Plant Populations. V. Evolution of Self-Fertility," *Heredity*, 23, 219–238.

Antonovics, J. (1975). "Metal Tolerance in Plants: Perfecting an Evolutionary Paradigm." In *International Conference on Heavy Metals in the Environment, Symposium Proceedings*, Vol. II, Toronto, pp. 169–186.

Antonovics, J. and Bradshaw, A. D. (1970). "Evolution in Closely Adjacent Plant Populations, VIII: Clinal Patterns n *Anthoxanthum odoratum* across a Mine Boundary," *Heredity*, 25, 349–362.

Antonovics, J., Bradshaw, A. D. and Turner, R. G. (1971). "Heavy Metal Tolerance in Plants," *Adv. Ecol. Res.*, 7, 1–85.

Ashida, J. (1965). "Adaptation of Fungi to Metal Toxicants," *Ann. Rev. Phytopathol.*, 3, 153–174.

Ashida, J., Higashi, N. and Kikuchi, T. (1963). "An Electron Microscopic Study on Copper Precipitation by Copper Resistant Yeast Cells," *Protoplasma*, 57, 27–32.

Baker, A. J. M. (1978). "Ecophysiological Aspects of Zinc Tolerance in *Silene maritima* With.," *New Phytol.*, 80, 635–642.

Baumeister, W. (1967). "Schwermetallpflanzengesellschaften und Zinkresistenz einiger Schwermetallpflanzen," *Angew. Bot.*, 40, 185–202.

Bjerrum, J., Schwarzenbach, G., and Sillèn, L. G. (1957). *Stability Constants*. The Chemical Society, London.

Bradshaw, A. D. (1952). "Populations of *Agrostis tenuis* Resistant to Lead and Zinc Poisoning," *Nature*, 169, 1098.

Bradshaw, A. D. (1975). "The Evolution of Metal Tolerance and Its Significance for Vegetation Establishment on Metal Contaminated Sites." In *International Conference on Heavy Metals in the Environment, Symposium Proceedings*, Vol. II, Toronto, pp. 599–622.

Bradshaw, A. D., McNeilly, T. S. and Gregory, R. P. G. (1965). "Industrialisation, Evolution and the Development of Heavy Metal Tolerance in Plants." In *Ecology and the Industrial* Society. Fifth British Ecological Society Symposium, Oxford, pp. 327–343.

Briggs, D. (1972). "Population differentiation in *Marchantia polymorpha* L. in Various Lead Pollution Levels," *Nature,* 238, 166–167.

Bröker, W. (1963). "Genetisch-physiologische Untersuchungen über die Zinkverträglichkeit von *Silene inflata* Sm.," *Flora,* B153, 122–156.

Clarkson, D. T. (1966). "Aluminium Tolerance in Species within the Genus *Agrostis*," *J. Ecol.,* 54, 167–178.

Cook, S. C. A., Lefébre, C., and McNeilly, T. (1972). "Competition between Metal Tolerant and Normal Plant Populations on Normal Soil," *Evolution,* 26, 366–372.

Coughtrey, P. J. and Martin, M. H. (1977). "Cadmium Tolerance of *Holcus lanatus* from a Site Contaminated by Aerial Fallout," *New Phytol.,* 79, 273–280.

Cox, R. M. and Thurman, D. A. (1978). "Inhibition by Zinc of Soluble and Cell Wall Acid Phosphatases of Zinc-Tolerant and Non-tolerant Clones of *Anthoxanthum odoratum*," *New Phytol.,* 80, 17–22.

Cram, W. J. (1974). "Effects of Cl^-, HCO_3^- and Malate Fluxes and CO_2 Fixation in Carrot and Barley Root Cells," *J. Exp. Bot.,* 25, 253–268.

Cram, W. J. and Laties, G. G. (1974). "The Kinetics of Bicarbonate and Malate Exchange in Carrot and Barley Root Cells," *J. Exp. Bot.,* 25, 11–27.

Ernst, W. (1974). "Schwermetallvegetation der Erde." In *Geobotanica selecta,* Vol. V. Stuttgart, pp. 1–194.

Ernst, W., Mathys, W., Salaske, J., and Janiesch, P. (1974). "Aspekte von Schwermetallbelastungen in Westfalen," *Abh. Landesmus. Naturk. Münster,* 36 pp. 1–31.

Ernst, W., Mathys, W., and Janiesch, P. (1975). "Physiologische Grundlagen der Schwermetallresistenz," *Forschungsber. Landes. Nordfhein-Westfalen.* 2496 21–38.

Foster, P. L. (1977). "Copper Exclusion as a Mechanism of Heavy Metal Tolerance in a Green Alga," *Nature,* 269, 322–323.

Gartside, D. W. and McNeilly, T. (1974a). "Genetic Studies in Heavy Metal Tolerant Plants. I: Genetics of Zinc Tolerance in *Anthoxanthum odoratum*," *Heredity,* 32, 287–297.

Gartside, D. W. and McNeilly, T. (1974b). "Genetic Studies in Heavy Metal Tolerant Plants. II: Zinc Tolerance in *Agrostis tenuis*," *Heredity,* 33, 303–308.

Gries, B. (1966). "Zellphysiologische Untersuchungen über die Zinkresistenz bei Galmeiformen und Normalformen von *Silene cucubalus* Wib.," *Flora,* B156, 271–290.

Hewitt, E. J. (1958). "The Role of Mineral Elements in the Activity of Plant

Enzyme Systems." In W. Ruhland, *Handbuch der Pflanzenphysiologie*, Vol. IV, pp. 427–481.

Horii, T., Higashi, T., and Okunuki, K. (1956). "Copper Resistance of *Mycobacterium tuberculosis avium*. II: The Influence of Copper Ion on the Precipitation of the Parent Cells and Copper-Resistant Cells," *J. Biochem.*, **42**, 491–498.

Jiracek, V. and Krulich, J. (1971). "Biosynthesis of Indole Glucosinolates in Rape Seedlings under the Influence of Some Sulphur Metabolism Inhibitors," *Experientia*, **27**, 1010–1013.

Jowett, D. (1959). "Adaptation of a Lead Tolerant Population of *Agrostis tenuis* to Low Soil Fertility," *Nature*, **184**, 43.

Jowett, D. (1964). "Population Studies on Lead Tolerant *Agrostis tenuis*," *Evolution*, **18**, 70–80.

Keller, T. (1974). "The Use of Peroxidase Activity for Monitoring and Mapping Air Pollution Areas," *Eur. J. For. Pathol.*, **4**, 11–19.

Khan, M. S. (1969). "The Process of Evolution of Heavy Metal Tolerance in *Agrostis tenuis* and Other Grasses." M. Sc. thesis, University of Wales.

Kjaer, A. (1960). "Naturally Derived Isothiocyanates (Mustard Oils) and Their Parent Glucosides," *Fortschr. Chem. Org. Naturst.*, **18**, 123–176.

Kutacek, M. and Kralova, M. (1972). "Biosynthesis of the Glucobrassicin Aglycone from ^{14}C and ^{15}N labelled L-Tryptophan Precursors," *Biol. Plant.*, **14**, 279–285.

Lefébre, C. (1970). "Self Fertility in Maritime and Zinc Mine Populations of *Armeria maritima* (Mill) Willd.," *Evolution*, **24**, 771–777.

Lefébre, C. (1975). "Evolutionary Problems in Heavy Metal tolerant *Armeria maritima*." In *International Conference on Heavy Metals in the Environment, Symposium Proceedings*, Vol. II, pp. 155–167.

Lips, S. H. and Beevers, H. (1966). "Compartmentation of Organic Acids in Corn Roots. II: The Cytoplasmic Pool of Malic Acid," *Plant Physiol.*, **41**, 713–716.

Longo, G. P. and Scandalios, J. G. (1969). "Nuclear Gene Control of Mitochondrial Malic Dehydrogenase in Maize," *Proc. Natl. Acad. Sci.*, **62**, 104–111.

Lüttge, U. (1973). *Stofftransport der Pflanzen. Springer*, Berlin, Heidelberg, New York.

Lyon, G. L., Peterson, P. J., and Brooks, R. R. (1969). "Chromium-51 Distribution in Tissues and Extracts of *Leptospermum scoparium*," *Planta*, **88**, 282–287.

Mathys, W. (1973). "Vergleichende Untersuchungen der Zinkaufnahme von resistenten und sensitiven Populationen von *Agrostis tenuis* Sibth.," *Flora*, **162**, 492–499.

Mathys, W. (1975a). "Physiologische Aspekte der Zinktoleranz bei *Silene cucubalus* Wib. und anderen Arten," Ph.D. thesis, University of Münster.

Mathys, W. (1975b). "Enzymes of Heavy Metal Resistant and Nonresistant Populations of *Silene cucubalus* and Their Interaction with Some Heavy Metals *in vitro* and *in vivo*," *Physiol. Plant.*, 33, 161–165.

Mathys, W. (1977). "The Role of Malate, Oxalate, and Mustard Oil Glucosides in the Evolution of Zinc-Resistance in Herbage Plants," *Physiol. Plant.*, 40, 130–136.

McNeilly, T. (1968). "Evolution in Closely Adjacent Plant Populations. III: *Agrostis tenuis* on a Small Copper Mine," *Heredity*, 23, 99–108.

Murayama, T. (1961). "Studies on the Metabolic Pattern of Yeast with Reference to Its Copper Resistance," *Mem. Ehime Univ.*, II(B4), 43–66.

Ohki, K. (1976). "Effect of Zinc Nutrition on Photosynthesis and Carbonic Anhydrase Activity in Cotton," *Physiol. Plant.*, 38, 300–305.

O'Sullivan, S. A. and Wedding, R. T. (1972). "Malate Dehydrogenase Isoenzymes from Cotton Leaves: Molecular Weights," *Plant Physiol.*, 49, 117–123.

Peterson, P. J. (1969). "The Distribution of Zinc-65 in *Agrostis stolonifera* L. and *Agrostis tenuis* Sibth. tissues," *J. Exp. Bot.*, 20, 863–875.

Preuss, P. W., Colavito, L., and Weinstein, L. H. (1970). "The Synthesis of Monofluoroacetic Acid by a Tissue Culture of *Acacia georginae*," *Experientia*, 26, 1059–1060.

Randall, P. J. and Bouma, D. (1973). "Zinc Deficiency, Carbonic Anhydrase, and Photosynthesis in Leaves of Spinach," *Plant Physiol.*, 52, 229–232.

Reilly, A., Rowel, J., and Stone, J. (1970). "The Accumulation and Binding of Copper in Leaf Tissue of *Becium homblei* (de Wild) Duvign.," *NewPhytol.*, 69, 993–997.

Reilly, C. (1972). "Amino Acids and Amino Acid Copper Complexes in Water-Soluble Extracts of Copper-Tolerant and Non-tolerant *Becium homblei*," *Z. Pflanzenphysiol.*, 66, 294–296.

Repp, G. (1963). "Die Kupferresistenz des Protoplasmas höherer Pflanzen auf Kupfererzböden," *Protoplasma*, 57, 643–659.

Rocha, V. and Ting, J. P. (1970). "Tissue Distribution of Microbody, Mitochondrial and Soluble Malate Dehydrogenase Isoenzymes," *Plant Physiol.*, 46, 754–756.

Rosenfeld, I. and Beath, O. A. (1964). *Selenium: Geobotany, Biochemistry, Toxicity, and Nutrition. Academic Press,* New York, London.

Rüther, F. (1967). "Vergleichende physiologische Untersuchungen über die Resistenz von Schwermetallpflanzen," *Protoplasma*, 64, 400–425.

Sasse, F. (1976). "Ökophysiologische Untersuchungen der Serpentinvegetation in Frankreich, Italien, Östereich und Deutschland." Ph. thesis, University of Münster.

Torii, K. and Laties, G. H. (1966). "Organic Acid Synthesis in Response to Excess Cation Absorption in Vacuolate and Non-Vacuolate Sections of Corn and Barley Roots," *Plant Cell Physiol.*, 7, 395–403.

Turner, R. G. (1970). "The Subcellular Distribution of Zinc and Copper within the Roots of Metal Tolerant Clones of *Agrostis tenuis* Sibth.," *New Phytol.,* **69**, 725–731.

Turner, R. G. and Gregory, R. P. G. (1967). "The Use of Radioisotopes to Investigate Heavy Metal Tolerance in Plants." In *Isotopes in Plant Nutrition and Physiology.* International Atomic Energy Agency/Food and Agriculture Organization Symposium, Vienna, pp. 493–509.

Turner, R. G. and Marshall, C. (1971). "The Accumulation of ^{65}Zn by Root Homogenates of Zinc-Tolerant and Non-tolerant Clones of *Agrostis tenuis* Sibth.," *New Phytol.,* **70**, 539–545.

Turner, R. G. and Marshall, C. (1972). "The Accumulation of Zn by Subcellular Fractions of Roots of *Agrostis tenuis* Sibth. in Relation to Zinc Tolerance," *New Phytol.,* **71**, 671–676.

Urquhart, C. (1971). "Genetics of Lead Tolerance in *Festuca ovina*," *Heredity,* **26**, 19–33.

Vallee, B. L. and Wacker, W. E. C. (1970). "Metalloproteins."In H. Neurath, *The Proteins,* Vol. 5, New York.

Veltrup, W. (1975). "Zink- und Kupferaufnahme durch Gerstenwurzeln." Ph. thesis, University of Münster.

Veltrup, W. (1978). "Der Einfluss von Cu^{2+} und Mg^{2+} auf die Aktivität von ATPasen aus Gerstenwurzeln," *Biochem. Physiol. Pflanz.,* **173**, 17–22.

Wainwright, S. J. and Woolhouse, H. W. (1977). "Some Physiological Aspects of Copper and Zinc Tolerance in *Agrostis tenuis* Sibth.: Cell Elongation and Membrane Damage," *J. Exp. Bot.,* **28**, 1029–1036.

Wakiuchi, N., Matsumoto, H., and Takahashi, E. (1971). "Changes of Some Enzyme Activities of Cucumber during Ammonium Toxicity," *Physiol. Plant.,* **24**, 248–253.

Wilkins, D. A. (1957). "A Technique for the Measurement of Lead Tolerance in Plants," *Nature,* **180**, 37–38.

Wilkins, D. A. (1960). "The Measurement and Genetical Analysis of Lead Tolerance in *Festuca ovina.*" *Rep. Scot. Res. Plant Breed. Sta.,* pp. 85–98.

Wood, I. G. and Silby, P. M. (1952). "Carbonic Anhydrase Activity in Plants in Relation to Zinc Content," *Aust. J. Sc. Res.,* **5**, 244–250.

Woolhouse, H. W. (1969). "Differences in the Properties of the Acid Phosphatases of Plant Roots and Their Significance in the Evolution of Edaphic Ecotypes." In *Ecological Aspects of the Mineral Nutrition of Plants,* Oxford, Edinburgh.

Wu, L. and Antonovics, J. (1975a). "Experimental Ecological Genetics in *Plantago.* I: Induction of Roots and Shoots on Leaves for Large Scale Vegetative Propagation and Metal Tolerance Testing in *P. lanceolata*," *New Phytol.* **75**, 277–282.

Wu, L. and Antonovics, (1975b). "Zinc and Copper Uptake by *Agrostis stolonifera,* Tolerant to Both Zinc and Copper," *New Phytol.,* **75**, 231–237.

Wu, L. and Bradshaw, A. D. (1972). "Aerial Pollution and the Rapid Evolution of Copper Tolerance," *Nature,* **238**, 167.

Wu, L., Bradshaw, A. D., and Thurman, D. A. (1975a). "The Potential for Evolution of Heavy Metal Tolerance in Plants. III: The Rapid Evolution of Copper Tolerance in *Agrostis stolonifera,*" *Heredity,* **34**, 165–187.

Wu, L., Thurman, D. A., and Bradshaw, A. D. (1975b). "The Uptake of Copper and Its Effect upon Respiratory Processes of Roots of Copper-Tolerant and Non-tolerant Clones of *Agrostis stolonifera,*" *New Phytol.,* **75**, 225–229.

18

ZINC TRANSPORT AND METABOLISM IN MICROORGANISMS

Mark L. Failla

Department of Biochemistry & Nutrition, Virginia Polytechnic Institute & State University, Blacksburg, Virginia

Eugene D. Weinberg

Department of Biology and Program in Medical Sciences, School of Medicine, Indiana University, Bloomington, Indiana

1. Introduction	439
2. Acquisition and Storage of Zinc	440
3. Functions	446
4. Summary	454
Acknowledgments	455
References	455

1. INTRODUCTION

The earliest demonstration that the transition metal zinc had a biological function occurred in 1869, when Raulin, a student of Pasteur, found that it was essential for the growth of *Aspergillus niger,* a filamentous fungus. Interestingly, a similar requirement for bacterial growth was not detected until 1947 (Feeney et al., 1947), well after the essentiality of zinc had been established for algae, plants, and mammals, in 1900, 1914, and 1934, respectively. Undoubtedly, the delayed discovery of the bacterial requirement was due to the submicromolar quantities needed since media prepared with reagent grade chemicals are generally contaminated with 0.5 to 1.0 μM zinc. Bacteria and most eucaryotic microorganisms achieve optimal growth without supplementation of growth media with additional zinc; some fungi require an increased quantity of the metal (Foster, 1949; Rawla, 1969). Moreover, attempts to prevent optimal cell growth by significantly reducing the zinc content of media by solvent extraction techniques or alumina absorption have often been unsuccessful.

The first comprehensive review of zinc accumulation and functions in microorganisms was published recently (Failla, 1977). This chapter extends that discussion and emphasizes recent developments in this rapidly expanding area of research. We will first consider how microorganisms acquire and store zinc and will then survey the various known and proposed roles that the metal has in these cells.

2. ACQUISITION AND STORAGE OF ZINC

Zinc is present in all natural environments. The bioavailability of this trace metal is largely determined by its physiochemical state. Zinc exists as a soluble, hydrated species in acidic environments. However, it may also exist as a component of insoluble complexes in neutral and alkaline soils (Jurinak and Inouye, 1962) and adsorbed to colloids in seawater (Zirino and Healy, 1970). Organic sequestering agents may be required for solubilization of the metal in such cases. For example, the addition of synthetic chelating agents, such as ethylenediamine tetraacetic acid (EDTA), as well as culture filtrates containing unidentified natural metal-binding substances, to systems containing low levels of available trace metals markedly enhanced microbial cell growth (Johnston, 1964; Lange, 1974). Microorganisms may have also developed the ability to synthesize, secrete, and reacquire high-affinity metal-binding molecules. Indeed, many microorganisms produce low molecular weight iron-sequestering agents, collectively referred to as siderophores (Byer and Arceneaux, 1977). Similar high-affinity zinc-binding substances of microbial origin have not been isolated thus far. Perhaps fungi, which generally proliferate in acidic rather than alkaline environments, have not been compelled to evolve a mechanism for zinc ionophore synthesis in order to succeed in most of their natural habitats. Nevertheless, it is reasonable to assume that some organisms inhabiting environments with higher pH have the ability to produce zinc-sequestering agents.

Zinc can interact with the numerous anionic sites on the microbial cell surface. Nonspecific absorption resembles cationic binding to ion exchange resins, that is, it is rapid, reversible, pH dependent, and energy independent (Gutknecht, 1963). Because of the critical roles zinc has in cellular metabolism (discussed in

Section 3), highly specific receptors that either directly or indirectly mediate translocation of the metal across the cell surface are probably also present on or exposed to the outer surface. The activity of such "permeases" would be expected to be subject to regulatory control(s), thereby assuring that the cell contained the necessary intracellular complement of zinc at all times. Below we summarize available data concerning the characteristics of microbial zinc transport systems.

In bacteria, energy-dependent zinc uptake has been reported for *Escherichia coli* (Bucheder and Broda, 1974; King et al., 1976; S. Silver and A. Weiss, personal communication), *Salmonella enteriditis* (Chipley and Edwards, 1972), and *Serratia marcescens* (Witney et al., 1977). Bucheder and Broda (1974) found that zinc uptake in *E. coli* exhibited saturation kinetics with an apparent K_m of 20 μM and V_{max} of 2.7 μmol/min g dry weight. These values are similar to those for the magnesium transport system, but are much higher than the ones for the uptake of manganese, which, like zinc, is needed by *E. coli* in only trace quantities (Silver, 1978). Moreover, magnesium was absent from the assay buffer, and ten-fold excess of cadmium reduced zinc uptake by only 50%. Thus we agree with Silver's suggestion that zinc was probably being acquired via the magnesium transport system in these studies. Zinc has been shown to be an alternative substrate for the magnesium uptake system (Webb, 1970).

Kinetic values for zinc uptake that are more consistent with trace nutrient requirements have been observed in *E. coli* more recently (Kung et al., 1976; Silver and Weiss, personal communication). Furthermore, when uptake was examined in synchronously dividing cells, the cellular zinc content increased in a steplike manner 10 to 15 min after each cell division. In contrast, Mg, K, and Ca uptake occurred gradually throughout the cell cycle (Kung et al., 1976).

In fungi, energy-dependent uptake of zinc has been observed in mycelia of *Neocosmospora vasinfecta* (Paton and Budd, 1972), but the absence of added magnesium in the assay system and the low affinity (K_m of 200 μM Zn^{2+}) again suggest that zinc was acquired via the magnesium transport system. Moreover, zinc uptake was inhibited by magnesium as well as by a nonphysiological quantity (500 μM) of manganese. In contrast, in a study with *Candida utilis*, when a 1000-fold excess of magnesium over zinc was employed (to simulate physiological conditions), zinc was transported by a highly specific energy-, pH-, and temperature-dependent system (Failla et al., 1976). Neither Mg, Na, Ca, Cr, Mn, Fe, Co, Ni, nor Cu suppressed zinc uptake; however, cadmium, the toxic congener of zinc, was an effective inhibitor. Transport was also inhibited by metabolic poisons such as cyanide, arsenate, and *m*-chlorophenylcarbonylcyanide hydrazone, as well as by *N*-ethylmaleimide, a sulfhydryl alkylating agent. Intact membranes were required for uptake inasmuch as the process was halted by addition of either a detergent (sodium dodecyl sulfate), an organic solvent (toluene), or a polyene antibiotic (nystatin).

At 30°C early- and mid-exponential phase cells of *C. utilis* harvested from low-zinc medium had a high-affinity system (apparent K_m of 2.0 μM Zn^{2+}) with a maximal rate of 0.22 nmol Zn^{2+}/min mg dry weight (Failla and Weinberg, 1977). Late-exponential phase cells had a similar apparent K_m (1.8 mM), but the V_{max} was increased 17-fold to 3.65 nmol/min mg dry weight (Figures 1 and 2). Cells in lag phase likewise had a high capacity for zinc uptake. Thus, during the batch culture growth cycle, a sequence of accumulation → dilution (as cells multiplied during exponential phase) → accumulation occurred. In recent studies on the kinetics of zinc uptake in *C. utilis* grown in continuous culture, the following values have been determined: apparent K_m of approximately 0.4 μM and V_{max} of approximately 3.0 nmol Zn^{2+}/min mg dry weight (Lawford et al., 1980).

While the pattern of uptake observed in *C. utilis* has not been detected for zinc in batch cultures of various bacterial species, cyclical uptake was observed in cultures of mouse fibroblast cells (Schwartz and Matrone, 1975). Since the characteristics of growth of *C. utilis* are identical in cells that contain low and high amounts of zinc, the purpose of "activated" zinc transport in late-

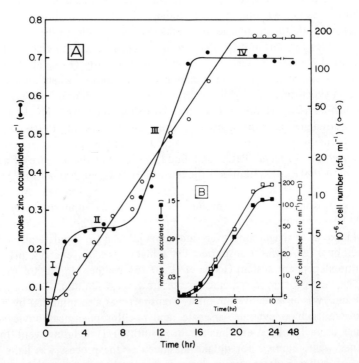

Figure 1. Accumulation of zinc and iron by *Candida utilis* in batch cultures. (Reproduced with permission from Failla and Weinberg, 1977.)

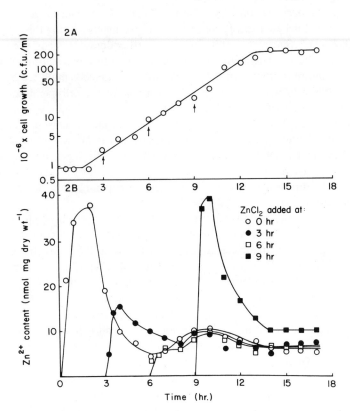

Figure 2. Effect of time of addition of ZnCl$_2$ on zinc uptake by *Candida utilis* in batch culture. At 0 hr, (0), 3 hr (●), 6 hr (□), and 9 hr (■), ZnCl$_2$ was added to batch cultures to a final concentration of 10 μM. When Zn^{2+} was not added until 9 hr, all the metal was taken up by the cells within the first hour. (Reproduced with permission from Failla and Weinberg, 1977.)

exponential phase organisms is not understood. The phenomenon perhaps is associated in some manner with the stringent quantitative requirements of zinc for such postexponential phase events as secondary metabolism and differentiation in fungi (discussed below).

Zinc exit-exchange reactions have not been observed in yeasts (Failla et al., 1976; Failla and Weinberg, 1977; Lawford et al., 1980; Ponta and Broda, 1970), filamentous fungi (Paton and Budd, 1972), or an alga (Parry and Haywood, 1973). EDTA caused partial release of the metal from *Chlorella* (Matzku and Broda, 1970); inasmuch as this chelator does not enter intact cells (Rahman and Wright, 1975), the zinc lost must have been derived from that bound to the surface. In *Escherichia coli* 25% of intracellular zinc exchanged with exogenous [65]Zn in 2 hr (Bucheder and Broda, 1974); unfortunately, during this period 50% of the cells died. Since, at least in fungi, zinc efflux does not occur, the transport system must be regulated through control of the influx mechanism. A possible means is transinhibition, in which uptake of a substance is inhibited by binding of the material to its carrier on the inner surface of the cell membrane. Microbial transport of various nutrilites is controlled by transinhibition (Becker and Lichstein, 1972; Bellenger et al., 1968; Cummins and Mitchison, 1967; Grenson, 1969; Hunter and Segal, 1973; Kadner, 1975; Morrison and Lichstein, 1976; Ring et al., 1970; Rytha, 1975; Yamamoto and Segal, 1966). However, the marked differences in zinc uptake in *Candida utilis* at different times in the batch culture growth cycle are difficult to explain solely on the basis of transinhibition.

During the period of activated transport in the late-exponential phase of batch cultures of *C. utilis,* the amount of zinc accumulated was as much as 115 to 130 μmol/g dry weight, which is equivalent to 0.7 to 0.85% of the dry weight of the cells (Failla and Weinberg, 1977). This high level of the metal was indeed located within the cytoplasm inasmuch as suspension of the cells in dilute HCl, EDTA, or nonradioactive Zn^{2+} after activated zinc uptake failed to reduce the quantity of [65]Zn associated with the viable microorganisms. Since the metal did not injure the cells in any observable way, it must have been packaged in innocuous form. Possibly intracellular zinc storage proteins and/or localization within a vacuole play such a role.

Our present understanding of zinc metabolism and storage after transfer across the cell surface is minimal in microorganisms. In contrast, this subject has received much attention in animal studies during the past decade. The isolation and characterization of soluble, low molecular weight zinc- and cadmium-binding proteins from equine (Kagi and Vallee, 1960, 1961) and human (Pulido et al., 1966) renal cortex provided the impetus for this work. Further investigations have shown that these soft metal-binding proteins, collectively referred to as metallothioneins (MT), are ubiquitous in the animal kingdom. The physiochemical and biological properties of MT are discussed in detail in Chapter 6 by Cousins and Failla. Below we briefly summarize the characteristics of MT and review data related to its possible presence in microorganisms.

Metallothioneins are single-chained, sulfur-rich polypeptides weighing approximately 6000 daltons. Cysteine accounts for about 30% of the total amino acid residues, while glycine, lysine, and serine comprise an additional 30%. The sulfur ligands of the cysteine residues can bind as many as 7 g-atoms of various metals (specifically Zn, Cu, Cd, Hg, Ag, and possibly Sn) per molecule. Both aromatic amino acids and intramolecular disulfide linkages are absent from the protein. The quantity of MT in tissues is directly proportional to the quantity of metal present; its synthesis is regulated at the transcriptional level. Many of the avail-

---

able data suggest that MT protects cells against both toxic (Cd, Hg. Ag) and elevated levels of essential (Zn, Cu), but nevertheless potentially toxic, metals. However, recent evidence strongly argues for a central role of MT in normal zinc (Bremner and Davies, 1975; Oh et al., 1978; Richards and Cousins, 1976; Sobocinski et al., 1978) and possibly copper (Ryden and Deutsch, 1978) metabolism in mammals.

Whether all microorganisms are able to synthesize MT-like metal storage proteins has not been resolved. A zinc-binding protein with chromatographic properties similar to those of MT was identified in *Anabena nidulans* (MacLean et al., 1972) and *Candida utilis* (Failla et al., 1976). However, such proteins were not found in cadmium-exposed *Escherichia coli* (Mitra et al., 1975) and *Saccharomyces cerevisiae* cultured in medium containing high levels of either zinc (Premakumar et al., 1975) or cadmium (Macara, 1978, Premakumar et al., 1975).

Low molecular weight copper-binding proteins have been isolated from yeasts (Naiki and Yamagata, 1976; Premakumar et al., 1975; Prinz and Weser, 1975; Weser et al., 1977). Like animal cell MT, these proteins weigh about 6000 daltons, lack aromatic amino acids, and have a high percentage of glycine, lysine, and serine (Table 1). However, the cysteine content ranges from 1.7 to 24.3%, and the glutamate and aspartate levels are elevated (Table 1). That all of these polypeptides are probably variant forms of copper-thionein is suggested by the

Table 1. Selected Amino Acid Compositions of Metallothionein-like Copper-Binding Proteins

Residue	CuBPI[a] Yeast (Naiki and Yamagata 1976)	Copper-thionein, Yeast (Weser et al., 1977)	Copper-chelatin, Yeast (Prema-Kumar et al., 1975)	Copper-thionein, Human Fetal Liver (Ryden and Deutsch, 1978)	Zinc-thionein, Rat Liver (Bremner and Davies, 1975)
Cysteine	7.1	24.3	13.0	24.3	27.2
Glycine	12.2	9.6	12.0	6.1	7.3
Serine	15.0	9.6	5.4	13.1	14.1
Lysine	13.0	8.9	8.8	13.6	12.5
Aspartic acid	17.3	14.6	11.1	5.3	6.7
Glutamic acid	21.3	19.2	16.3	6.0	6.3

[a]Copper-binding protein I.

data of Weser et al. (1977). As the copper concentration of the growth medium is increased, there is a concomitant increase in both the copper and cysteine contents of the intracellular metal-binding protein. Thus thioneins containing 1, 2, 8, and 10 g-atoms of copper and 1.7, 5.1, 20.0, and 24.3% cysteine, respectively, have been purified. Moreover, the percentage of cysteine residues is inversely proportional to the total percentage of alanine, valine, and leucine; the sum of these four amino acids in copper-thioneins is $23.4 \pm 2.3\%$, regardless of the metal content. It appears that the decreased percentage of cysteine residues in the copper-binding proteins isolated by Premakumar et al. (1975) and Naiki and Yamagata (1976) may have resulted from the different culture conditions and/or strains employed. Although the mechanisms related to the synthesis and function of such proteins in yeasts remain unknown, it is apparent that the ability to synthesize MT-like proteins developed early in evolution.

Do bacteria have the ability to synthesize MT-like proteins? Except for the report by MacLean et al. (1972) concerning a cyanobacterium, low molecular weight, soft-metal-binding proteins have not been isolated. It is possible that alternative mechanisms of "detoxification" have evolved in procaryotes. Cadmium resistance in *Staphylococcus aureus*, mediated through plasmids (Kondo et al., 1974; Novick and Roth, 1968) results from the ability of the bacterium to partially exclude the toxic metal from its intracellular environment (Chopra, 1970; Weiss et al., 1978). Adaptation of *Escherichia coli* to cadmium is likewise due to the decreased permeability of the cell membrane to the metal (Mitra et al., 1975). A systematic study of representative bacteria, including those isolated from soft-metal-contaminated soil and water, will be required to answer the question posed above.

3. FUNCTIONS

Many proteins require stoichiometric quantities of metal ions for catalytic and/or structural purposes. A list of microbial zinc metalloenzymes is given in Table 2. These zinc-dependent enzymes (*a*) are synthesized by both pro- and eucaryotes; (*b*) represent a diverse group of catalytic proteins (oxidoreductases, transferases, hydrolases, lyases, an isomerase, and a ligase); (*c*) are required for protein, carbohydrate, lipid, and nucleic acid metabolism; and (*d*) may have either extracellular (e.g., amylase and neutral protease) or intracellular functions.

Specific roles for zinc in nucleic acid biosynthesis have recently been elucidated with the discovery that microbial and viral nucleotidyl transferases require the metal for activity (Table 3). Sea urchin DNA polymerase (Slater et al., 1971) and wheat germ RNA polymerase II (Petranyi et al., 1977) contain 1 and 7 g-atoms of tightly bound zinc per mole of enzyme, respectively. In addition, rat liver RNA polymerases I and II, sea urchin RNA polymerases I, II, and III

Table 2. Microbial Zinc Metalloenzymes

Enzyme	Source[a]	References	Enzyme	Source[a]	References
Alcohol dehydrogenase	Y	Hock and Vallee (1955)	Lactate-cytochrome c reductase	Y	Cremona and Singer (1964)
Aldolase	B, M, Y	Harris et al. (1969), Jagannathan et al. (1956), Kobes et al. (1969), Kowal et al. (1969), Schwartz et al. (1974), Schwartz and Feingold (1973), Sugimoto and Noson (1971)	Lactate dehydrogenase	Y	Curden and Labeyrie (1961), Gregolin and Singer (1963), Iwotsubo and Curdel (1961)
Alkaline phosphatase	B	Simpson and Vallee (1968), Simpson et al. (1968), Yoshizumi et al. (1974)	Megateriopeptidase	B	Keay et al. (1971)
Alkaline protease	B	Nakajima et al. (1974)	Neutral protease	B	Feder et al. (1971), Holmquist (1977), Keay and Wildi (1970), McConn et al. (1967), Miyata et al. (1971)
Aminopeptidase	B	Griffith and Prescott (1970), Metz and Rome (1976), Rodriguez-Absi and Prescott (1978)	Neutral proteinase	B, M	Hiramatsu and Ouchi (1972), Sekino (1972)
Amylase	B	Isemura and Kokiuchi (1955), Stein and Fischer (1962)	Nuclease P_1	M	Fugimoto et al. (1974, 1975)

Table 2. Continued

Enzyme	Source[a]	References	Enzyme	Source[a]	References
Aspartate transcarbamylase	B	Nelbach et al. (1972)	Nucleotide pyrophosphatase	Y	Twu et al. (1977)
Carboxypeptidase	B	Chabner and Bertino (1972), Seber et al. (1976)	Phospholipase C	B	Little and Otnass (1975)
Collagenase	B	Seifter et al. (1970)	Phosphomannose isomerase	Y	Gracy and Noltmann (1968a, 1968b)
Dihydroorotase	B	Taylor et al. (1976)	Pyruvate carboxlase	Y	Scrutton et al. (1970)
Dipeptidase	B	Hayman et al. (1974)	RNA polymerase	A, B, Y	Auld et al. (1976), Falchuk et al. (1976, 1977), Halling et al. (1977), Lattke and Weser (1976), Scrutton et al. (1971), Wandzilak and Benson (1977, 1978)
DNA polymerase	B	Slater et al. (1971), Springgata et al. (1973)	Superoxide dismutase	M, Y	Gosin and Fridovich (1972), Misra and Fridovich (1972), Rapp et al. (1973)
Elastase	B	Morihara and Tasuzuki (1975)	Thermolysin	B	Latt et al. (1969), Levy et al. (1975)
Glyceraldehyde 3-phosphate dehydrogenase	Y	Keleti (1966)	Transcarboxylase	B	Northrop and Wood (1969)
Lactamase	B	Davies and Abraham (1974)			

[a] A = algae, B = bacteria, M = molds, Y = yeast.

Table 3.   Microbial and Viral Zinc-Dependent Nucleotidyl Transferases

Transferase	Source	Zinc (g-atoms/ mol)	Molecular Weight (daltons)	References
DNA polymerases	*Escherichia coli*	1.0	109,000	Slater et al. (1971), Springgate et al. (1973)
	Phage T4	1.0	112,000	Springgate et al. (1973)
RNA polymerases	*Bacillus subtilis*	2.0	395,000	Halling et al. (1977)
	Escherichia coli	2.0	395,000	Scrutton et al. (1971)
	Saccharomyces cerevisiae			
	I	2.4	650,000	Auld et al. (1976)
	II	1.0	460,000	Lattke and Weser (1976)
	III	2.0	380,000	Wandjilak and Benson (1977, 1978)
	Euglena gracilis			
	I	2.2	624,000	Falchuk et al. (1977)
	II	2.2	700,000	Falchuk et al. (1976)
	Phage T7	2.4	107,000	Coleman (1974)
Reverse transcriptases	Avian myeloblastosis virus	1.8	108,000	Auld et al. (1974), Poiesz et al. (1974)
	Woolly monkey type C virus	1.0		Auld et al. (1975)
	Murine type C virus	1.4		Auld et al. (1975)

(Valenzuela et al., 1973), calf thymus DNA terminal nucleotidyl transferase (Chang and Bollum, 1970), *Escherichia coli* tRNA nucleotidyl transferase (Williams and Schofield, 1975), and mammalian type C oncogenic virus reverse transcriptases (Auld et al., 1975) are probably zinc metalloenzymes.

It has been suggested that the tightly bound zinc ions present in these polymerases are required for template binding, as well as for the initiation and elongation of the newly synthesized chain (Auld et al., 1976; Coleman, 1974;

Falchuk et al., 1976; Halling et al., 1977; Scrutton et al., 1971; Slater et al., 1971; Springgate et al., 1973; Wandzilak and Benson, 1978). Elucidation of the specific role(s) of this metal would be greatly facilitated by replacement of zinc with a spectroscopically "active" transition metal. While the intrinsic zinc ions of *E. coli* RNA polymerase do not exchange with Mn, Co, and Cu *in vitro*, Speckhard et al. (1977) have demonstrated that such a substitution can be achieved *in vivo*. Cells were grown in cobalt-supplemented (5 μM), low-zinc (0.05 μM) medium without alteration of cell morphology, growth rate, and yield. Purified cobalt-substituted RNA polymerase contains 1.8 to 2.2 g-atoms metal/mol enzyme, is as active as the native enzyme on a variety of templates, and has physical and chemical properties almost identical with those of zinc-containing RNA polymerase. In contrast, reduction of the zinc concentration of the growth medium for *Euglena gracilis* from 10 to 0.1 μM results in early cessation of growth and alterations in cell morphology and chemical composition (Falchuk et al., 1977). That such "low"-zinc cells contain an unusual RNA polymerase, a population of mRNA with significantly altered base composition, an elevated percentage of mRNA, and a 35- and 5-fold increase in the intracellular manganese and magnesium concentrations, respectively, is of particular interest. Falchuk and associates (1977) have suggested that the altered composition of mRNA may result from a decreased fidelity of the transcription process, which in turn is due to the presence of manganese (or magnesium) in the zinc-binding site of the polymerase.

Escherichia coli alkaline phosphatase provides the classic example of an enzyme that requires zinc for both structural and catalytic functions. Two gram-atoms of the metal are present at the activie site, while another two stabilize interactions between monomeric units of this dimeric protein (Bosrow et al., 1975; Simpson and Vallee, 1968; Yosizumi and Coleman, 1974). When the zinc content of the medium was reduced to 0.2 μM, alkaline phosphatase activity decreased markedly; cell yield was not altered (Torriani, 1968). The addition of the metal to stationary-phase cultures of low zinc cells resulted in a rapid increase in alkaline phosphatase activity without concomitant cell growth. It was demonstrated that the additional complement of zinc was required for aggregation of the polypeptide monomers that had been synthesized and accumulated during the exponential growth period. Specific catalytic and structural binding sites for zinc have also been suggested for yeast alcohol dehydrogenase (Dickinson and Dickinson, 1976; Klinman and Welsh, 1976).

Examples of enzymes that require zinc strictly for stabilization of the native quaternary structure include *E. coli* asparate transcarbamylase (Nelbach et al., 1972) and *Bacillus subtilis* amylase (Isemura and Kakiuchi, 1955; Stein and Fischer, 1962). The binding of zinc to *Clostridium perfringens* lecithinase increases the resistance of this enzyme to proteolytic degradation (Sato et al., 1978). Finally, the distal surface of the hexagonal baseplate of T-even viruses contans 4 or 6 g-atoms of tightly bound zinc (Kozloff, 1978; Kozloff and Lute, 1977). The metal serves to bridge the baseplate and the tail fibers.

In light of the zinc requirements of various proteins and the known interactions between this metal and nucleic acids (Failla, 1977), might zinc have some role in the structure and/or function of ribosomes? In *Euglena gracilis* zinc is required for the structural integrity of ribosomes (Prask and Plocke, 1971). When the intracellular level of the metal decreased below a certain concentration, the ribosomes dissociated, unfolded, and were degraded by ribonucleases and proteases. Cysts of *Entamoeba invadens* condense ribosomes into crystalline chromatoid bodies. Electron microprobe analysis revealed that the zinc content of these bodies was extremely high; specifically, the number of zinc ions was approximately equal to the number of ribosomal RNA nucleotides (Morgan and Satillaro, 1972). It was suggested that the metal might be stabilizing the structure of this organelle either by neutralizing the negative charges of rRNA or by inhibiting localized RNase activity.

Relatively high levels of zinc are associated with cell membranes, and specific interactions between the metal and membrane components have been demonstrated (Chvapil, 1973; Failla, 1977). The binding of the peptide antibiotic bacitracin to the polar head groups of membrane phospholipids in gram-positive bacteria is enhanced maximally by zinc (Storm and Strominger, 1973). Although additional evidence suggesting various functions of this metal in microbial cell membrane structure, stabilization, and function has been discussed (Failla, 1977), the precise binding sites and role of zinc in this organelle need to be characterized. Zinc and other cations, principally magnesium and calcium, are also present in the cell walls of bacteria and probably all microorganisms. These metals mediate and/or stabilize interactions between various components of the wall and phage-cell and cell-cell attachment (Failla, 1977). For example, zinc-treated female cells of *Escherichia coli* can mate with Hfr cells lacking F pili (Ou, 1973).

Regulatory functions of zinc in various aspects of microbial cell metabolism have been proposed. For example, the increase in zinc uptake that occurs about 10 min before the initiation of a new round of DNA synthesis in *E. coli* may be a prerequisite for the onset of molecular events that lead to cell division (Kung et al., 1976). Zinc has long been known to have a role in heme synthesis. In *Ustilago sphaerogena* cytochrome formation occurred optimally at about 10 μM zinc and was depressed at both lower and higher concentrations (Grimm and Allen, 1954). Also, the synthesis of δ-aminolevulinate dehydratase by this organism is zinc dependent (Komai and Neilands, 1968). Finally, zinc porphyrins have been reported in yeast, but their significance is uncertain (Pretlow and Sherman, 1967).

In batch cultures of *Aspergillus niger,* medium zinc levels of 1 to 2 μM promoted the growth phase, whereas concentrations below 1 μM favored the stationary phase (Wold and Suzuki, 1976). When the medium zinc content of cultures of stationary-phase cells that were producing citric acid was elevated to 2 μM, the acidogenic process was suppressed after a brief lag period (Figure 3). Likewise, the levels of glycolytic and tricarboxylic acid cycle enzymes in aging cultures of *Aspergillus parasiticus* were found to vary in response to the zinc concentration of the medium (Gupta et al., 1976, 1977).

The efficient synthesis of secondary metabolites (e.g., toxins and antibiotics) and/or morphological differentiation by fungi and actinomycetes have a narrower tolerance for range of culture zinc concentration than does vegetative cell growth (Weinberg, 1970, 1977). Specific examples are provided in Table 4. Excellent vegetative growth generally occurs through a concentration range of two and three orders of magnitude, whereas the range of quantity of zinc that permits high yield of a secondary product or of differentiated cells is usually less than one order of magnitude (Figure 4). The concentration of iron in the culture medium also must be strictly controlled to obtain high yields of secondary substances in actinomycetes. In contrast with the above, the quantity of iron in the culture medium generally is much more important than that of zinc for secondary metabolic processes in procaryotes other than the actinomycetes.

Tejwani et al. (1976) have demonstrated that rat liver fructose-1,6-bis phosphatase is inhibited by 0.3 μM zinc and activated by 10 μM zinc at neutral pH. When present at lower concentrations, zinc binds solely to a high-affinity site and inhibits enzyme activation by either magnesium or manganese in a noncompetitive manner. When the zinc concentration is elevated, the metal replaces

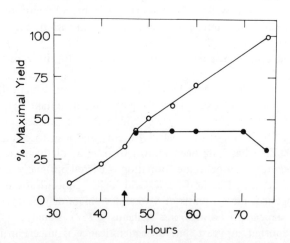

Figure 3. Effect of addition of zinc on citric acid formation by stationary phase cells of *Aspergillus niger*. At 45 hr Zn^{2+} was added to a final concentration of 1.5 μM. Top curve = culture without added zinc; bottom curve = culture with added zinc. (Figure redrawn from Figure 2 of Wold and Suzuki, 1976.)

Table 4. Secondary Metabolites or Differentiated Cell Forms Whose Synthesis Requires a Finer Adjustment of Zinc Concentration Than is Necessary for Maximal Vegetative Growth of the Producer Microorganisms

Product or Structure	References	Product or Structure	Reference
a. Fungi		b. Actinomycetes	
Aflatoxin	Marsh et al. (1975), Mateless and Ayde (1965)	Actinomycin	Katz et al. (1958)
		Candicidin	Acker and Lechavalier (1954)
Cynodontin	White and Johnson (1971)	Chloramphenicol	Gallichio and Gottlieb (1958)
Ergotamine	Stoll et al. (1957)	Kanamycin	Basak and Majumdar (1975)
Fusaric acid	Kalyanasundaraman and Saraswathi-Devi (1955)	Neomycin	Majumdar and Majumdar (1965)
		Streptomycin	Chester and Robinson (1951)
Gentisyl alcohol	Ehrensvaard (1955)		
Griseofulvin	Grove (1967)		
Lysergic acid	Rosazza et al. (1967)		
Malformin	Steenbergen and Weinberg (1968)		
Napthazarin	Kern et al. (1972)		
Oogonia (*Pythium*)	Lenny and Klemner (1966)		
Patulin	Brack (1974)		
Penicillin	Foster et al. (1943)		
Sclerotia (*Whetzelinia*)	Vega and Le Tourneau (1974)		
Vesicles (*Puccinia*)	Sharp and Smith (1952)		
Yeast → Mycelia (*Candida*)	Yamaguchi (1975)		
Zearalenone	Hidy et al. (1977)		

453

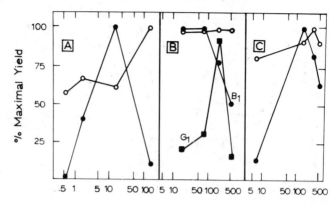

Figure 4. Effect of zinc on yield of (*A*) sclerotia, (*B*) aflatoxin B_1 and G_1, and (*C*) rubratoxin. 0———0. = Dry weight mycelia; ●———● = sclerotia or toxins (■————■ = aflatoxin G_1). [Figures drawn from data in (*A*) Table 3, Vega and LeTourneau, 1974; (*B*) Table 1A, Marsh et al., 1975; (*C*) Table 6, Emeh and Marth, 1976.

magnesium or manganese from the lower affinity activation site and the inhibitory effects of zinc are abolished. Likewise, removal of zinc from the high-affinity site by chelating agents results in enzyme activation. Might similar allosteric zinc-binding proteins have a role in regulating microbial secondary metabolic processes? Also, it is known that the concentrations of many precursor substances are altered as the cell enters the stationary phase (Weinberg, 1970). What effect do altered levels of low molecular weight zinc-complexing agents (e.g., histidine, cysteine, and nucleotides) have on the intracellular distribution of the metal and secondonary metabolic synthetase activity? These questions surely merit investigation.

4. SUMMARY

High-affinity uptake systems specific for zinc are just beginning to be characterized in microorganisms. The uptake of this metal oscillates during the cell cycle in *Escherichia coli* and during the culture cycle in *Candida utilis*. Zinc efflux apparently does not occur in eucaryotic microbial cells; thus they must possess either exclusionary mechanisms or methods of packaging intracellular zinc in an innocuous form. Preliminary reports concerning these two possibilities are cited. Finally, evidence is reviewed for four roles of zinc in microbial cells: catalytic, structural, stabilizing, and regulatory. Even casual reading of the material will lead to the conclusion that vast gaps in our understanding of zinc metabolism, homeostasis, and function in the microbial world presently exist.

ACKNOWLEDGMENTS

M. L. F. was a postdoctoral trainee of the National Institute of Arthritis, Metabolism, and Digestive Diseases (1 F32 AMO5726-01). E. D. W. was supported in part by a research grant from the National Science Foundation (BMS 75-16753).

REFERENCES

Acker, R. F. and Lechavalier, H. A. (1954). "Some Nutritional Requirements of Streptomyces griseus 3570 for Growth and Candicidin Production," Appl. Microbiol., 2, 152-157.

Auld, D. S., Kawaguchi, H., Livingston, D. M., and Vallee, B. L. (1974). "RNA-Dependent DNA Polymerase (Reverse Transcriptase) from Avian Myeloblastosis Virus: A Zinc Metalloenzyme," Proc. Natl. Acad. Sci., 71, 2091-2095.

Auld, D. S., Kawaguchi, H., Livingston, D. M., and Vallee, B. L. (1975). "Zinc Reverse Transcriptases from Mammalian RNA Type of C Viruses," Biochem. Biophys. Res. Commun., 62, 296-302.

Auld, D. S., Atsuza, I., Campino, C., and Valenzuela, P. (1976). "Yeast RNA Polymerase: A Eucaryotic Zinc Metalloenzyme," Biochem. Biophys. Res. Commun., 69, 548-554.

Basak, K., and Majumdar, S. K. (1975). "Mineral Nutrition of Streptomyces kanamyceticus for Kanamycin Formation," Antimicr. Agents Chemother., 8, 391-395.

Becker, J. M. and Lichstein, H. C. (1972). "Transport Overshoot during Biotin Uptake by Saccharomyces cerevisiae," Biochem. Biophys. Acta, 282, 409-422.

Bellenger, N., Nissen, R., Wood, T. C., and Segel, I. H. (1968). "Specificity and Control of Choline-o-Sulfate Transport in Filamentous Fungi," J. Bacteriol., 96, 1574-1585.

Bosron, W. F., Kennedy, F. S., and Vallee, B. L. (1975). "Zinc and Magnesium Content of Alkaline Phosphatase from Escherichia coli," Biochemistry, 14, 2275-2282.

Brack, A. (1974). "Isolierung von Gentisalkohol neben Patulin," Helv. Chim. Acta, 30, 1-14.

Bremner, I. and Davies, N. T. (1975). "The Induction of Metallothionein in Rat Liver by Zinc Injection and Restriction of Food Intake," Biochem. J., 149, 733-738.

Bucheder, F. and Broda, E. (1974). "Energy Dependent Zinc Transport by Escherichia coli," Eur. J. Biochem., 45, 555-559.

Byer, B. R. and Arceneaux, J. E. L. (1977). "Microbial Transport and Utilization of Iron." In E. D. Weinberg, Ed., *Microorganisms and Minerals,* Marcel Dekker, New York, pp. 215-249.

Chabner, B. A. and Bertino, J. R. (1972). "Activation and Inhibition of Carboxypeptidase G1 by Divalent Cations," *Biochim. Biophys. Acta,* **276,** 234-240.

Chang, L. M. S. and Bollum, F. J. (1970). "Deoxynucleotide-polymerizing Enzymes of Calf Thymus Gland, IV. Inhibition of terminal deoxynucleotidyl transferase by metal lignands," *Proc. Natl. Acad. Sci.,* **65,** 1041-1048.

Chester, C. G. C. and Rolinson, G. N. (1951). "Trace Elements and Streptomycin Production," *J. Gen. Microbiol.,* **5,** 559-565.

Chipley, J. R. and Edwards, H. M. (1972). "Cation Uptake and Exchange in *Salmonella enteriditis,*" *Can. J. Microbiol.,* **18,** 509-513.

Chopra, I. (1970). "Decreased Uptake of Cadmium by a Resistant Strain of *Staphylococcus aureus,*" *J. Gen. Microbiol.,* **63,** 265-267.

Chvapil, M. (1973). "New Aspects in the Biological Role of Zinc: A Stabilizer of Macromolecules and Biological Membranes," *Life Sci.,* **13,** 1041-1049.

Coleman, J. E. (1974). "The Role of Zn(II) in Transcription by T7 RNA Polyerase," *Biochem. Biophys. Res. Commun.,* **60,** 641-648.

Cremona, T. and Singer, T. P. (1964). "The Lactic Dehydrogenase of Yeast. V: Chemical Properties and Function of the Zinc Component of D-Lactic Cytochrome Reductase," *J. Biol. Chem.,* **239,** 1466-1473.

Cummins, J. E. and Mitchison, J. M. (1967). "Adenine Uptake and Pool Formation in the Fission Yeast *Schizosaccharomyces pombe,*" *Biochim. Biophys. Acta,* **136,** 108-120.

Curdel, A. and Labeyrie, F. (1961). "D-Lactic Dehydrogenase from Anaerobic Yeast: Combination of Apoenzyme with Zinc and Cobalt and Comparison of Reconstituted Enzymes," *Biochem. Biophys. Res. Commun.,* **4,** 175-179.

Davies, R. B. and Abraham, E. P. (1974). "Metal Cofactor Requirements of β-Lactamase II," *Biochem. J.,* **143,** 129-135.

Dickinson, C. J. and Dickinson, F. M. (1976). "Some Properties of an Alcohol Dehydrogenase Partially Purified from Baker's Yeast Grown without Added Zinc," *Biochem. J.,* **153,** 309-319.

Ehrensvaard, G. (1955). "Some Observations on Aromatic Biosynthesis in *Penicillium urticae* Bainer," *Exp. Cell Res.,* Suppl. 3, 102-109.

Emeh, C. O. and Marth, E. H. (1976). "Cultural and Nutritional Factors that Control Rubratoxin Formation," *J. Milk Food Technol.,* **39,** 184-190.

Failla, M. L. (1977). "Zinc: Functions and Transport in Microorganisms." In E. D. Weinberg, Ed., *Microorganisms and Minerals,* Marcel Dekker, New York, pp. 151-214.

Failla, M. L. and Weinberg, E. D. (1977). "Cyclic Accumulation of Zinc by *Candida utilis* during Growth in Batch Culture," *J. Gen. Microbiol.,* **99,** 85-97.

References   457

Failla, M. L., Benedict, C. D., and Weinberg, E. D. (1976). "Accumulation and Storage of Zn²⁺ by *Candida utilis*," *J. Gen. Microbiol.*, **94**, 23-36.

Falchuk, K. H., Mazus, B., Ulpino, L., and Vallee, B. L. (1976). "*Euglena gracilis* DNA Dependent RNA Polymerase. II: A Zinc Dependent Metalloenzyme," *Biochemistry*, **15**, 4468-4475.

Falchuk, K. H., Hardy, C., Ulpino, L., and Vallee, B. L. (1977). "RNA Polymerase, Manganese and RNA Metabolism of Zinc Sufficient and Deficient *Euglena gracilis*," *Biochem. Biophys. Res. Commun.*, **77**, 314-319.

Falchuk, K. H., Ulpino, L., and Vallee, B. L. (1977b). "*E. gracilis* RNA Polymerase. I: A Zinc Metalloenzyme," *Biochem. Biophys. Res. Commun.*, **74**, 1206-1212.

Feder, J., Keay, L., Garett, L. R., Circulis, N., Moseley, M. H., and Wildi, B. S. (1971). "*Bacillus cereus* Neutral Protease," *Biochim. Biophys. Acta*, **251**, 74-78.

Feeney, R. E., Lightbody, H. D., and Garibaldi, J. A. (1947). "Zinc as an Essential Element for Growth and Subtilin Formation by *Bacillus subtilis*," *Arch. Biochem. Biophys.*, **15**, 13-17.

Foster, J. W. (1949). *Chemical Activities of Fungi*. Academic Press, New York, pp. 251-281.

Foster, J. W., Woodruff, H. B., and McDaniel, L. E. (1943). "Microbiological Aspects of Penicillin. III: Production of Penicillin in Surface Cultures of *Penicillium notatum*," *J. Bacteriol.*, **46**, 421-433.

Fugimoto, M., Kuninaka, A., and Yoshino, H. (1974). "Identity of Phosphodiesterase and Phosphomonoesterase Activities with Nuclease P₁, a Nuclease from *Penicillium citrium*," *Agric. Biol. Chem.*, **38**, 785-790.

Fugimoto, M., Kuninaka, A., and Yoshino, H. (1975). "Some Physical and Chemical Properties of Nuclease P₁," *Agric. Biol. Chem.*, **39**, 1991-1998.

Gallichio, V. and Gottlieb, D. (1958). "The Biosynthesis of Chloramphenicol. III: Effects of Micronutrients on Synthesis," *Mycologia*, **50**, 490-496.

Goscin, S. A. and Fridovich, I. (1972). "The Purification and Properties of Superoxide Dismutase from *Saccharomyces cerevisiae*," *Biochem. Biophys. Acta*, **289**, 276-283.

Gracy, R. W. and Noltmann, E. A. (1968a). "Studies on Phosphomannose Isomerase. II: Characterization as a Zinc Metalloenzyme," *J. Biol. Chem.*, **243**, 4109-4116.

Gracy, R. W., and Noltman, E. A. (1968b). "Studies on Phosphomannose Isomerase. III: A Mechanism for Catalysis and for the Role of Zinc in the Enzymatic and the Nonenzymatic Isomerization," *J. Biol. Chem.*, **243**, 5410-5419.

Gregolin, C. and Singer, T. P. (1963). "The Lactic Dehydrogenase of Yeast," *Biochim. Biophys. Acta*, **67**, 201-218.

Grenson, M. (1969). "The Utilization of Exogenous Pyrimidines and the Recycling of Uridine-5'-Phosphate Derivatives in *Saccharomyces cerevisiae,* as Studied by Means of Mutants Affected in Pyrimidine Uptake and Metabolism," *Eur. J. Biochem.,* 14, 197–204.

Griffith, T. B. and Prescott, J. M. (1970). "Some Physical Characteristics of a Proteinase from *Aeromonas proteolytica," J. Biol. Chem.,* 245, 1348–1356.

Grimm, P. W. and Allen, P. J. (1954). "Promotion by Zinc of the Formation of Cytochromes in *Ustilago sphaeogena," Plant Physiol.,* 29, 369–377.

Grove, J. F. (1967). "Griseofulvin." In D. Gottlieb and P. D. Shaw, Eds., *Antibiotics II. Biosynthesis,* Springer-Verlag, Berlin, pp. 123–133.

Gupta, S. K., Maggon, K. K., and Venkitasubramanian, T. A. (1976). "Zinc-Dependence of Glycolytic Enzymes in an Aflatoxigenic Strain of *Aspergillus parasiticus," Microbios Lett.,* 3, 89–92.

Gupta, S. K., Maggon, K. K., and Venkitasubramanian, T. A. (1977). "Effect of Zinc on Tricarboxylic Acid Cycle Intermediates and Enzymes in Relation to Aflatoxin Biosynthesis," *J. Gen. Microbiol.,* 99, 43–48.

Gutnecht, J. (1963). "^{65}Zn Uptake by Benthic Marine Algae," *Limnol, Oceanogr.,* 8, 31–39.

Halling, S. H., Sanchez-Anzaldo, F. J., Fukuda, R., Doi, R. H., and Meares, C. F. (1977). "Zinc is Associated with the Subunit of DNA-Dependent RNA Polymerase of *Bacillus subtilis," Biochemistry,* 16, 2880–2884.

Harris, C. E., Kobes, R. D., Teller, D. C., and Rutter, W. J. (1969). "The Molecular Characteristics of Yeast Aldolase," *Biochemistry,* 8, 2442–2454.

Hayman, S., Gatmaitan, J. S., and Patterson, E. K. (1974). "The Relationship of Extrinsic and Intrinsic Metal Ions to the Specificity of a Dipeptidase from *Escherichia coli* B," *Biochemistry,* 13, 4486–4494.

Hidy, P. H., Baldwin, R. S., Greasham, R. L., Keith, C. L., and McMullen, J. R. (1977). "Zearalenone and Some Derivatives: Production and Biological Activities," *Adv. Appl. Microbiol.,* 22, 59–82.

Hiramatsu, A. and Ouchi, T. (1972). "A Neutral Protease from *Streptomyces naraensis.* III: An Improved Purification and Some Physiochemical Properties," *J. Biochem.,* 71, 767–781.

Hoch, F. L. and Vallee, B. L. (1955). "Zinc: a Component of Yeast Alcohol Dehydrogenase," *Proc. Natl. Acad. Sci.,* 41, 327–337.

Holmquist, B. (1977). "Characterization of the "Microprotease" from *Bacillus cereus.* A Zinc Neutral Endoprotease," *Biochemistry,* 16, 4591–4594.

Hunter, D. R. and Segel, I. H. (1973). "Control of the General Amino Acid Permease of *Pencillium chrysogenum* by Transinhibition and Turnover," *Arch. Biochem. Biophys.,* 154, 387–399.

Isemura, T. and Kakiuchi, (1955). "Association and Dissociation of Bacterial α-Amylase Molecule," *J. Biochem.,* 51, 358–392.

Iwatsubo, M. and Curdel, A. (1961). "D-Lactic Dehydrogenase from Anaerobic Yeast: Reversible Dissociation of FAD and Zn by Acid Treatment of the Holoenzyme," *Biochem. Biophys. Res. Commun.,* 6, 385–388.

Jagannathan, V., Singh, K., and Damodaran, M. (1956). "Carbohydrate Metabolism in Citric Acid Fermentation. 4: Purification and Properties of Aldolase from *Aspergillus niger*," *Biochem. J.*, **63**, 94–105.

Johnston, R. (1964). "Sea Water, the Natural Medium of Phytoplankton," *J. Mar. Biol. Assoc. U.K.*, **44**, 87–109.

Jurinak, J. J. and Inouye, T. S. (1962). "Some Aspects of Zinc and Copper Phosphate Formation in Aqueous Systems," *Proc. Soil Sci. Soc. Am.*, **26**, 144–147.

Kadner, R. J. (1975). "Regulation of Methionine Transport Activity in *Escherichia coli*," *J. Bacteriol.*, **122**, 110–119.

Kagi, J. H. R. and Vallee, B. L. (1960). "Metallothionein: a Cadmium and Zinc-Containing Protein from Equine Renal Cortex," *J. Biol. Chem.*, **235**, 3460–3465.

Kagi, J. H. R. and Vallee, B. L. (1961). "Metallothionein: a Cadmium- and Zinc-Containing Protein from Equine Renal Cortex. II: Physiochemical Properties," *J. Biol. Chem.*, **236**, 2435–2442.

Kalyanasundaraman, R. and Saraswathi-Devi, L. (1955). "Zinc in the Metabolism of *Fusarium vasinfectum* Atk.," *Nature*, **175**, 945–945.

Katz, E., Pienta, P., and Sivak, A. (1958). "The Role of Nutrition in the Synthesis of Actinomycin," *Appl. Microbiol.*, **6**, 236–241.

Keay, L. and Wildi, B. S. (1970). "Protease of the Genus *Bacillus*. I: Neutral Proteases," *Biotechnol. Bioeng.*, **12**, 179–212.

Keay, L., Feder, L., Garrett, L. R., Moseley, M. H., and Cirulis, N. (1971). "*Bacillus megaterium* Neutral Protease, a Zinc-Containing Metalloenzyme," *Biochim. Biophys. Acta*, **229**, 829–835.

Keleti, T. (1966). "Zn in Yeast D-Glyceraldehyde-3-phosphate Dehydrogenase," *Biochem. Biophys. Res. Commun.*, **22**, 640–643.

Kern, H., Naef-Roth, S., and Ruffner, F. (1972). "Der Einfluss der Ernahrung auf die Bildung von Naphthazarinderivaten durch *Fusarium Martii* var. *pisi*," *Phytopathol. Z.*, **74**, 272–280.

Klinman, J. P. and Welsh, K. (1976). "The Zinc Content of Yeast Alcohol Dehydrogenase," *Biochem. Biophys. Res. Commun.*, **70**, 878–884.

Kobes, R. D., Simpson, R. T., Vallee, B. L., and Rutter, W. J. (1969). "A Functional Role of Metal Ions in a Class II Aldolase," *Biochemistry*, **8**, 585–588.

Komai, H. and Neilands, J. B. (1968). "Effect of Zinc Ions on δ-Aminolevulinate Dehydratase in *Ustilago sphaerogena*," *Arch. Biochem. Biophys.*, **124**, 456–461.

Kondo, I., Ishikawa, T., and Nakuhara, H. (1974). "Mercury and Cadmium Resistances Mediated by the Penicillinase Plasmid in *Staphylococcus aureus*," *J. Bacteriol.*, **117**, 1–7.

Kowal, J., Cremona, T., and Horecker, B. L. (1966). "Fructose-1,6-diphosphate Aldolase of *Candida utilis*: Purification and Properties," *Arch. Biochem. Biophys.*, **114**, 13–23.

Kozloff, L. M. (1978). "Properties of T4D Bacteriophage Grown in Synthetic Media Containing Zn^{2+}, Co^{2+}, or Ni^{2+}," *J. Biol. Chem.*, 253, 1059–1064.

Kozloff, L. M. and Lute, M. (1977). "Zinc, an Essential Component of the Baseplates of T-even Bacteriophages," *J. Biol. Chem.*, 252, 7715–7724.

Kung, F. C., Raymond, J., and Glaser, D. A. (1976). "Metal Ion Content of *Escherichia coli* versus Cell Age," *J. Bacteriol.*, 126, 1089–1095.

Lange, W. (1974). "Chelating Agents and Blue-Green Algae," *Can. J. Microbiol.*, 20, 1311–1321.

Latt, S. A., Holmquist, B., and Vallee, B. L. (1969). "Thermolysin: a Zinc Metalloenzyme," *Biochem. Biophys. Res. Commun.*, 37, 333–339.

Lattke, H. and Weser, U. (1976). "Yeast RNA-Polymerase B: a Zinc Metalloenzyme," *FEBS Lett.*, 65, 288–292.

Lawford, H. G., Pik, J. R., Lawford, G. R., Williams, T., and Kligerman, A. (1980). "Hyper Accumulation of Zinc by Zinc-Depleted *Candida utilis* Grown in Chemostat Culture," *Canad. J. Microbiol.*, 26, in press.

Lenny, J. F. and Klemner, H. W. (1966). "Factors Controlling Sexual Reproduction and Growth in *Pythium graninicola*," *Nature*, 209, 1365–1366.

Levy, P. I., Pangburn, M. K., Burnstein, Y., Ericsson, L. H., Neurath, H., and Walsh, K. A. (1975). "Evidence of Homolgous Relationship between Thermolysin and Neutral Protease A of *Bacillus subtilis*," *Proc. Natl. Acad. Sci.*, 72, 4341–4345.

Little, C. and Otnass, A. (1975). "The Metal Ion Dependence of Phospholipase C from *Bacillus cereus*," *Biochem. Biophys. Acta*, 391, 326–333.

Macara, I. G. (1978). "Accommodation of Yeast to Toxic Levels of Cadmium Ions," *J. Gen. Microbiol.*, 104, 321–324.

MacLean, F. I., Lucis, O. J., Shaikh, Z. A., and Jansey, E. R. (1972). "The Uptake and Subcellular Distribution of Cd and Zn in Microorganisms," *Fed. Poc.*, 31, 699–699.

Mujumdar, M. K. and Majumdar, S. K. (1965). "Effect of Minerals on Neomycin Production by *Streptomyces fradiae*," *Appl. Microbiol.*, 13, 190–193.

Marsh, P. B., Simpson, M. E., and Trucksess, M. W. (1975). "Effects of Trace Metals on the Production of Aflatoxins by *Aspergillus parasiticus*," *Appl. Microbiol.*, 30, 52–57.

Mateles, R. I. and Adye, J. C. (1965). "Production of Aflatoxin in Submerged Cultures," *Appl. Microbiol.*, 13, 208–211.

Matzku, S. and Broda, E. (1970). "Die Zinkaufnahme in das Innere von *Chlorella*," *Planta*, 92, 29–40.

McConn, J. D., Tsuru, D., and Yasumohi, K. T. (1967). "*Bacillus subtilis* Neutral Protease. III: Exchange Properties of Zinc and Preparation of Some Metal Derivatives," *Arch. Biochem. Biophys.*, 120, 479–486.

Metz, G. and Rohm, K. H. (1976). "Yeast Aminopeptidase. I: Chemical Composition and Catalytic Properties," *Biochim. Biophys. Acta*, 429, 933–949.

Misra, H. P. and Fridovich, I. (1972). "The Purification and Properties of Superoxide Dismutase from *Neurospora crassa*," *J. Biol. Chem.*, 247, 3410–3414.

Mitra, R. S., Gracey, R. H., Chin, B., and Berstein, I. A. (1975). "Molecular Mechanisms of Accommodation in *Escherichia coli* to Toxic Levels of Cadmium Ions," *J. Bacteriol.*, 121, 1180–1188.

Miyata, K., Tomoda, K., and Isono, M. (1971) "*Serratia* Pretease," *Agric. Biol. Chem.*, 35, 460–467.

Morgan, R. S. and Satillaro, R. F. (1972). "Zinc in *Entamoeba invadens*," *Science*, 176, 929–930.

Morihara, K. and Tasuzuki, H. (1975). "*Pseudomonas aeruginosa* Elastase: Affinity Chromatography and Some Properties as a Metalloneutral Proteinase," *Agric. Biol. Chem.*, 39, 1123–1128.

Morrison, C. E. and Lichstein, H. C. (1976). "Regulation of Lysine Transport by Feedback Inhibition in *Saccharomyces cerevisiae*," *J. Bacteriol.*, 125, 864–871.

Naiki, N. and Yamagata, S. (1976). "Isolation and Some Properties of Copper-Binding Proteins Found in a Copper-Resistant Strain of Yeast," *Plant Cell Physiol.*, 17, 1281–1295.

Nakajima, M., Mizusawa, K., and Yoshida F. (1974). "Purification and Properties of an Extracellular Proteinase of Psychrophilic *Escherichia freundii*," *Eur. J. Biochem.*, 44, 87–96.

Nelbach, M., Pigiet, V., Gerhart, J., and Schachman, H. K. (1972). "A Role for Zinc in the Quaternary Structure of Aspartate Transcarbamylase from *Escherichia coli*," *Biochemistry*, 11, 315–327.

Northrop, D. B. and Wood, H. G. (1969). "Transcarboxylase. V: The Presence of Bound Zinc and Cobalt," *J. Biol. Chem.*, 244, 5801–5807.

Novick, R. P. and Roth, C. (1968). "Plasmid-Linked Resistance to Inorganic Salts in *Staphylococcus aureus*," *J. Bacteriol.*, 95, 1335–1342.

Oh, S. H., Deagen, J. T., Whanger, P. D., and Weswig, P. H. (1978). "Biological Function of Metallothionein. V: Its Induction in Rats by Various Stresses," *Am. J. Physiol.*, 234, E282–E285.

Ou, J. T. (1973). "Effect of Zn^{2+} on Bacterial Conjugation: Increase in Ability of F^- Cells to Form Mating Pairs," *J. Bacteriol.*, 115, 648–654.

Parry, G. D. R. and Haywood, J. (1973). "The Uptake of [65]Zn by *Dunaliella tertiolecta*," *J. Mar. Biol. Assoc. U.K.*, 53, 915–922.

Paton, W. H. N. and Budd, K. (1972). "Zinc Uptake in *Neocosmospora vasinfecta*," *J. Gen. Microbiol.*, 72, 173–184.

Petranyi, P., Jendrisak, J. J., and Burgess, R. B. (1977). RNA Polymerase II from Wheat Germ Contains Tightly Bound Zinc," *Biochem. Biophys. Res. Commun.*, 74, 1031–1038.

Poiesz, B. J., Seal, G., and Loeb, L. A. (1974). "Reverse Transcriptase: Correlation of Zinc Content with Activity," *Proc. Natl. Acad. Sci.*, 71, 4892–4896.

Ponta, H. and Broda, E. (1970). "Mechanismen der Aufnahme von Zink durch Backerhefe," *Planta*, **95**, 18-26.

Prask, J. A. and Plocke, D. J. (1971). "A Role for Zinc in the Structural Integrity of the Cytoplasmic Ribosomes of Euglena gracilis," *Plant Physiol.*, **48**, 150-155.

Premakumar, R., Winge, D. R., Wiley, R. D., and Rajagopolan, K. V. (1975). "Copper Chelatin: Isolation from Various Eucaryotic Sources," *Arch. Biochem. Biophys.*, **170**, 278-288.

Pretlow, T. P. and Sherman, F. (1967). "Porphyrins and Zinc Porphyrins in Normal and Mutant Strains of Yeast," *Biochim. Biophys. Acta*, **148**, 629-644.

Prinz, R. and Weser, U. (1975). "Cuprodoxin," *FEBS Lett.*, **54**, 224-229.

Pulido, P., Kagi, J. H. R., and Vallee, B. L. (1966). "Isolation and Some Properties of Human Metallothionein," *Biochemistry*, **5**, 1768-1777.

Rahman, Y. E. and Wright, B. J. (1975). "Liposomes Containing Chelating Agents: Cellular Penetration and a Possible Mechanism of Metal Removal," *J. Cell Biol.*, **65**, 112-122.

Rapp, U., Adams, W. C., and Miller, R. W. (1973). "Purification of Superoxide Dismutase from Fungi and Characterization of the Reaction of the Enzyme with Catechols by Electron Spin Resonance Spectrography," *Can. J. Biochem.*, **51**, 158-171.

Raulin, J. (1869). "Études Cliniques sur la Vegetation," *Ann. Sci. Nat. Bot.*, Ser. V, **11**, 93-299.

Rawla, G. S. (1969). "A Note on Trace Elements for the Growth of Nigrosora orzae (B. and Br. Petch)," *New Phytol.*, **68**, 941-943.

Richards, M. P. and R. J. Cousins. (1976). "Metallothionein and Its Relationship to the Metabolism of Dietary Zinc in Rats," *J. Nutr.*, **106**, 1591-1599.

Ring, K., Gross, W., and Heinz, E. (1970). "Negative Feedback Regulation of Amino Acid Transport in Streptomyces hydrogenans," *Arch. Biochem. Biophys.*, **137**, 243-252.

Rodriguez-Absi, J. and Prescott, J. M. (1978). "Isolation and Properties of an Aminopeptidase from Bacillus licheniformis," *Arch. Biochem. Biophys.*, **186**, 383-391.

Rosazza, J. P., Kelleher, W. J., and Schwarting, A. E. (1967). "Production of Lysergic Acid Derivatives in Submerged Culture. IV: Inorganic Nutrition Studies with Claviceps paspali," *Appl. Microbiol.*, **15**, 1270-1283.

Ryden, L. and Deutsch, H. F. (1978). "Preparation and Properties of the Major Copper-Binding Component in Human Fetal Liver: Its Identification as Metallothionein," *J. Biol. Chem.*, **253**, 519-524.

Rytka, J. (1975). "Positive Selection of General Amino Acid Permease Mutants in Saccharomyces cerevisiae," *J. Bacteriol.*, **121**, 562-570.

Sato, H., Yamakawa, Y., Ito, A., and Murata, R. (1978). "Effect of Zinc and Calcium Ions on the Production of alpha toxin and Proteases by Clostridium perfringens," *Infect. Immunol.*, **20**, 325-333.

Schwartz, N. B. and Feingold, D. S. (1973). "L-Rhamnulose 1-Phosphate Aldolase from *Escherichia coli*. III: The Role of Divalent Cations in Enzyme Activity," *Bioinorg. Chem.*, 2, 75–86.

Schwartz, F. J. and Matrone, G. (1975). "Methodological Studies on the Uptake of Zinc by 3T3 Cells," *Proc. Soc. Exp. Biol. Med.*, 149, 888–892.

Schwartz, N. B., Abram, D., and Feingold, D. S. (1974). "L-Rhamnulose 1-Phosphate Aldolase of *Escherichia coli*. IV: The Role of Metal in Enzyme Structure," *Biochemistry*, 13, 1726–1730.

Scrutton, M. C., Young, M. R., and Utter, M. F. (1970). "Pyruvate Carboxylase from Baker's Yeast: The Presence of Bound Zinc," *J. Biol. Chem.*, 245, 6220–6227.

Scrutton, M. C., Wu, C. W., and Goldthwait, D. A. (1971). "The Presence and Possible Roles of Zinc in RNA Polymerase Obtained from *Escherichia coli*," *Proc. Natl. Acad. Sci.*, 68, 2497–2501.

Seber, J. F., Toomey, T. P., Powell, J. T., Brew, K., and Awad, W. M. (1976). "Proteolytic Enzymes of the K-1 Strain of *Streptomyces griseus* Obtained from a Commercial Preparation (Pronase)," *J. Biol. Chem.*, 251, 204–208.

Seifter, S., Takahashi, S., and Harper, E. (1970). "Further Demonstration that Cysteine Reacts with the Metal Component of Collagenase," *Biochim. Biophys. Acta*, 214, 559–561.

Sekino, H. (1972). "Some Properties of Neutral Proteinases I and II of *Aspergillus sojae* as a Zinc-Containing Metalloenzyme," *Agric. Biol. Chem.*, 36, 2143–2150.

Sharp, E. L. and Smith, F. G. (1952). "The Influence of pH and Zinc on Vesicle Formation in *Puccinia coronata* avenae corda," *Phytopathology*, 42, 581–582.

Silver, S. (1978). "Active Transport of Cations and Anions." In B. Rosen, Ed., *Bacterial Transport*, Marcel Dekker, New York, pp. 221–324.

Simpson, R. T. and Vallee, B. L. (1968). "Two Differentiable Classes of Metal Atoms in Alkaline Phosphatase in *Escherichia coli*," *Biochemistry*, 7, 4343–4350.

Simpson, R. T., Vallee, B. L., and Tait, G. H. (1968). "Alkaline Phosphatase of *Escherichia coli*: Composition," *Biochemistry*, 7, 4336–4342.

Slater, J. P., Mildvan, A. S., and Loeb, L. A. (1971). "Zinc in DNA Polymerases," *Biochem. Biophys. Res. Commun.*, 44, 37–43.

Sobocinski, P. Z., Canterbury, W. J., Mapes, C. A., and Dinterman, R. E. (1978). "Involvement of Hepatic Metallothioneins in Hypozincemia Associated with Bacterial Infections," *Am. J. Physiol.*, 234, E399–E407.

Speckhard, D. C., Wu, F. Y. H., and Wu, C. W. (1977). "Role of Intrinsic Metal in RNA Polymerase from *Escherichia coli*: *In vivo* Substitution of Tightly Bound Zinc with Cobalt," *Biochemistry*, 16, 5228–5234.

Springgate, C. F., Mildvan, A. S., Abramson, R., Engle, J. L., and Loeb, L. A. (1973). "*Escherichia coli* Deoxyribonucleic Acid Polymerase I, a Zinc Metalloenzyme," *J. Biol. Chem.*, 248, 5987–5993.

Steenbergen, S. T. and Weinberg, E. D. (1968). "Trace Metal Requirements for Malformin Biosynthesis," *Growth*, **32**, 125–134.

Stein, E. A. and Fischer, E. H. (1962). *"Bacillus subtilis* α-Amylase, a Zinc-Protein Complex," *Biochim. Biophys. Acta*, **39**, 287–296.

Stoll, A., Brack, A., Hofmann, A., and Kobel, H. (1957). "Process for the Preparation of Ergotamine." U.S. Patent 2,809,920.

Storm, D. R. and Strominger, J. L. (1973). "Complex Formation between Bacitracin Peptides and Isoprenyl Pyrophosphate," *J. Biol. Chem.*, **248**, 3940–3945.

Sugimoto, S. and Nosoh, Y. (1971). "Thermal Properties of Fructose-1,6-diphosphate Aldolase from Thermophilic Bacteria, *Biochem. Biophys. Acta*, **235**, 210–221.

Taylor, W. H., Taylor, M. L., Balch, M. E., and Gilchrist, (1976). "Purification and Properties of Dihydroorotase, a Zinc-Containing Metalloenzyme in *Clostridium oroticum," J. Bacteriol.*, **127**, 863–873.

Tejwani, G. A., Pedrosa, F. O., Pontremoli, S., and Horecker, B. L. (1976). "Dual Role of Zn^{2+} as Inhibitor and Activator of Fructose-1,6-bisphosphatase of Rat Liver," *Proc. Natl. Acad. Sci.*, **73**, 2692–2695.

Torriani, A. (1968). "Alkaline Phosphatase Subunits and Their Dimerization *in vivo," J. Bacteriol.*, **96**, 1200–1207.

Twu, J. S., Haroz, R. K., and Bretthauer, R. K. (1977). "Nucleotide Pryophosphatase from Yeast: The Presence of Bound Zinc," *Arch. Biochem. Biophys.*, **184**, 249–256.

Valenzuela, P., Morris, R. W., Faras, A., Levinson, W., and Rutter, W. J. (1973). "Are All Nucleotidyl Transferases Metalloenzymes?" *Biochem. Biophys. Res. Commun.*, **53**, 1036–1041.

Vega, R. R. and LeTourneau, D. (1974). "The Effect of Zinc on Growth and Sclerotial Formation in *Whetzelinia sclerotium," Mycologia*, **66**, 256–264.

Wandzilak, T. M. and Benson, R. W. (1977). "Yeast RNA Polymerase III: A Zinc Metalloenzyme," *Biochem. Biophys. Res. Commun.*, **76**, 247–252.

Wandzilak, T. M. and Benson, R. W. (1978). *"Saccharomyces cerevisiae* DNA-Dependent RNA Polymerase III: A Zinc Metalloenzyme," *Biochemistry*, **17**, 426–431.

Webb, M. (1970). "Interrelations between the Utilization of Magnesium and the Uptake of Other Bivalent Cations by Bacteria," *Biochem. Biophys. Acta*, **222**, 428–439.

Weinberg, E. D. (1970). "Biosynthesis of Secondary Metabolites: Roles of Trace Metals," *Adv. Microb. Physiol.*, **4**, 1–44.

Weinberg, E. D. (1977). "Mineral Element Control of Microbial Secondary Metabolism. In E. D. Weinberg, Ed., *Microorganisms and Minerals*, Marcel Dekker, New York, pp. 289–316.

Weiss, A. A., Silver, S., and Kinscherf, T. G. (1978). "Cation Transport Alteration Associated with Plasmid-Determined Resistance to Cadmium in *Staphyloccus aureus,"* *Antimicrob. Agents Chemother.* **14**, 856–865.

Weser, U., Hartmann, H. J., Fretzdoff, A., and Strobel, G. J. (1977). "Homologous Copper (I)–(thiolate) 2-chromophores in Yeast Copper Thionein," *Biochem. Biophys. Acta,* **493**, 456–477.

White, J. P. and Johnson, G. T. (1971). "Zinc Effects on Growth and Cynodontin Production of *Helminthosporium cynodontis,"* *Mycologia,* **63**, 548–561.

Williams, K. R. and Schofield, P. (1975). "Evidence for Metalloenzyme Character of RNA Nucleotidyl Transferase," *Biochem. Biophys. Res. Commun.,* **64**, 262–267.

Witney, F. R., Failla, M. L., and Weinberg, E. D. (1977). "Phosphate Inhibition of Secondary Metabolism in *Serratia marcescens,"* *Appl. Environ. Microbiol.,* **33**, 1042–1046.

Wold, W. S. M. and Suzuki, I. (1976). "The Citric Acid Fermentation by *Aspergillus niger:* Regulation by Zinc of Growth and Acidogenesis," *Can. J. Microbiol.,* **22**, 1083–1091.

Yamaguchi, H. (1975). "Control of Dimorphism in *Candida albicans* by Zinc: Effect on Cell Morphology and Composition," *J. Gen. Microbiol.,* **86**, 370–372.

Yamamoto, K. A. and Segel, I. H. (1966). "The Inorganic Sulfate Transport System of *Penicillium chrysogenum,"* *Arch. Biochem. Biophys.,* **114**, 523–538.

Yoshizumi, F. K. and Coleman, J. E. (1974). "Metalloalkaline Phosphatases from *Bacillus subtilis:* Physiochemical and Enzymatic Properties," *Arch. Biochem. Biophys.,* **160**, 255–268.

Zirino, A. and Healy, M. L. (1970). "Inorganic Zinc Complexes in Seawater," *Limnol. Oceanogr.,* **15**, 956–958.

Index

Abiotic factors modifying zinc uptake, by
    marine biota, 327-334
Accelerators in zinc dental cements, 238-
    243
Accumulation of zinc:
    aquatic biota, 384-386, 389-390
    aquatic plants, 384-386
    laboratory populations of marine biota,
       319, 322-326
    marine biota, 259-351
Acrodermititis enteropathica, induced by
    abnormal zinc metabolism, 17, 38-39,
    80, 153
Actinomycetes, effect of zinc on synthesis
    of secondary metabolites by, 452-454
Active process of zinc uptake, by crops,
    403-404
Adaptation of aquatic plants, to high zinc
    levels, 377, 380
Additives in zinc dental cements, 238-243
Adolescents, zinc deficiency in, 14-15
Adrenocorticotropin, effect of zinc de-
    ficiency on, 77-78
Adults, zinc deficiency in, 15
Aflatoxin B_1, effect of zinc on microbial
    synthesis of, 453-454
Age
    effect on plasma zinc levels, 142
    effect on zinc uptake, by marine biota,
       328-334
Agricultural crops:
    diagnosis of zinc deficiency in, 408-409
    factors affecting zinc uptake by, 404-409
    mechanism of zinc uptake by, 403-404
    uptake of zinc by, 401-414
    zinc fertilization of, 409-411
Agricultural soils, zinc contents of, 402
Agrostis tenuis, uptake and localization of
    zinc by, 418-419

ALAD activity:
    activation by zinc, 174
    inhibitory effects of lead on, 172-175
Albumen, binding of zinc by, 45, 143
Alcohol, role in hyperzincuria, 33-34,
    148-149
Alcohol dehydrogenase activity, effect of
    zinc on, 48
Alcohol withdrawal, effect on zinc balance
    in humans, 112
Algae:
    adaptation to high zinc levels, 377, 380
    bioassay studies using, 388
    concentrations factors for, 320
    factors modifying zinc accumulations by,
       327-328
    laboratory studies of zinc accumulation
       by, 322
    levels of zinc in field collections of, 262-
       266
Alkaline phosphatase activity, effect of zinc
    deficiency on, 47, 81-85
Allium cepa, zinc fertilization of, 410-411
Alternaria, tolerance of zinc by, 370
Amino acids, in metallothionein, 444-445
Amphipods, distribution of zinc, in field
    samples of, 284
Analysis of zinc, in blood samples, 138-139
Animal studies of zinc therapy, in wound
    healing, 226
Annelida:
    concentrations of zinc in field samples of,
       294-295
    factors modifying zinc accumulation by,
       332-333
    laboratory study of zinc accumulation by,
       322-323, 326
Antagonistic effect of zinc, on lead poison-
    ing, 171-175

Antibiotics, effect of zinc, on microbial
    synthesis of, 452-454
Application of zinc fertilizers, methods and
    rates, 410-411
Aquatic plants:
    accumulations of zinc by, 262-266, 384-
        386
    adaptation to high zinc levels, 377, 380
    composition of, in lotic environments,
        370-376
    concentration factors for, 320
    diversity in polluted rivers and streams,
        370-376
    morphological adaptations to high zinc
        levels, 381-382
    physiological adaptations to high zinc
        levels, 381-382
    taxonomy of zinc resistant species, 377,
        378-379
    tolerance of zinc in lotic environments by,
        370-376
    use of, in monitoring zinc in rivers, 387-
        390
Arachnoidea, zinc in field samples of, 283
Areas zinc deficiency reported, 18-19
Areas with high zinc levels in streams and
    rivers, 366-368
Ascophyllum nodosum, accumula-
    tion of zinc by, 322, 326
Aspergillus niger, effect of zinc on growth
    of, 451-452
ATP activity, suppression by zinc, 207
ATPase in leaves, effect of metals on, 424-
    426
Availability of zinc from foods, 4, 45, 105-
    119

Bacillariophyta, zinc tolerance by, 378
Bacteria, see Microorganisms
Bacteria cultures, removal of zinc from
    seawater by, 328
Bacteriacidal action of zinc dental cements,
    244-245
Balanus balanoides, concentrations factors
    for zinc in tissues of, 319
Barnacles, distribution of zinc in field
    samples of, 285-286
Beans, response of genotypes to zinc, 405-
    406
Bile, release of zinc to intestinal excretions
    by, 113

Bioassay data for zinc, 355-356
Bioassays on zinc tolerance by aquatic
    plants, 388
Bioavailability of zinc in environment, 440
Biochemical roles of zinc, 3-4, 46-50, 89-93
Biochemistry of zinc, in biological systems,
    46-50
Biological factors influencing zinc uptake,
    by marine biota, 327-334
Biological materials, determination of zinc
    in, 138-139
Biomass production, by plants grown in zinc
    culture, 421-422
Biosynthesis pathways of heme, 172
Blood zinc:
    abnormal levels associated with hemato-
        logical disorders, 146-147
    analysis of, 138-139
    diurnal variations, 139-141
    during pregnancy, 145
    effect of hepatic disease on, 149
    effect of lead poisoning on, 171-181
    influence of medications on, 156-158
    levels in malignancy, 201-204
    mechanisms that produce changes in, 155-
        156
    normal concentrations, 141-143
    patients with
        infectious diseases, 151-152
        malignancies, 150-151
    reduction by oral contraceptive agents,
        146
    renal disease patients, 149-150
    toxic levels of, 148
    transplacental transfer of, 145-146
Blood of marine organisms, zinc contents
    of, 274-318
Bluegill, toxicity of zinc to, 355-356
Blue-green algae, see Algae
Body content of zinc, 2-3
Body fluids of marine organisms, zinc con-
    tents of, 269-318
Body weight, effect on zinc uptake by
    marine biota, 331-334
Bone, accumulation of zinc in, 122
Bridge retainers made of zinc cements,
    250
Bronchial carcinoma, blood zinc in patients
    with, 150
Bronchitis, blood zinc level conditioned by,
    154

Brook trout, toxicity of zinc to, 355-356
Bryophytes, accumulation of, zinc by, 389
Burns, zinc deficiency conditioned by, 35

Cadmium:
  activation of, ALAD by, 175
  cotolerance by plants, 422
  effect on :
    zinc toxicity to aquatic plants, 384
    zinc uptake by marine biota, 328-334
  enzymatic effects of, 422-426
  inhibition of microbial zinc uptake by,
    441
Calamine, medicinal use of, 215
Calcium:
  effect on zinc uptake by marine biota,
    328-334
  metabolic interactions with zinc, 114
  replacement by zinc in cell microskeleton,
    49
Calcium diet, effect on zinc balanace in human beings, 110
Cancerous tissues, zinc levels in, 202-203
Candida utilis:
  activated transport of zinc in, 444
  mechanism of zinc uptake by, 441-443
Carapaces of marine biota, zinc contents
  of, 269-318
Carbonate zinc complexes in water, 353-359
Carbonic anhydrase, effect of zinc deficiency on, 85-86, 426
Carboxypeptidase, effect of zinc deficiency
  on, 86
Carcinogenicity of zinc, 200-201
Castration, effect on zinc accumulation in
  prostate, 130
Cast restorations using zinc dental cements,
  247-250
Catalytic functions of zinc in microbial
  metalloenzymes, 450
Causes of zinc deficiency in humans:
  alcohol, 33-34
  burns and skin disorders, 35-36
  collagen diseases, 37
  diabetes, 36-37
  genetic disorders, 38-39
  gastrointestinal disorders, 34
  iatrogenic causes, 36
  liver disease, 34
  neoplastic diseases, 35
  nutritional causes, 30-33

parasitic infestations, 36
pregnancy, 37-38
renal diseases, 34-35
Cellular aspects of zinc metabolism and
  homeostasis, 121-135
Celiac disease, induced by zinc deficiency,
  40
Cellular distribution of zinc, in plants, 419
Cellular membrane effects of zinc, 223-224
Cerebral infarction patients, blood zinc
  levels in, 155
Chaetognatha, zinc content of, 294
Changes in enzyme activities, induced by
  zinc deficiency, 81-89
Chemical factors, influencing zinc toxicity
  to aquatic plants, 383-384
Childhood zinc requirements, 11-14
Children, zinc deficiency in, 12-14
Chlamydomonas sp. tolerance of zinc by,
  370-390
Chlamys opercularis, discrimination be-
  tween chemical species of zinc by, 319
Chlorella sp., adaptation to high zinc levels,
  380
Chlorophyta, zinc tolerance by, 370, 378-
  379
Chromatography of mustard oil glucosides in
  thlaspi alpestre, 430
Chronic alcoholism, effect on urinary zinc
  excretion, 115
Chronic diseases affecting zinc metabolism,
  123-124
Chrysophyta, zinc tolerance by, 370, 378
Cirrhosis of liver conditioned by zinc deficiency, 39-40
Citrus, application of zinc fertilizers to, 411
Cladophora:
  accumulation of zinc by, 389-390
  adaptation to high zinc levels, 370-385
  sensitivity to heavy metal pollution, 386-
    387
  tolerance of zinc by, 370-390
  use of, in monitoring zinc in rivers, 388
Climatic factors influencing zinc uptake by
  crops, 404-405
Clinical manifestations of zinc deficiency,
  8-9, 39-50
Clinical uses of zinc cements, 246-253
Cobalt:
  cotolerance by plants, 422
  effect on:

hepatic zinc accumulation, 127
zinc uptake by marine biota, 328-334
enzymatic effects of, 422-426
replacement of zinc in metalloenzymes
by, 90
Codfish, distribution of zinc in, 305-306
Coelenterata, zinc in field samples of, 267-268
Collagenase enzymes, role in wound healing, 219-222
Collagen diseases, zinc deficiency conditioned by, 37
Collagen synthesis in wound, 218, 219-222
Complexing agents, role in development of zinc tolerance by plants, 427-430
Complexing ligands for zinc in water, 353-359
Composition of lotic vegetation, 370-376
Compressive strengths of luting cements, 248-249
Concentration factors for zinc, in marine biota, 261, 319-321
Conjugatophytta, zinc tolerance by, 378
Contamination control in blood zinc analysis, 138-139
Copepods, distribution of zinc in field samples of, 286
Copper:
  cotolerance by plants, 422
  diurnal variations in blood concentration of, 140
  effect on
    hepatic zinc accumulation, 127
    zinc uptake by marine biota, 327-334
  enzymatic effects of, 422-426
  influence of zinc on plasma concentrations of, 155
  levels in cancer patients, 203
  replacement of zinc in RNA polymerase by, 450
Copper deficiency caused by excessive zinc intake, 52
Corals, zinc in field samples of, 267-268
Corn, response of genotypes to zinc, 405-406
Cotolerance of metals by plants, 421-422
Crab, distribution of zinc in field samples of, 287-289
Crassostrea virginica, accumulation of zinc by, 323, 326

Crustacea:
  concentration factors for zinc in, 320
  concentrations of zinc in field samples of, 283-293
  factors modifying zinc accumulations by, 330-332
  laboratory study of zinc accumulation by, 323, 326
Cryptogams, zinc tolerance by, 379
Ctenophora, zinc in field samples of, 268
Cutthroat trout, toxicity of zinc to, 355-356
Cysteic acid residues in intestinal metallo-thioneins, 125

Daily intakes of zinc by humans, 4-5, 46, 105-119
Decreased taste acuity attributable to zinc deficiency, 41-42
Deficiency of zinc in man:
  clinical manifestations, 8-9, 39-50
  etiology, 6-8, 30-39
  experimental production of, 42-45
Degenerative connective tissue diseases, blood zinc level conditioned by, 155
Degree of zinc tolerance by plants, determination of, 420-422
Dehydrogenases, effect of zinc deficiency on activity of, 86-87
Dental cements:
  clinical uses, 246-253
  composition of 238-243
  tissue response to, 243-246
Dentin, thermal conductivity of, 251
Determination of degree of zinc tolerance by plants, 420-422
Diabetic patients, blood zinc levels in, 154
Diabetes caused by zinc deficiency, 38
Diagnosis of zinc deficiency:
  human beings, 9-10, 39-50
  soils and plants, 408-409
Dialyzable zinc in oysters, 328
Dialysis, effect on blood zinc, 149-150
Diatoms, tolerance of combined zinc and copper in solution by, 371
Dicotyledons, zinc tolerance by, 379
Dietary allowances for zinc, 4-5
Dietary availabity and absorption of zinc during pregnacy, 185-186
Dietary perturbations of plasma zinc, 123

Disease conditions associated with zinc
abnormalities 1-27, 39-50, 137-169
Distribution of zinc in plant cells, 419
Diurnal variations in blood zonc, 139-141
DNA:
    effect of zinc on synthesis of, 203-206,
    220-222
    functions of zinc in microbial synthesis of,
    449-454
    role of zinc in synthesis of, 3, 40, 45,
    46, 87
Down's syndrome, zinc deficiency condi-
    tioned by, 39
DTPA test for zinc status of soils, 409
Dunaliella tertiolecta, light-dependent up-
    take of zinc by, 386
Dwarfism attributable to zinc defiency, 144

Echinodermata:
    concentration factors for zinc in, 320
    factors modifying zinc accumulations by,
    333
    levels of zinc in field samples of, 295-297
EDTA, release of zinc from chlorella by,
    443
Elasmobranchii, zinc in field samples of,
    298-301
Electrogustatometry, application of, 191
Emission spectra of whole blood, containing
    zinc protoporphyrin, 176
Endogenous secretion of zinc, 125-126
Energy poisons in liver, 127
Energy-dependent zinc uptake by micro-
    organisms, 441-442
Enolase in leaves, effect of metals on, 423-
    426
Enzymes, sensitivity to metals, 422-426
Enzyme-activating effect of zinc, on lead
    poisoning, 174-175
Enzyme activation, inhibitory effect of zinc
    on, 452, 454
Enzyme functions, affected by zinc, 217-
    218
Epidemiological aspects of zinc deficiency
    in humans, 1-27
Equilibria regulating zinc speciation in
    water, 354
Equilibrium zinc balance in human beings,
    113-114
Erythrocytes, normal zinc levels in, 141

Escherichia coli:
    acquisition of zinc by, 441
    functions of zinc in DNA synthesis by,
    449-451
Esophogeal cancer, blood zinc in patients
    with, 150-151
Ethambutol, chelation of metal ions by, 158
Etiological factors in zinc deficiency, 6-8,
    30-39
Euglenophyta, zinc tolerance by, 370, 378
Europe:
    areas with high values of zinc in streams
    and rivers, 376-378
    floristic composition of lotic vegetation
    in, 372-373
Excess zinc, suppression of turmor growth
    by, 206-207
Excretion of zinc, mechanisms for, 184
Exit-exchange reactions of zinc in micro-
    organisms, 443-444
Experimental production of zinc deficiency
    in human beings, 42-45
Experimental study of zinc influence on
    lotic vergtation, 370

Factors affecting zinc accumulation by
    aquatic plants, 386
Factors influence zinc toxicity to aquatic
    plants, 382-384
Factors modifying zinc accumulations by
    marine biota, 327-335
Fathead minnow, toxicity of zinc to, 355-
    356
Features of lotic environments with high
    zinc levels, 364-369
Fecal excretion of zinc, 108-115
Fertilizer nitrogen, enhancement of crop
    assimilation of zinc by, 407-408
Fetal alcohol syndrome, 40
Fetal malformations conditioned by
    maternal zinc deficiency, 188
Fetus, zinc deficiency in, 10
Fibroblast, effect of zinc on, 221-222
Field collections of marine biota, zinc
    levels in, 261-318
Field populations of aquatic plants, ac-
    cumulation of zinc by, 384-385
Fillers in zinc dental cements, 238-243
Fish:
    concentration factors for zinc in, 321

distribution of zinc in field samples of, 301-316
laboratory study of zinc accumulation by, 325
Fishmeals, zinc content of, 307
Flagfish, toxicity of zinc to, 355-356
Floristic composition of vegetation in lotic environments, 370-376
Flowing waters, see Lotic environments
Food groups, lead contents of, 5
Food web zinc concentration, 329-334
Forms of zinc:
    influence on zinc uptake by marine biota, 328-334
    in soils, 402-403
France, areas with high zinc levels in rivers and streams of, 366-367
Free radical reactions, effects of zinc on, 49
Frustules of marine biota, zinc contents of, 266-292
Functional relation of zinc to ALAD activity, 174-175
Functions of zinc in microorganisms, 446-454
Fundulus heteroclitus, accumulation of zinc by, 325, 326
Fungi:
    effect of zinc on synthesis of secondary metabolites by, 452-454
    energy-dependent zinc uptake by, 441

Gastrointestinal bleeding conditioned by zinc toxicity, 51
Gastrointestinal disorders:
    blood zinc level conditioned by, 155
    zinc deficiency induced by, 34
Genetic disorders, zinc deficiency conditioned by, 38-39
Genetics of zinc tolerant plants, 416
Gills of marine organisms, zinc contents of, 269-318
Glucocorticosteroid hormones, effect on hepatic zinc metabolism, 128-130
Glucose tolerance, lowering of, conditoned by zinc deficiency, 73-75
Goldfish, toxicity of zinc to, 355-356
Gonadal function, role of zinc in, 48
Growth hormones, effect of zinc deficiency on, 77-78

Growth phase, effect on zinc uptake by marine biota, 328-334
Growth retardation induced by zinc deficiency, 39-40
Guppies, toxicity of zinc to, 355-356

Habitat of zinc tolerant plants, 417-418
Hematofluorometer, applications of, 178
Hematological disorders, abnormal blood zinc levels associated with, 146-148
Hepatic diseases, blood zinc levels conditioned by, 148
Hepatic metallothionein biosynthesis, mechanisms of, 127
Hepatic zinc metabolism, 127-130
Hepatocysts, biosynthesis of metallothioneins in, 128-130
Herbaceous plants, development of zinc tolerance by, 431
Homarus vulgaris, accumulation of zinc by, 323, 326
Homeostatic control of zinc in mammalian system, 45-50, 121-135
Hormidium sp., adaptation to high zinc levels by, 377, 381, 383-384, 387, 389
Hormones, biochemical effects of zinc deficiency on, 72-81
Horses, effect of zinc on lead poisoning of, 171
Hospital diets, zinc contents of, 106-108
Human studies of zinc therapy, in wound healing, 226-227
Human zinc deficiency, epidemiological aspects of, 1-27
Hydroxy zinc complexes in water, 353-359
Hyperthyroidism, blood zinc level conditioned by, 154, 155
Hyperzincemia in human patients, 51
Hypogeusia induced by zinc deficiency, 41-42
Hypogonadism conditioned by zinc deficiency, 9, 144
Hypophysectomy, effect on prostate zinc, 130
Hypospadia glandis, attributable to maternal zinc deficiency, 190

Iatrogenic causes of zinc deficiency, 38

Immune system, effect of zinc deficiency on, 50
Impression materials made of zinc dental cements, 253
In vivo enzyme activity in leaves, effects of metals on, 425-426
Index of zinc tolerance by plants, 420
Infants, zinc deficiency in, 10-12
Infections, conditioned effect of zinc deficiency on, 50
Infectious diseases, blood zinc levels conditioned by, 151-152
Inflammatory cells in wounds, effects on zinc on, 224-225
Injury, tissue transport of zinc following, 225-226
Insecta, zinc contents of, 294
Insulin, effect of zinc deficiency on, 72-73, 75-77
Intake of zinc, by human beings, 5, 123, 185-186
Interactions between zinc and nutrients, affecting zinc uptake by plants, 406-408
Intestinal malabsorption of zinc, 8
Intestinal zinc, form and absorption of, 124-126
Intracellular ligands, role in metabolic zinc absorption, 126, 129
Intracoronal dressings with zinc cements, 246
Iron, effect on:
    hepatic zinc accumulation, 127
    zinc uptake by marine biota, 328-334
    zinc uptake by plants, 404, 408

Kidney of marine organisms, zinc contents of, 269-318

Laboratory populations of marine biota, accumulation of zinc by, 319, 322-326
Lactation, zinc requirement during, 16
Lead:
    dysfunctions caused by poisoning with, 171
    enzymatic effects of, 424
    functional relationship with zinc protoporphyrin in blood, 176-178
    hematofluorometric detection of human absorption of, 178

protective action of zinc on, 171-175
Lead poisoning, ameliorating effect of zinc on, 172-175
Lemanea sp., accumulation of zinc by, 389-390
Leptomitus, tolerance to zinc, 370
Leukemia, blood zinc levels conditioned by, 146
Leukemia patients, blood zinc levels in, 202-203
Leukocytes, normal zinc levels in, 141
Lewis lung carcinoma, suppression of growth by zinc deficiency, 205
Light:
    effect on zinc toxicity to aquatic plants, 383
    impact on zinc deficiency in plants, 404-405
Light intensity, effect on zinc uptake by algae, 327
Liver:
    metallothioneins in, 125
    zinc metabolism in, 126-130
Liver carcinomas, blood zinc levels in, 201-202
Liver disease, abnormal zinc metabolism induced by, 34
Liverworts, zinc tolerance by, 379
Lobster, distribution of zinc in, 289-293
Localization of zinc in plants, 418-419
Loss of zinc:
    from mussels, 329-330
    during starvation, 111
Lotic environments:
    accumulations of zinc by plants in, 384-386
    adaptation of plants to high zinc levels in, 370, 380
    chemical features of, 365-369
    floristic composition of vegetation in, 370-376
    geographic distribution of, 366-368
    occurrence of zinc in, 364-365
    physical features of, 365-368
    taxonomy of zinc resistant plants in, 377-379
    use of plants to monitor zinc pollution in, 387-390
Low calorie diet, effect on zinc loss in humans, 111
Luting agents made of zinc cements, 247-252

Lymphocytes, effect of zinc on, 223-225
Lymphoma, blood zinc levels conditioned by, 147
Lysl oxidase activity, suppression by zinc, 218

Macroglobulin, binding of zinc by, 143
Macrophages, effects of zinc on, 224
Macrophytes, levels of zinc in field samples of, 266
Magnesium, effect on zinc uptake by marine biota, 327-334
Major zinc species in test waters, 357
Malate dehydrogenase in leaves, effect of metals on, 424-426
Malignancies, blood zinc levels in patients with, 150-151
Malignant cells, behavior of zinc in, 203-204
Malignant systems, tissue and blood zinc in, 201-204
Mammals, metabolism and homeostasis of zinc in, 121-135
Manganese:
    enzymatic effects of, 422-425
    replacement of zinc in RNA polymerase by, 450
Manifestations of zinc abnormalities:
    animals, 61-70
    human beings, 29-59
Manipulation of zinc dental cements, 238-243
Marine biota:
    distribution of zinc in species of, 268-318
    zinc levels in field collections of, 260-318
Marine mammals, distribution of zinc in, 317-318
Measurement of nutritional status, 5-4
Measurement of zinc tolerance, by aquatic plants, 388-389
Meats, zinc contents of, 112
Mechanism of intestinal zinc absorption, 125
Mechanism of zinc tolerance by plants, 430-432
Mechanism of zinc uptake by crops, 403-404
Medication:
    effect on blood zinc, 156-158
    effect on zinc metabolism in humans, 111, 114-115

Membrane effects of zinc, 223-224
Metabolic balance of zinc during pregnancy, 186-188
Metabolic diets, zinc contents of, 108
Metabolism of zinc:
    effect on wound healing, 218-222
    human beings, 2-6, 45-50, 105-119
Metal:
    energy-dependent uptake by microorganisms, 441
    inhibition of enzyme activities by, 422-426
    levels in cancer patients, 203
    levels in plants growing on contaminated soils, 417
    tolerance by plants, 415-437
Metal fume fever due to zinc toxicity, 51
Metalloenzymes:
    biochemical functions of, 3-4, 46-50
    in microorganisms, 446-448
Metallothionein:
    amino acid compositions of, 444-445
    biosynthetic mechanisms of, 126-130
    microbial synthesis of, 445-446
Methods of applying fertilizer zinc to crops, 410-411
Microbial cell surface, transfer of zinc across, 441-445
Microbial zinc-dependent nucleotidyl transferases, 449-454
Microbial zinc metalloenzymes, 446-448
Microorganisms:
    acquisition of zinc by, 440-446
    elimination of toxic metals by, 445-446
    functions of zinc in, 446-454
    metabolism of zinc in, 444-454
    metalloenzymes in, 446-448
    storage of zinc by, 440-446
Microthamnion strictissimum, genetic adaptation to high zinc levels, 377, 380
Milk, zinc concentration in, 11
Miramichi, composition of lotic vegetation in, 370-372
Mobility of zinc to plant roots, 403
Model development of zinc tolerance in herbaceous plants, 431
Molecular aspects of zinc metabolism and homeostasis, 121-135
Mollusca:
    concentration factors for zinc in, 320-321

factors modifying zinc accumulations by, 327-330

laboratory study of zinc accumulation by, 323-326

zinc levels in field samples of, 268-282

Monocotyledons, zinc tolerance by, 379

Morphological adaptations of aquatic plants to high zinc levels, 381-382

Mosses, zinc tolerance by, 379

Mougeotia sp., genetic adaptation to high zinc levels by, 377, 380

Movement of zinc to plant roots, 403

Muscles of marine organisms, zinc contents of, 268-318

Mya arenaria, accumulation of zinc by, 324-326

Myocardial infarction, blood zinc level conditioned by, 154

Mytilus edulis, accumulation of zinc by, 325, 326

Mytilus sp., distribution of zinc in field samples of, 276-277

Myxophyta, zinc tolerance by, 370, 378

Neocosmospora vasinfecta, zinc uptake by, 441

Neonates:

blood zinc levels in, 145-146

zinc deficiency in, 11

Neoplastic diseases, zinc deficiency induced by, 35

Nephrotic syndrome, abornal blood zinc level associated with, 149

Nereis diversicolor, accumulation of zinc by, 322-323, 326

New Lead Belt, composition of lotic vegetation in, 374-375

Nickel:

cotolerance by plants, 422

effect on hepatic zinc accumulation, 127

enzymatic effects of, 422, 426

Nitrate ruductase, effect of metals on, 423-426

Nitrogen, effect on zinc uptake by plants, 407-408

Normal concentrations of blood zinc, 141-142

Nucleic acid synthesis in wound, effect of zinc on, 220-222

Nutritional causes of zinc deficiency, 30-33

Nutritional dwarfism induced by zinc deficiency, 13

Nutritional state, effect on human zinc metabolism, 111, 115

Nutritional status of zinc for humans, 5-6

Nutritional zinc deficiency in children, 10-14, 32-33

Occupationally exposed adults, zinc proto-porphyrin levels in, 176-178

Occurrence of zinc in lotic environments, 364-365

Octopus, concentration of zinc in, 278

Old mine wastes, metal contents of plants growing on, 417

Older children, zinc deficiency in, 13-14

Oral contraceptive agents, effects on blood zinc, 146

Organic matter in soils, binding of zinc by, 404

Orthodontic brackets and bands, luting agents for, 250-251

Osteosarcoma, blood zinc in patients with, 150

Ostrea sp., distribution of zinc in field samples of, 278-279

Oxalate:

ecotype variations in production of, 427-430

role in development of zinc tolerance by plants, 427-430

Oxygen concentration, effect on zinc uptake by marine biota, 327-334

Oysters, distribution of zinc in field samples of, 280-281

Packed cell zinc, normal concentrations of, 142

Pancreatic carboxypeptidase, effect of zinc deficiency on, 86

Parasitic infestation, zinc deficiency conditioned by, 36

Parental zinc nutrition, 16-17

Pathways of zinc excretion in human beings, 113

Pelagic fish, distribution of zinc in, 312

Perfused rat intestinal system, absorption of zinc by, 125-126

Periodental packs containing zinc cements, 252-253

Pesticides, effect on zinc uptake by marine biota, 327-334

pH, effect on:
 zinc speciation in water, 353-359, 369
 zinc toxicity to aquatic plants, 383
 zinc toxicity to fish, 355-359
 zinc uptake by marine biota, 327-334
Phaeodactylum tricornutum, diffusion controlled zinc uptake up by, 386
Phaseolus vulgaris, zinc fertilization of, 410-411
Phosphatase activity in leaves, effect of metals on, 423-426
Phosphates of zinc, use in dental cement, 238-239
Phosphorus:
 antagonistic effect on zinc uptake by plants, 406-407
 effect on zinc metabolism, 114
 effect on zinc toxicity to aquatic plants, 384
 physiological effect on zinc metabolism in plants, 407
Phosphorus diet, effect on zinc balance in humans, 110
Photosynthesis, effect of zinc on, 382
Photosynthetic organisms highly tolerant to zinc, 370-390
Phylogenetic groups, different concentrations factors for, 319
Physical factors influencing zinc toxicity to aquatic plants, 382-383
Physical features of lotic environments with high zinc levels, 365
Physiological adaptations of aquatic plants to high zinc levels, 381-382
Physiological role of zinc, 3-4
Physiology of zinc tolerance by plants, 420-430
Phytoplankton, zinc concentrations in, 265-266
Phytosociology of aquatic plants, 373
Piturary-gonadal axis, zinc metabolism in, 131
Plaice, distribution of zinc in, 312-315
Plant:
 concentrations of zinc in, 266, 408
 diagnosis of zinc deficiency in, 408
 ecotypes susceptible to zinc, 415-432
 localization of zinc in, 418-419
 mechanism of zinc tolerance by, 430-432
 mechanism of zinc uptake by, 403-404

 metallic inhibition of enzyme activity in, 422-426
 responses of genotypes to zinc, 405-406
 uptake of zinc by, 418-419
 zinc tolerant ecotypes, 416-432
 see also Agricultural crops; crops; Aquatic plants
Plant roots, movement of zinc to, 403
Plasma, zinc concentrations in, 123-124
Plasma zinc:
 binding of, by protein molecules, 143-144
 diurnal variations, 139-141
 half-life of, 124
 lowering of, during pregnancy, 187
 monitoring of response of tuberculosis to chemotherapy using, 152
 normal concentrations of, 141-142
 see also Blood zinc
Plasticizers in zinc dental cements, 238-243
Plectonema gracillimum, nitrogen fixation by, 382
Potassium, levels in cancer patients, 203
Porifera, zinc in field samples of, 267
Pregnancy:
 blood zinc during, 145
 congenital malformations conditioned by zinc deficiency, 188
 dietary zinc availability and absorption during, 185-188
 requirements of zinc during, 184-185
 zinc disorders conditioned by, 37-38
 zinc requirement during, 15-16
 zinc supplementation during, 188-191
Premature infants, zinc requirement of, 12
Preschool children, zinc deficiency in, 12-13
Properties of zinc dental cements, 238-243
Prostate, zinc accumulation in, 130-131
Prostatic carcinoma, blood zinc in patients with, 150
Protective action of zinc, on lead poisoning, 171-181
Proteins:
 binding of plasma zinc by, 143-144
 metabolic binding of zinc by, 46-50
Protein deficiency, role of zinc in, 144-145
Protein synthesis in wound, effect of zinc on, 218-220
Protista, factors modifying zinc accumulations by, 328

Protoplasmic resistance test of plant sus-
    ceptibility to zinc, 420
Protozoa, levels of zinc field samples of, 267
Psoriasis, zinc deficiency conditioned by,
    35-36
Pteropods, zinc contents of, 282
Pups of *zalophus californianus*, zinc content
    of, 317-318

Rainbow trout, toxicity of zinc to,
    355-356
Rat:
    biochemical changes conditioned by zinc
        deficiency in, 71-97
    effect of zinc on lead poisoning of, 171-
        175
Rates of zinc fertilizer application, 410-411
Reactants in zinc dental cements, 238-243
Recommended daily dietary dose for zinc,
    4-5
Redistribution of zinc in wounds, 225-226
Regulatory functions of zinc, in microbial
    cell metabolism, 451
Relationships between reduced enzyme
        activity and zinc deficiency symptoms,
        91-93
Renal disease:
    effect on zinc absorption, 34-35
    patients, blood zinc levels in, 149-150
Reproducibility of zinc balances, in human
    beings, 108-109
Requirements of zinc:
    by humans, 4-5, 105-119
    during pregnancy, 184-185
Residual effect of applied zinc in soils, 409
Responses of individual zinc metallo-
    enzymes to zinc depletion, 89-91
Response of plant genotypes to zinc, 405-
    406
Restorative use of zinc dental cements, 246-
    247
Rheumatoid arthritics, blood zinc level con-
    ditioned by, 154
Rhizoclonium, sensitivity to heavy metal
    pollution, 386-387
Rhodophyta, zinc tolerance by, 378
Rice, zinc deficiency in, 403
Rice cultivars, response of genotypes to
    zinc, 406
Rivers, *see* Lotic environments

RNA:
    functions of zinc in microbial synthesis of,
        449-454
    role of zinc in synthesis of, 3, 45, 46, 88,
        203-206, 220-222
Role of zinc therapy, in tissue repair, 226-
    227
Root canal sealers made of zinc cements,
    253
Rooting test of plant sensitivity to zinc, 420
Root interception of zinc, 403
Rubratoxin, effect of zinc on microbial
    synthesis of, 453-454

Salinity, effect on zinc uptake by marine
    biota, 329-334
Salmonella enteriditis, zinc uptake by, 441
Salmonelli typhi, blood zinc in patients in-
    fected with, 151
Salmonid, toxicity of zinc to, 355-356
Sclerotia, effect of zinc on microbial syn-
    thesis of, 453-454
Scope of zinc deficiency, in human beings,
    18-19
Seasonal variations in zinc uptake by
    crustacea, 331
Seaweeds, zinc levels in, 265-266
Secondary metabolites, effect of zinc on
    synthesis of, 452-454
Secondary zinc deficiency, in humans, 17-
    18
Serum zinc:
    diurnal variations, 139-141
    normal concentrations, 84, 142-143
    reduction of, during pregnancy, 187
Seston, zinc contents of, 293
Setting reactions of zinc dental cements,
    238-243
Sewage-fungus communities in rivers, 376
Sex hormones, effect of zinc deficiency on,
    79-81
Shells of marine organisms, zinc contents
    of, 268-318
Shrimp, zinc contents of, 293
Sickle-cell disease:
    patients, blood zinc levels in, 147-148
    zinc deficiency conditioned by, 38
Silene sp., uptake of zinc by, 419
Silicophosphate dental cements, 241-
    243

Skin disorders, zinc disorders conditioned by, 35-36

Small intestine, absorption of zinc by, 122

Snails, concentration of zinc in, 282

Soft parts of marine organisms, zinc contents of, 268-318

Soils:
contamination with heavy metals, 417-418
diagnosis of zinc deficiency in, 408-409
forms of zinc in, 402
influence on zinc uptake by crops, 404-405

Sources of fertilizer zinc, 409-410

Soybean, response of genotypes to zinc, 405

Speciation of zinc in waters, 353-359

Starvation, effect on zinc loss in humans, 111

Stigeoclonium:
adaptation to high zinc levels, 370-385
biomonitoring of zinc in rivers using, 388
sensitivity to heavy metal pollution, 387

Streams, *see* Lotic environments

Structural role of zinc, in microbial metalloenzymes, 449-454

Suppression of tumor growth, by zinc deficiency, 204-206

Suspended matter in streams, zinc content of, 365, 369

Sweat, loss of zinc via, 184

Symptons of human zinc toxicity, 51

Taste abnormalities induced by zinc deficiency, 41-42

Tzxonomy of zinc resistant plants, in lotic environments, 377-379

Teleostei:
accumulation of zinc in field samples of, 301-316
concentration factors for zinc in, 321
factors modifying zinc accumulations by, 333-334
laboratory study of zinc accumulation by, 325

Temperature:
effect on zinc toxicity to aquatic plants, 382
effect on zinc uptake by marine biota, 327-334
role in zinc deficiency in plants, 405

Temporal restorations using zinc cements, 246-247

Testicular teratoma attributable to zinc, 200

Therapeutic uses of zinc, 114

Therapy for zinc deficiency during pregnancy, 188-191

Thermal conductivities of dental basing and restorative materials, 251-252

Thermal insulation of dentine using zinc cements, 251-252

Thymidine kinase activity, effect of zinc deficiency on, 89, 205

Tissue repair, role of zinc in, 215-236

Tissue response to dental cements, 243-246

Tissue transport of zinc following injury, 225-226

Tissue zinc levels, in malignancy, 201-204

Tolerance of lotic vegetation to zinc, 370-376

Tomatoes, zinc deficiency in, 405

Total parental zinc nutrition, 16-17

Toxicity of zinc:
aquatic plants, 363-392
fish, 353-354
human beings, 50-51, 148
terrestial plants, 415-432

Transferrin complex, transport of zinc by, 143

Trauma, effect on tissue zinc distribution, 225-226

Treatment of zinc deficiency in man, 9-10; 42-45

Trout, toxicity of zinc to, 355-356

Tuberculosis patients, blood zinc in, 151-152

Tumor:
blood zinc modification by, 150
growth, suppression by zinc, 204-207

Tumorgenesis and zinc, 199-213

Tuna, accumulation radioactive zinc by, 260

Tunicata, zinc in field samples of, 297

Typha latifolia, trace metal tolerance by, 380

United States:
areas with high zinc levels in rivers and streams of, 368
zinc deficiency in children of, 32-33

Uptake of zinc by plants, 418-419

Urinary excretion of zinc in human beings, 8, 109-115

Use of plants to monitor zinc in rivers, 387-390

Variability of zinc contents of meals, 107
Viola calaminaria, zinc tolerance by, 387-388
Viral zinc-dependent nucleotidyl transferases, 449-454

Water chemistry, effect on zinc toxicity to aquatic plants, 383-384
Water hardness, influence on zinc toxicity to fish, 353-359
Wheat, application of zinc fertilizer to, 410-411
Worldwide problem of zinc deficiency, in humans, 18-19
Wound healing, role of zinc in, 215-236

Xanthophyta, zinc tolerance by, 378

Yeast, copper-binding proteins in, 445

Zebrafish, toxicity of zinc to, 355-356
Zinc:
    absorption and balance during pregnancy, 185-188
    accumulation:
        by agricultural crops, 401-414
        by marine biota, 259-351
    adaptations of aquatic plants to high levels of, 377-382
    areas with high levels of in rivers and streams, 365-369
    balances in human beings, 108-109
    blood concentrations of, in health and disease, 137-179
    carcinogenesis by, 200-201
    cellular membrance effects of, 223-224
    deficiency in man, 6-10, 29-50
    dental cements containing, 238-258
    enzyme functions affected by, 217-218
    epidemiological aspects of deficiency in humans, 1-27
    as essential element, 2-6
    factors influencing zinc toxicity to aquatic plants, 382-384
    floristic composition of vegetation, in streams and rivers with high levels of, 370-376
    functions in microorganisms, 446-454
    homeostasis in mammalian system, 121-135
    intestinal absorption of, 105-112, 124-126
    localization by plants, 418-419
    manifestations
        of abnormalities in human beings, 29-59
        of deficiency in humans, 89, 39-50
    measurement of plant susceptibility to, 388-389
    metabolism:
        in human beings, 2-6, 45-50, 105-119
        in mammals, 121-135
        in microorganisms, 439-465
    protective action on lead poisoning, 171-175
    requirements during pregnancy, 184-185
    role in wound healing, 215-236
    speciation in water, 353-361
    supplementation in diets, 106, 110-111
    therapy:
        during pregnancy, 188-191
        in wound healing, 236-237
    tolerance by plants, 415-437
    toxicity of, 51-52; 353-361
    transport in microorganisms, 439-465
    tumorgenesis by, 199-213
    uptake by crops, 401-414
Zinc carbonate, precipitation in water, 358
Zinc deficiency:
    biochemical changes induced by, 3-4, 45-50, 71-103
    changes in enzyme activities induced by, 81-89
    in children, 7-16
    clinical manifestations, 8-9
    diagnosis and treatment, 9-10, 39-50
    epidemiologic considerations, 1-27
    etiology, 6-8, 30-39
    hormone changes conditioned by, 72-81
    manifestations in humans, 29-50
    metalloenzymatic changes conditioned by, 81-93
    role in protein deficiency, 144-145
    suppression of tumorgenesis and tumor growth by, 204-206
    symptoms, 8-19, 39-50, 91-93
    syndrome in pregnancy, 187
Zinc dental cements:
    clinical uses of, 246-253

formulations of, 238-243
  tissue response to, 243-246
Zinc fertilizers, application rates and
  methods, 409-411
Zinc metalloenzymes in animal and human
  body, 81-93
Zinc oxide in dental cements, 239-241
Zinc phosphate cement in dentistry, 238-
  239

Zinc polyacrylate dental cements, 243
Zinc protoporphyrin, indicator of lead
  poisoning, 175-178
Zinc silicophosphate dental cement, 241-
  243
Zinc therapy in tissue repair, 226-227
Zooplankton, zinc in field samples of, 293-
  294

FOOD, CLIMATE AND MAN
Margaret R. Biswas and Asit K. Biswas, Editors

CHEMICAL CONCEPTS IN POLLUTANT BEHAVIOR
Ian J. Tinsley

RESOURCE RECOVERY AND RECYCLING
A.F.M. Barton

ATMOSPHERIC MOTION AND AIR POLLUTION
Richard A. Dobbins

INDUSTRIAL POLLUTION CONTROL – Volume I: Agro-Industries
E. Joe Middlebrooks

BREEDING PLANTS RESISTANT TO INSECTS
Fowden G. Maxwell and Peter Jennings, Editors

NEW TECHNOLOGY OF PEST CONTROL
Carl B. Huffaker, Editor

COPPER IN THE ENVIRONMENT, Parts I and II
Jerome O. Nriagu, Editor

ZINC IN THE ENVIRONMENT, Parts I and II
Jerome O. Nriagu, Editor